Principles of Molecular Photochemistry

An Introduction

Principles of Molecular Photochemistry

An Introduction

Nicholas J. Turro
COLUMBIA UNIVERSITY

V. Ramamurthy
UNIVERSITY OF MIAMI

J. C. Scaiano
UNIVERSITY OF OTTAWA

University Science Books

University Science Books
www.uscibooks.com

PRODUCTION MANAGER   Paul C. Anagnostopoulos
MANUSCRIPT EDITOR   Jeannette Stiefel
DESIGN   Windfall Software
ILLUSTRATORS   John Choi and LineWorks
PROOFREADER   Yonie Overton
COMPOSITOR   Windfall Software, using ZzTEX
COVER DESIGN   Genette Itoko McGrew
PRINTER AND BINDER   Victor Graphics

This book is printed on acid-free paper.

LIBRARY OF CONGRESS CATALOGING-IN-PUBLICATION DATA

Turro, Nicholas J., 1938–
    Principles of molecular photochemistry : an introduction / Nicholas J. Turro,
V. Ramamurthy, J.C. Scaiano.
        p.   cm.
    Includes index.
    ISBN 978-1-891389-57-3 (alk. paper)
    1. Photochemistry.   I. Ramamurthy, V.   II. Scaiano, J. C. (Juan C.), 1945–   III. Title.
    QD708.2.T97 2009
    541′.35—dc22

                                                                        2008040312

Printed in the United States of America
10 9 8 7 6 5 4 3

To my one and only Muse and Soul Mate Sandy,
To my two Foxes, Cindy and Claire,
To my five Munchkins, Nicholas, Charlie, Catherine, Maria, and Julia.

N.T.

To my Jyothi: Jayalakshmi, Rajee, and Pradeep.

V.R.

To my wife Elda,
For 40 years of love, friendship, and patience.

J.C.S.

Contents

CHAPTER 4 Radiative Transitions between Electronic States   169

CHAPTER 5 Photophysical Radiationless Transitions   **265**

CHAPTER 7   Energy Transfer and Electron Transfer   383

Preface

It has been over three decades since the publication of Modern Molecular Photochemistry (MMP) in 1978. During this period, the concepts and paradigms described in MMP have become part and parcel of modern synthetic and mechanistic photochemistry, and have been absorbed as routine intellectual tools in a wide range of fields, including physical organic chemistry, chemical biology, polymer chemistry, materials science and nanoscience. Remarkably, most of the basic paradigms of MMP remain the bed rock of current mechanistic analyses, investigations and application of photochemical reactions. However, the elaboration of these paradigms by the effect of spin and electron transfer was not covered in MMP. It was decided that a primer, which included these factors and integrated them with the general successful pedagogical philosophy of MMP, would be of use and interest to not only practicing photochemists and their students, but also to those in a variety of other fields, such as biological scientists, polymer scientists, materials scientists and nanoscientists who integrate photochemistry and photophysics in their research and teaching.

The primer, *Principles of Molecular Photochemistry: An Introduction,* introduces photochemical and photophysical concepts from a small set of principles that are familiar and understood by students of chemistry and other sciences. An initial paradigm is introduced that relates the photon and a reactant molecular structure to photochemistry through the structure of electronically excited states, reactive intermediates and products. The same paradigm is readily adapted to incorporate the photon and a reactant molecular structure to photophysics. The role of electronically excited states, and electronic–vibrational and electronic–spin interactions are clearly described in pictorial terms that can be readily understood and applied to systems of interest.

For the first time in any photochemical text, a fundamental description of electronic spin and its impact on photochemical and photophysical processes is described with an intuitive, pictorial and powerful vector model. The mysterious processes of spin–orbit coupling, intersystem crossing and magnetic effects on photochemical and photophysical processes are readily handled with this model.

Also, for the first time in any photochemical text, the concepts of electronic energy transfer and electron transfer are integrated and treated from a common foundation and set of concepts. The tremendous progress in theoretical and experimental electron transfer is covered in depth and will be of great assistance as an introduction to these two critical aspects of molecular photochemistry.

Level and Approach

The goal of this primer is to familiarize both students and researchers with the critical concepts and methodology involved in the investigation of organic photochemical reactions. A simple paradigm at the start of each chapter is elaborated with exemplars that provide the student with an understanding through examples of the underlying principles. The material can be understood easily by students with the fundamental knowledge of college general chemistry, organic chemistry and physics. An important feature of the primer is avoidance of complex mathematics and the translation of all concepts into familiar visualized representations, providing a complete and unified theoretical background for an understanding of light absorption and emission, or radiationless processes and of photochemical reactions. For example, the concept of electronic energy surfaces coupled with simple molecular orbital theory is used to visualize the triggers of photochemical processes and the analogies that intellectually reduce the thousands of organic photochemical reactions to a handful of fundamental primary photochemical reactions from electronically excited states.

This primer will be of use to students interested in a qualitative, pictorial interpretation of spectroscopic processes involving the absorption and emission of light by organic molecules and the photochemical processes that result from light absorption. In this regard, in addition to students of chemistry, the text will be useful to nonchemistry majors pursing interests in biology, biochemistry, materials science or chemical engineering.

Acknowledgments

This text is an outgrowth of courses and lectures on Organic Photochemistry. The authors are indebted to the many students who have assisted in the development of the text through their probing and stimulating questions as they sought an understanding of molecular organic photochemistry. We are also thankful to many colleagues who allowed their "brains to be picked" and thereby enabled the authors to produce a translation of abstruse mathematical concepts into concrete representations that provide students with an understanding of the subject. Special thanks to Professor David Schuster of New York University for a careful and critical reading of the early drafts of the book. Thanks also to the following photochemists who have read and provided critical comments: R.S.H. Liu, F.D. Lewis, J.R. Scheffer, L. Johnson, D.I. Schuster, and A. Griesbeck.

The authors express their appreciation to the many students and colleagues who participated in discussions that were very useful and important in the editing and improvement of the manuscripts at various levels. In particular, thanks go to J. Chen, M. Chrétien, M. Ivan, Steffen Jockusch, L. S. Kaanumalle, M. Laferrière, J. Lancaster, L. Mikelsons. A. Moscatelli, A. Natarajan, T. Poon, E. Park, J. Sivaguru. K. Sivasubramanian.

Molecular Photochemistry of Organic Compounds: An Overview

1.1 What Is Molecular Organic Photochemistry?

Molecular organic photochemistry is a science concerned with the structures and dynamic processes that result from the interaction of light with organic molecules. The field of molecular organic photochemistry can be conveniently classified in terms of the *photophysics of organic compounds* (the interactions of light and organic molecules resulting in net physical changes) and the *photochemistry of organic compounds* (the interactions of light and organic molecules resulting in net chemical changes). The molecular photochemistry of organic molecules is a rather broad and interdisciplinary topic embracing the fields of chemical physics, molecular spectroscopy, physical organic chemistry, synthetic organic chemistry, computational organic chemistry, and supramolecular organic chemistry.

In simplest terms (Scheme 1.1), molecular organic *photochemistry* involves the overall process $R + hv \rightarrow {}^*R \rightarrow P$, where R is an organic molecule that absorbs a photon (hv), whose frequency (v) is correct for light absorption by R; $^*R$ is an electronically excited molecule; and P is an isolated product (or products). Organic *photophysics*, on the other hand, involves the overall process $R + hv \rightarrow {}^*R \rightarrow R$, where R absorbs a photon, but does not undergo any net chemical change. In general, R will stand not only for the reactant molecule (R) that absorbs the photon, but also any other molecules (M) that are required for production of the product (P). If not stated explicitly, it should be assumed that the reactions described in this text are conducted in a solution of an inert solvent at or near room temperature ($\sim 25\,°C$). The electronically excited molecule ($^*R$) is the essential species that is universal to all photochemical and photophysical processes.

The text describes how the overall photochemical process $R + hv \rightarrow P$ and the overall photophysical process $R + hv \rightarrow R$ can be visualized in structural, mechanistic, theoretical, and experimental terms. For example, Scheme 1.1 describes a global paradigm for understanding the possible paths for the *photochemical* process

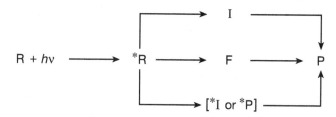

Scheme 1.1   A global paradigm for organic photochemical reactions. Photophysical processes that return *R to R are not included at this point, for simplicity. The *R → R photophysical processes are shown in Scheme 2.1.

R + $h\nu$ → P. The holy grail of molecular organic photochemistry uses Scheme 1.1 and its plausible elaborations for *the achievement of a complete structural and dynamic mechanistic description of all of the physical and chemical steps that occur as the result of the absorption of a photon by an organic molecule* (R) *and eventuate in the formation of an isolated product* (P) *or the regeneration of the starting material* (R). The nature of the species I and F will be discussed in Sections 1.9 and 1.10.

Scheme 1.1 and its elaboration for photophysical processes (R + $h\nu$ → R) provide a *paradigm* of remarkable versatility and scope that serves as a basis *for analyzing all organic photochemical reactions and all photophysical processes.* We shall employ the term "molecular photochemistry" to include *both* the photophysics of *R and the photochemistry of *R, since the concepts and laws of photophysics and photochemistry are intimately interwoven. Indeed, we shall see that it is impossible to have a proper understanding of the photochemical processes of *R without a corresponding understanding of the photophysical processes of *R.

Scheme 1.1 shows that there are *three* fundamentally distinct pathways, termed *primary photochemical processes,* that *R may follow on the way to P:

1. A pathway, *R → I → P, that leads to the formation of a discrete reactive intermediate (I) that can typically be described as having the characteristics of a radical pair (RP), a biradical (BR), or a zwitterion (Z).
2. A pathway, *R → F → P, that does not involve a discrete reactive intermediate (I), but instead proceeds through a "funnel" (F). This pathway takes R to P and can be described in the language of energy surfaces as a "conical surface intersection" or as a minimum produced by surface-avoided intersections.
3. A pathway, *R → *I → P or *R → *P → P, that involves the formation of an electronically excited *intermediate* (*I) or an electronically excited product (*P).

Of these three possibilities, *R → I (RP, BR, or Z) is the most commonly observed pathway for organic photochemical reactions.

The "molecular" part of molecular photochemistry emphasizes the use of molecular structure and its implied dynamics (transitions between states) and molecular substructure (electron configuration, nuclear configuration, and spin configuration) as the crucial and unifying intellectual units for organizing and describing the possible, plausible, and probable pathways of photochemical reactions from "cradle" (the absorption of a photon by a reactant, R, to form *R) to "grave" (the isolation of a product, P, produced by one of the three pathways from *R shown in Scheme 1.1).

1.2   Learning Molecular Organic Photochemistry through the Visualization of Molecular Structures and the Dynamics of Their Transformations

This text attempts to teach effective cognitive and contextual strategies for learning molecular organic photochemistry. Physical organic chemistry has thrived and progressed rapidly because of a tradition of correlating molecular structures with reaction mechanisms and with chemical reactivity. Molecular structure provides a powerful and effective visual means of coupling molecular dynamics to the change of molecular structure. We strive to provide an understanding of photochemical reactions through the visualization of molecular structure and the molecular dynamics of the processes described in Scheme 1.1.

1.3   Why Study Molecular Organic Photochemistry?

Scheme 1.1 displays schematically, at an elementary level, the *structural and dynamic content* that is important for the study and understanding of modern molecular organic photochemistry. *Every* organic photochemical reaction can be understood and described in terms of the paradigm of Scheme 1.1 or some plausible and straightforward modification or elaboration of Scheme 1.1. The motivation for studying the molecular photochemistry of organic compounds depends on the *context* in which the student views the content of the field, and many different motivations can provide such a context.

For example, there is the pure intellectual satisfaction of understanding how to visualize the ways in which two of the most fundamental components of the universe, photons ($h\nu$) and molecules (R), interact with one another to produce an electronically excited molecule (*R), which eventually is transformed to an isolated product (P). In particular, there can be a special intellectual delight in learning how to integrate different fields, such as spectroscopy, quantum mechanics, reaction mechanisms, molecular structure, magnetic resonance, and chemical dynamics. A qualitative understanding of each of these fields is important for an understanding of molecular organic photochemistry.

The intellectual structure of this field is inherently interdisciplinary and therefore requires a practitioner to seek a commonality and integration of the ideas and methods of many scientific disciplines. Such a process presents a challenge to the student. To a student who starts to learn a scientific subject, theories may appear to be based on disparate and seemingly conflicting concepts and laws, such as the theory of waves and the theory of particles. Molecular organic photochemistry must integrate theories from different fields. In this text, the required integration of theories and concepts is achieved by providing a visualization of the molecular structures, energetics, and dynamics involved in molecular organic photochemical reactions.

Other motivations for studying organic molecular photochemistry are found in its importance in modern technologies, in molecular and chemical biology, in medical applications, and in solar energy resources. For example, molecular organic photochemistry provides an understanding of the mechanism of photosynthesis, the fundamental process by which nature harnesses the sun's energy by the absorption of solar photons to produce food and energy for our planet. Photosynthesis is initiated by a primary photochemical process involving an electron transfer reaction (Chapter 7). Vision, our most important sense for observing and surviving in the external world, is triggered by a remarkably simple primary process of cis–trans isomerization, which triggers a cascade of physiological events that result in the sensation of vision in the brain.

During the late 1900s, the advent of laser technology revolutionized the field of telecommunications by making it possible to transmit information using light (through glass fibers), rather than electrons (through metal wires). New technologies, termed photonics, employ light to perform tasks that were originally relegated to the domain of electronics. Photochemistry is also attaining an increasingly important role in the health sciences, curing some forms of cancers (through phototherapy), repairing tissues, and performing microsurgery using lasers. Other important applications of photochemistry include the use of photolithography to manufacture computer chips and photopolymerization to produce protective coatings for a variety of high-value materials, such as optical fibers. One of the "holy grails" of photochemistry is the discovery of practical ways to convert sunlight into high-grade fuels to replace fossil fuels. Photophysics, especially the use of fluorescence as a sensor, is currently of enormous importance for applications in the materials sciences and in the biological sciences. All of these applications require an understanding of the essential players in molecular organic photochemistry, namely, the photon ($h\nu$), the molecule (R), and the electronically excited molecule (*R) as outlined in Scheme 1.1.

Among the most exciting developments in photochemistry during the past four decades has been the ever-increasing speed at which "pictures" of reacting molecules, such as *R, can be taken. Lasers can now routinely produce pulses of light whose duration are on the order of a few femtoseconds (fs; $1\,\text{fs} = 10^{-15}$ s). With such short pulses it is possible to take "snap shots" of actual atomic motions in real time down to periods as short as vibrational time scales. Typical atomic motions (i.e., bond stretching and bending) occur on the scale of $\sim 10^{12}$ nm s^{-1} (10^{13} Å s^{-1}). Thus, for a bond stretching and breaking of ~ 1 nm (10 Å), the time scale is on the order of 1000 fs. Pulses on the order of 10 fs, therefore, are able to follow such fast atomic processes. Femtosecond lasers are now routine in the laboratories of physicists and chemical physicists.[1]

What will be the eventual limit of short laser pulses that will be of interest to chemists? Since chemistry involves the movement of electrons, we can define the time scale for electron motion as the lower limit for events of interest to the chemist.

An electron in a Bohr hydrogen atom makes a complete orbit in $\sim 10^{-16}$–10^{-17} s, so we can use this time scale as the ultimate limit of time scales of interest to the chemist. Pulses on the order of 100 as (attoseconds; 1 as = 10^{-18} s) have been produced.[2] If the past is any guide, chemists will someday be taking snap shots of electrons moving in their orbitals. Chemists can probably agree that the zeptosecond (zs) time scale (1 zs = 10^{-21} s) will probably be the exclusive domain of physicists, who will take snap shots of excited nuclei exploding!

A unique property of lasers is the coherence (or phase alignment) of the light that is emitted. This coherence has the promise of controlling the course of chemical reactions by "steering" reactions toward specific pathways.

1.4   The Value of Pictorial Representations and Visualization of Scientific Concepts

Molecular photochemistry employs a number of theories and representations of molecular structure to describe the interaction of light and organic molecules to produce electronically excited states (i.e., R + $h\nu$ → *R in Scheme 1.1) and to describe the dynamics of the overall pathways from electronically excited states to products (i.e., *R → P in Scheme 1.1). Organic chemists are accustomed to analyzing ground-state, thermally induced reactions of R in terms of molecular structure, molecular energetics, and molecular dynamics. We show that the familiar molecular structural theory of organic chemistry provides an effective and powerful starting point for understanding mechanistic organic photochemistry. However, we have to make some important modifications to the theory of ground-state reactions as we proceed. We need to develop a theory of light and of the interaction of light with molecules where the usual structural theory of organic chemistry is replaced by a theory of interacting waves, and where the classical continuum of states and energies is replaced by quantized states and quantized energies. Consequently, we shall seek to understand and visualize the paradigms of wave and quantum mechanics, which have evolved as an authoritative and powerful means for understanding all structural and dynamic aspects of molecular organic photochemistry.

In order to understand molecular organic photochemistry, in addition to the familiar chemical representations involving molecular structure and dynamics, we must develop an understanding of the concepts of electron spins, electromagnetic radiation, and photons. The latter concepts are described quantitatively and most effectively by the mathematics of wave and quantum mechanics. However, this text is directed at students who do not possess the mathematical background necessary for a quantitative computation of molecular properties through quantum mechanics. Instead, we show that there are classical representations that are readily visualizable and capture the spirit and essence of most of the critical features of quantum mechanics that are needed to understand molecular organic photochemistry. These visualizable classical representations will provide the student with a *quantum intuition* for an understanding of the qualitative details of the pathways given in Scheme 1.1. For those who plan to proceed more deeply into the mathematics of quantum mechanics, we hope that the pictorial representations will provide a useful framework for the more quantitative mathematical aspects. To delve as deeply into the mathematical aspects as desired, the interested and able student can proceed to standard textbooks and references on quantum mechanics.[3]

1.5   Scientific Paradigms of Molecular Organic Photochemistry

Consensus exists among scientists on how to perform research and how to describe experimental observations when *authoritative scientific paradigms* exist that provide an accepted process on how to deal with important questions such as What are the fundamental entities that exist in the universe, and what are their properties? and What are the legitimate theoretical concepts and experimental tools that are required to understand and to measure the properties of the entities that exist? Authoritative paradigms allow the practicing scientist to perform everyday research and enable a student to be readily initiated into a mature field of science by studying, learning, and mastering the paradigms of the field.

Now, we examine briefly the concept of scientific paradigms and how it relates to the development of a paradigm for molecular organic photochemistry. The simple paradigm of modern molecular photochemistry shown in Scheme 1.1 helps answer the question What are the fundamental entities that exist along a photochemical or photophysical pathway? We will also answer questions, such as What are the structural, energetic, and dynamic properties of the entities shown in Scheme 1.1? and What are the legitimate theoretical concepts and experimental tools that are required to understand and to measure the properties of these entities? We use the word "paradigm" throughout the text because of its importance in science. We digress briefly now to describe how the word has evolved in the scientific community.

In a book entitled *The Structure of Scientific Revolutions*,[4] Thomas Kuhn, a philosopher of science, defined a *scientific paradigm* as a complex set of intellectual and experimental structures consisting of assumptions, concepts, strategies, methods, and techniques that provide a framework for performing scientific research in a field and for organizing and interpreting observable phenomena of the universe in a systematic and organized manner. According to Kuhn, *the accepted paradigms of a field provide the authority to which scientists appeal in deciding on the course of everyday, normal scientific activities and in recognizing expected results, exceptional results, and likely errors or artifacts*. A scientific paradigm sets the expectations and coordinates the benchmarks for what a scientific community considers legitimate concepts, laws, theories, and research within the field over which the paradigm governs. In effect, a scientific community is defined by the paradigm that directs the everyday research efforts of the practitioners. *This text is concerned with the description and development of the scientific paradigms of modern molecular organic photochemistry.*

The authority of the currently reigning paradigm prevents practitioners in a field from wasting time arguing over fundamentals, irrelevancies, errors, or artifacts. *Because they share the same paradigm, practitioners can proceed rapidly to advanced levels of inquiry without arguing over fundamental issues.* For example, the paradigm of atomic and molecular structure is so authoritative and widely accepted that no modern chemist or physicist argues whether molecules can be usefully represented by three-dimensional (3D) models of atoms connected by bonds that result from the interactions of electrons and nuclei. However, a little over 150 years ago the paradigm of describing molecular structures in terms of 3D geometry was hotly debated by the scientific community, and before 1955, there were no authoritative paradigms governing the description of organic photochemical reactions. Yet, today photochemists are convinced that all observable photochemical phenomena, no matter how complex, can be understood and investigated based on the paradigm of molecular structure and dynamics implied in Scheme 1.1 and its plausible elaborations. The paradigm of organic photochemistry is now considered to be mature.

Because of the maturity of the paradigms of modern molecular photochemistry, organic photochemists do not argue whether the paradigm of Scheme 1.1 is correct in any essential way. Thus, the critical entities of interest to the organic photochemist are immediately defined by Scheme 1.1 as R, hv, *R, I, F, *I, *P, and P. The structures, energetics, and dynamics of these entities are therefore of vital interest to the photochemist. This text will help to develop an understanding of the structures, energetics, and dynamics of these entities and the dynamics of their transformations through the paradigms that currently are the basis of modern molecular organic photochemistry.

In closing this section on paradigms, the student must be warned that the ruling paradigms are by no means permanent but are always subject to change. This is true for at least three reasons, namely, the tentative nature of theories, the incompleteness of experimental information, and the inevitable possibility of completely novel and unanticipated results that may be observed in the future as new techniques are developed and perfected. The history of science over the past two centuries has shown that paradigms that were considered to be absolute and indisputable authorities were eventually overturned as paradigm shifts occurred and the once reigning paradigms were replaced by new governing paradigms. For example, the classical paradigm of light as an electromagnetic wave has been replaced by the quantum mechanical paradigm in which light is viewed as a quantized entity possessing both wave and particle characteristics. The electron, considered a classical particle at the turn of the nineteenth century, is now considered a quantized entity with both wave and particle characteristics (both paradigm shifts are described in Sections 4.2–4.5).

1.6 Exemplars as Guides to the Experimental Study and Understanding of Molecular Organic Photochemistry

Typically, a textbook will describe the paradigms that constitute the assumptions, concepts, strategies, methods, and techniques of the field of interest. An important cognitive tool in learning a field's paradigms is the explicit consideration of specific informative and well-tested examples called *exemplars*. Exemplars provide pedagogical tools that introduce students to new fields of science. For example, in organic chemistry the concept of functional groups provides a familiar set of exemplars for the understanding of organic structures, organic mechanisms, and organic syntheses. The carbonyl, olefinic, enone, aromatic, and other functional groups are all exemplars of a broad scope of chemical and physical properties that can be extended broadly to cover many actual examples. Thus, we can use a functional group as an exemplar to predict the types of reactions and the properties of an extremely wide range of molecules.

An exemplar can also be loosely defined as a universally recognized scientific accomplishment or a set of accomplishments that for a time provides a theoretical and experimental framework for the scientist on how to investigate a new system. For example, the photoreaction of benzophenone with alcohols has served as an exemplar of how to investigate the mechanisms of organic photochemical reactions. Exemplars of the entities and processes shown in Scheme 1.1 are widely used in this text and they provide a basis for the understanding of molecular organic photochemistry.

1.7   The Paradigms of Molecular Organic Photochemistry

Molecular organic photochemistry integrates the paradigms of structure–energy–reactivity correlations, which are the domain of physical organic chemistry, with the paradigms describing the interaction of electromagnetic radiation (photons) with matter (the electrons and nuclei of organic molecules). The paradigm of organic chemistry employs the structure of the molecule (with its implied electronic, nuclear, and spin configurations) as the key organizing concept; the paradigm of electromagnetic radiation employs photons or oscillating electromagnetic waves as the key organizing concept. Thus, *the field of molecular photochemistry is concerned with the interactions of light (represented by photons or oscillating electromagnetic waves) and matter (represented by the electrons and nuclei of molecules) that lead to the formation of* *R *and its eventual conversion to* P *(photochemistry) or* R *(photophysics) through pathways that are elaborations of Scheme 1.1.*

1.8   Paradigms as Guides for Proceeding from the Possible to the Plausible to the Probable Photochemical Processes

The paradigm of Scheme 1.1 provides the organic photochemist with guides for proceeding from the *possible*, to the *plausible*, to the *probable* when considering how to study and interpret photochemical and photophysical processes. How do you characterize a reaction pathway, such as *R → P? For any reaction pathway to be *possible*, molecules (and their vibrational and spin substructures) must obey *all* four of the conservation laws of chemical reactions: (1) the conservation of energy, (2) the conservation of momentum (linear and angular), (3) the conservation of mass (the number and kinds of atoms), and (4) the conservation of charge. As we shall see (Chapter 8), these conservation laws place considerable restrictions on the number of *a priori possible* structures (*R, I, *I, P, *P, and F) and *a priori possible* pathways (Scheme 1.1) that a photochemical reaction can follow. *Only the set of structures and pathways that obeys the conservation laws is considered possible and all others are ruled out, absolutely, with no exceptions!*

However, even when the conservation laws are fully obeyed, the paradigm constrains the actual number of *plausible* pathways for a photochemical reaction by the consideration of the details of molecular structure and implied energies and reorganization associated with structural transformations. In addition, one needs to consider available interactions that couple structures, and available mechanisms of momentum and energy exchange. These considerations lead to a set of "selection rules" that indicate the *plausible* (at some assumed level of approximation) reactions that should be considered from the initial set of *possible* reactions.

To move from the plausible to the *probable*, you must consider specific details of the structure and the available interactions, reorganization energy, and time scales available to the plausible structures. These considerations determine the *kinetics* (or rates) of each of the steps in Scheme 1.1. After eliminating pathways based on kinetic considerations, the remaining (much smaller) set of plausible pathways, which occur at the fastest rates, is considered to be the *set of most probable reaction pathways* of the plausible processes—that is, those that proceed at the fastest rates will win the race from *R to P and are therefore the most probable. We present paradigms that show how to generate selection rules for plausible sets of pathways by employing structures, energetics, and interactions that cause transitions between structures to decide whether a pathway is possible, plausible, or probable. We also describe the experimental and computational methods available to photochemists to experimentally "prove" which of the probable pathways is actually the one that occurs under a given set of conditions.

In attempting to understand an overall photochemical transformation, $R + h\nu \rightarrow P$, it is very useful first to list *all of the plausible pathways* that are available to *R after the absorption of a photon by R (e.g., from Scheme 1.1, the formation of I, the passage through a funnel F, or the formation of *I or *P) and then to qualitatively predict, based on selection rules described in Section 4.13, the relative rates of the plausible pathway(s) to P compared to the rate of all other plausible pathways available to *R that do not lead to P. Predicting an observed or most probable pathway of a photochemical reaction under a given set of conditions requires the ability to use the paradigm of molecular organic photochemistry shown in Scheme 1.1 to make informed judgments based on a knowledge of known, measured rates, exemplars, or theoretically estimated rates based on structure, interactions, energy, and dynamics for a given set of conditions.

The goal of this text is to teach, and for the student to learn, the global and everyday working paradigms that relate, from cradle to grave, the structure, energetics, and dynamics of molecules and photons to photochemical transformations, such as the overall photochemical process $R + hv \to P$ and the overall photophysical process $R + hv \to R$.

1.9 Some Important Questions that Will Be Answered by the Paradigms of Molecular Organic Photochemistry

Now, let us consider in detail one of the possible paths of the global paradigm given in Scheme 1.1, the $^*R \to I \to P$ sequence, which involves the following steps:

1. The *absorption of a photon* (hv) by a reactant molecule (R) to produce an electronically excited state ($^*R$).
2. The *primary photochemical reaction* of the electronically excited state ($^*R$) to produce a thermally equilibrated ground-state reactive intermediate (I).
3. The *thermally induced reaction* of I to produce the observed product(s) (P).

The paradigm of Scheme 1.1 suggests that a photochemist should always ask and attempt to answer a number of standard questions concerning the details of an overall photochemical reaction, $R + hv \to P$. For example,

1. How do we visualize a photon interacting with the electrons of R to induce absorption of a photon to produce $^*R$, and how does this interaction of a photon and the electrons of R relate to theoretical and experimental quantities, such as extinction coefficients, radiative lifetimes, and radiative efficiencies?
2. What are the possible and plausible *structures, energetics, and dynamics* available to $^*R$ and I that occur along the reaction pathway from $^*R \to P$?
3. What are the *possible* and *plausible* sets of *primary photochemical processes* corresponding to the $^*R \to I$ process?
4. What are the legitimate *theoretical approaches, experimental design strategies, experimental techniques, and computational strategies* for experimentally "observing" or validating the occurrence of the species $^*R$ and I that are postulated to occur along the reaction pathway from $^*R \to P$?
5. What is the most *probable* pathway from $^*R \to I$?
6. How is the most probable pathway determined by the competing *kinetic* pathways for the photophysics and photochemistry of $^*R$?
7. What are the absolute rates (rate constants) at which each elementary step occurs along the reaction pathway from $^*R \to P$?
8. What sorts of *structures, energetics, and dynamics* correspond to $^*R$ and I in typical organic photoreactions?

Questions such as these and many more that are implicitly posed by the paradigm of Scheme 1.1 (and its elaborations) can be handled by establishing a more detailed working paradigm and by referencing exemplars that serve as benchmarks for the analysis of photochemical reactions.

1.10   From a Global Paradigm to the Everyday Working Paradigm

In solving normal scientific puzzles, we save a great deal of time by employing an "everyday working paradigm" that is based on considerable experience or precedent and that is found to be generally applicable to a wide range of commonly encountered situations. This shortcut of using a working paradigm is a sort of mechanistic "Occam's razor," relieving the photochemist from always starting from scratch and examining a large number of hypothetically plausible, but historically improbable, situations each time a photochemical reaction is analyzed, an experiment is designed, or a theoretical point is discussed. The paradigm discourages the scientist from wasting time by considering theoretical or experimental situations that are expected to be outside the paradigm. In our study of molecular organic photochemistry, we shall always start with Scheme 1.1 as the global paradigm and determine how we can continuously elaborate it into an ever more specific everyday working paradigm for molecular organic photochemistry. A very effective method for refining Scheme 1.1 is the appeal to exemplars.

As mentioned in Section 1.6, students of organic chemistry are familiar with the effectiveness of the exemplar approach through the study of *functional groups*, where a functional group is an atom or group of atoms that possess qualitatively similar reactivities, spectroscopic properties, and physical properties that are independent of the molecule in which the functional group is found.

Coupling the functional group approach with exemplars from molecular orbital (MO) theory provides a powerful means of predicting chemical reactivity at a qualitative level and will be used extensively in this text to advance an understanding of the photochemistry of exemplar systems. We show that, to a good starting approximation, having an understanding of the photochemistry of the common functional groups of organic chemistry (carbonyl, olefinic, enone, aromatic compounds, etc.) means the working paradigm needs only to consider two things: (1) the electron configurations of two MOs [the highest occupied molecular orbital (HOMO), abbreviated as HO, and the lowest unoccupied molecular orbital (LUMO), abbreviated as LU] and (2) the electron spin configurations of the electrons in the HO and LU for the key structures (i.e., R, *R, I, and P) shown in Scheme 1.1.

Scheme 1.2, an elaboration of Scheme 1.1, includes the energy levels of the HO and LU of R, *R, I, and P as a working paradigm for the examination of molecular organic photochemical reactions that proceed through the path R $+ hv \rightarrow$ *R $\rightarrow$ I $\rightarrow$ P. Scheme 1.2 displays qualitatively the energies of the HO and LU, and at this level, electron spin is not explicitly considered. The energies of the HO and LU for R, *R, and P are assumed to be far apart (typically > 40 kcal mol^{-1}), whereas *the energies of the HO and LU molecular orbitals for* I *are assumed to be very similar and may often be approximated as nonbonding* (NB) *orbitals*. It is assumed in the working paradigm of Scheme 1.2 that all of the remaining electrons that are not shown in this scheme are spin paired (according to the aufbau and Pauli exclusion principles) in orbitals of lower energy and are of secondary importance in determining the course of

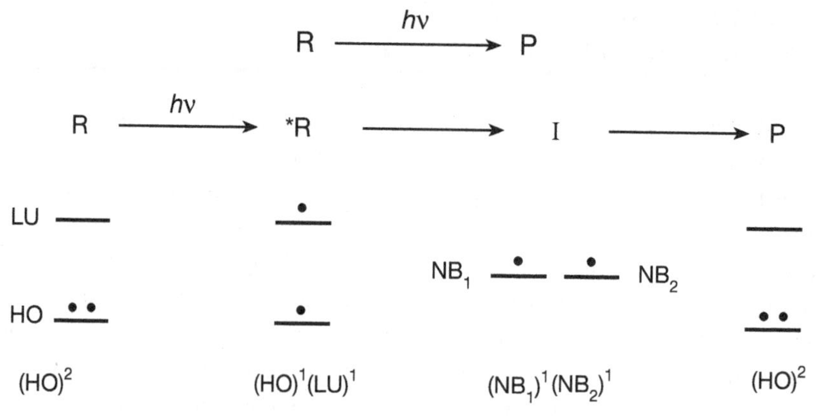

$$R \xrightarrow{\enspace h\nu \enspace} P$$

$$R \xrightarrow{\enspace h\nu \enspace} {}^*R \longrightarrow I \longrightarrow P$$

LU ——   •——

                                      NB$_1$ •—— •—— NB$_2$ ——

HO •• ——   •——   •• ——

(HO)^2 (HO)^1(LU)^1 (NB$_1$)^1(NB$_2$)^1 (HO)^2

Scheme 1.2   The global paradigm of organic photochemical reactions displaying orbital configurations of R, *R, I, and P.

the photochemical and photophysical processes (because these lower-energy electrons are difficult to perturb, even in photochemical processes).

The starting point for the analysis of a photochemical or photophysical process is the assignment of the electronic nature of the HO and LU for R and *R. This amounts to assigning a specific electron configuration to R(HO)^2 and *R(HO)(LU). The reactive intermediate I is generally a species possessing two nonbonding orbitals that are produced by the primary photochemical process *R → I(NB)^1(NB)^1. There are no HO or LU of very different energies in the latter case, but instead there are two NB orbitals of similar energy. Thus, the chemistry of I will be determined by the electronic configuration of two electrons in two NB orbitals (and, as we shall see in Section 1.11, by the spin configuration of the two electrons, too).

When the two NB orbitals are located mainly on carbon atoms, the lowest-energy orbital configuration of I corresponds to one electron in each NB orbital [i.e., I(NB)^1(NB)^1], thus producing a radical pair I(RP) or biradical (BR). (The terms "biradical" and "diradical" are sometimes used interchangeably in the photochemical literature, but we use the term "biradical" solely for the situation in which two NB orbitals each contain one electron and both NB orbitals are contained in the same molecular structure. We use the symbol D to mean a more general "diradicaloid" species that could be a RP, BR, or some related structure. The definition of a diradicaloid is discussed in detail in Chapter 6.)

In some cases, when the energies of the NB orbitals are significantly different, the reactive intermediate I may possess an electron configuration that places both electrons in the lower-energy orbital, a situation that requires the two electrons to be spin paired. Such electronic configurations correspond to species called *zwitterions*, Z(NB)^2, which for simplicity we are ignoring at this point.

In all cases the orbitals assigned to the HO and LU will be simple one-electron orbitals that are familiar to the student from courses in organic and physical chemistry.

In this approximation, moreover, we ignore electron–electron repulsions that would lead to different energies of the HO and LU in R and *R. This approximation is discussed in more detail in Chapter 2 and is employed throughout the text.

The working paradigm of Scheme 1.2 suggests a number of questions that need to be answered when studying or analyzing any organic photochemical reaction:

1. What are the electronic characteristics of the HO and LU involved in the R + $h\nu$ → *R process?
2. What is the electronic configuration of *R (i.e, the orbital occupancy of the HO and LU)?
3. What are the plausible primary photochemical and photophysical processes typical of *R based on its electron configuration $(HO)^1(LU)^1$?
4. What are the electronic natures of the NB orbitals of I?
5. What are the plausible secondary thermal reactions of I that lead to P?

Scheme 1.2 requires one more level of structural elaboration before it can be employed as an everyday working paradigm. (For simplicity, at this stage, we are ignoring both the *R → F processes and the *R → *I processes.) The nature of the funnel (F) will be considered in detail in Chapter 6. This final level of detail includes not only the electronic configurations of *R and I but also the electronic *spin* configurations of *R and I, as shown in Scheme 1.3. Now, we consider the role of spin in a photochemical reaction of the type *R → I → P.

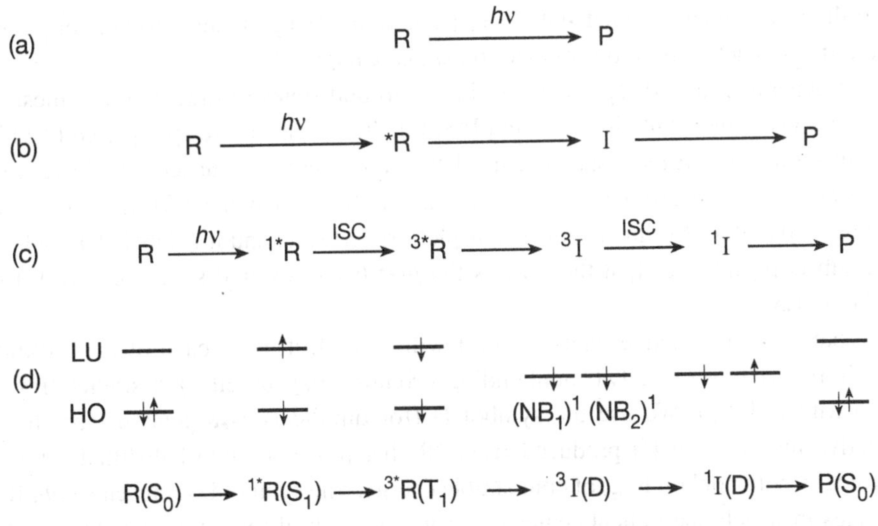

Scheme 1.3 Exemplar paradigm for an organic photochemical reaction that proceeds through a triplet state.

1.11   Singlet States, Triplet States, Diradicals, and Zwitterions: Key Structures Along a Photochemical Pathway from *R to P

Scheme 1.3 describes an elaboration of the orbital and spin structural detail for the exemplar photochemical reaction $R + h\nu \rightarrow {}^*R \rightarrow I(D) \rightarrow P$. First, we consider the elaboration of the orbital description of the species along the reaction path (b), then we consider an elaboration of the spin description of the species along the reaction path (c).

The electronic configurations of the ground states of R and P are generally $(HO)^2(LU)^0$ for ordinary organic molecules. According to the Pauli exclusion principle, the spins of two electrons in the same orbital must be paired (the spins will be symbolized as $\uparrow\downarrow$, termed "antiparallel spins," and correspond to a singlet spin configuration or a singlet state). The electronic configuration of *R and I, both of which typically possess one electron in each of the two key orbitals (HO and LU or the two NB orbitals) shown in Scheme 1.3d, are not required by the Pauli exclusion principle to be spin paired, so the two key electrons in the half-filled orbitals can be either paired ($\uparrow\downarrow$, singlet states) or unpaired (symbolized as $\uparrow\uparrow$ and referred to as "parallel spins," corresponding to a triplet spin configuration or a triplet state).

The *singlet states* of a molecule are given the symbol S_n, where the subscript n ranks the energy of the singlet state. The subscript 0 is reserved for the lowest-energy electronic ground state, which is always a singlet state (i.e., S_0) for ordinary organic molecules. The first excited singlet state is S_1, the second excited singlet state is S_2, and so on. When *R (or I) possesses two orbitally unpaired electrons and the electron spins are spin unpaired ($\uparrow\uparrow$), the structure is termed a *triplet* state and labeled T_n, where the subscript n ranks the energy of the triplet state. [Since the subscript 0 is reserved for the lowest-energy electronic ground state (i.e., S_0), $n = 1, 2, \ldots$ for triplet states, and the lowest energy triplet state is T_1.] The terms "singlet" and "triplet" originate from the magnetic properties of electron spins (Chapter 2).

In general, R and P represent the singlet ground states of organic molecules, so they are given the symbols $R(S_0)$ and $P(S_0)$. If the electrons are spin paired ($\uparrow\downarrow$) in *R, this is a singlet *excited* state and it is labeled S_1, where the subscript indicates that the state is the *first excited* singlet state [i.e., ${}^*R(S_1)$ in Scheme 1.3]. If the electron spins are parallel ($\uparrow\uparrow$) in *R, this is a triplet *excited* state and it is labeled T_1, where the subscript indicates that the state is the *first* triplet excited state [i.e., ${}^*R(T_1)$ in Scheme 1.3].

Likewise, the reactive intermediate I with one electron in each of two orbitals of similar energy (e.g., two nonbonding orbitals) may be either a singlet ${}^1I(\uparrow\downarrow)$ or a triplet ${}^3I(\uparrow\uparrow)$. We use the symbol D (for diradical) as a general label for a reactive intermediate (I) produced from *R that possesses two half-filled orbitals (typically both nonbonding) of comparable energy. The symbol D represents *both* RP, species in which one radical center is located on each of two molecular fragments, and BR, species in which the two radical centers are located on a single molecular structure. Thus, the symbol I(D) refers to a reactive intermediate that possesses diradical character and for which the two half-filled orbitals are of similar energy. The I(D) species differ from *R, because *R species possess two half-filled orbitals that are of *very different energies*. This distinction will be of particular importance when we consider the role of electron–electron interactions in changing the energies of orbitals from the values for one-electron orbitals.

The symbols ^{1}I(D) and ^{3}I(D) represent singlet and triplet diradical intermediates, respectively. The superscript indicates the spin state of the intermediate, and the D means the intermediate possesses two electrons in half-filled orbitals. Then, it follows that the symbols ^{1}I(RP) and ^{3}I(RP) represent singlet and triplet radical pairs, respectively, and the symbols ^{1}I(BR) and ^{3}I(BR) represent singlet and triplet biradicals, respectively.

If I is in a singlet state, it is also possible for the two electrons to be in *one* NB orbital, and for no electrons to be in the other, that is, $I(NB)^2(NB)^0$. Such a species is referred to as a *zwitterion* and given the symbol I(Z). The I(Z) species are involved in the $^{1*}R \rightarrow {}^1I(Z)$ and the $^{1*}R \rightarrow F$ steps of photoreactions involving certain singlet states, whereas D species are always involved in the $^{3*}R \rightarrow {}^3I(D)$ step of photochemical reactions involving photochemical processes initiated in $^{3*}R$. The rules for D or Z formation and the chemical properties of these species are described in Chapter 6.

Scheme 1.3 represents a working *exemplar paradigm* for all photochemical reactions of organic molecules that proceed through a triplet excited state, $*R(T_1)$. For any given reaction, R may be a carbonyl, an olefinic, an enone, an aromatic compound, or so on. We need to know the nature of the HO and LU of each of these structures to deduce the electronic configuration of *R. Given the electronic configuration of *R, we can generate "selection rules" for the plausible *primary photochemical reactions* *R →I. Predicting and understanding photochemical reactions requires a knowledge of the structures of the entities shown in Scheme 1.3, namely, $R(S_0)$, $*R(S_1)$, $*R(T_1)$, ^{3}I, ^{1}I, and $P(S_0)$, and of the probabilities of the transitions between the structures connected by the pathways shown in Scheme 1.3(c).

In Chapter 6, we see that when the electronic configuration of T_1 is HO = n (i.e., a nonbonding MO) and LU = π^* (i.e., an antibonding MO), which is the case for acetone, benzophenone, and many other ketones, there is only a small set of primary photochemical processes of the type $*R(T_1) \rightarrow {}^3I(D)$ that are plausible. The important role of electron spin rears its head in the overall reaction because (Scheme 1.3) the reactive intermediate ^{3}I(D) must be converted to a singlet intermediate, ^{1}I(D), before the final product $P(S_0)$, which is a singlet state, can be formed.

How do the ISC processes $*R(S_0) \rightarrow *R(T_1)$ and $^3I(D) \rightarrow {}^1I(D)$, which require a change in electron spin, occur? A useful vector representation of electron spin is presented in Chapter 2 to describe how electron spin operates to control the steps that interconvert singlets and triplets. In Chapters 3 and 6, this vectorial representation of electron spin is used to answer questions pertaining to the interconversion of spin states. Now that we have introduced the important global and exemplar paradigms for analyzing organic photochemical reactions, we can develop the *state energy diagram*, which makes it possible to use and manipulate the working and exemplar paradigms.

(a) Norrish Type I reaction: I(D) = radical pair, I(RP)

$$R \xrightarrow{h\nu} {}^*R \xrightarrow{k_{PP}} I(RP) \xrightarrow{k_{SP}} P_1$$

(b) Norrish Type II reaction: I(D) = biradical, I(BR)

$$R \xrightarrow{h\nu} {}^*R \xrightarrow{k_{PP}} I(BR) \xrightarrow{k_{SP}} P_1 \longrightarrow P_2$$

Figure 1.1   (a) An example of a primary photochemical $^*R \rightarrow I(RP)$ process, the "Type I" α-cleavage of ketones. (b) An example of a primary photochemical $^*R \rightarrow I(BR)$ process, the "Type II" intramolecular hydrogen abstraction of ketones with alkyl side chains.

Figure 1.1 presents two concrete exemplars of $^*R \rightarrow I(D)$ primary photochemical processes. In the first example, the Norrish Type I reaction, $^*R$ undergoes an α-cleavage of the C—C bond of the C=O function to produe a radical pair, I(RP). In the second example, the Norrish Type II reaction, $^*R$ undergoes an intramolecular hydrogen abstraction to produce a biradical, I(BR). These two types of $^*R \rightarrow I(D)$ primary photochemical processes are very common and provide excellent exemplars for the analysis of the photochemical primary processes of a wide range of organic molecules.

1.12   State Energy Diagrams: Electronic and Spin Isomers

According to Scheme 1.3, our exemplar paradigm of organic photochemistry, there are three important molecular states, $R(S_0)$, $^*R(S_1)$, and $^*R(T_1)$, that must always be considered when starting an analysis of a photochemical reaction involving organic molecules. A *state energy diagram* (Scheme 1.4) provides a compact working exemplar for displaying the relative energies and keeping track of the ground state (S_0), the lowest-energy excited singlet state (S_1), and the lowest-energy triplet state (T_1) of an organic molecule (where E_S is the energy of S_1 and E_T is the energy of T_1). The electronic configurations of the S_0, S_1, and T_1 states are also shown. Higher-energy singlet states (S_2, S_3, etc.) and higher-energy triplet states (T_2, T_3, etc.) can also be included as desired, but need not be explicitly included in the working state diagram, because experience has shown that *excitation of these higher-energy excited states generally results in deactivation to S_1 and T_1 faster than any other measurable process* (Kasha's

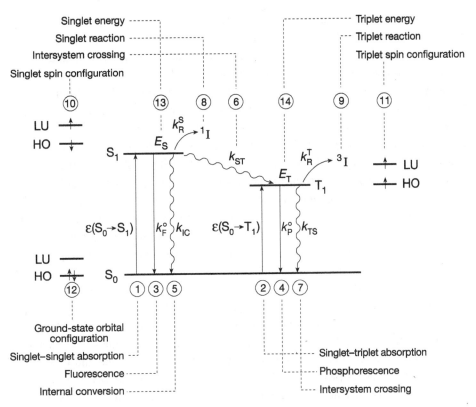

Scheme 1.4   The state energy diagram is a fundamental paradigm of modern molecular photochemistry. The intersystem-crossing rate constants k_{ST} and k_{TS} are sometimes referred to as k_{ISC}.

rule, Chapter 4). In a state energy diagram, the y-(vertical) coordinate represents the potential energy (PE) of the system. The x-(horizontal) coordinate has no physical meaning (it is *not* a reaction coordinate or potential energy surface). The lines representing the state energies of S_1 and T_1 are displaced for convenience and to avoid congestion. Thus, a state energy diagram shows the ranking of the energies of the S_0, T_1, and S_1 states and is most useful if actual values of these energies can be associated with the T_1 and S_1 states. How the energies of S_1 and T_1 are determined experimentally is described in Chapter 4.

State energy diagrams are sometimes referred to as Jablonski diagrams[5] in honor of Aleksander Jablonski, a Polish physicist who used a schematic portrayal of the relative positions of the electronic and vibrational levels of R and *R without any attempt to indicate the relative nuclear geometries. For simplicity, vibrational levels are omitted in this introduction to state energy diagrams. Vibrational levels, which are very important in determining the rates of photophysical processes, are introduced into state energy diagrams in Chapter 2.

In state energy diagrams, it is assumed that the equilibrium nuclear geometries of R and *R are similar and that these geometries represent minima of R and *R. Since all of the structures in the state energy diagram have the same composition (i.e., numbers and kinds of atoms) and the same constitution (i.e., atom connectivity) as S_0, but are chemically different from S_0, all of the states in the state diagram (S_0, S_1, and T_1) are formally isomers of S_0 and also isomers of each other!

What is the basis of the isomerism? The isomerism results from differences in the *electronic configurations* (*electronic isomers*) or in the *spin configurations* (*spin isomers*) between the displayed states. The S_n and T_n states are electronic isomers of each other. The differences in electronic isomers are due to differences in the orbital configurations (i.e., in the different occupancies of the HO or LU) or to differences in the electronic spin configurations of each state (i.e., ↑↓ or ↑↑). The S_n and T_n states are related to each other as *spin–electronic isomers*, that is, singlet spin configuration (↑↓) or triplet spin configuration (↑↑). In addition to the spin–electronic isomerism, the states in the state energy diagram may also be stereoisomers of one another (i.e., they may have the same constitutions and the same spin–electronic configurations, but different spatial arrangements of their atoms).

The state energy diagram provides a handy and useful way to organize and systematize the state electronic structures, the state electronic energies, and the dynamics of interstate transitions corresponding to *all possible* photophysical processes that interconnect S_0, S_1, and T_1. Transitions between any two electronic states in the diagram correspond to the *possible* connections between the states indicated and may be radiative or radiationless processes. The *plausibility* and the *probability* of a transition between any two states, however, requires knowledge of specific molecular structures and reaction conditions, which can be varied at will by the experimenter. The photophysical processes are defined as transitions in the energy diagram that interconvert excited states with each other or that interconvert excited states *R with the ground-state *R. All possible photophysical transitions from S_1 and T_1 must be considered in an overall *R $\rightarrow$ R photochemical analysis, since photophysical processes will, in principle, be competitive with the photochemical processes from these two key states. If the photophysical processes are very fast compared to the photochemical processes, the competing *photochemical* processes may be plausible, when considered as an isolated process, but will be inefficient and *improbable* because plausible competing photophysical processes occur at a faster rate.

As an exemplar, let us see how the state diagram describes the *possible* photophysical *radiative* processes (processes 1–4 in Scheme 1.4), which involve the absorption or emission of a photon.

1. The spin-allowed singlet–singlet absorption of photons ($S_0 + h\nu \rightarrow S_1$), characterized experimentally by an extinction coefficient $\varepsilon(S_0 \rightarrow S_1)$.
2. The spin-forbidden singlet–triplet absorption of photons ($S_0 + h\nu \rightarrow T_1$), characterized experimentally by an extinction coefficient $\varepsilon(S_0 \rightarrow T_1)$.
3. The spin-allowed singlet–singlet emission of photons ($S_1 \rightarrow S_0 + h\nu$), called *fluorescence*, characterized by a rate constant, k_F.
4. The spin-forbidden triplet–singlet emission of photons ($T_1 \rightarrow S_0 + h\nu$), called *phosphorescence*, characterized by a rate constant, k_P.

The *plausible* photophysical *radiationless* processes are processes 5–7 in Scheme 1.4:

5. The spin-allowed radiationless transitions between states of the same spin ($S_1 \to S_0$+ heat), called *internal conversion*, characterized by a rate constant, k_{IC}.

6. The spin-forbidden radiationless transitions between excited states of different spin ($S_1 \to T_1$ + heat), called *intersystem crossing*, characterized by a rate constant, k_{ST}.

7. The spin-forbidden radiationless transitions between the triplet and the ground state ($T_1 \to S_0$ + heat), also called intersystem crossing, characterized by a rate constant, k_{TS}.

All of the structures in the state energy diagram refer to a single fixed equilibrium (minimum) nuclear geometry of R; the geometry of *R is assumed to be very similar to that of R in the state energy diagram. As a useful extension of the state energy diagram, *primary photochemical processes* can be defined as transitions from an electronically excited state *R(S_1 or T_1) that yield molecular structures of different constitution or geometry from that of *R. These chemically different molecular structures are the reactive intermediates I of Schemes 1.1–1.3 and are produced by either process 8 or 9 in Scheme 1.4:

8. A photochemical reaction from S_1 to produce a reactive intermediate, $S_1 \to {}^1I$, called a *primary photochemical reaction*, characterized by a rate constant, k_R^S.

9. A photochemical reaction from T_1 to produce a reactive intermediate, $T_1 \to {}^3I$, also called a primary photochemical reaction, characterized by a rate constant, k_R^T.

The final, isolated product of a photochemical process results from the thermal chemistry of I under the reaction conditions. The thermal I $\to$ P processes are called *secondary thermal reactions* and are expected to occur in exactly the same manner as when the reactive intermediate I(D) is produced by a ground-state thermolysis. Although not a photochemical process, an understanding of the I $\to$ P pathway that occurs completely in a ground state is crucial, however, in order to be able to completely describe the overall process R + hv $\to$ P. A more complete description of the *R $\to$ P process is provided by the working paradigm of a *potential energy surface*, which is described qualitatively in the next sections and in detail in Chapters 3 and 6.

In order to determine which of the *plausible* processes are most *probable* from S_1 or T_1, we need information on the *relative rates* of all of the plausible photochemical and photophysical processes that compete for deactivation of these states. The values of these rates are available if the *rate constants* (k) for the various processes shown in the energy diagram of Scheme 1.4 are known or can be estimated from experiment, via an appeal to exemplars, or through computation. The relative rates of the transitions from a given state determine the probability of the various plausible processes that can occur from the state. These relative rates depend on a number of structural and energetic factors that are discussed in Chapters 2–6. At this stage, the exemplar working paradigm is incomplete, since for simplicity we have not explicitly considered other possible, but less common, pathways of Scheme 1.1, such as *R → F → P or *R → (*I, *P) → P. These possibilities are described in Chapters 4–6.

1.13   An Energy Surface Description of Molecular Photochemistry

In proceeding from the state energy diagram (that assumes a fixed nuclear geometry of R) to a complete analysis of a photochemical reaction (that creates a different nuclear geometry than that of R to first form I and then proceed to P), it is necessary to keep track of a number of structures, energies, and dynamics of transitions. Keeping track of all of these features of a photochemical reaction involves a complicated energetic, structural, and dynamic bookkeeping that is nicely handled by the paradigms associated with *potential energy curves and surfaces* (which are discussed in detail in Chapters 3 and 6). For now, we preview how the paradigm of energy surfaces handles the problem of simultaneously integrating the structure, energetics, and transition dynamics involved in photochemical and photophysical processes.

A potential energy (PE) surface displays the PE of a molecular system (the y-coordinate) versus the varying molecular structure of the system (the x-coordinate). The *lowest PE path* along a given potential energy surface is called the *reaction coordinate*. Strictly speaking, PE surfaces are multidimensional mathematical objects that are difficult to visualize. However, as a reasonable "zero-order" (i.e., working) approximation to an energy surface, we can use two-dimensional (2D) "potential energy curves"; for simplicity we use the term "energy surface" to describe these curves.

A PE curve extends the concept of a state energy diagram to describe how the PE of the states of a system changes as the nuclear geometry of *R changes from one that is very similar to that of R to one that begins to resemble the geometry of the possible structures (e.g., I) involved in the photochemical transformation of *R → I → P. Consider the *hypothetical* example of the energy surfaces of the ground and excited states shown in Scheme 1.5. For simplicity, both surfaces are assumed to be singlet states. This exemplar energy surface is intended to display important common features of photochemical reactions but is not representative of any particular class of photoreactions. Whereas in the state energy diagram a nuclear geometry similar to that of the ground state (R) is assumed for all of the structures considered, each point on the PE curve represents a different nuclear geometry (specified on the x-axis) and an associated PE (specified on the y-axis). For a given nuclear configuration, the PE of a molecule is determined mainly by its electronic orbital configuration and its spin configuration.

As an exemplar, the lower-energy curve shown in Scheme 1.5 corresponds to the *reaction coordinate* (the lowest-potential-energy path) for the hypothetical *thermal* transformation R → P. When more than one energy surface is involved, for simplicity, we assume that the reaction coordinate that refers to the pathway involving the

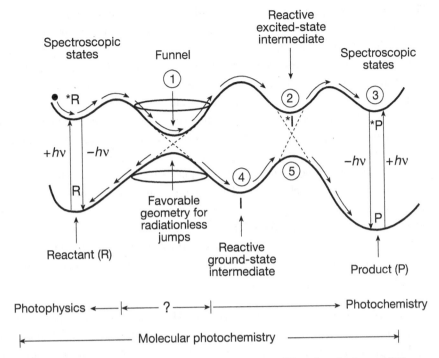

Scheme 1.5   Schematic representation of a ground-state (R) and excited-state (*R) energy surface. The arrows on the surface indicate the motion of a point representing a molecule whose nuclear geometry is moving along the reaction coordinate from left to right. The "?" indicates a "twilight zone" region where the distinction between photochemistry and photophysics is fuzzy.

ground surface starting from R is the same reaction coordinate that *R follows. In general, however, the lowest-energy pathway for the ground-state transformation for the R → P process need not be the lowest-energy pathway from *R → P. In Chapter 6, we discuss the theory of photochemical reactions, which will make it possible to qualitatively predict the reaction coordinates of excited-state reactions.

Scheme 1.5 shows hypothetical surfaces for an overall reaction *R → P for two different starting electronic configurations, a ground-state R and an excited-state *R. The lower-energy surface is called the *ground-state electronic surface,* and the higher-energy surface is called an *excited-state electronic surface.* Any point, r, of interest on either surface is a *representative point* of the PE (the y-axis) of the system for each nuclear geometry along the reaction coordinate (the x-axis). In this way, we can envision photophysical and photochemical processes in terms of the *motion* of a representative point on a PE curve, where each representative point corresponds to the energy of a specific nuclear configuration on the R → P pathway for one surface or the other. Starting from R, we can track the representative point r starting from R and imagine the trajectory of r moving along the excited surface or along the ground surface (propelled along the reaction coordinate by collisions with other molecules in the environment).

The representative point of an electronically excited molecule *R spends its time on either the excited or ground surface on the way from *R → P, except for the short periods of time when it finds a "funnel" (F) between the excited- and ground-state surface through which the representative point can "jump" from one surface to the other (the time scales for these jumps are very short for singlet states). To the extent that these hypothetical surfaces are valid, they make it possible to visualize or map *all plausible pathways* for the *R → P transformation. Now, we consider some of the *a priori* plausible pathways for the electronically excited-state *R based on the nuclear-geometry changes that are mapped out by the two curves in Scheme 1.5.

First, consider the important topological (qualitative) features of the two hypothetical surfaces shown in Scheme 1.5. These features include the nuclear geometries for the maxima and minima on each surface, the nuclear geometries for which the surfaces are far apart in energy, the relative disposition of the maxima and minima to each other, and the geometries for which two surfaces come close to one another in energy. We focus on the following important features of the maxima and minima of the two surfaces:

1. *Spectroscopic (Franck–Condon) Minima.* The *absorption* of a photon (the R + *hv* → *R step) involves a jump from a minimum in the ground state (e.g., R or P) to a minimum on the excited surface (e.g., *R or *P). The *emission* of a photon involves a jump from a minimum on the excited-state surface to a minimum on the ground-state surface (e.g., *R → R + *hv* or *P → P + *hv*). In Chapters 3 and 4, we will see that radiative jumps occur with the highest probability between surfaces for which there is a minimum and similar nuclear geometry in *both* the excited- and ground-state surfaces. Such minima are called "spectroscopic" or "Franck–Condon" minima. Small maxima (energy barriers of a few kilocalories per mole) may separate such spectroscopic minima from other regions of the PE surfaces, as shown on the left in Scheme 1.5. If the barriers are small, thermal energy from collisions with neighboring molecules will be sufficient to propel the representative point "to the right" of the energy surface toward region 1. If the barriers are high, then the representative point will be "trapped" in the excited-state minimum for *R until the point returns to R by either emission of a photon (fluorescence) or radiationless deactivation (internal conversion).

2. *Surface-Crossing Minima.* Excited energy surfaces possess minima in energy corresponding to excited-state energy surface crossings with lower-energy surfaces. Crossings of this type, depending on the available electronic interactions, may be true or weakly avoided crossings (the dotted lines in region 1 of Scheme 1.5) or may be strongly avoided crossings (the dotted lines in region 2 of Scheme 1.5). These crossings are described in detail in Chapter 6. If the representative point approaches a weakly avoided crossing, a very fast and therefore probable transition to the ground state will occur. When very fast transitions occur from the excited to the ground surface, a "funnel" is said to exist on the excited surface (region 1). In Chapter 6, we will see that passage

through such funnels can occur on the fastest time scales possible. In addition, we see why such funnels are given the special name of "conical intersections."

3. *Ground-State Maxima Due to Surface Crossings.* Barriers that exist on the ground-state energy surface can often be viewed as having an approximate surface-crossing origin before a weak electronic interaction is taken into account. A surface crossing or a weakly avoided crossing leads to the same result; namely, the excited surface comes close in energy to the ground-state surface at some geometry along the reaction coordinate. Weakly avoided surface crossings may be identified with large barriers along the reaction coordinate in the ground state (excited-state region 1), and strongly avoided surface crossings may be identified with low barriers along the reaction coordinate in the ground state (excited-state region 2). These crossings are described in detail in Chapter 6.

The $R + h\nu \rightarrow {}^*R$ process places the system on the excited-state surface, as an electron jumps from a HO to a LU and the electron configuration of negative charges due to the electrons changes "instantaneously" from that of R to that of *R. Consequently, the positively charged nuclei feel a different negative electrical force field whose direction and magnitude are given by the shape (i.e., the gradient) of the excited surface (*R) and no longer by the ground state (R). The new force on the nuclei is due to the different configuration of electrons for *R (i.e., one electron in the HO and one in the LU) compared to R (i.e., two electrons in the HO and none in the LU). The new electrical forces resulting from the HO $\rightarrow$ LU electronic jump cause the nuclei to rearrange to better accommodate the new electronic distribution. The impulse of the newly created electronic distribution causes the nuclei to move, generating kinetic energy of the nuclei (kinetic energy generated in this way is called vibrational energy), which is rapidly transferred intramolecularly and then intermolecularly to the surrounding solvent so that *R reaches the minimum vibrational energy in a few picoseconds (ps). The rates of these vibrational energy-transfer processes are described in Chapter 3 and Chapter 5.

The motion of a representative point, r, on a PE surface completely controls the nuclear motion of molecules, except for regions of the surface that come close to one another. When two surfaces do happen to come close together, each surface has a chance to "compete" for control of the motion of the representative point and, therefore, the control of the nuclear motion of the reacting system. In such regions, the nuclear system is "confused" as to which surface will control its motion.

Let us follow some possible trajectories of a representative point along the energy surfaces of Scheme 1.5. Begin with R on the ground surface (the spectroscopic minimum at the bottom left of Scheme 1.5). Absorption of a photon is extremely fast relative to vibrational motion, and therefore the representative point makes a "vertical jump" (with no change in the nuclear geometry of R) from the ground (R) to the excited-state surface to produce *R, which is assumed to be formed in a relatively shallow energy spectroscopic (Franck–Condon) minimum. A radiative transition is possible for *R, which can emit a photon of fluorescence and return to the ground state (R). A radiationless path that moves the system toward the structure of P is also possible: As a result of thermal collisions with surrounding molecules, the representative point r corresponding to *R may overcome the small barrier along the excited surface and proceed to region 1, which happens in this hypothetical example to be a weakly avoided "crossing" of the excited- and ground-state surfaces. Such a situation is highly favorable for a very rapid jump from the excited to the ground surface (we explain why this occurs in Chapters 3 and 6), so that such regions on the excited state serve as "funnels" (F), which can take the representative point from the excited to the ground-state surface. These funnels are the same species (F) first encountered in the *R → F processes of Scheme 1.1.

After reaching region 1, the representative point r has two options. First, the point may jump to the ground surface and "spill" into the R minimum (resulting in a net "photophysical" cycle, R + hv → *R → F → R). These jumps are *internal conversions*, since the initial and final states are both singlet states. Internal conversion from *R → R occurs inefficiently when there is a large gap separating the two states, as is the case near region 2 (the rules concerning the factors controlling the rates and efficiencies of internal conversions are explained in Chapter 5), and becomes more and more rapid as the gap between the states undergoing internal conversion decreases.

In the second option, the representative point may jump from region 1 of the *R surface to the right of the maximum on the ground surface and form the reactive intermediate I. Such jumps correspond to *primary photochemical processes* (i.e., *R → I transitions). Since I is a reactive intermediate, it may live long enough to achieve thermal activation and proceed over the barrier (region 5 in Scheme 1.5) to yield the product P. For the pathway R* → (1) → (4) → (5) → P, the nuclear motion is controlled by the excited-state surface for part of the reaction, *R → (1), and by the ground-state surface, (4) → (5) → P, for another part of the reaction. This situation, although hypothetical in the example given, is typical of many photoreactions, as described in Chapter 6.

Because of the rapid rate of passing through the funnel (F) to the ground-state surface, only a few, if any, *R molecules moving on the excited surface will be able to gather enough thermal energy and proceed to region 2 of the excited surface, especially if there is a significant barrier for proceeding to *I. The latter is a minimum on the excited surface that corresponds to an *electronically excited reactive intermediate* *I. Note that unlike the weakly avoided surface crossings of region 1, *I is separated from a maximum ground state by a relatively large amount of energy. This sort of minimum–maximum/excited-surface–ground-surface correspondence is a signature of a strongly avoided surface crossing.

In some rare cases, the representative point may make it from region 2 and pass over an energy barrier to region 3, which possesses a minimum that corresponds to *P, an excited state of the product (P). When *I is formed, a true photoreaction has occurred, since a reactive intermediate (I) and its excited state (*I) possess a nuclear geometry that is quite distinct from that of R. Note that the minimum corresponding to *P on the excited surface possesses a corresponding minimum on the ground surface. This means that the nuclear geometries of *P and P are similar, as is the case for *R and R on the right side of Scheme 1.5. The representative point in region 3 may jump to the ground state with the emission of a photon (fluorescence) or with the release of heat (internal conversion).

As a rule, the absorption and emission of light occur near spectroscopic minima corresponding to the nuclear geometry of the reactants and products. (This rule, called the *Franck–Condon principle*, is discussed in Chapter 4.) Thus, there is both a radiative (*R $\to$ R + $h\nu$) and a radiationless (*R $\to$ F $\to$ R) pathway for *R to return to R. Pathways that return the system back to R after the absorption of light are called *photophysical* pathways and are of great importance because they generally compete with the *photochemical* pathways that carry *R to I and eventually to P. It is also possible that *R may proceed to an electronically excited *I and *P, although this pathway (called an adiabatic photoreaction) is rarely found. Thus, these processes are considered possible, but not plausible, except in special circumstances.

Although Scheme 1.5 represents an arbitrary and hypothetical overall photochemical reaction (R + $h\nu$ $\to$ P), the pathways and processes shown represent an exemplar for most of the important photochemical and photophysical processes and allow the following generalizations to be made based on considerable theoretical and experimental experience:

1. Absorption (R + $h\nu$ $\to$ *R) and emission (*R $\to$ R + $h\nu$ and *P $\to$ P + $h\nu$) of photons tend to occur at nuclear geometries corresponding to *spectroscopic minima* in both the ground and the excited surface.
2. Radiationless jumps from one surface to another are most probable for nuclear geometries at which two surfaces, a minimum and a maximum, come close together in energy (*R $\to$ R and *R $\to$ I).
3. The location and heights of energy barriers on *both* the excited- and ground-state surfaces may determine the specific pathway of a photoreaction.
4. Some minima on excited surfaces (e.g., funnels, F) may not be readily detected by conventional absorption and emission techniques.
5. The course of a photoreaction depends on competing photophysical, as well as photochemical, processes.

In Chapter 3, we explain how to use PE curves to describe photochemical and photophysical transitions, and then apply this knowledge to many situations in the subsequent chapters.

1.14 Structure, Energy, and Time: Molecular-Level Benchmarks and Calibration Points of Photochemical Processes

The most powerful paradigms in all of chemistry are derived from the representation of molecules as particles possessing various levels of internal structure (i.e., atoms, nuclei, electrons, and spins). Both a qualitative and a quantitative appreciation of molecular dimensions, molecular dynamics, and molecular energy are important for visualizing events and estimating their rates at the electronic and molecular levels. An understanding of the sizes of molecules and the time and energy required for electrons and nuclei to move in space is at the heart of mechanistic descriptions of molecular and spectroscopic phenomena. The ability to achieve a transformation at the molecular level depends on the energy of the initial state, the energy of the final state, the amount of thermal energy available to do work to cause the transition, and the time required to execute the transformation relative to the interactions (forces) that drive the structural changes of interest. Intuitively, the rate of the transformation depends on the efficiency of getting energy into the correct modes or degrees of freedom that cause a motion that can change the structure in the appropriate fashion.

In order to calibrate the energy, distance, and time scales, we now consider some benchmark values of energy and time of great importance to photochemistry.

1.15 Calibration Points and Numerical Benchmarks for Molecular Energetics

Organic chemists are accustomed to counting molecules and using the mole and Avogadro's number (6.02×10^{23}) as a benchmark for the number of molecules contained in 1 mole of molecules. The measurable mass of a pure molecular substance (in grams, g) can be translated into the number of moles of the substance divided by the molecule's molecular weight (in grams per mole, $g\,mol^{-1}$).

Photochemists, however, are interested in counting not only molecules but also the number of photons in a light source (the intensity of the light source is the number of photons emitted per second at a specific wavelength, λ). If we consider the photon as a "massless reagent," then the intensity of a light that is absorbed from a source to produce *R in a given volume is related to the concentration of molecules R in a solution. The number of photons absorbed, by a given concentration of molecules through a given path length, is a measure of the "cross section" for absorption (at a given wavelength) that a molecule presents to a stream of passing photons corresponding to the wavelength of absorption (Chapter 4). The number of molecules of I or P produced per photon absorbed is called the *quantum yield* (Φ) of the formation of a reactive intermediate (I) or a product (P).

We now try to understand some of the quantities that are important in all chemical transformations but which are particularly important in photochemical transformations. In this section we consider some *calibration points and numerical benchmarks for molecular and photonic energetics*, and in Section 1.16, we consider some calibration points for molecular and photonic sizes and dynamics.

In general, in photochemistry we are concerned with the *difference in energy of the energy gap*, (ΔE), *between states of a molecule* (Eq. 1.1), rather than the absolute energy of a state.

$$\Delta E = |E_2 - E_1| \qquad \text{The energy gap between } E_2 \text{ and } E_1 \qquad (1.1)$$

Absorption of a photon by a molecule ($R + hv \rightarrow {}^*R$) transforms light energy (a photon, hv) into the electronic excitation energy (*) of a molecule. The photon uses its energy to do work by changing the structure of the orbiting electrons, or vibrating nuclei, or "precessing spins" of a molecule (see Section 2.28). The absorption of light not only provides the molecule with energy that it can employ to make or break chemical bonds but also changes the electronic configuration, and therefore the electronic distribution about the nuclei. The change in the electronic configuration generally promotes a change in the configuration of the positively charged nuclei in response to the change in the electronic distribution. The change in electronic and nuclear configuration may also assist in changing the electron spin configuration.

The energy required to produce an electronically excited state ($R + hv \rightarrow {}^*R$) is obtained by inspecting the absorption or the emission spectrum of the molecule in question (see Chapter 4), as well as applying Einstein's *resonance condition* for the absorption of light (Eq. 1.2):

$$\Delta E = |E_2 - E_1| = |E_2({}^*R) - E_1(R)| = hv = hc/\lambda \qquad (1.2)$$

where h is Planck's constant (1.58×10^{-34} cal s $= 1.58 \times 10^{-37}$ kcal s), v is the frequency (commonly given in units of $s^{-1} = $ Hz), λ is the wavelength at which absorption occurs (commonly given in units of nanometers, nm), c is the speed of light (3×10^8 cm s^{-1}), and E_2 and E_1 are the energies of a molecule in an excited (*R) and an initial state (R), respectively.

Equation 1.2 is fundamentally important to spectroscopy and photochemistry, since it relates the energy gap (ΔE) between two states to measurable properties, namely, the frequency (v) and the wavelength (λ) of an absorbed photon. Knowing the absolute energies, E_2 and E_1, is not required in these kinds of energy analyses, since *it is the difference in energy between the two states that is required when applying Eq. 1.2.*

In the energy diagram of Scheme 1.4, the two most important values of ΔE are the energy gap between S_1 and S_0 (called the singlet energy, E_S) and the energy gap between T_1 and S_0 (called the triplet energy, E_T). These energies reflect the available energy that can serve as a driving force for these two states to do work on the making and breaking of bonds in photochemical processes. Both E_S and E_T correspond to excess electronic energy that can be converted into free energy to drive bond making and breaking in primary photochemical processes. In Chapter 7, for example, the values of E_S and E_T play critical roles in photoinduced electron- and energy-transfer processes. E_S and E_T are similarly important in overcoming thermodynamic endothermicity in bond-breaking processes.

Since photochemistry is concerned with the making and breaking of chemical bonds after the absorption of a photon, it is useful to have calibration points and numerical benchmarks for the energy of absorbed photons and to compare these energies to the energies required to break bonds that commonly occur in organic molecules. It is also important to relate bond energies to the frequency (v) and wavelength (λ) of light. In Chapter 4, we develop a model of light as a photon that treats light as consisting of "particles" (or quanta) of energy. Just as a solvent molecule consists of a "field of particles" whose collisions with reactant molecules provide a source of activation energy for a reaction, a beam of light provides a "field of photons" whose collisions with reactant molecules cause the absorption of energy that can serve as activation energy for a reaction.

1.16 Counting Photons

How do photochemists count photons that are emitted from a light source or absorbed by a sample during a photoreaction? Equation 1.3, the second of Einstein's light–energy relationships, relates the energy of a *single* photon to the wavelength (or frequency) of light. Thus, we can use Eq. 1.3 to "count" the photons *emitted* if we know the energy (E) of the light source. Similarly, we can "count" the photons *absorbed* by a sample if we know the energy of the light absorbed by the sample. In other words, a mole of photons of light for a given wavelength (λ) or frequency (ν) corresponds to a definite energy (E), so that if we know the energy contained in the light source at a given ν (or λ), we can compute the number of photons of that ν (or λ) through Eq. 1.3. Since we will be dealing with E in kilocalories per mole, with λ in nanometers, and ν in hertz (Hz = s^{-1}), we need to use 1.58×10^{-37} kcal s for Planck's constant (h) and 3.00×10^{17} nm s^{-1} for the speed of light (c).

$$E = h\nu = h(c/\lambda) \qquad \text{The energy of a single photon} \qquad (1.3)$$

A mole ($N_0 = 6.02 \times 10^{23}$) of photons is called an *einstein* in honor of the intellectual father of the photon. (*It is important to distinguish between Eq. 1.3, which relates the energy of a single photon to a light wave's frequency and wavelength, and Eq. 1.1, the resonance condition that relates the energy gap (ΔE) between two states and the frequency of the light wave that corresponds to a photon whose energy is exactly equal to ΔE.*) According to Eq. 1.3, the energy contained by an arbitrary number of photons or by an einstein of photons depends on the wavelength (frequency) of the corresponding light wave, which leads directly to Eqs. 1.4a and b, where n is an arbitrary number of photons and N_0 is 1 mol of photons.

$$E = nh\nu = nh(c/\lambda) \qquad \text{The energy of } n \text{ photons} \qquad (1.4a)$$

$$E = N_0 h\nu = N_0 h(c/\lambda) \quad \text{The energy of } N_0 \text{ photons (an einstein)} \quad (1.4b)$$

The energy of 1 mol (N_0) of photons given by Eq. 1.4b provides a direct relationship between the amount of light energy absorbed by a system and the number of photons absorbed. Thus, *by measuring the energy of light absorbed (E) and knowing the wavelength (or frequency) of the absorbed light, we have a way to count photons!*

1.17   Computing the Energy of a Mole of Photons for Light of Wavelength λ and Frequency ν

By using Eq. 1.4b, the energy of 1 mol of photons in kilocalories per mole may be computed from the frequency associated with the photon (Eq. 1.5a), or the wavelength associated with the photon (Eq. 1.5b).

$$E(\text{kcal mol}^{-1}) = (9.52 \times 10^{-14}\ \text{kcal mol}^{-1}\ \text{s})\nu \qquad (1.5a)$$

$$E(\text{kcal mol}^{-1}) = (2.86 \times 10^4\ \text{kcal mol}^{-1}\ \text{nm})/\lambda \qquad (1.5b)$$

The data in Table 1.1 show how the energy of 1 mol of photons (an einstein) is related to the corresponding wavelength of light (λ in nm) and frequency of light (ν in $\text{s}^{-1} = \text{Hz}$) for the range of wavelengths of greatest photochemical interest ($\lambda = 200$–1000 nm). These values were calculated using Eqs. 1.5a and b. Historically, because different energy units were used for investigating light in different regions of the electromagnetic spectrum, a number of different energy units are commonly used in both spectroscopy and photochemistry. Values of the energies corresponding to 1 mol of photons of varying ν or λ are commonly given in terms of kilocalories per mole (kcal mol^{-1}), kilojoules per mole (kJ mol^{-1}), reciprocal centimeters (cm^{-1}), and electronvolts (eV). For the most part, however, kilocalories per mole are used in this text, since this unit is commonly employed in chemistry and is associated with bond energies and reaction activation energies. The conversion units are $1\ \text{kcal mol}^{-1} = 4.18\ \text{kJ mol}^{-1} = 350\ \text{cm}^{-1} = 0.0434\ \text{eV}$.

Table 1.1   Relationship among energy, wavelength, and frequency[a]

Type of Radiation	Wavelength ($\lambda = $ nm)	Energy $E = $ kcal mol^{-1}	Frequency ($\nu = $ Hz $= $ s^{-1})
Ultraviolet (UV)	200–400	140–70	1.5×10^{15}–7.50×10^{14}
Violet	~400	70	7.50×10^{14}
Green	~500	60	6.00×10^{14}
Red	~700	40	5.00×10^{14}
Near-Infrared (NIR)	~1000	30	3.00×10^{14}

a. The violet–green–red portion of the spectrum (400–700 nm) corresponds to the visible portion of the spectrum. See Scheme 1.6 for a schematic representation of these data. For a comprehensive list of conversion factors, see the front end endpaper.

1.18   The Range of Photon Energies in the Electromagnetic Spectrum

The range of electromagnetic radiation extends from gamma (γ) rays (the high-frequency, short-wavelength limit) to radiofrequency (rf) waves (the low-frequency, long-wavelength limit). The highest-energy photon in this range corresponds to a γ ray (for $\lambda = 0.0001$ nm and $\nu = 3.0 \times 10^{21}\ \text{s}^{-1}$, the energy of an einstein of γ-ray photons $\cong 3 \times 10^8$ kcal mol^{-1}!). The lowest-energy photons of interest to chemists correspond to a rf wave (for $\lambda = 1 \times 10^{10}$ nm and $\nu = 3.0 \times 10^6$ s^{-1}, the energy of an einstein of radiowave photons $\cong 3 \times 10^{-6}$ kcal mol^{-1}). Thus, the range of energies corresponding to 1 mol of photons spans ~ 13 orders of magnitude, from 3×10^8 kcal mol^{-1} for γ rays to 3×10^{-6} kcal mol^{-1} for radio waves!

However, the range of wavelengths (and therefore of energies) of interest to the organic photochemist is just a tiny region of the electromagnetic spectrum, corresponding to ~ 200–1000 nm (143–30 kcal mol^{-1}). This range corresponds to the UV (200–400 nm), visible (vis, 400–700 nm), and NIR (700–1000 nm) regions of the electromagnetic spectrum. The cutoff at short wavelengths (200 nm $\cong 140$ kcal mol^{-1}) is determined by practical considerations such as the need for a transparent material (quartz or Pyrex glass) from which to construct photolysis vessels. The most transparent material commonly available is quartz, which becomes strongly absorbing at wavelengths shorter than 200 nm, thereby setting the practical short-wavelength cutoff for organic photochemical reactions at $\lambda > 200$ nm. The cutoff at long wavelengths (1000 nm $\cong 29$ kcal mol^{-1}) is somewhat arbitrary and corresponds to the longest practical wavelength for electronic excitation of organic molecules to produce *R. Light with wavelengths longer than 1000 nm tends to excite vibrations rather than electrons.

Equations 1.5a and b offer a convenient formula for the conversion of the wavelength of an electromagnetic wave into the energy (kcal) of 1 mol of photons. Thus, we can use Eq. 1.5a to convert an einstein (1 mol of photons) possessing a wavelength of 700 nm (red light) into its energy equivalent in kilocalories per mole as shown in Eq. 1.6a (Eq. 1.5a $\times 6.02 \times 10^{23}$).

$$E(\text{kcal mol}^{-1} \text{nm}) = 2.86 \times 10^4 / 700 \text{ nm} = 40.8 \text{ kcal mol}^{-1} \qquad (1.6a)$$

Likewise an einstein of light possessing a wavelength of 200 nm (UV light) can be converted into its energy equivalent in kilocalories per mole, as shown in Eq. 1.6b (Eq. 1.5b $\times 6.02 \times 10^{23}$).

$$E(\text{kcal mol}^{-1} \text{nm}) = 2.86 \times 10^4 / 200 \text{ nm} = 143 \text{ kcal mol}^{-1} \qquad (1.6b)$$

For comparison with photochemical excitation energies in the UV–vis region, some typical bond energies are shown in Scheme 1.6. The weakest single bonds commonly encountered in organic molecules have strengths of ~ 35 kcal mol^{-1} (e.g., an O—O bond) and the strongest single bonds have strengths on the order of ~ 100 kcal mol^{-1} (e.g., an O—H bond). A photon with a wavelength of ~ 820 nm carries sufficient energy (~ 35 kcal mol^{-1}) to break an O—O bond, whereas a photon with a wavelength of ~ 290 nm (~ 100 kcal mol^{-1}) would be required to break an O—H bond.

Does absorption of 250 nm light (114 kcal mol^{-1}) lead to the random rupture of any of the single bonds of an organic molecule? No, it does not. In fact, many photoreactions proceed with remarkable selectivity, even if UV light corresponding to photons of energy much greater than that of the strongest bonds of organic molecules

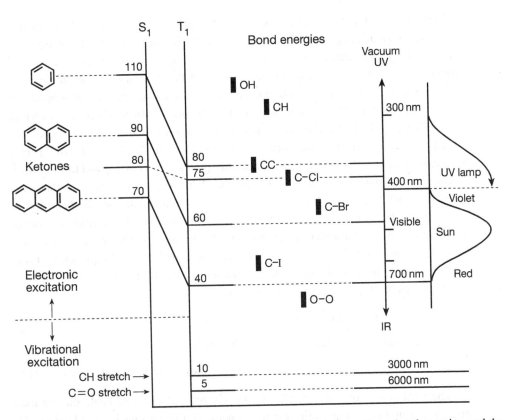

Scheme 1.6 A comparison of energies involved in photochemical reactions, bond energies, and the emission of the sun. Vibrational energies are shown for comparison at the bottom of the scheme. Energies of S_1 and T_1 are kcal mol^{-1}.

is absorbed. Indeed, only certain bonds are made or broken even when the energy per photon absorbed is higher than the energy of most of the individual bonds of a molecule. The reasons for this selectivity include a combination of the rapid deactivation of excess vibrational energy by electronically excited states, the localization of electronic excitation on certain atoms, and the specificity with which this electronic excitation is employed to make or break bonds. In other words, specific mechanisms exist for the conversion of electronic excitation energy into the nuclear motion that results in a net chemical reaction (e.g., *R → I). In this text, we seek to explain these mechanisms in order to understand photoreactions.

Light of wavelength in the range of 1000–10,000 nm (3×10^{14}–3×10^{13} s^{-1}, 29–2.9 kcal mol^{-1}) corresponds to photons in the near infrared (NIR) and the infrared (IR) region of the electromagnetic spectrum. Photons in this energy region excite fundamental and overtone vibrations (stretches and bends) of organic molecules. For example, a photon corresponding to $\lambda = 3000$ nm corresponds to an energy of ~ 10 kcal mol^{-1} (i.e., the energy required to stretch a C—H bond) and a photon corresponding to $\lambda = 10,000$ nm corresponds to an energy of ~ 3 kcal mol^{-1} (i.e., the energy required to stretch a C—C bond).

Light of wavelength in the region of 1×10^6 nm (1 cm, 3×10^9 s^{-1}, 0.029 kcal mol^{-1}) corresponds to photons in the microwave region of the electromagnetic spectrum, and light of wavelength in the region of 1×10^{10} nm (10 m, 3×10^6 s^{-1}, 0.0000029 kcal mol^{-1}) corresponds to photons in the rf region of the electromagnetic spectrum. The value of ΔE for electron and nuclear spin states depends on the size of the magnetic field in which they are placed. In fields on the order of $\sim 10,000$ G, typical electron spin energies correspond to microwave frequencies ($\sim 10^9$–10^{10} s^{-1}), which correspond to energies of $\sim 10^{-4}$–10^{-5} kcal mol^{-1}). In a magnetic field of $\sim 10,000$ G, typical nuclear spin energies correspond to radiowave frequencies ($\sim 10^6$–10^7 s^{-1}), which correspond, in turn, to energies on the order of $\sim 10^{-6}$–10^{-7} kcal mol^{-1}. Finally, it is worthwhile to relate the *number* of photons (n) and the moles of photons ($N = n/N_0$) that correspond to a given amount of light energy. As an exemplar, let us compute the number of photons corresponding to 100 kcal mol^{-1} of energy for light of different wavelengths (frequencies). From Eq. 1.4a, the value of n (the number of photons) is given by Eq. 1.7 and the value of N (the moles of photons) is given by Eq. 1.8.

$$n \text{ (number of photons)} = E\lambda/hc \tag{1.7}$$

$$N \text{ (moles of photons)} = n/N_0 = E\lambda/N_0hc$$

$$= E\lambda/(2.86 \times 10^4 \text{ kcal mol}^{-1} \text{ nm}) \tag{1.8}$$

The total energy of 1 mol of photons of 350 nm light is ~ 82 kcal, and the total energy of 2 mol of 700 nm light (the total energy of the photons is ~ 41 kcal mol^{-1}) is also ~ 82 kcal. However, absorption of one photon of 350 nm light instantaneously provides a single molecule with the equivalent of the entire 82 kcal; that is, this energy, in principle, could be employed to break a bond whose energy is ~ 82 kcal mol^{-1} in a single molecule. Absorption of one photon of 700 nm light provides only the equivalent of ~ 41 kcal of energy to a single molecule. The simultaneous absorption of two photons with ordinary lamps by the same molecule is implausible (having two photons and a molecule together in the same space is analogous to the improbable simultaneous collision or reaction of three molecules), so it would be improbable to efficiently break bonds whose dissociation energy is ~ 82 kcal mol^{-1} with 700 nm light, no matter how intense the beam. Thus, the total energy is not as important as the energy per photon; that is, an intense red lamp with a large total energy of photons would be useless to efficiently break 82-kcal-mol^{-1} bonds, but a weak blue lamp could do the job. This relationship, in which a threshold energy is required to break a bond in an organic molecule, is completely analogous to the photoelectric effect (Chapter 4), for which there is a threshold of photon energy to remove an electron from a metal. Indeed, Einstein's interpretation of the photoelectric effect was the first interpretation of light in terms of quantized photons and was made in analogy to Planck's interpretation of the quantization of energy (Chapter 4).

The final calibration point in Table 1.1 is the relationship between the number of photons corresponding to 100 kcal mol^{-1} of energy for different values of λ or ν. A beam of 0.1 nm (X-ray) light corresponding to this energy contains 3×10^{-4} mol of photons; a beam of 286 nm light contains 1 mol of photons; a beam of 1000 nm (NIR) light contains 3.5 mol of photons; a beam of 10^8 nm (microwave) light contains 3.3×10^4 photons; and a beam of 10^{10} nm (radiowave) light contains 3.3×10^6 photons.

1.19   Calibration Points and Numerical Benchmarks for Molecular Dimensions and Time Scales

Chemists often think of molecules in terms of "ball-and-stick" models that are useful for evaluating many static (time-independent) properties of molecules, such as molecular geometries (bond lengths and bond angles), but microscopic particles (electrons, nuclei, and spins) are never at rest. Because of the uncertainty principle, nuclei undergo vibrational motions even at temperatures close to 0 K. In addition to vibrating nuclei, electrons in orbits and their electromagnetic spin moments execute characteristic zero-point motions. Indeed, even the electromagnetic field has a zero-point motion (which corresponds to the absence of photons in the field). In Chapter 3, we discuss physical and chemical radiationless transitions, such as the reorganization of the nuclear, electronic, or spin structure of a molecule, may be viewed as changes in zero-point motions. Understanding how this reorganization of structure over distances (on the order of the dimensions of molecules) occurs as a function of time is critical for an understanding of photophysical and photochemical processes. Thus, we need some numerical benchmarks for dimensions and time scales.

First, consider the dimensions of typical *chromophores*, the groups of atoms that are responsible for the absorption of light. The standard functional groups of organic chemistry (carbonyl, olefinic, enone, aromatic, etc.) correspond to simple chromophores. If we consider the typical atoms or groups for organic molecules involved in the absorption of light ($R + h\nu \rightarrow$ *R), the "size" of these groups is generally on the order of 2–6 Å (0.2–0.6 nm) and involves a relatively small number of connected atoms. A photon travels at the speed of light ($c = 3 \times 10^{10}$ cm s^{-1} = 3×10^{17} nm s^{-1}). In other words, a photon travels 1 cm(10^7 nm) in 33×10^{-12} s (33 picoseconds, ps)!

If we associate the wavelength of light (λ) with the "length" or "dimension" (d) of a photon, then photons corresponding to blue light have a dimension on the order of 400 nm (Table 1.1). We may interpret the dimension or length of photons in terms of their ability to collide (interact) with a molecule. Thus, the time it takes a "blue" photon with a wavelength of 400 nm to pass a point is $\tau = d/c = 400$ nm/3×10^{17} nm s$^{-1} \sim 10^{-15}$ s, one femtosecond (1 fs). Crudely, this corresponds to an order of magnitude for the "interaction time" available for the absorption of a photon by a molecule. If absorption does not occur in this time period, the photon zips past the chromophore and absorption does not occur.

Can an electron jump from one orbital to another or from one atom to another in this period, or does the photon zip by a molecule too rapidly? Let us use a concrete physical model, the Bohr atom, to estimate the time required for an electron to jump from the orbital of one atom to the orbital of an adjacent atom. The time it takes an electron to make one complete circuit in the lowest-energy Bohr orbit of a hydrogen atom (the radius of the lowest-energy orbit of a hydrogen atom is ~ 0.05 nm or 0.5 Å) is $\sim 10^{15}$ Å s^{-1}. Thus, an electron may move on the order of 0.1 nm in 10^{-16} s and 1 Å in 10^{-15} s. Since 0.1–0.3 nm (1–3 Å) is on the order of common bond lengths of organic molecules, the orders of magnitude of the time scales of photon interaction and electron motion overlap.

For absorption of light to occur and to cause an electron to jump from one orbital to another (the $R + h\nu \rightarrow {}^*R$ process), the frequency of the light must match a possible frequency of motion of an electron; that is, the resonance condition of Eq. 1.1 ($\Delta E = h\nu$) must be satisfied. Thus, if the resonance condition is met, the energy of the photon may be absorbed and an electron may be excited. In the wave picture, when light is absorbed, energy is transferred from the oscillating electromagnetic field to the electrons, which are simultaneously sent into oscillation due to excitation. In Chapter 4, we discuss the quantum mechanical selection rules that make the absorption of light by a molecule plausible. The time period of $\sim 10^{-15}$ s sets an upper limit to the scale of chemical events, since no chemistry can occur before electron motion has occurred (i.e., before an electron has changed its position in space.) *Thus, 10^{-15} s (1 fm) serves as a numerical benchmark time for the fastest events of chemical or photochemical interest.* Remarkably, modern laser techniques make it possible to measure processes occurring on the time scale of 10^{-15} s. For his work in developing these techniques, Ahmed Zewail[1] was awarded the Nobel Prize in Chemistry in 1999.

Now, let us obtain a feeling for the magnitudes of the rates (or lifetimes of processes) that can occur during the lifetime of an excited state (*R), that is, the lifetimes of the processes shown in the state energy diagram of Scheme 1.4 and the energy surfaces of Scheme 1.5. What are the calibration points or benchmark rates of the slowest and fastest processes available to *R? What limits the maximum lifetime of *R?

Radiative processes limit the maximum lifetimes of electronically excited states (*R). In other words, *R cannot live longer than its natural radiative lifetime; if no other process deactivates *R, it will eventually emit a photon, and the $^*R \rightarrow R + h\nu$ process will take the excited molecule back to its ground state. Thus, any radiationless transition (photophysical or photochemical) from S_1 or T_1 must occur at a rate faster (in a time scale shorter) than the natural rate of emission, or emission will be the "default" deactivation process. That is, the molecule will deactivate by emitting a photon faster than undergoing a photophysical or photochemical event.

What, then, are the benchmark limits for the fastest and the slowest pure radiative processes? In Chapter 4, we explain that the largest *fluorescence* ($S_1 \rightarrow S_0 + h\nu_F$) rates of organic molecules are on the order of 10^9 s^{-1} and the smallest fluorescence rate constants are on the order of 10^6 s^{-1}. This finding puts the time scale for competitive processes from S_1 in a time period shorter than the range of 10^{-6}–10^{-9} s. In other words, a radiationless process that takes 10^{-5} s or longer from S_1 will be inefficient, even for the longest-lived S_1 states, and a radiationless process that takes $< 10^{-10}$ s will compete with even the fastest radiative rate.

On the other hand, the largest *phosphorescence* ($T_1 \rightarrow S_0 + h\nu_P$) rate constants for organic molecules, k_P, are on the order of $10^3 \, s^{-1}$ and the smallest are on the order of $10^{-2} \, s^{-1}$. This means that the time scale for competitive processes from T_1 must occur in a period that is shorter than the range of 10^{-3}–100 s. Thus, a radiationless process taking place in the period of 10^{-5} s (which is far too long to compete with fluorescence from S_1) or longer may be quite efficient for a T_1 state. The discussion in Chapter 4 shows that the values of k_F and k_P are related to the structure of *R, but for now we have some numerical benchmarks for the limits of the rates of processes that can occur competitively from S_1 or T_1. All other factors being similar, photoreactions of triplet states are more likely to occur than photoreactions of singlet states, based on lifetime considerations alone.

Now, we compare the time scales for the emission of light to the time scales for the internal nuclear motions of molecules (i.e., vibrations). The fastest vibrations of organic molecules occur with a frequency of $10^{14} \, s^{-1}$ (C—H stretching vibrations), and the slowest occur with a frequency of $\sim 10^{12} \, s^{-1}$ (C—Cl stretching vibrations). This means that it takes somewhere between $\sim 10^{-12}$ and $\sim 10^{-14}$ s to complete a zero-point vibration for the bonded groups in typical organic molecules. Since the inherent lifetime of the fluorescence of most organic molecules falls in the range of 10^{-6}–10^{-9} s, S_1 states undergo thousands to millions of vibrations before emitting a photon! The T_1 state, which takes 10^{-3} s or longer to emit, can execute $\sim 10^{11}$–10^{14} vibrations before emitting photons! Thus, there is plenty of time for nuclear motion to become equilibrated during the lifetime of an electronically excited molecule that is deactivated through a radiative process.

Electron spin plays an important role in many photochemical reaction pathways and is the key structural feature of all singlet–triplet interconversions, radiative or radiationless. In general, the rates of spin interconversions vary over many orders of magnitude but are slow relative to vibrational motions and relatively slow compared to electronic motions. The fastest spin interconversions for organic molecules composed of H and atoms of the elements of the first full row of the periodic table occur at a rate of $\sim 10^{12} \, s^{-1}$. The slowest spin interconversions occur at a rate of $\sim 10^{-1} \, s^{-1}$. The rates of spin interconversions are determined by an interaction between an electron's spin motion and its orbital motion. This interaction is called spin–orbit coupling and is discussed in Chapter 3.

The rates of photoreactions (k_R, Scheme 1.4) vary over enormous ranges, from $\sim 10^{14}$ to $\sim 10^{-2} \, s^{-1}$. The fastest reactions are limited by vibrational motion (passing through funnels) and electron transfer (ionization), and the slowest reactions are limited by the slowest phosphorescence rates. Whether the photoreaction occurs from S_1 or T_1 depends on both the rate constant (k_R) of the *R $\rightarrow$ I process and Σk (where Σk represents the sum of the rates of all deactivating pathways of the excited state).

Scheme 1.7 compares the spread of time scales for events of photochemical interest, which range from $\sim 10^{-16}$ s (1/10 fm) to ~ 1 s, with the same spread of history going back into the past (from ~ 1–10^{16} s; 10^{16} s = 10 ps = 3×10^8 years). When compared in this manner, the history of a photoreaction passes through about as many "decades" of time as the history of the earth!

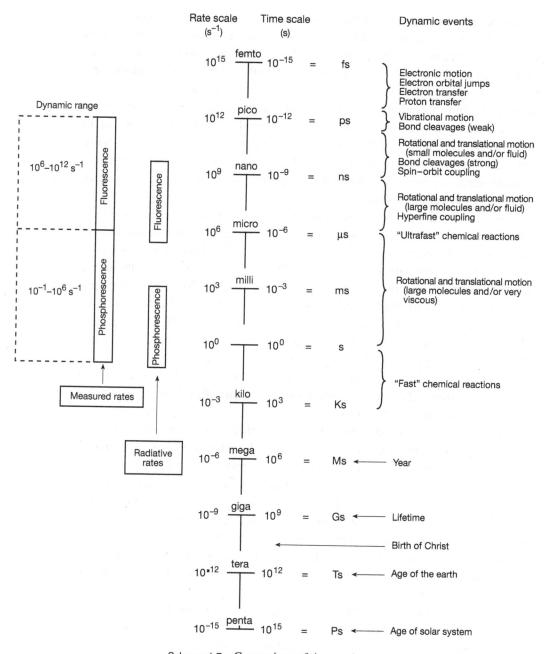

Scheme 1.7   Comparison of time scales.

1.20   Plan of the Text

Now that we have described a broad overview of modern molecular organic photochemistry, indicated the paradigm-and-exemplar approach that will be adopted in the text, and reviewed some calibration points and benchmarks for thinking about molecular organic photochemistry, we conclude this chapter with a review of the plan of the text.

The concepts of structure, energetics, and dynamics are crucial for understanding molecular photochemistry. To get started, in Chapter 1, we need to understand the structures of the species R, *R, I, and P in the paradigms of Schemes 1.1–1.3. Chapter 2 (Electronic, Nuclear, and Spin Configurations of Electronically Excited States), describes how we can visualize the electronic, vibrational, and electronic spin structures of R, *R, I, and P to a zero approximation. Each stable configuration of electrons and electron spins corresponds to a stable nuclear geometry and possesses an associated energy. The enumeration, classification, and visualization of molecular states of electronically excited molecules and their relative energies in terms of orbital configurations, nuclear configurations, and spin configurations are the topics of Chapter 2.

Knowing the plausible structures we will encounter in organic photochemistry, we next consider the problem of the transition of an initial structure into a different final structure, that is, processes, such as $R + hv \rightarrow {}^*R$, ${}^*R \rightarrow I$, $I \rightarrow P$, and ${}^*R \rightarrow P$. These issues are addressed in Chapter 3 (Transitions between Molecular States: Photophysical Processes), which ties together the concepts of structure, dynamics, and energetics in terms of PE surfaces that allow an effective and concrete visualization of the plausible pathways by which molecular states may be interconverted.

Chapter 4 (Radiative Transitions Between Electronic States) describes how radiative transitions (i.e., absorption, $R + hv \rightarrow {}^*R$, and emission, ${}^*R \rightarrow R + hv$) can be understood and visualized, as well as how these radiative transitions are qualitatively and quantitatively related to molecular electronic structure and the structure of the electromagnetic field. Chapter 5 (Photophysical Radiationless Transitions) describes the mechanisms of radiationless transitions between excited states (${}^*R \rightarrow {}^*R' + \text{heat}$) and each other and between excited states and ground states (${}^*R \rightarrow R + \text{heat}$). The transitions considered in Chapters 4 and 5 are termed "photophysical," because they occur between initial and final molecular states of very similar *nuclear geometry* and do not correspond to traditional chemical processes in which bonds are clearly broken or formed to create different nuclear configurations.

In Chapter 6 (A Qualitative Theory of Molecular Photochemistry), we consider radiationless transitions corresponding to chemical reactions, and we develop a theory and paradigms for the understanding and visualization of photochemical reactions in terms of energy surfaces. In addition, we describe the theoretical aspects of photochemical reactions after the electronically excited state (*R) has been formed by the absorption of a photon (hv). The primary photochemical processes ${}^*R \rightarrow I$ and ${}^*R \rightarrow F$ are considered in theoretical terms of orbital interactions and orbital (and state) correlation diagrams.

Chapter 7 (Energy and Electron Transfer) describes the relationship and scope of two closely related and very important processes involving *R. The common orbital interaction relationship between electron and electronic energy transfer will be discussed, and a number of exemplars of each process will be reviewed within the current paradigms.

To close this chapter, we reference a number of useful sources of information on organic photochemistry.[5-14]

References

1. (a) A. H. Zewail, *Pure App. Chem.* **72**, 2219 (2000). (b) A. H. Zewail, *Angew. Chem., Int. Ed. Engl.* **39**, 2586 (2000).

2. (a) P. M. Paul et al. *Science* **292**, 1689 (2001). (b) D. Labrador, *Sci. Am.* **287**, 56 (2002).

3. P. W. Atkins and R. Friedman, *Molecular Quantum Mechanics,* 5th ed., Oxford University Press, Oxford, UK, 2005.

4. T. Kuhn, *The Structure of Scientific Revolutions,* 2nd ed., The University of Chicago Press, Chicago, 1970.

5. A useful guide to the terminology of photochemistry. S. E. Braslavsky and K. N. Houk, *Pure & Appl. Chem.* **60**, 1055, (1988).

6. A useful handbook on photochemical data of all types (energies, rate constants, spectral data, etc.). *Handbook of Photochemistry,* 3rd ed., M. Montalti, A. Credi, L. Prodi, M. T. Gandolfi, eds., CRC Taylor and Francis, Boca Raton, 2006.

7. An excellent text on photochemistry that integrates spectroscopy and quantum mechanics. M. Klessinger and J. Michl, *Excited States and Photochemistry of Organic Molecules,* VCH Publishers, New York, 1995.

8. A comprehensive annual review of all areas of photochemistry since 1969. *Specialist Periodical Reports,* The Chemical Society, London.

9. Reviews of a range of topics in all areas of photochemistry appearing every year or two, from 1963. *Advances in Photochemistry,* Wiley & Sons, Inc., New York.

10. Reviews of a range of topics in organic photochemistry in 11 volumes during the period 1967–1991. *Organic Photochemistry,* Marcel Dekker, New York.

11. Reviews of a range of topics in molecular and supramolecular photochemistry in 14 volumes during the period 1997–2006. *Molecular and Supramolecular Photochemistry,* Marcel Dekker, New York.

12. Reviews of a range of topics in organic photochemistry. (a) W. Horspool and F. Lensi, eds., *CRC Handbook of Organic Photochemistry and Photobiology,* 2nd ed., CRC Press, Boca Raton, FL, 2004. (b) J. C. Scaiano, ed., *CRC Handbook of Organic Photochemistry,* CRC Press, Boca Raton, FL, 1989.

13. The previous versions of this text. (a) N. J. Turro, *Molecular Photochemistry,* Benjamin, New York, 1965. (b) N. J. Turro, *Modern Molecular Photochemistry,* University Science Press, Menlo Park, CA, 1991.

14. Reviews of organic syntheses. (a) *Photochemical Key Steps in Organic Synthesis,* A. G. Griesbeck and J. Mattay, eds., VCH, Weinheim, 1994. (b) *Synthetic Organic Photochemistry,* A. G. Griesbeck and J. Mattay, eds., Marcel Dekker, New York, 2005. (c) A. Schonberg, *Preparative Organic Photochemistry,* Springer-Verlag, New York, 1968. (d) *Organic Photochemical Synthesis,* Vols. 1 and 2, R. Srinivasan and T. D. Roberts, eds., Interscience, New York, 1972.

Electronic, Vibrational, and Spin Configurations of Electronically Excited States

2.1 Visualization of the Electronically Excited Structures through the Paradigms of Molecular Organic Photochemistry

Let us review what we have learned from the working paradigms of Schemes 1.1–1.3. Electronically excited states, *R, are the critical initial transient structures in the photochemical reactions of *all* organic molecules. Any *chemical* process that is produced directly from *R is called a *primary photochemical process*. Any *physical* process that proceeds directly from *R is called a *primary photophysical process*. One of the most common primary photochemical process for organic molecules is *R → I, as shown in Scheme 2.1. Scheme 2.1 expands on Scheme 1.1, with the addition of the primary photophysical processes from *R. These photophysical pathways, *R → R, generally compete with any primary photochemical pathways, *R → I. The less common primary photochemical process that takes *R to P by way of a funnel (*R → F → P) or through electronically excited intermediates (*R →*I) are not included at this point for simplicity.

The reactive intermediate (I) undergoes conventional *thermal processes* leading to the observed product (P). The structure I represents the common reactive intermediates (radical pairs, biradicals, carbonium ion–carbanion pairs, zwitterionic species, carbenes, etc.) that are produced in the primary photochemical step *R → I. The typical electronic structures associated with I are discussed in Chapter 6. The chemical pathways from I →P are called *secondary thermal processes* to contrast with the *primary photochemical pathways* *R → I. Primary photochemical pathways proceeding directly through funnels, *R → F → P, and pathways involving formation of an electronically excited state of the product (e.g., *R →*I and *R →*P) are known but are relatively less common experimentally and are not considered explicitly until Chapters 5 and 6.

The symbols given in Scheme 2.1 (i.e., R, *R, I, and P) are shorthand notations and need not necessarily be restricted to a single molecular species. In general, *R

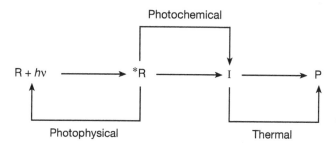

Scheme 2.1   The global paradigm for the overall photochemical
pathways of *R → P and the overall photophysical pathways
from *R → R.

symbolizes the system of reactants that would include any molecule (M) in addition
to *R that are involved in the primary photochemical process. *All of the species
in Scheme 2.1, including *R, can be represented by traditional Lewis structures
or discussed in terms of conventional one-electron molecular orbitals* (MOs). It is
assumed that the student is familiar with Lewis structural and MO representations of
R, P, and I. The latter possess the ordinary organic functional groups contained in
organic molecules (ketones, ethylenes, enones, benzenes, etc.) that are discussed in
all basic courses in organic chemistry (Chapters 9–12). One of the goals of this text is
to make the student comfortable with applying knowledge of the Lewis structures and
MOs (associated with the ground-state structures R, P, and I) toward an understanding
of the detailed *electronic, vibrational, and spin structure* of the excited-state structures
(*R) and of the reactive intermediates (I).

Quantum mechanics[1] provides the most general, powerful, and effective paradigm
of physics and chemistry for the description of the molecular structure and dynamics
of organic molecules. However, a proper understanding of molecular structure and
dynamics through quantum mechanics requires a mathematical sophistication that is
well beyond the intended scope and the expected readership of this text. Nevertheless,
this chapter and Chapter 3 exploit the power of the concepts and methods of quantum
mechanics by translating the mathematics that are required for a quantum mechanical
treatment of organic photochemistry into visual pictures and structures that students,
who may not have a background or interest in formal mathematics, can comprehend
readily. *We translate the mathematics of quantum mechanics into pictures that can,
in turn, be translated into molecular structures (this chapter) and molecular dynamics (Chapter 3).* This approach recognizes that the mathematical ideas of quantum
mechanics can be interpreted in a *qualitative* manner through easily visualizable, pictorial representations and that these representations are almost always, to a certain
degree, acceptable approximations to the real systems.

The language of organic chemistry employs pictorial objects called *molecular
structures* (e.g., Lewis structures and MOs) to make correlations with molecular
function, properties, and reactivity. The language of quantum mechanics uses mathematical objects called *wave functions* (a term we describe in detail in this chapter) and
the idea of quantized electronic, vibrational, and spin energies. Since the molecule is

the *same object* being described by both molecular structures and wave functions, there must be methods to transform molecular structures into wave functions and to transform wave functions into the more familiar visualizable molecular structures described by Lewis structures and MOs. In Scheme 2.1, R, *R, I, and P may be described as ordinary molecular structures or as wave functions. Visualizations of the wave functions of R, *R, I, and P, although understood to be only approximate models, allow the photochemist's imagination to create a "motion picture" of the choreography of structural, energetic, and dynamic events occurring at the molecular level for the overall photochemical transformation $R + h\nu \rightarrow {}^*R \rightarrow I \rightarrow P$.

As in all analogies and models, such visualizations of wave functions are an approximate and imperfect representation of the more "correct" mathematics of quantum mechanics; nevertheless, such pictures are valid and useful qualitative approximations for introducing the concepts of a new field. Indeed, as long as the pictures provided capture the correct nature and qualitative form of the fundamental structures, fundamental forces, and fundamental dynamics of the processes of Scheme 2.1, a valuable introductory and operational knowledge and "quantum intuition" can be achieved that can also be helpful as a basis for probing deeper into the subject.[1]

The pictorial approach to understanding chemical systems is most effective when the "pictures" themselves are, or represent, mathematical objects, such as those derived from topological or Euclidean geometry.[2] For example, Lewis electron line–dot structures possess the properties of a mathematical or topological object, called a *graph*; the Lewis structure displays the geometric relationships between point vertices (atoms) and the connections of the vertices with lines (bonds). The symbols R, *R, I, and P from Scheme 2.1 can be represented as Lewis structures (topological objects or chemical graphs) or as three-dimensional (3D) representations of the position of the atoms in space (i.e., Euclidean objects or chemical structures).[2] Organic chemists are familiar with the topological and Euclidian representation of the structure of organic molecules. Even though such structures are mathematical abstractions, to the chemist they take on a reality that leads to powerful structure–function, structure–reactivity, and structure–property correlations. Understanding that traditional molecular structures employed to describe organic molecules are actually mathematical objects (just as mathematically abstract as quantum mechanical wave functions) may allow the student to more easily accept the use of the mathematics of wave functions as an alternate description of organic molecules.

Although wave functions are formulated completely in mathematical terms, the physical basis for constructing wave functions[1] typically starts with the classical electrostatic systems of physics (i.e., negatively charged electrons and positively charged nuclei), which chemists are accustomed to visualizing. Thus, from this standpoint much of quantum mechanics does not really provide an entirely new way of understanding and investigating natural phenomena; in many respects quantum mechanics represents an evolution of classical mechanics that introduces wave functions and quantization for the description of microscopic particles. *In this sense, using the visualizations provided by classical mechanics as a starting point, we can develop a "quantum intuition" that is based on the classical mechanics of the world around us.* Thus, classical mechanics can be usefully employed to provide the pictures we use to understand processes that occur in the quantum mechanical world. There are some features of quantum mechanics, however, that cannot at all be considered a natural evolution of classical mechanics, such as the consequences of electron exchange (the Pauli principle) and the indeterminacy of measuring quantities precisely at the microscopic level (the Heisenberg uncertainty principle).

This chapter *explains how to use electronic orbitals to visualize electronic wave functions, how to use vibrating masses connected by elastic springs to visualize vibrational wave functions, and how to use precessing vectors to visualize electron spin configuration wave functions.* Each of these visualizations can be associated with the mathematical paradigms of quantum mechanics in a qualitative manner that makes it possible to usefully interpret and understand a very wide range of photochemical phenomena.

2.2 Molecular Wave Functions and Molecular Structure[1]

According to the principles of quantum mechanics,[1] the wave function of any atomic or molecular system is a mathematical function that contains all of the information that is necessary to determine any measurable property, static or dynamic, of the atomic or molecular system "simply" by performing appropriate mathematical operations on the function to produce a solution that includes the value of the property of interest. Conventionally, a complete wave function of any molecular system is composed of a rather complicated and abstruse mathematical function represented by the symbol Ψ. Quantum mechanics provides an understanding of molecular structure, molecular energetics, and molecular dynamics based on computations that "operate" mathematically on the wave function Ψ. According to the principles of quantum mechanics, *if the mathematical form of Ψ is known precisely for a given molecule, it is possible, in principle, not only to compute the electronic, nuclear, and spin configurations of a molecule but also to compute the average value of any experimental observable property (electronic energy, dipole moment, nuclear geometry, electron spin energies, probabilities for transitions between electronic states, etc.) of any state of the molecule for an assumed set of initial conditions and interactions provided by internal and external forces.* The application of quantum mechanics to molecules boils down to solving an appropriate equation and using the mathematical solution as the wave function for the property or transition of interest. Once the wave function is available, all of the properties of the molecule can in principle be computed, because quantum mechanics teaches us how to extract the values of all molecular properties.

According to the laws of quantum mechanics,[1] a wave function Ψ is produced by solving the Schrödinger "wave equation" (Eq. 2.1).

$$H\Psi = E\Psi \tag{2.1}$$

In words, Eq. 2.1 is a mathematical expression of the fundamental *laws* of nature, according to quantum mechanics, in the same sense as Newton's laws provide a mathematical expression of the fundamental laws of nature, according to classical mechanics. The laws of quantum mechanics tell us that in nature there can exist only certain "allowed" stable states, and that these allowed states possess specific (quantized) energies. Each of these stable states corresponds to a specific wave function (Ψ) that is a solution of Eq. 2.1. Each Ψ has an associated energy (E) that is produced by the solutions of Eq. 2.1. In the wave equation, H is called the "Hamiltonian" and corresponds to a mathematical "operator" for the *possible* energies (E) of the system. These energies may be the electronic energies of a molecule, the vibrational energies of the atoms of a molecule, or the spin energies of the electrons. The special properties of Eq. 2.1 are that the "allowed" (stable) wave functions (Ψ) have the remarkable property that when the mathematical operator H is multiplied by Ψ the result, for certain values of E, is $H\Psi = E\Psi$ (Eq. 2.1). The wave functions (Ψ) are called *eigenfunctions,* and the solutions (E) are called *eigenvalues* of the operator H. The prefix "eigen-" derives from the German for "proper" (i.e., both Ψ and E are "proper" solutions of Eq. 2.1). According to the laws of quantum mechanics, only the eigenvalues that correspond to the solutions of Eq. 2.1 are the energies that are "allowed" for any stable state, that is, the energies of all molecular states (electronic, vibrational, and spin) are *quantized* as a result of the wave features of Eq. 2.1. Thus, quantization is a natural consequence of the mathematical properties of the wave equation (Eq. 2.1) and is seen as the result of the wavelike properties of matter at the atomic and molecular levels.

In the following discussion, students who are unfamiliar with the mathematics of quantum mechanics may want to think of a complete molecular wave function (Ψ) as a mathematical representation of the entire molecular structure (i.e., electronic, vibrational, and spin). Thus, for the molecular structural representations of R, *R, I, and P in Scheme 2.1, there exist corresponding wave functions $\Psi(R)$, $\Psi(^*R)$, $\Psi(I)$, and $\Psi(P)$, respectively. A system whose wave function is an eigenfunction of some operator that represents a measurable property of a system is said to be in a (stable) *eigenstate* for the measured property. The explicit mathematical form of Ψ does not appear in this text, but we use pictorial representations or graphs of Ψ, and we relate these representations to the appropriate features of the familiar molecular structures.

Quantum mechanics can be used to answer the following questions about the species *R and I in Scheme 2.1:

1. What are the detailed electronic, vibrational, and spin structures and energies of *R and I?
2. What are the electron distributions about the atoms of *R and I?
3. What is the probability of light absorption by R, and what is the probability of light emission by *R?
4. What are the rates of the photochemical and photophysical transitions of *R?
5. What is the role of electronic spin in determining the properties of *R and I?

We take as accepted, from this point on, that by employing quantum mechanics and Eq. 2.1 (and related equations), we can compute the outcome of an observation of a particular property by performing the appropriate mathematical operation on the wave function Ψ (e.g., the operation of the Hamiltonian operator, H, on Ψ produces the allowed energies). But how exactly do we visualize this mathematical "operator" in Eq. 2.1? A mathematical operator is related to the forces or interactions that determine the measurable properties of a system (the energy, dipole moment, bond angle, angular momentum, transition probability, etc.). In fact, a quantum mechanical operator generally has a form similar to the mathematics that are used to compute the classical property. For example, the quantum mechanical operator for the repulsive interaction between two electrons has the form e^2/r (where e is the negative charge on an electron and r is the separation between the electrons), which is exactly the form of the expression for computing interactions between classical negatively charged particles using Coulomb's laws of electrical attraction and repulsion.

We finish this introductory section with the following important conclusions and generalizations derived from the laws of quantum mechanics:[1]

1. According to the principles of quantum mechanics, the only possible "allowed" values of a measurement must be eigenvalues of equations of a form similar to Eq. 2.2, which is a generalization of Eq. 2.1. For example, for every measurable property of a molecular system (P), there is a mathematical function P that operates on the eigenfunction Ψ to produce an eigenvalue P that corresponds to an experimental measurement of that property of the system.

$$P\Psi = \mathrm{P}\Psi \qquad (2.2)$$

2. Equation 2.2 refers to a single measurement in an experiment on a single molecule. A large number of experiments are actually performed on a huge number of molecules in any real experiment, so that an average (ave) value of the property (P_{ave}) is obtained in a laboratory experiment. According to the laws of quantum mechanics, P_{ave} (called the expectation value of the measurement) is obtained by a mathematical integration of the form $\int \Psi P \Psi$ (Eq. 2.3), where $\int \Psi P \Psi$ is shorthand for the actual mathematics, which involve complex wave functions and mathematical "normalization," features that we ignore for simplicity. An alternate symbol for the integral corresponding to the expectation value $< \Psi | P | \Psi >$ is called a "matrix element" (Eq. 2.3). The important point is that information on a molecular system is extracted from a wave function Ψ by mathematically "acting" on it with an operator P and computing a matrix element $< \Psi | P | \Psi >$ that corresponds to an expectation value of an experimental measurement.

$$\underbrace{P_{ave}}_{\text{Expectation value}} = \int \Psi P \Psi = \underbrace{< \Psi | P | \Psi >}_{\text{Matrix element}} \qquad (2.3)$$

Now, we describe a common and convenient approximation for Ψ that can be employed to solve Eq. 2.1 and produce a set of allowed *approximate* wave functions and *approximate* electronic energies of a system. Then, we use approximate wave functions to visualize the electron configurations, the nuclear configurations, and the spin configurations of R, *R, I, and P from Scheme 2.1. Next, we describe how to use these approximate wave functions to estimate some properties, such as the electronic distributions and energies of R, *R, I, and P, starting with Eq. 2.3. In these cases, in addition to visualizing the approximate wave functions, we need to identify and visualize the appropriate operator (P) corresponding to the property of interest, and then visualize the matrix element of Eq. 2.3, in which the appropriate operator P is mathematically "sandwiched" between the approximate wave functions.

2.3   The Born–Oppenheimer Approximation: A Starting Point for Approximate Molecular Wave Functions and Energies

For organic molecules, the most important method for determining molecular wave functions and associated energies is the Born–Oppenheimer approximation.[1,2] According to this approximation, *the motions of electrons in orbitals are much more rapid than nuclear vibrational motions.* The Born–Oppenheimer approximation assumes that the low-mass, rapidly moving, negatively charged electrons can immediately adjust their distribution to the positive potential of slowly moving, heavy, massive nuclei. The important consequence of this approximation is that it allows electronic and nuclear motions to be treated mathematically, but independently. This feature greatly simplifies the solution to Eq. 2.1 because it allows the electronic wave function to be computed for any selected *static* nuclear framework (i.e., for any nuclear geometry). This finding has the practical effect, when making calculations of wave functions, that any arbitrary, fixed nuclear framework (the frozen nuclei approximation) can be selected and the electronic wave function can then be calculated. The energy obtained from such a calculation is "potential energy" (PE) in the sense that it corresponds to the total electronic energy of a molecule less the kinetic energy (KE) of the nuclei (which are assumed to be motionless for each geometry so that kinetic energy is not included in the computation).

The nuclei, in principle, can be moved to all possible configurations and the electronic energies computed for each nuclear configuration. After all of the possible computations have been made, the lowest-energy configuration can be determined and identified as the equilibrium configuration, thus providing the most stable shape of the molecule. It is possible within the Born–Oppenheimer approximation to compute the energies of ground-state reactions as the system proceeds from R → P or excited-state reactions as the system proceeds from *R → P (e.g., that shown in the hypothetical example of Scheme 1.5). In this case, the *lowest energy path* for a ground- or excited-state reaction can be selected, and this special lowest-energy path is called the *reaction coordinate.*

For singlet states, which are the only states of interest for organic molecules (R and P) in their ground states, the net electron spin is zero, so spin is not involved in the solution to Eq. 2.1. For *R and I, however, spin is often a critical feature in determining the overall pathway to products. Electron spin motion is due to a magnetic interaction (Section 2.33), and since magnetic and electronic phenomena interact only weakly for most organic molecules, the spin motion of electrons may also be treated independently from both the motion of electrons in orbitals and the motion of nuclei in vibrations and rotations.

The Born–Oppenheimer approximation is generally excellent for a stable molecular state, such as the ground state of organic molecules, that does not interact significantly with other excited (electronic, vibrational, or spin) states. The Born–Oppenheimer approximation breaks down when a single nuclear configuration corresponds to two or more electronic states of similar energy. (This is a situation that is commonly encountered for *R and I but is rarely encountered for R or P.) When two states of equal energy have the same nuclear geometry, the conditions for resonance are met and the wave functions of two or more states compete for control of the nuclear motion as the system approaches this special geometry.

The Born–Oppenheimer approximation makes it possible to compute a good first guess of Ψ, the "true" molecular wave functions of a molecule. For example, the Born–Oppenheimer approximation allows an approximate Ψ to be computed in terms of *three* independent and approximate wave functions (i.e., Ψ_0, χ, and S in Eq. 2.4). The function Ψ_0 describes the electron configuration, χ describes the nuclear configuration, and S describes the spin configuration.

$$\underbrace{\Psi}_{\substack{\text{"True" molecular wave function} \\ \text{Exact solution to Eq. 2.1}}} \sim \underbrace{\Psi_0 \chi S}_{\substack{\text{(orbitals)(nuclei)(spin)} \\ \text{Approximate solution to Eq. 2.1}}} \qquad (2.4)$$

The wave function Ψ_0 in Eq. 2.4 represents an approximate *electronic* wave function (the subscript zero refers to a "zero-order," or initial working approximation) for the electron position and the electronic orbital motion in space about the positively charged (positionally frozen) nuclear framework. The wave function χ represents the approximate *vibrational* wave function (which is discussed in detail in Section 2.18). The wave function S represents the approximate *spin* wave function (which is discussed in detail in Section 2.22). Thus, in a zero-order approximation, the "true" molecular wave function, Ψ, is approximated in terms of three separate, approximate wave functions, Ψ_0, χ, and S. *This approximation breaks down whenever there is a significant interaction between the electrons and the vibrations* (called *vibronic* coupling) *or between the spins and the orbiting electrons* (called *spin–orbit* coupling). In essence, it is the Born–Oppenheimer approximation that justifies the visualization of Ψ in terms of the approximate wave functions, Ψ_0, χ, and S, and provides the Lewis structures with a certain validity. Now, we describe a way to visualize each of the three approximate wave functions in detail.

First, consider the approximate *electronic* wave function, Ψ_0, which we visualize in terms of electrons distributed in orbitals about a fixed field of the positively charged nuclear structure. The detailed nature of Ψ_0 depends on the level of sophistication and accuracy desired. A common approach to approximate Ψ_0 is not to solve Eq. 2.1 for all of the electrons of a molecule (which is impossible because of the complications of electron–electron interactions) but to solve for a fictitious molecule that contains only "one electron" (which is much simpler for computations because of the absence of any electron–electron interactions). The wave functions generated by solving Eq. 2.1

by this procedure for the one-electron molecule are called "one-electron orbitals," ϕ_i, each of which possesses an eigenvalue of energy, E_i. For many qualitative analyses of molecular phenomena, Ψ_0 is usefully approximated as a product or overlap of one-electron molecular orbitals, ϕ_i (see Eq. 2.5).

$$\underbrace{\Psi_0}_{\substack{\text{Approximate}\\\text{electronic wave function}}} \sim \underbrace{\phi_1\phi_2\cdots\phi_n}_{\substack{\text{Overlapping one-electron}\\\text{orbitals}}} \tag{2.5}$$

where $\phi_i\,(i = 1, 2, 3, \ldots, n)$ is a solution of the wave equation (Eq. 2.1) for a one-electron molecule. For a discussion of these approximations, refer to any one of a number of elementary texts.[1] Because a fictitious molecule possessing only one electron does not experience any electron–electron repulsions, this model only approximates a real molecule; nevertheless, it is still a remarkably useful zero-order approximation for qualitative purposes and for the visualization of the orbitals about a nuclear framework. Indeed, the one-electron wave function is the level of approximation for orbitals that is conventionally given in introductory texts on organic chemistry (the so-called Hückel orbitals). These are the familiar orbitals that organic chemistry students have usually studied, and so they will provide a certain comfort level and intuitive appeal for students who are unfamiliar with quantum mechanics. Furthermore, in many cases consideration of two one-electron orbitals, the highest-energy occupied molecular orbital (called the HOMO or HO) and the lowest-energy unoccupied molecular orbital (called the LUMO or LU), is a sufficiently useful starting approximation for discussing photochemical reactions.

A description of the wave functions corresponding to the nuclear (vibrational) configurations (χ) and the electronic spin configurations (**S**) of Eq. 2.4 is generally not covered in introductory chemistry course or texts and is thus unfamiliar to many students. In this chapter, then, in addition to the critical one-electron wave functions (ϕ_i), we develop detailed pictorial representations of the wave functions χ and **S** in order to provide you with a degree of "quantum intuition" concerning vibrational wave functions of spin wave functions that will make it possible to visualize molecular vibrations and electron spins as they obey the laws of quantum mechanics.

2.4   Important Qualitative Characteristics of Approximate Wave Functions

An important feature of quantum mechanics is that although the wave function Ψ contains all of the information needed to determine any observable property of a molecular system, the information must be extracted from the wave function by acting on it with a mathematical operator and calculating an *expectation value* **P** for the observable (Eq. 2.3). The following qualitative characteristics of the approximate molecular wave functions Ψ_0, χ, and **S** are of considerable general importance for an understanding of the organic photochemical paradigm of Scheme 2.1 and its elaborations in the remaining chapters.

1. The properties of a wave function (Ψ_0, χ, or S) are not subject to direct experimental observation; however, the properties of the *square* of the wave function [Ψ_0^2 (and ϕ_i^2), χ^2, and S^2] *are* subject to direct experimental observation.

2. The square of a wave function [Ψ_0^2 (and ϕ_i^2), χ^2, and S^2] relates to the *probability* of finding the electrons, nuclei, and spins, respectively, at particular points in space in a molecular structure and therefore *provides a means of pictorially representing electrons, nuclear geometries, and spins as geometric structures that represent a molecule.*

3. The wave functions Ψ_0, χ, and S *may be visualized as structures in three dimensions* with respect to a given nuclear configuration of a molecule (e.g., the nuclear configurations of R, *R, I, or P).

4. For molecules possessing local or overall symmetry elements, Ψ_0 (and ϕ_i), χ, and S often possess useful symmetry properties that can be related to the motion and spatial positions of the electrons, nuclei, and spin, which *provide a pictorial basis for the selection rules governing transitions between two states.*

5. Waves can undergo "resonance" with other waves under proper conditions. The wave function of any molecular state may always be considered to consist, to a greater or lesser extent, of a superimposition (mixture) of other wave functions. The degree of mixing of the wave function of a state with other states is always small when the molecular state under examination possesses a *large* separation in energy from the other possible interacting states. However, the mixing of the wave function of a state with the wave functions of other states may be large when a molecular state under examination possesses an energy separation that is *small* relative to other possibly interacting states. We need to clarify what the terms "large energy" and "small energy" mean, but when two states are *degenerate* (i.e., possess exactly the same energy), the necessary conditions for *resonance* exist and the mixing of states is probable if sufficient interactions for mixing are available to couple the two states.

In this chapter, we are concerned with characteristics 1–5, above, and we learn how qualitative pictorial knowledge of Ψ_0, χ, and S can lead to a visualization of the *key electronically excited structures*, *R(S_1) and *R(T_1), in the important working paradigms of organic photochemistry (e.g., Scheme 1.3). In particular, we seek to qualitatively estimate two important equilibrium (static) properties of molecular states: (1) the various state electronic, nuclear, and spin *configurations*, and (2) a qualitative ranking of the *energies* corresponding to the various state electronic, nuclear, and spin configurations. From a knowledge of these two properties, we can show how to readily construct energy diagrams (Scheme 1.4), which allow the energetic ranking of the important low-energy electronic states of a given molecule. Chapter 3 describes how to extend our procedure to visualize operators corresponding to forces that trigger *transitions* between states of a state energy diagram (Scheme.1.4), all of which have a very similar nuclear configuration (e.g., photophysical processes, *R → R). We will present many experimental examples of radiative (Chapter 4) and radiationless (Chapter 5) photophysical processes. Chapter 6 considers how to use approximate wave functions to pictorially describe transitions that cause chemical changes in the nuclear configuration (e.g., primary photochemical processes, $^*R \rightarrow I$ and $^*R \rightarrow F \rightarrow P$).

2.5 From Postulates of Quantum Mechanics to Observations of Molecular Structure: Expectation Values and Matrix Elements[1,2]

One of the most important postulates of quantum mechanics is that the average value (or *expectation value*) P, of any observable molecular property (the energy of a state, the dipole moment of a state, the transition probability between two states, the magnetic moment associated with an electron spin, etc.) can be evaluated mathematically in terms of a so-called *matrix element*, $< \Psi|P|\Psi >$ (Eq. 2.6),

$$\underbrace{P}_{\substack{\text{Expectation value} \\ \text{Expected from experiment}}} = \underbrace{< \Psi|P|\Psi >}_{\substack{\text{Matrix element} \\ \text{Obtained by computation}}} \qquad (2.6)$$

where P is a mathematical operator representing the forces or interactions operating on Ψ to produce the property, P. As a matter of convention, when the property of interest (P) is the energy (E) of the system, the operator P is given the special symbol H, for the "Hamiltonian" operator (or energy operator, Eq. 2.1). The quantitative evaluation of matrix elements is an important procedure in theoretical chemistry, but the mathematical details of the computation of matrix elements, which can be found in standard texts,[1] are beyond the scope of this text.

For our purposes, the matrix element $< \Psi|P|\Psi >$ of Eq. 2.6 *is the quantum mechanical representation of the magnitude of an observable property of a molecular system.* Instead of computing the matrix element mathematically, we attempt to qualitatively evaluate the matrix element by devising methods to *visualize* the components of the matrix element (i.e., the wave function, Ψ, and the operator, P) by attributing to these mathematical objects concrete structural properties that can be associated with classical mechanics, which elegantly provides us with clear pictorial representations. Starting with these pictures, we can seek qualitative conclusions based on the approximate, but *useful*, classical model. The magnitude of a matrix element provides the basis for estimating the energy values of states and for estimating "selection rules" for the plausibility and probability of transitions between states.

The approximate Born–Oppenheimer wave functions are employed to compute matrix elements. Thus, by combining Eqs. 2.3 and 2.4, the approximate expectation value of P is given by the matrix element of Eq. 2.7.

$$\underbrace{P_{ave}}_{\text{Approximate expectation value}} \sim \underbrace{< \Psi_0 \chi S|P|\Psi_0 \chi S >}_{\text{Approximate matrix element}} \qquad (2.7)$$

To understand the factors determining the magnitude of P_{ave} and to be able to qualitatively evaluate its magnitude, we need to visualize each of the wave functions (Ψ_0, χ, and S) and the operator P and then to perform qualitative "mathematical" operations on the picture of the wave functions that we have devised. From Eqs. 2.5 and 2.7, the approximate value of P_{ave} is given by the matrix element of Eq. 2.8, which involves one-electron orbitals (ϕ_i).

$$P_{ave} \sim \; < (\phi_1 \phi_2 \cdots \phi_n) \chi \mathbf{S} \, | \, P \, | \, (\phi_1 \phi_2 \cdots \phi_n) \chi \mathbf{S}) > \qquad (2.8)$$

This level of approximation of P_{ave} is a working approximation, or so-called *zero-order approximation*. In other words, we are within the Born–Oppenheimer approximation (i.e., the separation of electronic, nuclear, and spin motion) and we are dealing with *one-electron orbital wave functions*, ϕ_i. In the next highest level of approximation, the so-called first-order approximation, *we challenge some aspect of the zero-order approximation and note its effect on the magnitude of* P. For example, we might begin to "mix" the one-electron wave functions and introduce electron–electron interactions. The mixing of wave functions (ϕ_i) is nothing more than a mathematical process that leads to a better approximation of the wave function in first order than we had in zero order. The new wave function, if the mixing is done properly, is closer to the "true" wave function, Ψ.

In this approximation, we only use the one-electron wave functions that come from solving the simple one-electron molecule model. For example, we can let the electron's orbital motion mix with its magnetic spin motion (i.e., spin–orbit coupling), or we can introduce electron–electron repulsion (called electron correlation, or configuration interaction), or we can consider how vibrations mix electronic states (i.e., vibronic interactions), and note how the magnitude of P is influenced by these "first-order mixing" effects. If we have selected a good zero-order approximation, the extent of mixing will be small and the mixing can be handled by "perturbation theory," which is discussed in Section 3.5. If the mixing is strong, then the zero-order approximation is poor and should be replaced by a better zero-order approximation as a starting point to describe the approximate wave functions.

2.6 The Spirit of the Use of Quantum Mechanical Wave Functions, Operators, and Matrix Elements

Classical mechanics deals with observables, such as the position and momentum of particles, as functions and postulates in which all of the information required to describe a *classical mechanical system* is available, if the functions describing the position and momentum of the entities that make up the system are known. Newton's laws of motion enable these functions to be described in a concise and economical mathematical form.

In a similar fashion, quantum mechanics postulates that all of the information about a *molecular system* is contained in its wave function Ψ and that in order to extract the information about the value of an observable, some mathematical operation (P) must be performed on the function (Eq. 2.3). This finding is analogous to the necessity of doing an act (an experiment) on the system in order to make a measurement of its state. Problems in quantum mechanics often boil down to selecting the proper approximation to solve for Ψ and/or making the correct selection of the operator (P) corresponding to the appropriate interaction. Thus, the selection of the correct operators (mathematical operations) or interaction leading to observable properties is as crucial to the explicit solution of a problem in quantum mechanics as is knowledge of the exact mathematical form of Ψ.

Two of the key operators of quantum mechanics are related to the classical dynamical quantities of the *momentum* and *position* of a particle (electrons, nuclei, and spins), which are also critical in Newton's equations. These operators allow the determination of the *energies*, and therefore the *ranking of the allowed energy levels* of the system for the electrons, vibrations, and spins. Once the operators for these dynamical variables have been selected, it usually turns out that operators for many observables can be set up in terms of the two fundamental variables of momentum and position. In the same manner that zero-order wave functions Ψ_0, χ, $\mathbf{S}$ are employed to approximate Ψ, in a quantitative evaluation of a matrix element, we almost always begin with a workable zero-order approximate Hamiltonian (or energy) operator, H_0, which is postulated to be related to the dominant set of interactions responsible for the energies of the electronic orbitals of the molecule. In the first-order approximation, we consider interactions, H_i, that are "weak" (e.g., interactions of electrons with vibrations and/or interactions of electrons with spins) relative to those described by H_0 but that can be important in causing transitions between the zero-order states. Chapter 3 shows that weak interactions are especially important when the zero-order states are close in energy.

2.7 From Atomic Orbitals, to Molecular Orbitals, to Electronic Configurations, to Electronic States

We propose to visualize a representation of the electronic portion of "exact" wave function Ψ by approximating its electronic part, Ψ_0, as a structure, namely, a configuration of overlapping, but noninteracting, occupied one-electron orbitals, ϕ_i (Eq. 2.5). A molecular orbital (MO) is approximated as a sum (or superposition) of atomic orbitals (AOs). A primary goal of this chapter is to develop a protocol for generating a state energy diagram (Scheme 1.4) for a given stable ground-state nuclear geometry of an organic molecule (R) and for the low-lying electronically excited states (*R) of the molecule by employing an electron configuration of low-energy *filled* "one-electron" orbitals.

The general procedure is as follows. For a given nuclear geometry of a molecule, the appropriate one-electron MOs are filled with the available electrons to generate a set of molecular electronic configurations. Only the lowest-energy configuration (the ground configuration, $R = S_0$) and the first or second low-energy excited configuration (*$R = S_2$, S_1, T_1, or T_2) generally need to be considered explicitly for the overwhelming number of organic photochemical reactions. This situation corresponds to a *zero-order molecular electronic configuration* (i.e., the Born–Oppenheimer approximation, filled one-electron orbitals). In order to generate proper *molecular electronic states* we would have to take into account electron correlation in some manner in the *first-order approximation*. Electronic correlation accounts in some manner for electron–electron interactions that are missing in the one-electron molecule (only one electron; no electron–electron interactions).

2.8   Ground and Excited Electronic Configurations

An *electronic configuration* of an organic molecule is defined as a listing of the MOs that are occupied with electrons in the configuration. Thus, an electron configuration tells us how the electrons are distributed among the available orbitals and also provides a description of the electronic distribution of a molecule from the knowledge of the spatial distribution of the orbitals (i.e., the one-electron wave functions). The ground configuration is defined as the configuration for which the orbitals that are occupied produce the state of lowest energy (R). All other electronic configurations correspond to electronically excited states (*R). According to the working state energy diagram (Scheme 1.4), in addition to the ground state (S_0), we are interested mainly in the lowest-energy electronically excited states, particularly the lowest-energy excited singlet state (S_1) and the lowest triplet state (T_1).

As an example of the procedure for constructing the electronic configuration for the ground state and the most important low-energy excited states, we select the formaldehyde molecule ($H_2C{=}O$) as an *exemplar*. This exemplar captures the important features of the ground- and excited-state configurations for many organic molecules of photochemical interest that possess the carbonyl functional group, and the lessons learned from this exemplar can be readily extended to more complicated systems. The energies of the one-electron MOs (ϕ_i) for $H_2C{=}O$ increase in the order shown in Eq. 2.9. For simplicity, the O atom is assumed to be unhybridized. In this description, the highest-energy occupied orbital is an unhybridized p orbital on oxygen (n_O), which is located in the plane of the molecule. In Eq. 2.9, $1s_O$, $2s_O$, and $1s_C$ refer to the essentially atomic, nonbonding "core" MOs localized on oxygen (subscript O) and carbon (subscript C), and the other MOs refer to conventional representations of bonding and antibonding orbitals (the superscript "*" indicates an antibonding orbital). Electrons in orbitals of higher energy than the core electrons are called "valence" electrons. The carbon atom of $H_2C{=}O$ is assumed to be sp^2 hybridized.

$$1s_O < 1s_C < 2s_O < \sigma_{C-H} < \sigma_{C-O} < \pi_{C=O} < n_O < \pi^*_{C=O} < \sigma^*_{C-O} < \sigma^*_{C-H}$$

$$(2.9)$$

Note, from Eq. 2.9, that there also exists a series of antibonding orbitals ($\pi^*_{C=O}$, σ^*_{C-O}, and σ^*_{C-H}) that are unoccupied in the ground state. The antibonding π^*_{CO} orbital is LU in the ground state of $H_2C{=}O$. The unoccupied σ^*_{C-H} and σ^*_{C-O} orbitals lie at very high energies and can be ignored in zero order. The HO in the ground state of $H_2C{=}O$ is the nonbonding n_O orbital. The n_O orbital is approximated to be an unhybridized p orbital on the oxygen atom.

A simple and useful approximation to visualize the radiative process $R + h\nu \rightarrow {}^*R$ is to consider the absorption of light as resulting from a transition between MOs (e.g., a HO $\rightarrow$ LU orbital transition) that is induced by the absorption of a photon. Similarly, the emission process ${}^*R \rightarrow R + h\nu$ can be viewed as a LU $\rightarrow$ HO transition. Of all the MOs, the HO and LU are generally of greatest importance in organic photochemistry, both in picturing the photophysical $R + h\nu \rightarrow {}^*R$ and the photochemical ${}^*R \rightarrow I$ processes. The initial photophysical and photochemical aspects start from ${}^*R$, which can be characterized as having one electron in the HO and one electron in the LU. Now, we address the issue of how to picture and describe the electronic distribution of occupied orbitals in S_0, S_1, and T_1, the key states of the state energy diagram (Scheme 1.4), and how to estimate the energy ranking of these states.

What is the electronic configuration of the lowest-energy electronic state (S_0) for our exemplar molecule $H_2C{=}O$? The molecule $H_2C{=}O$ has a total of 16 electrons, which must be distributed to the available MOs given in Eq. 2.9. In general, the energetic order of the MOs, the number of available electrons, and the Pauli and aufbau principles are employed to determine the ground-state electronic configuration (S_0) of a molecule. The electronic configuration of S_0 is constructed by adding electrons two at a time to the lowest-energy orbitals until all 16 of the available electrons have been assigned to orbitals. According to the Pauli exclusion principle, no more than two electrons may occupy one orbital and, if two electrons are present in an orbital, their spins must be paired. According to the aufbau principle, the ground or lowest-energy state of a molecule results from placing electrons in the lowest-energy orbitals first in accordance with the Pauli principle. The configuration of the ground state (S_0) of $H_2C{=}O$ constructed according to these configuration building rules is represented by Eq. 2.10, where the label in the parentheses designates the kind of orbital that is occupied, the superscripts designate the number of electrons in each orbital, and the subscript designates the electronic nature of the orbital. The empty high-energy σ^*_{CO} and σ^*_{CH} orbitals are not shown for simplicity. All of the bonding and nonbonding orbitals of $H_2C{=}O$ are filled in the ground state (S_0).

$$\Psi_0(H_2C{=}O) = (1s_O)^2(1s_C)^2(2s_O)^2(\sigma_{CH})^2(\sigma'_{CH})^2(\sigma_{CO})^2(\pi_{CO})^2(n_O)^2(\pi^*_{CO})^0 \quad (2.10)$$

The total electronic distribution of an electronic configuration may be approximated as a superposition of each of the occupied MOs that make up the configuration (Eq. 2.5). However, as in the case of writing Lewis structures, it is a useful and valid simplification to explicitly consider *only the valence electrons* when considering chemical or photochemical processes, and to consider the "core" electrons as being so close to the nuclei that they are too stable to be perturbed during chemical or photochemical processes. In a further simplification, it usually suffices to consider explicitly only the highest-energy valence electrons. In the case of $H_2C{=}O$, we consider the valence electrons in both $\pi_{C{=}O}$ and n_O because they are of comparable energy and both turn out to be potential HOs in many substituted carbonyl compounds. We also include the $\pi^*_{C=O}$ orbital explicitly, even though it is unoccupied in S_0, since the $\pi^*_{C=O}$ orbital is the LU for $H_2C=O$ and also for many carbonyl compounds. With the latter approximations in mind, a shorthand notation (eliminating the occupied orbitals that are too low in energy to be perturbed and only retaining the LU) for the ground-state electronic configuration of $H_2C=O$ is given by Eq. 2.11:

$$\Psi_0(H_2C=O) = K(\pi_{C=O})^2(n_O)^2(\pi^*_{C=O})^0 \quad \text{Ground state } (S_0) \quad (2.11)$$

In Eq. 2.11, K represents all 12 of the core electrons (electrons in σ or lower-energy orbitals displayed in Eq. 2.10) that are "tightly bound" to the molecular framework; that is, they are close to, and are strongly stabilized by, the positive nuclear charge and difficult to perturb.

Following an analogous protocol to that followed for the $H_2C=O$ exemplar, we ignore the lower-energy MOs and describe the ground-state electronic configuration of a second exemplar, ethylene ($CH_2=CH_2$), by Eq. 2.12:

$$\Psi_0(CH_2=CH_2) = K(\pi_{C=C})^2(\pi^*_{C=C})^0 \quad \text{Ground state } (S_0) \quad (2.12)$$

In the case of $CH_2=CH_2$, it is sufficient to consider only $\pi_{C=C}$ and $\pi^*_{C=C}$ explicitly, because the electrons in the σ orbitals are of much lower energy than those in the π orbital, and the σ^* orbitals are of very high energy compared to the π^* orbital. For $CH_2=CH_2$, the $\pi_{C=C}$ orbital is the HO and the $\pi^*_{C=C}$ orbital is the LU. *From these two exemplars, we have established a protocol for describing the orbital configuration of R and *R of any organic molecule for which the MOs are known or for which they can be approximated.*

We are now in a position to describe the *electronically excited states* (*R) of $H_2C=O$ and $CH_2=CH_2$ in terms of orbital configurations. The *lowest-energy electronic states* (*R), which are most important in organic photochemistry (see the state energy diagram, Scheme 1.4), possess an electronic configuration for which an electron has been removed from the HO of the ground-state configuration and placed into the LU of the ground-state configuration; that is, they possess a $(HO)^1(LU)^1$ electronic configuration for the two half-occupied orbitals of *R. As examples, the lowest excited states of formaldehyde and ethylene possess the electronic configurations shown in Eqs. 2.13 and 2.14. The wave function for an electronically excited state *Ψ is labeled with a superscript "*" to emphasize that the state is electronically excited. In Eq. 2.13b, we show the electronic configuration of the π,π^* state of $H_2C=O$. For $H_2C=O$, the π,π^* electronic configuration corresponds to an upper electronically excited state, **R (e.g., S_2 or T_2). In some carbonyl compounds, because of either stabilization of the nonbonding electrons or destabilization of the π electrons, the HO may be a π orbital rather than a nonbonding orbital. In these situations, the lowest-energy excited state will be a π,π^* excited state.

$$*\Psi(H_2C=O) = K(\pi_{C=O})^2(n_O)^1(\pi^*_{C=O})^1 \quad n,\pi^* \text{ excited state,* R} \quad (2.13a)$$

$$*\Psi(H_2C=O) = K(\pi_{C=O})^1(n_O)^2(\pi^*_{C=O})^1 \quad \pi,\pi^* \text{ excited state,* R} \quad (2.13b)$$

$$*\Psi(CH_2=CH_2) = K(\pi_{C=C})^1(\pi^*_{C=C})^1 \quad \pi,\pi^* \text{ excited state,* R} \quad (2.14)$$

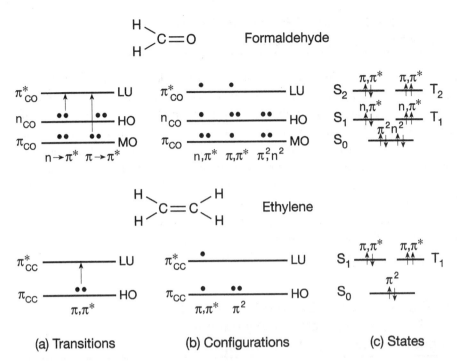

Figure 2.1 Schematic representation of (a) electronic transitions, (b) electronic configurations, and (c) electronic states for the lowest excited states of formaldehyde (top) and ethylene (bottom). The arrows for the states refer to the relative orientation of the two electrons in the HO and LU.

We can use electron orbital configurations not only to enumerate, to visualize, to energetically rank, and to conveniently classify electronically excited states (*R), but also to describe transitions between electronic states in electronic structural terms. For example, if the energies of the n and π orbitals of formaldehyde (Fig. 2.1a) are not too different in energy, $H_2C=O$ will possess two relatively low-energy electronic transitions (symbolized as $n \rightarrow \pi^*$ and $\pi \rightarrow \pi^*$). These two transitions will in turn produce two corresponding electronic excited-state configurations from excitation of the ground state: *R(n,π^*) and *R(π,π^*), respectively. For carbonyl compounds, which of these two states is actually lowest in energy depends on both structural and environmental factors (Chapter 4). On the other hand, ethylene will have only one low-energy electronic transition ($\pi \rightarrow \pi^*$) and one corresponding lowest-lying electronic excited-state configuration, *R(π,π^*), since the σ orbitals are very low in energy and the σ^* orbitals are very high in energy.

Electronically excited *configurations* are described in the remainder of the text in terms of the two singly occupied MOs, usually the HO and the LU (e.g., n,π^* and π,π^*). Electronic *transitions* are described only in terms of the orbitals undergoing a change in electronic occupancy (e.g., $n \rightarrow \pi^*$ and $\pi \rightarrow \pi^*$). This convenient shorthand notation for electronic configurations also provides a useful method to represent *excited states* (e.g., S_1 and T_1) as we show in the following sections.

2.9   The Construction of Electronic States from Electronic Configurations

The goal of this chapter is the development of ideas of molecular structure to enumerate, classify, energetically rank, and qualitatively visualize the electronic, vibrational, and spin nature of molecular states that appear in the state energy diagram of an organic molecule (Scheme 1.4). Now, we make an important distinction between the structure of excited singlet (S_1) and triplet (T_1) states that may be derived from the same electronic orbital and nuclear geometric configuration when the same two orbitals are half-occupied (i.e., the n,π^* and π,π^* configurations). The structural difference for the same orbital occupancy for a given nuclear configuration resides in differences in the configuration of the spins of the electrons in the two half-filled orbitals and the operation of the Pauli principle, which requires the electrons to behave *very* differently, depending on the relative orientation of the two electron spins.

2.10   Construction of Excited Singlet and Triplet States from Electronically Excited Configurations and the Pauli Principle

The Pauli *exclusion* principle states that no more than two electrons may occupy any given electronic orbital and that if two electrons do occupy an orbital, the electrons must have paired spins. The Pauli principle demands that any ground-state configuration (e.g., that for formaldehyde), in which all of the orbitals are filled with two electrons, must also have the two electrons spin paired in each orbital. Thus, organic molecules are ground-state *singlets*; that is, when two electrons occupy a single orbital and are orbitally paired, they must also be *spin paired* (we discuss the origin of the terms "singlet" and "triplet" in Section 2.27). In the excited state (*R), two electrons are *orbitally unpaired*; that is, each electron is in a different orbital, one in a HO and the other in a LU. The Pauli principle allows, *but does not require*, the spins of two electrons to be paired if they do not occupy the same orbital. As a result, either a *singlet* excited state, in which the two spins cancel and there is no net spin [i.e., the two electrons that are orbitally unpaired, HO and LU singly occupied, and with electron spins antiparallel ($\uparrow\downarrow$) or paired as in the ground state], or a *triplet* excited state [the two electrons that are orbitally unpaired, HO and LU singly occupied, and with electron spins parallel ($\uparrow\uparrow$) or unpaired] may result from the same electronic configuration of two electrons in half-occupied orbitals. This means that each of the *same* excited electronic $(HO)^1(LU)^1$ *configurations* given in Fig. 2.1b can refer to either an excited singlet (S_1) or to a triplet (T_1) electronic state, as shown in Fig. 2.1c.

Thus, *four low-energy excited states*, two singlets and two triplets, result from the two lowest-energy electronic configurations of formaldehyde. For $H_2C=O$, the lower-energy n,π^* state may be either a singlet (S_1) or a triplet (T_1) configuration, and similarly, the higher-energy π,π^* state may be either a singlet (S_2) or a triplet (T_2) configuration. These states are displayed in Fig. 2.1c (see a more detailed description

Table 2.1   States, Characteristic Orbitals, Characteristic Spin Configurations, and Shorthand Description of Low-Lying States of Formaldehyde ($CH_2{=}O$)[a]

State	Characteristic Orbitals	Characteristic Spin Electronic Configuration	Shorthand Description of State
S_2	π,π^*	$(\pi \uparrow)^1(n)^2(\pi^* \downarrow)^1$	$^1(\pi,\pi^*)$
T_2	π,π^*	$(\pi \uparrow)^1(n)^2(\pi^* \uparrow)^1$	$^3(\pi,\pi^*)$
S_1	n,π^*	$(\pi)^2(n \uparrow)^1(\pi^* \downarrow)^1$	$^1(n,\pi^*)$
T_1	n,π^*	$(\pi)^2(n \uparrow)^1(\pi^* \uparrow)^1$	$^3(n,\pi^*)$
S_0	π,n	$(\pi)^2(n)^2(\pi^*)^0$	$^1[(\pi)^2(n)^2]$

a. When an orbital contains two electrons, the spins must be paired, so the spins are not shown explicitly for filled orbitals.

Table 2.2   States, Characteristic Orbitals, Characteristic Spin Configurations, and Shorthand Description of Low-Lying States of Ethylene ($CH_2{=}CH_2$)

State	Characteristic Orbitals	Characteristic Spin Electronic Configuration	Shorthand Description of State
S_1	π,π^*	$(\pi \uparrow)^1(\pi^* \downarrow)^1$	$^1(\pi,\pi^*)$
T_1	π,π^*	$(\pi \uparrow)^1(\pi^* \uparrow)^1$	$^3(\pi,\pi^*)$
S_0	π^2	$(\pi)^2(\pi^*)^0$	π^2

of the orbital configuration and spin occupancy in Table 2.1). For $CH_2{=}CH_2$, only two electronically excited states, an $S_1(\pi,\pi^*)$ and a $T_1(\pi,\pi^*)$ state, are expected to be low lying in energy (Table 2.2).

2.11   Characteristic Configurations of Singlet and Triplet States: A Shorthand Notation

Now, let us briefly review the conventions described in Chapter 1 for the state energy diagram of a molecule (Scheme 1.4), which serve as a basis for a convenient notation to describe electronic configurations and states. We always label the ground state S_0. We label electronically *excited singlet* states S_1, S_2, and so on, where the subscript refers to the *energy ranking* of the state relative to the ground state (S_0). The state S_0 is arbitrarily given the rank of energy equal to zero, so all other states possess *positive* energy relative to the ground state. Thus, S_1 is the *first* electronically excited *singlet* state located energetically above S_0, and S_2 is the *second* electronically excited *singlet* state located energetically above S_0. Similarly, T_1 is the *first* electronic *triplet* state located above S_0, and T_2 is the *second* electronic *triplet* state above S_0.

As a shorthand, we describe electronic and spin *configurations* by indicating only the key molecular *orbitals* that are expected to dominate the energy and/or chemistry of the configuration. Thus, for formaldehyde we need to explicitly consider only the $\pi_{C=O}$, n_O, and $\pi^*_{C=O}$ orbitals when discussing electronic configurations, and hence when discussing electronic states, in a state energy diagram. This approximation assumes that consideration of the lower-lying nonbonding orbitals, the σ orbitals, and the higher-energy σ^* orbitals is unnecessary for an understanding of the photochemistry of formaldehyde. Consequently, *each electronic state may be described in terms of a characteristic electronic configuration, which in turn may be described in terms of the* HO *and* LU *and in terms of a characteristic spin configuration, either singlet* ($\uparrow\downarrow$) *or triplet* ($\uparrow\uparrow$).

A single electron *configuration* is often adequate to approximate the electronic characteristics of an electronic *state*. In some cases, however, a combination of two or more electronic configurations are required to achieve a good approximation of a state. This situation occurs *when two states in zero order have similar energies and can be coupled by some interaction, such as vibrational motion.* In this case, a single electron orbital is inadequate as a description because these orbitals are "mixed by vibrations"; that is, a nonbonding orbital may mix in some π character as the result of a vibration (Section 3.7). Later, we see examples of "mixed" n,π^* and π,π^* states, where neither "pure" (zero-order) configuration provides an adequate representation of the actual state but the mixed or first-order representation works well. Perturbation theory teaches the rules for such mixing and is discussed in Chapter 3.

In summary, the use of electronic and spin configurations to describe the S_0, S_1, S_2, T_1, and T_2 states of formaldehyde and the S_0, S_1, and T_1 states of ethylene may be described, as shown in Tables 2.1 and 2.2, respectively. In addition, based on the individual orbital energies, we can rank the energy of the n,π^* state of a given spin as being lower in energy than the corresponding π,π^* states of the same spin, for example, $S_1(n,\pi^*)$. The ranking of states of different electronic configurations may shift from the order given in the table as the result of substitution and environmental effects (Chapters 3 and 4).

2.12 Electronic Energy Difference between Molecular Singlet and Triplet States of *R: Electron Correlation and the Electron Exchange Energy

Hund's rule must be followed when filling the atomic orbitals of atoms with the available electrons to determine an atomic electronic configuration. That is, when electrons are added to atomic orbitals of *equal* energy (those that are degenerate in energy), they must first half-fill every orbital of equal energy, with the spins unpaired, before spin pairing in any orbital. In other words, *when filling two orbitals of equal energy, the triplet state is lower in energy than the singlet state for the same orbital occupancy.*

For organic photochemistry, Hund's rule can be rephrased for MOs as follows: *For molecules possessing two half-filled orbitals, one a HO and the other a LU, the triplet state ($\uparrow\uparrow$) is always of lower energy than the energy of the corresponding singlet state ($\uparrow\downarrow$) derived from the same electronic (HO)^1(LU)^1 configuration.* For example, $E_S(n,\pi^*) > E_T(n,\pi^*)$ and $E_S(\pi,\pi^*) > E_T(\pi,\pi^*)$, where E_S is the electronic energy of a singlet state in its lowest equilibrated vibrational level and E_T is the electronic energy of a triplet state in its lowest equilibrated vibrational level. No general statement can be made concerning a comparison of the energies of singlet and triplet states of different configuration. For example, a π,π^* triplet may be higher or lower in energy than an n,π^* singlet, and an n,π^* triplet may be higher or lower in energy than a π,π^* singlet, depending on substituents and environments.

A physical and theoretical basis for Hund's rule for the relative energies of singlet and triplet states for systems with two half-filled orbitals is available from consideration of the implications of the Pauli exclusion principle. That is, *no more than two electrons may occupy a given orbital, and if two electrons occupy a given orbital, they must be spin paired.* The Pauli principle may be viewed as a requirement that the electrons of a molecule must *correlate* their motions and positions in space, because two electrons with parallel spins ($\uparrow\uparrow$) are absolutely forbidden to occupy the same space at the same time. The Pauli exclusion principle is based on a very deep quantum mechanical law[1] that requires the overall wave function of a system to change its mathematical sign under pairwise *exchange* of identical odd spin particles, e.g., electrons.

The electrons in a singlet state are not compelled to obey the Pauli exclusion principle and both electrons may on occasion approach the same region of space; however, the Pauli principle forbids the two spin-unpaired electrons of a triplet from getting too close to one another and occupying the same region of space. This remarkable feature of the Pauli principle has a profound effect on the relative energies of a singlet and a triplet derived from the same configuration of half-filled orbitals: The energy of the singlet state (E_S) will generally be higher than that of the triplet state (E_T) for the *same* (HO)^1(LU)^1 configuration. Thus, by being instructed by the Pauli principle to avoid being in the same place, electrons in a triplet state are able to minimize electron–electron repulsions relative to a singlet state, in which the electrons are allowed to approach without the Pauli restriction. The basis of the energy difference E_S and E_T is perhaps even more interesting. The Pauli principle actually requires electrons of the same spin to tend to cluster more than expected from classical electron–electron repulsion. These features lead to differences in the energies of the singlet and triplet states that are derived from the same half-filled orbital configuration. An imperfect and somewhat whimsical analogy is available from observations of a busy traffic intersection. Cars that are required to follow the traffic rules can be considered to be in a low-energy state that avoids collisions, whereas cars that are not required to follow the traffic rules can be considered to be in a higher-energy state that allows collisions. Thus, the Pauli principle operates as a sort of quantum mechanical "traffic cop for electron motion" that instructs the two key orbitally unpaired electrons in triplet states how to avoid one another.

The difference in *electronic energy*, $E_S - E_T = \Delta E_{ST}$, between singlet and triplet states (that are derived from the same electron orbital configuration) results from the "better" energy-lowering correlation of electron motions in a triplet state that produces lower electron–electron repulsions. In order to visualize how this singlet–triplet energy separation (ΔE_{ST}) arises for a given orbital configuration of a molecule, and to gain an appreciation of how the magnitude of ΔE_{ST} depends upon the orbitals composing the characteristic configuration, we attempt to visualize the magnitude of *matrix elements* that correspond to the electronic energy of orbital configurations.

2.13   Evaluation of the Relative Singlet and Triplet Energies and Singlet–Triplet Energy Gaps for Electronically Excited States (*R) of the Same Electronic Configuration

The zero-order energy (e.g., E_S or E_T) of an electronically excited state, *R, is defined as the summation of all of the zero-order energies of all the *occupied* one-electron orbitals (which assumes no electron–electron interactions). The electron–electron repulsion energies are included in the first-order approximation. In the Born–Oppenheimer approximation, the nuclear geometry is fixed; the attractive forces between the negative electrons and the fixed positive nuclear framework are assumed to contribute a certain stabilization energy based on classical electrostatic attractions. *The differences in energy between different states in this approximation are due entirely to electron–electron repulsions.* Thus, when we discuss electronic energy differences between two electronic states, we are referring only to the differences between electron–electron repulsions in the two states. *The state with the smaller electron–electron repulsions always has the lower energy.* The classical part of the operator, H, for the energy increase due to electron–electron repulsions is defined by the *electrostatic repulsion,* or *coulombic repulsion, between electrons*: $H_{ee} = e^2/r_{12}$, where e is the charge on an electron and r_{12} is the separation between the electrons. For a fixed positive nuclear configuration, the coulombic interactions between two electronic distributions *are always repulsive,* and therefore will *always increase* the energy of the system compared to the zero-order approximation (which ignores electron–electron repulsions).

The magnitude of the electronic repulsion may be computed by integrating the repulsive interactions over the entire volume of a molecule, assuming a fixed positive molecular framework (i.e., by computing matrix elements, Eq. 2.3). These integrals correspond to the matrix elements for the *total* electron–electron repulsion energies and are broken down into the computation of mathematical integrals for two types: (1) the repulsions between electrons due to *classical mechanical electrostatic interactions between negative charge distributions of electrons* (mathematically equal to the *Coulomb integral* and given the symbol K), and (2) the first-order *correction,* J, to the value of the electron–electron repulsions between electrons when quantum mechanical interactions due to the Pauli exclusion principle (which is a corollary of the more fundamental Pauli principle that requires the electronic wave function to change sign upon *exchange* of any two electrons). This correction, J, is called the *electron exchange energy* and is equal to ΔE_{ST}, the energy difference between a singlet and a triplet state for the *same* half-filled orbital configuration. We interpret J as the *correction to the classical electron–electron repulsion K that is required to get the correct answer for a state energy.* Note that J is still a repulsion energy and therefore a positive quantity. As an example, the matrix element for J for the difference in energy between S_1 and T_1 derived from the same HO and LU would have the form $J \sim < HO|e^2/r_{12}|LU >$.

The quantum mechanical integral for J is referred to as the *electron exchange integral* because the Pauli principle is derived from a *postulate* on the symmetry property of electrons when two electrons *exchange* positions in space with one another. *The exchange integral is a purely quantum mechanical phenomenon since it amounts to a quantum mechanical correction to the classical electronic distribution, which does not take into account the influence of electron spin on electron–electron correlation and therefore electron–electron repulsion.*

As an exemplar for visualizing and qualitatively estimating the matrix element $< HO|e^2/r_{12}|LU >$, and therefore the differences in singlet and triplet energies *for electronically excited states, *R, of the same electronic configuration,* let us consider the energies of the ground state (S_0), and the lowest excited S_1 and T_1 states of $H_2C{=}O$ based on the computation of the electronic energies of the n and π^* orbitals of Fig. 2.1. The energies of S_0, S_1, and T_1 are defined by Eqs. 2.15–2.17, respectively. Since we are interested only in the energy differences, ΔE, between states and not their absolute energies, the energy of S_0 is conveniently defined (in Eq. 2.15) as 0 and the excited-state energies are considered to be positive quantities relative to this standard energy.

$$E_0 = 0 \text{ (by definition)} \tag{2.15}$$

$$E_S = E_0(n,\pi^*) + K(n,\pi^*) + J(n,\pi^*) \tag{2.16}$$

$$E_T = E_0(n,\pi^*) + K(n,\pi^*) - J(n,\pi^*) \tag{2.17}$$

In Eqs. 2.16 and 2.17, $E_0(n,\pi^*)$ is the zero-order energy (for a fixed nuclear framework) of the *one-electron orbital derived* excited state; $K(n,\pi^*)$ is the first-order Coulombic correction due to classical electron–electron correlation; and $J(n,\pi^*)$ is the first-order quantum mechanical correction of the electron–electron repulsion energy due to the Pauli principle. Since the charges are the same for electron–electron electrostatic interactions that lead to repulsions, the values for both of the energy integrals (K and J) in Eqs. 2.16 and 2.17 are defined as mathematically positive (positive mathematical quantities are energy-raising and negative mathematical quantities are energy-lowering).

A simple picture of how the Pauli exclusion principle operates is available by assuming that quantum mechanics requires that electrons of the same (unpaired, parallel) spin ($\uparrow\uparrow$) avoid approaching one another, but that electrons of opposite (paired, antiparallel) spin ($\uparrow\downarrow$) actually have an enhanced probability of being found near one another. Thus, from this assumption we build up the "quantum intuition" that if two electrons have the parallel spin orientation, the average repulsion energy will be less than the repulsion computed from the classical model *because of the tendency of electrons with parallel spins to avoid each other, and thus reduce electron–electron repulsions.* On the other hand, if two electrons have the antiparallel spin orientation, the average repulsion energy between them will actually be greater than the repulsion computed from the classical model *because of the tendency of two electrons of the opposite spin to "stick" to one another.* As a result, for the T_1 state, the average repulsive energy is reduced from the expected value for the classical case $[E(n,\pi^*) = E_0(n,\pi^*) + K(n,\pi^*)]$ to a lower value $[E_0(n,\pi^*) + K(n,\pi^*) - J(n,\pi^*)]$. Note that we are lowering the overall classical electron–electron repulsions $K(n,\pi^*)$ by putting a negative sign in front of $J(n,\pi^*)$.

Now, let us estimate the magnitude of the difference in energy between a singlet and triplet state of the same electronic $(HO)^1(LU)^1$ configuration by visualizing the *overlap of the orbitals*, an important factor in determining the magnitude of the integrals in Eqs. 2.16 and 2.17. As above, we start with the zero-order approximation (one-electron orbitals) for an n,π^* state, and let $E_0(n,\pi^*)$ be the value of the matrix element for the zero-order one-electron orbital energy. In this case, the electrons are assumed not to interact, so that the singlet and triplet energies would be identical (i.e., both J and $K = 0$). Now, we let $K(n,\pi^*)$ be the matrix element that measures the classical electron repulsion due to Coulombic interactions between electrons, and let the matrix element $J(n,\pi^*)$ refer to the correction to the value of the electron repulsion when the electron exchange is taken into account. We can obtain ΔE_{ST}, the energy difference between $S_1(n,\pi^*)$ and $T_1(n,\pi^*)$, by subtracting Eq. 2.17 from Eq. 2.16, thus obtaining Eq. 2.18a. The values of $E_0(n,\pi^*)$ and $K(n,\pi^*)$ drop out and $\Delta E_{ST} = E_S - E_T = 2J(n,\pi^*)$. Thus, ΔE_{ST} depends only on the value of $2J(n,\pi^*)$, as shown in Eq. 2.18b.

$$\Delta E_{ST} = E_S - E_T = E_0(n,\pi^*) + K(n,\pi^*) + J(n,\pi^*)$$

$$- [E_0(n,\pi^*) + K(n,\pi^*) - J(n,\pi^*)] \tag{2.18a}$$

$$\Delta E_{ST} = E_S - E_T = 2J(n,\pi^*) > 0 \tag{2.18b}$$

Since $J(n,\pi^*)$ corresponds to electron–electron repulsion (and is therefore positive), we have the general result of Eq. 2.18b, that is, $E_S - E_T = 2J(n,\pi^*) > 0$. We conclude, within this approximation, that E_S *must* be higher in energy than E_T *in general* for all n,π^* states. In addition, the value of ΔE_{ST} is precisely equal to $2J(n,\pi^*)$; that is, the energy gap between the singlet and triplet equals twice the value of the electron exchange energy, J. Since the details of the orbitals involved in determining the matrix element were not considered in making the argument, we can extend the above exemplar to predict the relative values of E_S and E_T for any $(HO)^1(LU)^1$ configuration. For example, we can conclude that for a π,π^* electronic configuration, $S(\pi,\pi^*)$ and $T(\pi,\pi^*)$, $\Delta E_{ST} = 2J(\pi,\pi^*)$ and that the $S(\pi,\pi^*)$ state will be higher in energy than the $T(\pi,\pi^*)$ state. This qualitative result can be generalized: *The energy gap between a singlet and a triplet state of the same electronic configuration of half-filled orbitals (i.e., the same orbital occupancy) is purely the result of electron exchange and is responsible for the observation that the energy of a triplet state is generally lower than that of a singlet state of the same electronic configuration (the same orbital occupancy) for organic molecules.* It follows, therefore, that we can estimate the value of ΔE_{ST} if we can estimate the value of J. Which is larger, in general, $\Delta E_{ST}(n,\pi^*)$ or $\Delta E_{ST}(\pi,\pi^*)$? We answer this question qualitatively by using a pictorial approach.

2.14   Exemplars for the Singlet–Triplet Splittings in Molecular Systems

Now, we consider our first attempt to use a pictorial representation of electronic orbitals as a means of qualitatively evaluating the value of a matrix element, namely, the matrix elements for the values of the energy differences $\Delta E_{ST}(n,\pi^*)$ and $\Delta E_{ST}(\pi,\pi^*)$. The estimation of ΔE_{ST} requires the qualitative evaluation of the magnitude of the matrix elements corresponding to the mathematical exchange integrals, $J(n,\pi^*)$ and $J(\pi,\pi^*)$ given in Eqs. 2.16 and 2.17. For this estimation and comparison of the relative values of $J(n,\pi^*)$ and $J(\pi,\pi^*)$, we consider the (n,π^*) and (π,π^*) states of a carbonyl group as an exemplar functional group. The energy diagram for $H_2C=O$ shown in Fig. 2.1c serves as an exemplar for the carbonyl functional group. In the exemplar $H_2C=O$, the $S_1(n,\pi^*)$ and $T_1(n,\pi^*)$ states are the two lowest-energy excited states, as is the case for many carbonyl compounds, such as ketones, with the $S_2(\pi,\pi^*)$ and $T_2(\pi,\pi^*)$ being somewhat higher in energy. From quantum analysis, the value of $J(n,\pi^*)$ represents the Pauli correction to the electrostatic repulsion, due to electron exchange, between an electron in a nonbonding orbital and an electron in a π^* orbital. The parameter $J(\pi,\pi^*)$ represents the Pauli correction to the electrostatic repulsion, due to electron exchange, between an electron in a π orbital and an electron in a π^* orbital.

The magnitude of $J(n,\pi^*)$ is given by the value of the matrix element shown in Eq. 2.19, where n and π^* represent the wave functions for the n and π^* orbitals, respectively. The numbers (1) and (2) refer to the two electrons occupying these orbitals, and e^2/r_{12} is the operator (where e is the charge on an electron and r_{12} is the distance separating the electrons) that represents the repulsion between the exchanging electrons. The e^2/r_{12} term may be factored out of the integral for Eq. 2.19 so that the value of $J(n,\pi^*)$ can be seen to be directly proportional to the mathematical integral of the wave functions for the overlap of the n orbital and π^* orbitals (Eq. 2.20). The magnitude of this orbital overlap is represented by the symbol $< n|\pi^* >$ and is called an *orbital overlap integral*. The quantum mechanical *mathematical* overlap integral corresponds to the degree of physical overlap of the orbitals in space. The smaller the overlap, the smaller the value of the overlap integral; and the larger the overlap, the larger the value of the overlap integral. The overlap integral may be visualized as a measure of the mutual resemblance of two wave functions. If the overlap integral is large, the two wave functions "look alike"; if the overlap integral is small, the two wave functions "look different." In Chapter 3, we show that transitions between states are always faster between states that "look alike" than between states that "look different." Thus, the overlap integral is a useful guide to the probability or rate of transitions between two states.

$$J(n,\pi^*) = < n(1)\pi^*(2)|e^2/r_{12}|n(2)\pi^*(1) > \qquad (2.19)$$

$$J(n,\pi^*) \sim e^2/r_{12} < n(1)\pi^*(2)|n(2)\pi^*(1) \sim < n|\pi^* > \quad \text{Overlap integral} \quad (2.20)$$

Thus, the magnitude of the exchange energy, $J(n,\pi^*)$, is proportional to the value of the overlap integral $< n|\pi^* >$ (Eq. 2.20). Since the magnitude of $< n|\pi^* >$ is proportional to the degree of overlap for the n and π^* orbitals, we can readily estimate qualitatively the value of the integral for Eq. 2.20 simply by pictorially estimating the overlap of orbitals as shown in Eq. 2.21a. Similarly, the value $< \pi | \pi^* >$ is proportional to the degree of overlap for the π and π^* orbitals; that is, $J_{\pi,\pi^*} \sim < \pi | \pi^* >$, as shown in Eq. 2.21b.

Small orbital overlap between n and π^*

$$\langle n | \pi^* \rangle \text{ Small}$$

(2.21a)

Large orbital overlap between π and π^*

(2.21b)

$$\langle \pi | \pi^* \rangle \text{ Large}$$

An overlap integral, such as that of Eq. 2.20, may be pictured as a measure of the mutual physical or mathematical resemblance for two wave functions in space relative to the nuclear framework of a molecule. If the two wave functions are identical with respect to their space occupancy, then the normalized value of the overlap integral, $< \phi_i | \phi_j > = 1$. In other words, $< n|n >$ and $< \pi | \pi >$ are both integrals whose normalized mathematical overlap is assigned a value of 1. They correspond to the overlap of an orbital with itself, which corresponds, in turn, to the maximum possible overlap, defined mathematically as having a maximum normalized value of 1. On the other hand, if the two wave functions do *not* overlap at all, then the normalized value of $< \phi_i | \phi_j > = 0$. For the purposes of qualitatively visualizing orbital overlap, the one-electron orbitals ϕ_n, ϕ_π, and ϕ_{π^*} are adequate. We may visualize the overlap integral $< n | \pi^* >$ if we replace the symbol n (which represents the one-electron wave function, ϕ_n, of the n_O orbital of formaldehyde) with a picture of the n orbital and replace the symbol π^* (which represents the one-electron wave function, ϕ_{π^*}, of the π^* orbital of a ketone) with a picture of the $\pi_{C=O^*}$ orbital (as is done in Eq. 2.21a). *Through visualization of the n and π^* orbitals, we can picture the common regions of overlap for these orbitals in the space about the nuclei of formaldehyde.* Then, we estimate qualitatively the degree of overlap between the n and π^* orbitals. By inspection, we immediately note that the mathematical overlap integral $< n | \pi^* >$ must be small because the n and π^* orbitals do not occupy very much of the same region of space. An electron in the π^* orbital has its density above and below the plane of the molecule (assuming a planar structure). Furthermore, the lobe of the π^* orbital on the carbon atom lies in the nodal plane of the n orbital (Eq. 2.21a), which means that at this level of approximation there is an exact mathematical canceling of the wave functions and zero overlap of the two orbitals n and π^*; that is, $J(n,\pi^*) \sim <n|\pi^*> = 0$ in zero order. In this approximation, there is no difference between the energies of the $S_1(n,\pi^*)$ and the $T_1(n,\pi^*)$, since the overlap integral is precisely equal to 0. This exercise is equivalent to determining a selection rule for the value of the singlet–triplet energy gap for the n,π^* states, where the selection rule allows a qualitative estimation of J by simply inspecting the overlap integral $< HO|LU >$.

Now, let us compare this result to the overlap of orbitals for a π,π^* configuration. If we employ the ideas leading to Eq. 2.20, we obtain for a π,π^* configuration the pictorial representation shown in Eq. 2.21b.

From a comparison of the orbitals, we see in Eq. 2.21b that there is significant orbital overlap at both the C and the O atoms. Thus, in contrast to the conclusion that the value of $J(n,\pi^*)$ was generally ~ 0, the value of $J(\pi,\pi^*)$ is finite and large. In other words, J_{π,π^*} will be greater, *in general*, than J_{n,π^*} because the overlap integral $< \pi|\pi^* >$ will usually be greater than the overlap integral $< n|\pi^* >$. From the above pictorial arguments, *we conclude that $\Delta E_{ST}(n, \pi^*) < \Delta E_{ST}(\pi,\pi^*)$, in general, because the overlap of a π with a π^* orbital will usually be greater than the overlap of an n and a π^* orbital.*

Now, we test the qualitative pictorial conclusion with some quantitative data to determine its validity. Experimentally, values of $\Delta E_{ST}(n,\pi^*)$ for ketones are 7–10 kcal mol^{-1}, while values of $\Delta E_{ST}(\pi,\pi^*)$ for aromatic hydrocarbons are on the order of 30–40 kcal mol^{-1}. Table 2.3 lists some experimental values of ΔE_{ST} and shows that states derived from n,π^* configurations consistently have smaller singlet–triplet splitting energies than states derived from π,π^* configurations. We note that the conclusions for the value of ΔE_{ST} refer to comparison of states of the same orbital symmetry.

Table 2.3   Some Examples of Singlet–Triplet Splittings

Molecule	Configuration of S_1 and T_1	ΔE_{ST}(kcal mol^{-1})
$CH_2{=}CH_2$	π,π^*	~ 70
$CH_2{=}CH{-}CH{=}CH_2$	π,π^*	~ 60
$CH_2{=}CH{-}CH{=}CH{-}CH{=}CH_2$	π,π^*	~ 48
	π,π^*	25[a] (52)[b]
	π,π^*	31[a] (38)[b]
	π,π^*	~ 34
	π,π^*	30
$CH_2{=}O$	n,π^*	10
$(CH_3)_2C{=}O$	n,π^*	7
$(C_6H_5)_2C{=}O$	n,π^*	5

a. ΔE_{ST} between states of different orbital symmetry.
b. ΔE_{ST} between states of the same orbital symmetry.

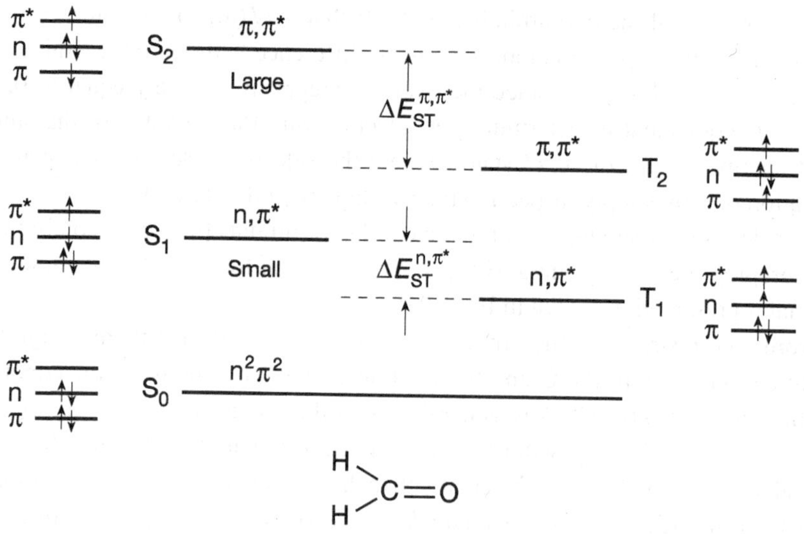

Figure 2.2   A qualitative state energy diagram for formaldehyde, including singlet–triplet splittings and electronic configurations of states.

The zero-order *state energy-level diagram* for formaldehyde (Fig. 2.1c) may now be modified to include the first-order singlet–triplet splittings (Fig. 2.2). From this point on, the states shown in a state energy-level diagram will be assumed to be at this level of approximation.

2.15   Electronic Energy Difference between Singlet and Triplet States of Diradical Reactive Intermediates: Radical Pairs, I(RP), and Biradicals, I(BR)

Now, we consider the electronic energy difference between singlet and triplet states of reactive intermediates, I(D), possessing a "diradical" character (Section 1.11).[2,3] Section 2.14 was concerned with the singlet and triplet energies and singlet–triplet energy gaps of *electronically excited* organic molecules (*R) possessing the electronic configuration $(HO)^1(LU)^1$. The energy difference between the HO and LU of organic molecules is typically on the order of 40–80 kcal mol^{-1} or more. The HO is a strongly bonding MO, which means that two spin-paired electrons will be very stable in such an orbital. The energy separation between the HO and LU is very *large* compared to vibrational or electron spin energy gaps. On the other hand, the reactive intermediate, I(D), produced in the *R →I(D) primary photochemical transformation of Scheme 2.1, is usually a diradical species, that is, a radical pair, I(RP) or a biradical I(BR).

In contrast to the situation for $*R(HO)^1(LU)^1$, which possesses two half-filled HO and LU with a large energy separation, a diradical intermediate I(D) is typically characterized by *two half-filled nonbonding* (NB) *orbitals of comparable energies.*

These two NB orbitals are often each localized on a carbon atom of the I(RP) or I(BR). We call these two nonbonding orbitals NB_L (the lower-energy nonbonding orbital) and NB_U (the upper-energy nonbonding orbital). The energy gap between NB_L and NB_U, and therefore of $I(NB_L)^1(NB_U)^1$, is generally small, often on the order of $< 1 \, \text{kcal mol}^{-1}$; indeed, this energy gap may be smaller than the energy gap between vibrational levels and sometimes even smaller than that between electron spin energy values. The characteristically small energy gap between the two NB orbitals causes a mixing of the possible states for I(D) and has a profound influence on the energy ranking of singlet and triplet states and singlet–triplet energy gaps as a function of the separation and orientation in space of the NB_L and NB_U orbitals for I(D). In Chapter 3, we show how these small energy gaps and their dependence on the separation and orientation of the NB_L and NB_U orbitals also have a profound influence on the rate of singlet–triplet transitions.

We have seen how the exchange interaction (J) dominates the value and direction of ΔE_{ST} for S_1 and T_1 of *R. In this case, the large energy gap between the HO and LU insures there is little mixing of $^*R(S_1)$ and $^*R(T_1)$ with the ground-state $R(S_0)$, which possesses two spin-paired electrons in a strongly bonding HO. Thus, for the ground-state R, for which there is a large energy gap between HO and LU, the electronic configuration $(HO)^2(LU)$ is strongly favored and the ground state is *always* an electron-paired singlet.

For I(D), however, where there is a very small energy gap between the NB_L and NB_U orbitals, there is no significant energy advantage to putting both electrons in the NB_L orbital [electronic configuration $(NB_L)^2$] because this will cause electron–electron repulsions that are absent when the two electrons are separated in space in the NB_L and NB_U orbitals [electronic configuration $(NB_L)^1(NB_U)^1$]. Now, let us examine the interplay of factors that determine whether the $(NB_L)^2$ or the $(NB_L)^1(NB_U)^1$ configuration is of lower energy.

As a first approximation,[4] we assume that the interactions between the two NB orbitals are weak, but finite. In this approximation, the value of the singlet–triplet gap for I(D), ΔE_{ST}, may be estimated by the two terms in Eq. 2.22.

$$\Delta E_{ST} = J - B \qquad (2.22)$$

The parameter J is proportional to the electron exchange integral for the $(NB_L)^1(NB_U)^1$ configuration, and B is proportional to the overlap of the two orbitals and the contribution of the bonding $(NB_L)^2$ configuration. Both J and B are positive quantities, and B corresponds to the bonding contribution of $(NB_L)^2$. Since bonding is an energy-lowering interaction (note the minus sign in front of B in Eq. 2.22), the spin-paired, singlet $(NB_L)^2$ configuration of I(D) will be lower in energy than the triplet $(NB_L)^1(NB_U)^1$ configuration when $B > J$. When this is the case, the ground state of I(D) will be a singlet (S) not a triplet (T). When $B < J$, however, Hund's rule applies and the ground state of I(D) will be a triplet (T), not a singlet (S).

When, however, is $B > J$ or $B < J$ for I(D)? We will consider a pictorial representation of Eq. 2.22 in terms of the overlap of two nonbonding orbitals to see how the singlet and triplet energies for I(D) vary with orbital orientation, using a simple model of two weakly interacting p orbitals on two spatially separated carbon atoms as an exemplar (Fig. 2.3). From the above discussion, the triplet state of I(D) will be of lower energy and will be the ground state when there is no bonding due to orbital overlap (i.e., $B = 0$, so $J > B$). The singlet state of I(D) will be of lower energy when there is significant bonding due to orbital overlap (i.e., when $B > J$).

Now, let us analyze the influence the orbital orientation of the two NB orbitals has for the following limiting cases: (1) the two p orbitals are oriented so that one orbital is in the nodal plane of the other (called the 90^0 orientation because the p orbitals are perpendicular to each other, as shown in Fig. 2.3a, left), and (2) the two p orbitals are "pointing at one another" (called the 0^0 orientation because the p orbitals are "head to head," as shown in Fig. 2.3a, right). For orientations close to (1), the two orbitals are close to orthogonal to each other and there is essentially no net overlap of the orbitals, in which case the B term in Eq. 2.22 is close to zero. For this situation, the exchange interaction can be more significant than the bonding interaction (i.e., $B < J$). When the latter situation holds, $\Delta E_{ST} \sim J$ (Eq. 2.22) and the T state of I(D) will be the ground state (Fig. 2.3, lower left). For orientations close to the head-to-head situation (2), however, the value of the overlap becomes significant and favors the bonding configuration, which puts two electrons into the NB_L orbital. In this case, $B > J$, so $\Delta E_{ST} \sim -B$ (Eq. 2.22) and the S state of I(D) will be the ground state (Fig. 2.3, lower right).

The following general conclusions apply when there is a small amount of overlap between NB_L and NB_U: (1) the T state of I(D) will be lower in energy than the S state when the p orbitals are close to the perpendicular orientation; (2) the S state of I(D) will be lower in energy when the p orbitals are close to the parallel orientation; and (3) the value of ΔE_{ST} for I(D) will vary as a function of both the separation and the orientation of NB_L and NB_U. As the energy difference between NB_L and NB_U increases, NB_L becomes a HO and NB_U becomes a LU. In this case, the $(HO)^2$ configuration becomes much lower in energy than the $(HO)^1(LU)^1$ configuration, so the singlet state is lower in energy than the lowest energy triplet. Note, however, that in this case we are comparing a $(HO)^2$ configuration and a $(HO)^1(LU)^1$ configuration. When we compare a singlet and a triplet $(HO)^1(LU)^1$ configuration, the triplet is always lower in energy than the singlet.

This analysis of the possible states for I(D) is oversimplified since, when we include spin, there are actually *four* possible electronic configurations for the two NB orbitals, namely, three singlet states, $NB_L(\uparrow)NB_U(\downarrow)$, $NB_L(\uparrow\downarrow)$, and $NB_U(\uparrow\downarrow)$, and a triplet state, $NB_L(\uparrow)NB_U(\uparrow)$. Since the energies of these states are similar, they may be strongly mixed with one another. These four states are discussed in more detail in Chapter 6.

In summary, whether the S or T state is lower in energy in a diradical depends on the energy gap between the orbitals. For nonbonding orbitals separated by distances greater than that for covalent bonding, J dominates and T is lower in energy than S. As the energy gap increases, the nonbonding orbitals split into a HO and LU and the energy features of repulsion are compensated by the decrease in energy due to covalent bonding, another quantum mechanical manifestation of the exchange interaction.

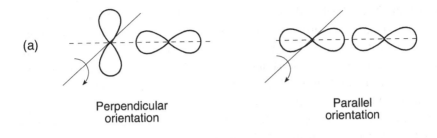

Perpendicular
orientation

Parallel
orientation

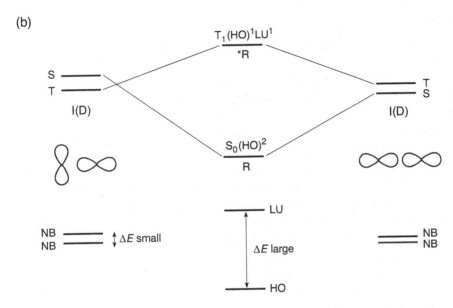

Figure 2.3   The electronic energy levels of a diradical (a radical pair or biradical). (a) The perpendicular (left) and parallel (right) orientations of two p orbitals. (b) The relative energies of the S and T states of I(D) as a function of orbital orientation and the correlation of S_0 and T_1 with S and T.

2.16   A Model for Vibrational Wave Functions: The Classical Harmonic Oscillator[5]

We used one-electron wave functions (ϕ_i, Eq. 2.5), and the configurations of one-electron orbitals ($\phi_1 \phi_2 \phi_3 \ldots \phi_i$) as models for the approximation and visualization of the electron orbital of the *electronic* portion of the complete molecular wave function, Ψ_0. Now, we describe how to visualize χ, an approximation to the wave function of the *vibrational* portion of the total molecular wave function (Ψ), in order to obtain some quantum mechanical intuition about the properties of vibrations at the molecular level. Chemists often employ a simple classical model of molecular vibrations in which the positive nuclei of a molecule are viewed as oscillating back and forth in the potential field of the electrons (again, in the Born–Oppenheimer approximation). These back-and-forth oscillations of the nuclei are approximated by the motions of a *classical harmonic oscillator*.[4] The classical harmonic oscillator is a very useful starting point for visualizing the wave function χ and for deducing the relative energy ranking of vibrational levels and some of the peculiar quantum characteristics of molecular vibrations. Chapter 3 shows how the probability of transitions between quantized vibrational levels can be elegantly interpreted in terms of the overlap of vibrational wave functions χ. Chapter 4 shows how vibrational wave functions χ are critical in determining the probability of *radiative* transitions between electronic states in terms of the Franck–Condon principle, and Chapter 5 shows that vibrational wave functions are important in determining the probability of *radiationless transitions* between electronic states.

Before describing the quantum harmonic oscillator, we need to review the basic features of the classical harmonic oscillator. The motions of many important quantum particles can be approximated starting with the mathematical form of a harmonic oscillator as an exemplar. Some examples are the oscillating vibrations of an electromagnetic field, the orbital motion of an electron about a nucleus, the vibrations between the two atoms of a chemical bond, and the precessional motion of an electron spin about a coupled magnetic field. The importance of the harmonic oscillator is found in the fact that *the same mathematics can be employed to treat an initial, approximate description of many physically distinct systems, all of which respond in a similar dynamic manner to a distorting or perturbing force.* By employing the mathematics of harmonic motion as an exemplar, we may deal with many apparently "different looking" systems (e.g., oscillating photons, orbiting electrons, vibrating nuclei, and precessing spins) that may be approximated as a harmonic oscillator even as we focus on one concrete exemplar.

A harmonic oscillator is defined as any physical system that, when perturbed from its equilibrium state, experiences a restoring force (F) that is proportional to the displacement from its equilibrium position. For a pair of masses connected by a flexible elastic spring as an exemplar, with r_e as the equilibrium separation and r as any arbitrary separation of the masses that is different from the equilibrium separation, the restoring force (F) to reposition the system to the equilibrium separation is given by Eq. 2.23 (Hooke's law).

$$F = -k\Delta r = -k|r - r_e| \tag{2.23}$$

where $\Delta r = |r - r_e|$ is the absolute distance of displacement from the equilibrium separation, F is the restoring force, and k is the proportionality constant that relates the force required to achieve a certain displacement. The proportionality constant (k) is called the force constant for oscillation. An oscillating pendulum, a vibrating string of a violin, and an oscillating tuning fork are common exemplars of classical harmonic oscillators. A classical harmonic oscillator that is originally at rest in its equilibrium position may be set into oscillation by some impulse or perturbation that is applied suddenly. After a certain period of time, called the relaxation time, the system returns to its equilibrium position by giving up the excess energy provided by the perturbation to the surroundings.

From classical physics, force (F) is defined as the negative of the slope of the potential energy of the system (Eq. 2.24a). From calculus, this relationship implies the existence of a potential-energy function relating the potential energy to the force constant and the displacement from equilibrium (Eq. 2.24b).

Eq. 2.24b provides the very important result that for a harmonic oscillator, *the potential energy* (PE) *varies directly with the magnitude of the force constant (k) and directly with the magnitude of the square of the degree of displacement of the oscillator from its equilibrium position (Δr)*. Thus, harmonic oscillation can be characterized by a parabolic PE curve (see Fig. 2.4) and given by Eq. 2.24b.

$$F = -d\mathrm{PE}/d\,\Delta r \tag{2.24a}$$

$$\mathrm{PE} = \tfrac{1}{2}k\Delta r^2 \tag{2.24b}$$

The common vibrations of a typical organic molecule may be classified as stretching vibrations between two bound atoms (e.g., C—H single-bond or C=O double-bond stretching) or as bending or twisting vibrations between three (or more) bound atoms (e.g., H—C—H or H—C—C bending). Essentially, all of the important features of the harmonic oscillator model for vibrations are incorporated into the characteristics of the stretching motion for a diatomic molecule as an exemplar (where only a stretching vibration is possible). Thus, we can analyze a generalized diatomic molecule as an exemplar and then apply the general results for *all* of the vibrations of an organic molecule.

The harmonic oscillator is a good zero-order approximation of a vibrating system when *small* displacements from the equilibrium positions of the atoms are considered. Consider a vibrating (oscillating) diatomic molecule, X—Y. For a classical harmonic oscillator, the relative vibrational motion of X with respect to Y is a periodic, oscillating function of time t (with one period occurring in the time τ). The frequency of oscillation, ν, (τ^{-1}), follows Eq. 2.25a, where k is the force constant (Eq. 2.23) of the chemical bond between the atoms X and Y, and μ is the reduced mass of the two atoms forming the bond (Eq. 2.25b). Based on Eqs. 2.25a and b, we can deduce the following for any classic harmonic oscillator: (1) the frequency of vibration (ν) of two masses, m_1 and m_2, is determined only by the reduced mass (μ) and the force constant (k), not by the amplitude of the motion (which does not appear in the equation); (2) the larger the force constant (k) for a given pair of masses of reduced mass μ, the higher the frequency (ν); and (3) the smaller the reduced mass (μ) for a given force constant (k), the higher the frequency (ν). In quantum mechanics, the frequency of motion of a quantum particle is related to its energy by the relationship $E = h\nu$, so $\nu = E/h$. As a result, higher-frequency vibrations correspond to higher energies and lower-frequency vibrations correspond to lower energies. Since we are concerned with the energy of states in this chapter, we will concentrate on the factors that determine the frequency (ν) of vibrations of a classical harmonic oscillator as given by Eq. 2.25, and then compute the energy through the $E = h\nu$ relationship.

$$\nu\ (\text{frequency}) = (k/\mu)^{1/2} \tag{2.25a}$$

$$\mu = [(m_1 + m_2)/m_1 m_2]^{1/2} \tag{2.25b}$$

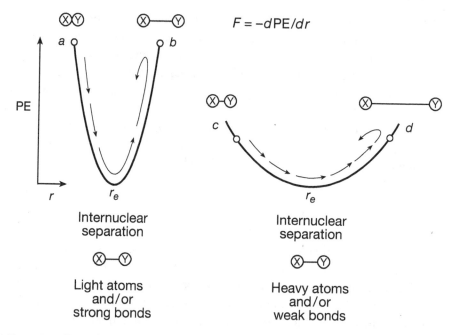

Figure 2.4   Potential energy curves for a classical diatomic molecule. (Left) A diatomic molecule with a strong bond and/or light atoms. (Right) A diatomic molecule with a weak bond and/or massive atoms.

According to Eq. 2.24b, a plot of the PE of a vibrating diatomic molecule as a function of internuclear separation (r) should give a *classical parabolic* PE *curve* (see Fig. 2.4). At some particular internuclear separation r_e the PE of the system is at a minimum (i.e., at the separation r_e, the nuclei possess their equilibrium configuration). If the separation is decreased to a value $< r_e$ (to the left in the curves of Fig. 2.4), the PE of the system increases rapidly as a result of internuclear and electronic repulsions; if the internuclear separation is increased to a value $> r_e$ (to the right in the curves of Fig. 2.4), the PE also increases due to the stretching and weakening of the X—Y bond. At the equilibrium position (r_e), there is no net restoring force operating on the atoms at r_e ($F = -d\text{PE}/dr = 0$, since $\Delta r = 0$, Eq. 2.24a), but at any displacement from r_e, a restoring force ($F = -d\text{PE}/dr \neq 0$, Eq. 2.23) exists that drives the system back toward r_e. The restoring force is a vector quantity that varies in *magnitude* and in *direction* in a periodic fashion at frequency ν in equal intervals of time called periods, τ. A pair of atoms undergoing classical harmonic motion stretch and contract back and forth through a point of separation, r_e, at which the PE of the system is at a minimum. Since generally we will be interested in energy differences between vibrational levels, we arbitrarily assign the PE minimum a value of $E_0 = 0$. All other vibrational energies are defined as positive (higher energy, less stable) relative to E_0.

The critical and general features of the classical harmonic oscillator may be applied to organic molecules using Eq. 2.25 and lead to two important conclusions: (1) C—H bonds tend to be of the highest frequency for organic molecules because they possess *both* large force constants k (strong bonds on the order of $\sim$ 90–100 kcal mol^{-1}, which are analogous to a stiff spring holding the atoms together) and a small reduced mass μ

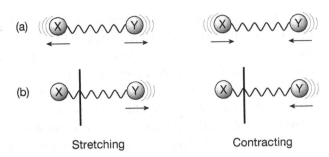

Stretching   Contracting

Scheme 2.2   Schematic of the stretching and contracting motions of a diatomic molecule. The masses are shown at the beginning of a vibration, either stretching or contracting. The line in (b) represents a heavy atom that is like a wall for a vibrating Y atom.

(H is the lightest atom), and (2) weak bonds ($\sim$ 60–80 kcal mol^{-1}) between two heavy atoms, such as a C—Cl bond, will be of very low frequency because they possess both small force constants (weak bonds, which are analogous to a soft spring holding the two atoms together) and a large reduced mass (the heavy Cl atom dominates the value of μ). Fig. 2.4 (left) shows an exemplar of the shape of the potential curve of a high-frequency vibration (e.g., a C—H bond), and Fig. 2.4(right) shows an exemplar of the shape of the potential curve for a low-frequency vibration (e.g., a C—Cl bond).

The stretching frequencies of C—H bonds, which combine a strong bond and a small reduced mass, are on the order of $\sim 10^{14}$ s^{-1}. Such a frequency is among the highest found for organic molecules. This frequency corresponds to a wavelength of $\sim$ 3000 nm (a vibrational energy gap of $\sim$ 9.5 kcal mol^{-1}, or 3323 cm^{-1}). On the other hand, the stretching frequency of a C—Cl bond, which combines a relatively weak bond and a large reduced mass, is relatively low, $\nu \sim 10^{13}$ s^{-1}. This frequency corresponds to a wavelength of $\sim$ 30,000 nm (a vibrational energy gap of $\sim$ 0.95 kcal mol^{-1} or 333 cm^{-1}). Indeed, the classical harmonic oscillator provides powerful quantum insights into the properties of vibrating atoms.

Scheme 2.2 depicts how a classical vibration between atoms X and Y can be visualized via a model in which an atom (mass) X is attached to another atom (mass) Y by a spring (representing the bond between X and Y). If the two atoms have a similar mass, then X and Y vibrate back and forth with respect to one another along the bond axis at the frequency, ν (Scheme 2.2a). If one of the two atoms is much lighter than the other (e.g., Y = H in HCl), nearly all of the motion in space is due to the lighter mass. In this case, we can assume the heavier mass is essentially stationary during the vibration. In effect, the heavy atom serves as a "fixed wall" to which the lighter mass is attached by a spring (Scheme 2.2b). The lighter mass vibrates back and forth relative to the wall, which does not move significantly during the vibration.

At any point r on the PE curves of Fig. 2.4 (except the minimum of the curve), the vibrating atoms experience a restoring force F that attracts the system toward the equilibrium geometry, r_e. The magnitude of the restoring force is given by $F = -d\text{PE}/dr$ (Eq. 2.24a); that is, the restoring force is the ratio of the magnitude of the PE of the system at any point to the value of the displacement from equilibrium, $\Delta r = |r - r_e|$. Mathematically, $d\text{PE}/dr$ is the value of the *slope* at any point r on a PE curve. Thus, the value of the slope provides a measure of the energetic resistance of the system to distortion from the equilibrium position. From Eq. 2.24b, we see that for a strong bond (i.e., one with a large k) a large change in PE will occur with only a small displacement because PE is a quadratic function of the displacement, Δr. A strong bond corresponds, therefore, to a PE curve with steep walls (Fig. 2.4, left), such that even small displacements cause a large increase in the PE. Such oscillators are relatively difficult to deform from their equilibrium geometry.

On the other hand, a weak bond corresponds to a PE curve with shallow walls (Fig. 2.4, right), such that relatively large displacements do not cause a relatively large increase in PE. Since the frequency of vibration (ν) is directly proportional to the bond strength and inversely proportional to the reduced mass of the oscillator (Eq. 2.25a), a steep PE curve implies a higher-frequency vibration and a shallow PE curve implies a lower-frequency vibration. Thus, we expect the PE curve on the left of Fig. 2.4 to be characteristic of a strong bond and the PE curve on the right of Fig. 2.4 to be characteristic of weak bonds, *if we compare similar reduced masses*. For example, the frequency of vibration of a C—C single bond will be much less than the frequency of vibration of a C=C double bond because the double bond has a much stronger bond strength and the reduced masses of the atoms involved in the vibration are similar.

The term "representative point" (r) on a PE curve may now be defined. Any particular point on the curves of Fig. 2.4 is a representative point corresponding to a specific displacement, $|\Delta r = r - r_e|$, along the x axis. As the representative point moves along a PE curve, r "mathematically tracks" the motion (i.e., the kinetic energy, or KE), the displacement from the equilibrium separation (Δr), and the PE of the pair of atoms XY. For example, when the representative point is at r_e, the molecule possesses zero PE (Eq. 2.24b, where $\Delta r = |r - r_e| = |r_e - r_e| = 0$). In the absence of any KE, the classical oscillator is motionless and the separation of X and Y is equal to r_e. For any value of r other than r_e on the PE curve, the molecule possesses some excess PE. The molecular structures corresponding to any representative points except r_e are kinetically unstable; that is, they are attracted toward r_e by the restoring force, $F = -d\text{PE}/dr$, of the harmonic oscillator, just as a classical particle on a downward-sloping surface is attracted toward a PE minimum. Therefore, any representative point at a position r has a tendency to move *spontaneously* toward r_e.

The degree of displacement (Δr) of the representative point from its equilibrium position during a vibration is called the *amplitude* (A) of the vibration. Equation 2.25a, which expresses the frequency of a vibration, depends only on the values of the force constant and the reduced mass *and is independent of the amplitude for the vibration*. For larger amplitudes, therefore, which correspond to greater displacements (Δr), the representative point must proceed with a greater velocity in order to maintain a constant frequency (ν) for the vibration. Larger amplitudes are associated with greater displacements from the equilibrium position, so the representative point experiences higher potential energies. Thus, the amplitude of a vibration, but not the frequency, increases as the total energy (PE + KE) of the oscillator increases.

Consider the points a and b (Fig. 2.4, left) and c and d (Fig. 2.4, right), which correspond to an excess of vibrational (potential) energy and are the "turning points" of vibration. At the turning point, the direction of the vibration must change, so the vibration must stop and the atoms are motionless for an instant. Suppose the representative point is initially at turning point a and has completed its stretching motion.

The representative point will then be attracted spontaneously to the equilibrium position due to the restoring force operating on it. In moving toward r_e, the representative point begins moving more and more rapidly and picks up more and more KE. The total vibrational energy (E_v) of the system equals PE + KE (Eq. 2.26). The KE of the representative point was 0 at point a, but is a maximum at r_e, where the PE is zero. At a turning point, on the other hand, the KE of a representative point is zero, so the total energy is equal to the maximum PE of the system at points a and b in Fig. 2.4. In the absence of friction, the representative point could oscillate back and forth between a and b forever. This oscillation is represented by the arrows on the curves in Fig. 2.4 and corresponds to the amplitude of the vibration.

Similar concepts apply for oscillations between points c and d in Fig. 2.4 (right). However, these oscillations occur at a lower frequency and larger amplitude than the vibrations between points a and b because the turning points correspond to a smaller maximum PE.

$$E_v = \text{PE} + \text{KE} \qquad (2.26)$$

2.17   The Quantum Mechanical Version of the Classical Harmonic Oscillator

With the classical harmonic oscillator of a diatomic molecule approximated as two vibrating masses held together by a flexible elastic spring as an exemplar, we now develop a model for the quantum mechanical vibrating diatomic molecule. The vibrating diatomic molecule is described by a vibrational wave function, χ. For a given electronic state (S_0, T_1, and S_1), the wave function χ describes the instantaneous shape of the nuclei (i.e., their positions in space relative to the electron cloud) and the motion of the nuclei, just as Ψ_0 describes the shape (i.e., their positions in space relative to the nuclei) and the motion of the electrons. We have discussed the orbital model for visualizing the one-electron wave functions, ϕ_i (Eq. 2.5). How can we visualize χ, the vibrational wave function, and what new aspects do quantum mechanics impose on the classical harmonic oscillator?

Solving the analogue of the electronic wave equation (Eq. 2.1) for a harmonic oscillator (with an energy operator that has the form of Eq. 2.24b for the PE of the oscillator)[5] yields a set of vibrational wave functions χ_i, each of which possesses a unique PE_v, where v is a vibrational quantum number. Examination of the form properties for the vibrational wave functions leads to a picture of the vibrating diatomic molecule that is strikingly different from that for the classical harmonic oscillator for low-energy vibrations, but becomes more analogous to the classical harmonic oscillator for high-energy vibrations. Now, let us consider how to visualize (1) the quantum mechanical features of a vibrating diatomic molecule, (2) the quantization and relative energy ranking of the vibrational energy levels, and (3) the wavelike character of the vibrations. The visualization of χ can be conveniently achieved by beginning with a *classical* PE curve (Fig. 2.4), then imposing quantization on the energy levels (Fig. 2.5), and finally describing the appearance of the vibrational wave functions of the quantized energy levels (Fig. 2.6).

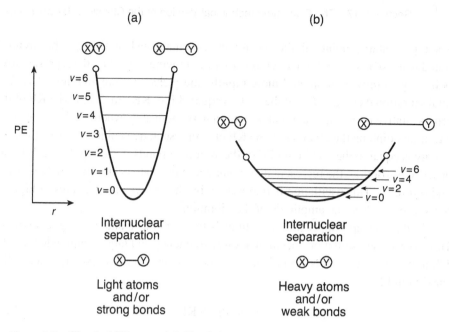

Figure 2.5   Classical PE curve (cf. Fig. 2.4) with quantized levels superimposed. See text for discussion.

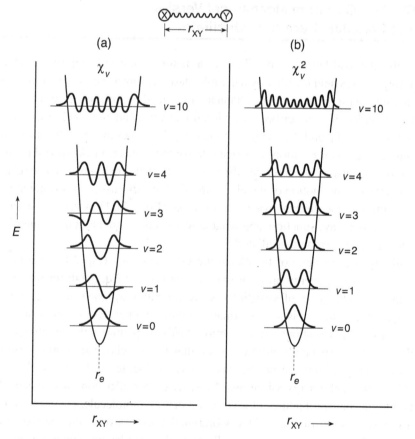

Figure 2.6   Quantum mechanical description of a vibrating diatomic molecule. See text for discussion.

2.18   The Vibrational Levels of a Quantum Mechanical Harmonic Oscillator

In quantum mechanics, as in classical mechanics, the starting model for a vibrating molecule is a harmonic oscillator that obeys Eq. 2.23. Solving the wave equation (Eq. 2.1) for a harmonic oscillator[5] obeying Eq. 2.23 reveals that the energy levels, E_v, are quantized and characterized by a quantum number, v (Fig. 2.5). The PE of each vibrational level, PE_v, is given by Eq. 2.27, where v is the vibrational quantum number (which can take only integral values, $v = 0, 1, 2, \ldots$), ν is the vibrational frequency of the classical oscillator, and h is Planck's constant. [Note that the literature and textbooks use two very similar looking symbols for the frequency of a vibration (i.e., ν, the Greek lowercase "nu") and for the quantum number of a vibration (i.e., v, an italicized Roman lowercase "vee").]

$$PE_v = h\nu(v + 0.5) \qquad (2.27)$$

The important results derived from the quantum mechanical solution for the harmonic oscillator, *which contrast sharply with the results for the classical harmonic oscillator*, are as follows:

1. Only the quantized PE values given by Eq. 2.27 are possible (stable) for the allowed values of the quantum number v for the quantum mechanical harmonic oscillator. The values of the PE for a classical harmonic oscillator, on the other hand, are continuous and can take on *any* value depending on the energy available in the system. In addition, the vibrational energy levels for a quantum mechanical harmonic oscillator are *equally spaced* in units of $h\nu$ above the $v = 0$ level.

2. The potential energy of the lowest possible vibration of a quantum mechanical harmonic oscillator (PE_0) is not zero (as it would be classically for a vibrating spring at rest) but is equal to $h\nu/2$ (Eq. 2.27, with $v = 0$). Thus, *the zero-point vibrational energy of a quantum mechanical harmonic oscillator is $h\nu/2$.* Since the harmonic oscillator is used as a model for oscillating electrons, vibrations, spins, and the electromagnetic field, the motion of each of these oscillations possesses a zero-point motion.

3. Because of the inherent, irreducible zero-point motion of all quantum particles, the vibrational motion of a quantum mechanical harmonic oscillator cannot be made to cease and the KE of vibration can never be zero. The classical harmonic oscillator, on the other hand, can be at rest in its equilibrium position (i.e., at $r = r_e$). *Unstoppable zero-point kinetic energy and zero-point motion are essential features of every quantum particle. Quantum particles are never at rest but always vibrate (or oscillate) to a certain degree as a consequence of the uncertainty principle.*

4. The energy of a quantum mechanical harmonic oscillator is directly proportional to the frequency (ν) of the vibration (i.e., $E = h\nu$) and *not* the amplitude (A) of the vibration. For a classical harmonic oscillator, the energy of a vibration is proportional to the square of the amplitude.

Figure 2.5 represents the quantum mechanical modification of the classical harmonic oscillator (Fig. 2.4) displaying the allowed energy levels whose energies are E_v. In Fig. 2.5, the energies corresponding to the representative values of v are indicated by horizontal lines (for $v = 0-6$) for the two PE curves depicted in Fig. 2.4 for the classical system. The amplitude of the corresponding classical vibrational motion (i.e., the extent of the motion of a representative point that is following the vibration from one extreme to the other) is obtained in Fig. 2.5 from the intersection of the PE curve with the corresponding energy level.

The classical turning points of a vibration are the positions defining the maximum PE during a period of vibration. *At the turning points, the total energy of the oscillator is pure potential energy, because the two masses have stopped vibrating in one direction and are starting to vibrate back in the other direction. The KE is therefore zero at the classical turning points.* Because the representative point is moving very slowly near the turning points, the classical harmonic oscillator tends to spend most of its time at the turning points. Even though the KE and PE are continually changing during the stretching and compressing motions of the nuclei, their sum, the total vibrational energy (E_v, Eq. 2.26), is constant. For the same value of v, though, the PE is much larger for the potential well in Fig. 2.5a, because the spacings between adjacent vibrational levels are larger for vibrations with higher frequencies (Eq. 2.27). Notice, for example, that the vibrating system in Fig. 2.5a has a larger zero-point energy ($v = 0$). This larger value occurs because the frequency of the PE in Fig. 2.5a is greater (Eq. 2.25a).

2.19 The Vibrational Wave Functions for a Quantum Mechanical Harmonic Oscillator: Visualization of the Wave Functions for Diatomic Molecules

The pictorial forms of the vibrational wave functions for a quantum mechanical harmonic oscillator near the equilibrium separation are very nonclassical and are, therefore, very nonintuitive based on classical physics. Thus, a closer inspection of some vibrational wave function pictures can provide some "quantum intuition" that is extremely valuable in the interpretation of both radiative and radiationless transitions between vibrational levels of states (Chapters 3–5). The mathematical form of the vibrational wave functions χ_v for $v = 0$, 1–4, and 10 are plotted qualitatively in Fig. 2.6a on the allowed (quantized) energy levels, which are indicated by horizontal lines. Recall that any wave, such as a sine wave, has an amplitude that continuously oscillates from *mathematically* positive to negative values, so the wave must pass through a value of 0 (since a change of mathematical sign must pass through 0 in order to take place). The mathematical value of χ_v *above* the horizontal line is arbitrarily considered to be mathematically positive; the value of χ_v *below* the horizontal line, therefore, is considered to be mathematically negative. The value of χ_v on the line is zero; that is, the horizontal line is a nodal line for the wave function for all levels except $v = 0$.

The number of times that χ_v passes through zero equals v, the vibrational quantum number. The wave function for $v = 0$ does not possess a node so this vibrational wave function is analogous to the wave function for a 1 s atomic orbital, which also does not possess a node. The mathematical sign of χ_v (positive above the energy-level line and negative below the energy-level line) corresponds to the *phase* of the wave function at some point in space, r, during a vibration. A critical distinction between particles and waves is that overlapping waves can interfere with one another constructively or destructively depending on their relative phases. If two wave functions have the same phase or the same mathematical sign (i.e., both plus or both minus) in the same region of space, the waves interfere *constructively* and the system is stabilized energetically; if two wave functions have a different phase or different mathematical sign in the same region of space (i.e., one plus and the other minus), the waves interfere *destructively* and the system is destabilized energetically. The process of constructive or destructive interference is an example of *resonance* that "mixes" two waves. A key feature of resonance between two waves that we see over and over is the requirement for the two waves to "look alike" in the same region of space and to have the same frequency. The same frequency feature assures that the two waves will possess the same energy (from $E = hv$). The "look-alike" feature is related to the "structure" of the wave, which includes its phase. Thus, if one wave has a positive phase and another has a negative phase in the same region of space, the two waves interfere destructively; however, if one wave has a positive phase in the same region of space where another wave also has a positive phase (or if one wave has a negative phase in the same region of space where another wave also has a negative phase), the two waves interfere constructively.

In quantum mechanics, it is the square of the wave function (χ_v^2), not the wave function (χ_v) that is directly related to laboratory observations. Since the *square* of a wave function represents the probability of finding particles (e.g., electrons, nuclei, or spins) in space, the function χ_v^2 (Fig. 2.6b) represents the *probability* of finding the nuclei at a given value of r during a vibration in a given energy level (i.e., at a given value of v). As you go up the vibrational ladder in Fig. 2.6, the wave function tends to have its highest probability near the turning points of the classical vibration ($v = 0$ is an exception). In the limit of very high quantum numbers, the quantum mechanical harmonic oscillator behaves similarly to the classical harmonic oscillator, which spends most of its time at the turning points. An important feature of Fig. 2.6 is the finite, non-negligible probability that the atoms will vibrate *outside of the region defined by the* PE *curve of the classical harmonic oscillator*. This peculiarity is the result of superimposing the wave description of the vibration and the classical picture for particles but is fully expected as a wave property. *Waves have a tendency to spread out in space and a wave description of a vibration allows the nuclei to "spread out" of the boundary for the classical* PE *curve and explore a region of space greater than the classical description would allow.* This leakage of the wave function outside of the classical PE well is responsible for the quantum mechanical phenomenon of "tunneling."

Another peculiarity of the quantum mechanical model is apparent through inspection of the probability of finding the nuclei at a given separation for upper vibrational levels (Fig. 2.6b). In addition to the broad maxima of the probability distribution in the vicinity of the classical turning points that is expected from the classical model, a number of maxima (for $v > 1$) exist in between the extremes of the vibration. For

$v = 1$, for example, there is a low probability of finding the vibrating atoms near r_e; whereas for $v = 2$, there is a relatively high probability of finding the vibrating atoms near r_e. Although the behavior of the harmonic oscillator is peculiar relative to the classical model for small values of v, as we go to higher and higher values of v, the quantum mechanical situation approximates the classical situation more closely (Bohr's correspondence principle); that is, the atoms tend to spend more time near the turning points of vibration and very little time in the region $\sim r = r_e$. In other words, a classical representative point has maximum KE near r_e (and therefore whizzes past this point) and minimum KE at turning points (and therefore moves slowly when it is in the vicinity of these points).

The probability distribution curve for the $v = 0$ level in Fig. 2.6b contrasts most dramatically with the classical picture. Instead of two turning points (as expected from the classical model), there exists one broad probability maximum at $r = r_e$. Classically, the vibrational state of lowest energy corresponds to the state of rest (point r_e in Fig. 2.6). This situation is impossible for the quantum mechanical model, since the position and velocity of this state would then be exactly defined (a violation of the uncertainty principle). As a result, the vibrating quantum system always possesses a certain amount of zero-point motion. *As explained in Chapters 4 and 5, radiative and radiationless transitions in condensed phases originate from thermally equilibrated vibrational states, which means that the $v = 0$ level will be the starting vibration for organic spectroscopy and photochemistry.* Furthermore, the zero-point motions of quantum mechanical systems, such as orbiting electrons, vibrating nuclei, precessing spins, and oscillating electromagnetic fields, can interact and stimulate transitions between the electronic, vibrational, and spin energy levels within a single molecule or between two molecules that are in each other's vicinity.

Based on these descriptions, we have built up an important quantum intuition for understanding the behavior of quantum mechanical vibrations, based on the quantization of vibrational energy levels and on visualizing the wave functions of the quantum mechanical harmonic oscillator shown in Fig. 2.6. Quantum intuition allows the smooth transfer from the classical description of a "dynamic" representative point moving on the surface to the quantum mechanical description of a "static" probability of finding the nuclei in a certain region of space with a given motion and phase. Chapter 3 employs these ideas to generate a model for understanding the factors determining the probability of transitions between vibrational levels of different electronic states.

2.20   A First-Order Approximation of the Harmonic-Oscillator Model: The Anharmonic Oscillator

The harmonic oscillator is a good zero-order approximation for a vibrating diatomic molecule X—Y (and for the vibrations of an organic molecule) for vibrations near the bottom of the PE curve (i.e., for small values of v). However, the model becomes less reliable and eventually fails for nuclear geometries corresponding to severe compressions and severe elongations of the X—Y bond. In these cases the restoring

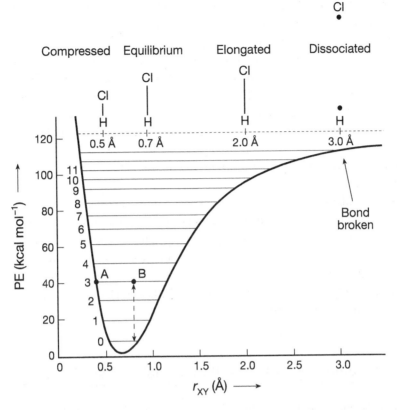

Figure 2.7 An exemplar of an anharmonic oscillator. The PE curve for HCl.

force does not obey Eq. 2.23. For example, when a bond has been extended to two or three times its normal length (i.e., from 1–2 Å to 5–6 Å), the atoms X and Y experience very little restoring force; that is, the bond is essentially broken, and the atoms may fly apart. The restoring force is zero for longer separations after the bond has broken. For an ideal classical harmonic oscillator, the restoring force increases indefinitely and smoothly with increasing or decreasing distance from the equilibrium position (see the PE curves of Figs. 2.4, 2.5, and 2.6). For a real molecule, however, the PE will rise more gradually than predicted by PE $= 0.5k \Delta r^2$ (Eq. 2.24b) when the bond is stretched because of a weakening of the X—Y bond at large r.

Consider Fig. 2.7 for the diatomic molecule hydrogen chloride (HCl) as an exemplar. The *anharmonic oscillator* considers first-order corrections to the zero-order harmonic oscillator model.[6] As the X—Y bond stretches, the PE for the anharmonic oscillator reaches a limiting value (the PE of the bond or the bond energy), the restoring force disappears, and the HCl bond breaks. Thus, the energy of the asymptote in Fig. 2.7 corresponds to the dissociation energy of the HCl molecule, which is an asymptote of the PE curve. If the energy of the system just corresponds to the asymptote, the atoms at a great distance from one another will have zero velocity. For a point above the asymptote, the KE of the two atoms (which is essentially continuous) is increased. On the other hand, compression of the nuclei results in a more rapid increase in PE than is predicted by the harmonic oscillator model, because of the sudden rise of nuclear–nuclear and electron–electron electrostatic repulsions with decreasing nuclear separation (Fig. 2.7a).

Recall from Eq. 2.26 that the sum total vibrational energy of the system equals the sum of the kinetic and potential energies. For example, in the $v = 3$ vibrational level, the anharmonic oscillator HCl shown in Fig. 2.7 possesses ~ 40 kcal mol^{-1} at point A (all PE), while at point B the system possesses ~ 35 kcal mol^{-1} of KE and 5 kcal mol^{-1} of PE energy (relative to an arbitrary energy of zero for the lowest point in the curve).

Another important feature of the anharmonic oscillator is that its vibrational levels, although quantized, are not equally separated in energy, which contrasts with harmonic oscillator (Fig. 2.6), where the separation of vibrational energy levels is constant and independent of v. Instead, the energy separations decrease slowly with increasing v, as shown in Fig. 2.7. For HCl, for example, the energy separation between $v = 0$ and $v = 1$ is ~ 12 kcal mol^{-1}, while the energy separation between $v = 10$ and $v = 11$ is only ~ 5 kcal mol^{-1}. Nevertheless, for small values of v, the energy levels are nearly equally spaced, and for these levels the harmonic oscillator approximation is still a reasonably good one.

2.21   Building Quantum Intuition for Using Wave Functions

Although the understanding and use of the paradigm of quantum mechanics requires a sophisticated mathematical background, the wave functions of electrons, vibrations, and spins may be approximated by pictorial representations that are useful for understanding the qualitative aspects of organic photochemistry. The electronic wave function and the energies of orbital configurations and associated states have been visualized in terms of the distribution of one-electron wave functions, ϕ_i, in space (i.e., MOs). The vibrational wave function (χ) has been visualized in terms of a PE curve with quantized vibrational levels whose energies depend on the quantum number, v. Upon these quantized energy levels, we picture the form of the wave function as the representative point moves along the PE curve. Chapter 3 uses this quantum intuition to examine how vibrational wave functions influence the rates of transitions between the vibrational levels of different electronic states. We have one wave function left to visualize, namely, the electron spin wave function, **S**. In the following sections, we will see that our quantum intuition for the energetic characteristics of the spin wave function will come from the visualization of classical vectors in space.

2.22   Electron Spin: A Model for Visualizing Spin Wave Functions

Electron spins play an important role in many photochemical reactions (e.g., Scheme 1.3). Consequently, in addition to visualizing the orbital nature of electrons and vibrating nuclei, we also need to visualize the nature of the electron's spin wave function, **S**, in order to determine the expectation values for important quantities, such as the energies of the spin quantum system and the ranking of the relative energies of spin states in a magnetic field. In Chapter 3, we use our model of **S** to develop an understanding of the mechanisms of intersystem crossing (electron spin change) transitions between singlet states (S_n) and triplet states (T_n), and vice versa. Since electron spin is generally covered only briefly, if at all, in elementary chemistry texts, we develop a relatively detailed model of the electron spin wave function (**S**) that is in the pictorial spirit of the electronic orbital model, Ψ_0, and the harmonic oscillator model for the vibrational wave function, χ. Our model[7] of electron spin is based on the properties of classical *vectors* and should provide a great deal of quantum intuition about the properties of the spin wave function (**S**). The language of the vector model for electron spin is readily transferable for the description of structures that are probably more familiar, namely, nuclear spins, the key objects of interest for nuclear magnetic resonance (NMR) spectroscopy.

In quantum mechanics, spin is a manifestation of an unremovable property of electrons that is at the same fundamental level as the electron's mass and charge. *Spin corresponds to the inherent angular momentum of an electron.* As we have done previously for orbiting electrons and vibrating atoms, we develop a quantum mechanical picture of electron spin angular momentum by first appealing to a classical model of angular momentum and then modifying this model with the appropriate quantum mechanical and wave characteristics. An important connection between an electron's spin and its magnetic moment resulting from the electron's angular momentum provide a way to rank the relative energies of spin states and to thereby produce a state energy diagram for electron spins.

The two most important classes of classical angular momentum involve (1) the circular "spin" motion of a point on the surface of a rotating sphere (or a rotating cylinder) about a defined axis (arbitrarily called the z-axis), and (2) the circular "orbital" motion of a particle rotating at a fixed distance from the center of a circle that contains an axis (arbitrarily called the z-axis) perpendicular to the plane of the circle. We begin to describe the electron spin by considering the angular momentum of a rotating sphere (case 1), then discuss the angular momentum of a rotating particle (case 2) in Section 2.31.

We can use the classical model of the angular momentum of a rotating sphere as a concrete physical model to describe the electron spin. Although electron spin, like electron exchange, is fundamentally a quantum mechanical phenomenon that has no classical analogue, chemists nevertheless accept the convenient fiction that electron spin is the angular momentum arising from the electron behaving like a negatively charged sphere that rotates without friction about an axis. Remarkably, *this concrete physical model allows chemists to visualize most of the critical quantum mechanical properties of electron spin.* In particular, *this model makes it possible to understand why an electron possesses not only a spin but also a magnetic moment, which is expected of a charged particle that is spinning.* The model of a magnetic moment, in turn, allows chemists to consider the energies of two interacting spins through their magnetic moments.

If we take a concrete physical model of an electron as a sphere of negative electricity, the *spin* of an electron is the angular momentum resulting from the electron's spin motion about the z-axis. Although a classical spinning electron would possess a continuous range of the angular momentum values, depending on the velocity and direction of its spinning motion, quantum mechanics demands that *electron spin has a fixed and fundamental characteristic value of exactly* $\hbar/2$. Recall that $\hbar$ is a fundamental quantity of angular momentum and is equal to Planck's constant (h), the fundamental quantum mechanical unit of angular momentum, divided by 2π (a characteristic mathematical quantity that occurs when you are dealing with circular motion). The value of the electron's spin is $\hbar/2$, whether it is a "free" electron that is in interstellar space and is unassociated with any nucleus (i.e., not bound to any positive nucleus), or whether the electron is associated with a nucleus in an atom, a molecule, an electronically excited state, or a free radical. Furthermore, the spin angular momentum of an electron is *always* the same, *exactly* $\hbar/2$, regardless of the orbital (e.g., n, π, or π^*) that the electron happens to occupy. Thus, an electron in the n orbital of an n, π^* state has the same spin as the electron in a π^* orbital of an n, π^* state: both have a spin of $\hbar/2$. If two electrons occupy the same orbital, they both still possess a spin of $\hbar/2$; however, because the two spins must be spin paired to satisfy the Pauli exclusion principle, the angular momentum of the two spins cancel each other for a net angular momentum of 0 and a net spin of 0. (*Note: For the remainder of the text, when dealing with spin,* $\mathbf{S}$, *the units of* $\hbar$ *will be assumed and may not be shown explicitly except when required for clarity.*)

Recall from Section 2.8 that electronically excited states (*R) that possess orbital configurations consisting of two half-filled orbitals may exist with different spin configurations of the orbitally unpaired electrons (i.e., a singlet-spin or a triplet-spin configuration). For a $(HO)^1(LU)^1$ electron configuration, for example, there exists an S_1 state in which the electron spins are "antiparallel" ($\uparrow\downarrow$), and for which there is no net spin (i.e., one electron spin cancels the other). There is also a T_1 state in which the electron spins are "parallel" ($\uparrow\uparrow$), for which there is a net spin of $1\hbar$ (i.e., each electron spin contributes $\hbar/2$ to the total spin of the state). In the following sections, we explain how to visualize the antiparallel and parallel configurations of two electron spins *by appealing to the classical vectorial properties of an object executing a spinning motion about an axis, such as a top or a gyroscope.* We review the fundamental classical properties of vectors that are important for an understanding of electron spin and then describe the quantum mechanics of spin in terms of this vector description.

Since we are interested in the relative energies associated with different spin configurations and states, we also describe the interactions between electron spins as being analogous to those between *spin magnetic moments* (given the symbol $\boldsymbol{\mu}$), which we directly relate to the *spin angular momentum* ($\mathbf{S}$). The idea behind this strategy is that we can build a classical intuition for the energy of interactions of two magnets and then relate that energy to the interactions of two electron spins. For simplicity, we use the same symbol ($\mathbf{S}$) to describe the spin wave function and the associated quantity, the spin angular momentum. In the case of the spin angular momentum, the unit $\hbar$ is assumed.

2.23   A Vector Model of Electron Spin

Physical quantities are classified as *scalars* if they can be completely described by a magnitude (i.e., a single number and a unit). Examples of scalar quantities are energy, mass, volume, time, wavelength, temperature, and length. Physical quantities that require both a magnitude and a *direction* in order to be fully defined, on the other hand, are called *vectors*. Examples of vector quantities are velocity, electric dipoles, angular momentum, magnetic fields, and magnetic dipoles. If required by the context, we represent *scalar* quantities in *italics* and **vector** quantities in **boldface.**

Spin angular momentum[7] is a vector quantity (as is any form of angular momentum) and is represented by the boldface symbol **S**. The *magnitude* of the spin angular momentum is a scalar quantity and is represented by the italicized symbol, S. The scalar S is mathematically equal to the absolute value of the vector **S**; that is, $S = |\mathbf{S}|$. The spin quantum number is a pure number (no units) and is represented by the roman symbol, S. The parameter S is the symbol for the *total* spin quantum number of a collection of two or more coupled spins (the most common in organic photochemistry being collections of two, three, or four spins), and the symbol s is to be used for the quantum number of a single spin. For example, an electron possessing a spin quantum number s = 1/2 (pure number) is described by a spin vector **S**, which is the symbol for the mathematical representation of the spin angular momentum of the electron, $\mathbf{S} = \hbar/2$ (vector). In experimental measurements, the angular momentum along the z-axis, $\mathbf{S}_z$, is measured. The value (length) of the spin angular momentum on the arbitrary z-axis is $S_z = |\mathbf{S}_z|$. The uppercase letter "es" (S) is used for a number of representations in organic photochemistry, including S_n, which symbolizes the singlet states of molecules. The meaning, though, should be clear from the context of the discussion for each case.

2.24   Important Properties of Vectors

In this section we review briefly some properties of vectors that are used repeatedly to describe the properties of electron spin and the interactions of electron spins. Vectors are conveniently represented by an arrow. The arrowhead (the "head") of a vector indicates its direction relative to a reference axis (conventionally designated as the Cartesian z-axis), and the length of the arrow represents the magnitude of the physical quantity represented by that vector. The orientation of the vector in space is defined in terms of the angle (θ) that the vector makes with the z-axis. Interacting vectors must all relate to the same z-axis system. *If the vectors do not interact, each will have its own arbitrary axis system; when they do interact, a single arbitrary z-axis must be used to describe all vectors.* Thus, for two uncoupled spins the vectors representing each spin can be oriented to any arbitrary z-axis; whereas for two spins that are coupled by some interaction (e.g., electron exchange or magnetic moments), each spin must possess a well-defined orientation relative to each other and to the reference z-axis.

2.25   Vector Representation of Electron Spin

In this section we consider some important properties of all vectors, with a vector representation of the spin angular momentum **S** as a concrete exemplar. Figure 2.8 summarizes the trigonometric relationships of a spin vector, **S**, that makes an angle θ with the z-axis. The important vectorial properties are (1) the spin vector **S** possesses a component $\mathbf{S}_z$ on the z-axis; (2) the spin vector **S** possesses a component $\mathbf{S}_{x,y}$ in the x–y plane; and (3) the magnitude of the component $\mathbf{S}_z$ on the z-axis is related to the magnitude of the total spin, **S**, by the trigonometric relationship given in Eq. 2.28, where $|\mathbf{S}_z|$ is the magnitude (the absolute value) of the vector on the z-axis and $|\mathbf{S}|$ is the absolute magnitude of the spin vector.

$$\cos \theta = |\mathbf{S}_z|/|\mathbf{S}| \tag{2.28}$$

According to the uncertainty principle, the precise value of the vector **S** cannot be precisely measured in any experiment. There is a way, however, to circumvent the uncertainty principle. *The uncertainty principle allows one component ($\mathbf{S}_z$ or $\mathbf{S}_{x,y}$) of the spin vector to be specified exactly (typically the z-component is the component selected) if the other components are completely unknown.* In this way, if the position of **S** in the x–y plane (i.e., $\mathbf{S}_{x,y}$, called the azimuth of the vector **S**) is completely undetermined, then the magnitude of $\mathbf{S}_z$ on the z-axis can be determined precisely in an experiment. This finding explains why only the z-axis is employed and specified when discussing the energies of interacting spin states, which are, by convention, defined as the value of **S** on the z-axis.

Now, let us consider the magnitude of an arbitrary spin vector **S** and its value on the z-axis, $\mathbf{S}_z$. Recall from quantum mechanics that it is the square of the wave function, Ψ^2, and not Ψ, that corresponds to the magnitude of physical quantities. Thus, it is $\mathbf{S}^2$ that is the starting point for the discussion of the measurement of the magnitude of electron spin. From the laws of quantum mechanics, the possible values of $\mathbf{S}^2$ are

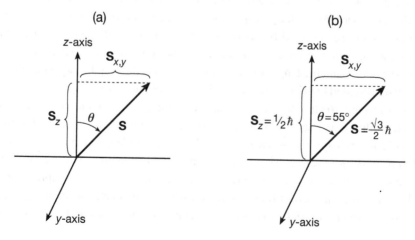

Figure 2.8   Trigonometric relationships between any spin vector, **S**, and its orientation θ and value, $\mathbf{S}_z$, with respect to the z-axis (left).

given by applying the allowed spin quantum numbers, S (i.e., the eigenvalues from the solution of the appropriate wave equation). The allowed values of $\mathbf{S}^2$ are limited to those given by Eq. 2.29a. Thus, for a given value of S, there is an associated square of the spin angular momentum ($\mathbf{S}^2$) that can be computed from Eq. 2.29a. The values of $\mathbf{S}$ are given by Eq. 2.29b.

$$\mathbf{S}^2 = [S(S+1)] \tag{2.29a}$$

$$\mathbf{S} = [S(S+1)]^{1/2} \tag{2.29b}$$

As a concrete example, consider the spin quantum number of a single electron: $S = 1/2$ (S is a pure number). According to Eq. 2.29a, the value of the magnitude of $\mathbf{S}^2$ (in units of $\hbar$) is $1/2(1/2 + 1) = 3/4$, and according to Eq. 2.29b, the value of $\mathbf{S}$ for a single electron is given by $[1/2(3/2)]^{1/2} = (3/4)^{1/2}$. Electrons possess the same value of the spin angular momentum, exactly 1/2, as other spin 1/2 particles, such as a ^{1}H or a ^{13}C nucleus, so the vector representation of the electron spin as an exemplar can also be applied to discussions of nuclear spins. Although the magnitude of the spin is identical for electrons and all nuclei of spin 1/2, the magnitude of the magnetic moments associated with electron and nuclear spin are very different (see Section 2.33).

Figure 2.8 shows the vector model for the case of $S = 1/2$. Although the value of the vector $\mathbf{S} = (3/4)^{1/2}$, the value of $\mathbf{S}_z$ must equal 1/2 on the z-axis. As a result, from Eq. 2.28, the value of θ must be 55°.

Now, we consider the vector representation of *quantized* electron spin in some detail for the two most important cases in photochemistry, namely, a single spin and two interacting spins that are coupled and therefore must behave as parallel ($\uparrow\uparrow$) or antiparallel ($\uparrow\downarrow$) spins. The latter description is essentially two-dimensional (2D), so for the three-dimensional (3D) representation of interacting spins, these two cases require a modification and reinterpretation of the terms "parallel" and "antiparallel."

2.26   Spin Multiplicities: Allowed Orientations of Electron Spins

The allowed values of the electron spin (or coupled electron spins) are given by Eq. 2.29, and the corresponding values of the electron spin on the z-axis are given by Eq. 2.28; that is, $\mathbf{S} = [S(S+1)]^{1/2}$ and $|\mathbf{S}|\cos\theta = |\mathbf{S}_z|$. The allowed values for the quantum number for electron spin (S) are 0, 1/2, 1, 3/2, 2, and so on, so the values of $|\mathbf{S}_z|$ are $0\hbar, \hbar/2, \hbar, 3\hbar/2, 2\hbar$, and so on. A remarkable principle of quantum mechanics is that angular momentum can take up a number of quantized orientations in space. It turns out that this spatial orientation of the angular momentum causes a very important spatial orientation of the magnetic moment associated with an electron spin in a magnetic field.

Now, we need to expand our model of electron spin to take into account the fact that quantum mechanics only allows certain orientations of the electron spin vector in space in a magnetic field. The *multiplicity* (M) of a given state of spin angular momentum is the *number* of quantum mechanically allowed orientations of a spin of magnitude **S** in a magnetic field. The multiplicity of a spin state is computed from S, the spin angular momentum quantum number:

Spin multiplicity $= M = 2S + 1 =$ number of allowed spin orientations in space

Each allowed orientation is assigned a spin orientation quantum number, M_S, where the subscripted S is related to the quantum number for the orientation of the spin and the value of the projection for the spin on the z-axis. The value of the spin remains the same for a given multiplicity, but the value of the spin projection can take on the values of M_S.

The multiplicities for a single electron and for two coupled electrons are of greatest interest in organic photochemistry. For a single electron, the spin quantum number is 1/2, so $M = [2(1/2) + 1] = 2$. Since $M = 2$, there are *two* and *only two* allowed orientations of the electron spin vector in a magnetic field, corresponding to $M_S = +1/2$ and $M_S = -1/2$. These two allowed orientations of the electron spin are shown in Fig. 2.9. One of the orientations has a value of $S_z = +1/2$ (referred to as the α orientation of a single electron spin), and the other orientation has the value of $S_z = -1/2$ (referred to as the β orientation of a single electron spin). Because there are two and only two allowed orientations of a single electron in a magnetic field, the $S = 1/2$ state is called a *doublet* state and is given the symbol D. The doublet state corresponds to the spin state of a radical that possesses a single unpaired electron. Once again we have redundancy in the use of a symbol. The symbol D that is used for a doublet spin state is also used to denote a reactive diradical intermediate I(D) (i.e., a radical pair, or a biradical) that possesses two half-filled orbitals, each orbital being occupied by a single electron. [Multiple uses of the same symbol for different entities are an unavoidable fact of the scientific literature. You must be sure of the context when a symbol that refers to different entities is used. In this text we try to be careful to indicate such contexts.]

Figure 2.9 shows the angular momentum vector pointing perpendicular to the plane of rotation of a spinning spherical electron, a feature based on analogy to the classical angular momentum of a spinning sphere. Although a specific arbitrary orientation

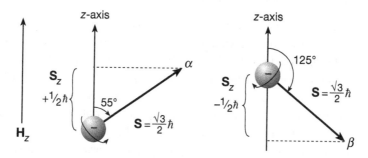

Figure 2.9　Vector representation of a spin $-1/2$ particle (e.g., an electron, a proton, or a ^{13}C nucleus). The symbol α refers to the spin wave function of a spin with $M_S = +1/2$, and the symbol β refers to the spin wave function of a spin with $M_S = -1/2$.

of the spin 1/2 vector is shown in Fig. 2.9, the vector may lie anywhere in a cone of possible orientations (see Section 2.28) that make an angle θ of either $55°$ (α-orientation) or $125°$ (β-orientation) with the z-axis.

2.27   Vector Model of Two Coupled Electron Spins: Singlet and Triplet States[7]

Perhaps the most important coupled spin systems in organic photochemistry correspond to those situations involving the coupling of two electron spins, each of which occupies a separate orbital. The quantum mechanical rule for the possible couplings of two 1/2 spins of the two orbitally uncoupled electrons is as follows: The final *total* spin angular momentum (in units of $\hbar$) of a two-spin system is either 0 or 1. For total spin S = 0, the multiplicity of the state is M = (2S + 1) = 1, and for total spin S = 1, the multiplicity of the state is M = (2S + 1) = 3. We are now in a position to understand the origin of the terms "singlet state" and "triplet state," which refer to the multiplicity of a state based on its spin and which has its origin in the study of the number of spin states of molecules in a magnetic field. Recall from Scheme 1.3 that both *R and I(D) can be either singlet or triplet states.

Now, we use the vector model of spin to visualize the singlet and triplet states that result from the coupling of two electron spins occupying separate orbitals. The singlet state results from the coupling of two electron spins in such a way that the spin vectors are antiparallel ($\uparrow\downarrow$) and collinear; the angle between the heads of the vectors is $180°$ (Fig. 2.10a) for a singlet state. Such an orientation of vectors results in the *exact cancellation of the net spin angular momentum of each of the spin vectors*, so the net spin vector length $|\mathbf{S}|$ is 0. This means that the net spin of the singlet state, as well as its projection on the z-axis, is exactly equal to $0\ \hbar$ (Fig. 2.10a). Although a specific arbitrary orientation of the two spin 1/2 vectors is shown in Fig. 2.10a, there is no preferred orientation in the cone of possible orientations (see Section 2.28) at all, since all orientations of the two spins that are $180°$ apart yield zero-spin angular momentum.

We conclude, therefore, that the singlet state has no net spin because of vectorial cancellation of the spin momentum of the two spins, and no preferred orientation of the component spin vectors in a magnetic field, even though the singlet state is composed of two electrons, each of which possesses a spin of 1/2 and each of which possesses a magnetic moment. This situation is analogous to the cancellation of two bond dipole moments in a linear molecule, such as $O\!=\!C\!=\!O$, which has two polar bonds but no net dipole moment. All electrons that are paired in an orbital are required by the Pauli exclusion principle to be in a singlet state, and therefore will have their spin vectors oriented as shown in Fig. 2.10a. Since nearly all organic molecules possess two electrons in each orbital in their ground states, organic molecules typically possess singlet ground states.

The triplet state (S = 1) also results from the coupling of two electron spins, but this time the coupling is such that the spin vectors make an angle of $\sim 70°$ between the two spin 1/2 vectors (Fig. 2.10b). The angle of $70°$ is required by trigonometry

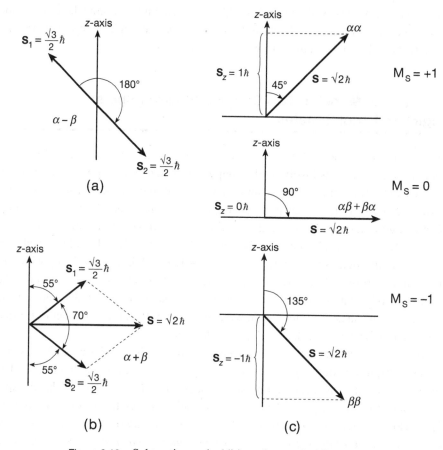

Figure 2.10   Subtraction and addition of two spin 1/2 particles.

so that the two spin 1/2 vectors will add vectorially to produce a single spin vector, $\mathbf{S} = (2)^{1/2}$. A vector of this length produces a value of $\mathbf{S}_z = 1$ on the z-axis when the vector is at an angle of $45°$ relative to the z-axis (Fig. 2.10c). Since $M = 3$ for an $S = 1$ state, there are three values for the quantum numbers for the orientation of the spin on the z-axis of an $S = 1$ system: $\mathbf{S}_z = +\hbar$, $\mathbf{S}_z = 0$, or $\mathbf{S}_z = -\hbar$ (Fig. 2.10c). These three spin states are assigned the quantum numbers $M_S = +1$, $M_S = 0$, and $M_S = -1$, respectively, and are called the sublevels of a triplet state, T. Thus, the origin of the term "triplet state" is based on the quantum mechanical requirement that an $S = 1$ spin state must possess three energy sublevels corresponding to the three quantum numbers, $M_S = +1$, $M_S = 0$, and $M_S = -1$, when the spin system is placed in a magnetic field. From trigonometry (Eq. 2.28), the angle θ that the vector $\mathbf{S}$ for an $S = 1$ spin state must make with the z-axis to achieve the values of $+\hbar$, 0, $\hbar$, or $-\hbar$ is $45°$, $90°$, or $135°$, respectively (Fig. 2.10c). The spin vectors $\mathbf{S}$ for $M_S = +1$ and $M_S = -1$ are shown as lying in some arbitrary position in the possible cone of orientations. For $M_S = 0$, the spin vector $\mathbf{S}$ is shown as a *circle* of possible orientations lying in the x–y plane. The possible cone of orientations that the spin vector may assume is discussed in Section 2.28.

We can now see how the vector model and trigonometry provide a very convenient and vivid means of visualizing singlet and triplet spin states based on the rules of quantum mechanics. For $S = 0$, two coupled spins are collinear and oriented at $180°$ relative to one another and exactly cancel each other's spin angular momentum. As a result, there is no net spin anywhere in space; only *one* spin state, a *singlet* state (quantum number $M_S = 0$) results, and this state remains a singlet state in a magnetic field because it lacks a magnetic moment. For $S = 1$, the coupled spins add to one another in a manner (with a relative angle in the cone of orientation of $70°$) to form a vector of length $|S| = 2^{1/2}$. The coupled spins have a fixed value of $(2^{1/2})\hbar$, but can possess one of three allowed orientations in a magnetic field (Fig. 2.10), corresponding to the quantum numbers $M_S = +1$, $M_S = 0$, and $M_S = -1$.

Let the spin wave function α represent an "up" spin ($M_S = +1/2$), and the spin wave function β that represents a "down" spin ($M_S = -1/2$). *Both* the singlet state (Fig. 2.10a) and the triplet component with $M_S = 0$ possess one α spin and one β spin (Fig. 2.10c, middle). Thus, a simple 2D up–down "arrow" notation ($\uparrow\downarrow$) would make the singlet state ($M_S = 0$) and the $M_S = 0$ level of the triplet state appear to be identical in terms of spin characteristics, even though the singlet has no net spin and the $M_S = 0$ level of the triplet possesses a net spin of $\hbar$. This distinction between the $M_S = 0$ for the singlet and triplet is made only in a vector representation relative to the z-axis (Fig. 2.10) or, even better, in the 3D representation in Section 2.28.

In closing this section, we comment on an interesting mathematical modification of our α and β label of spins that must be made to provide proper wave functions for both the singlet state and the $M_S = 0$ level of the triplet state. This modification is imposed by the Pauli exclusion principle and electron exchange (Section 2.9), which states that all of the properties of exchanged electrons must be identical. Suppose, for example, that we labeled the wave functions of the two states (i.e., the singlet and triplet with $M_S = 0$) with one spin up and one spin down as either $\alpha_1\beta_2$ or $\beta_1\alpha_2$ (where the subscripts refer to electrons 1 and 2). However, such an assignment would imply that we could distinguish electron 1 and electron 2 as being in a specific and differentiable spin state. *This distinction violates the Pauli exclusion principle, which states that two electrons must be indistinguishable upon exchange.* Quantum mechanics allows that an acceptable modification of the spin wave function for the singlet state is $S = (\alpha_1\beta_2 - \beta_1\alpha_2)$, where the minus sign signifies that the two spins are exactly $180°$ *out of phase* [a mathematical "normalization" factor of $(1/2)^{1/2}$ is of no interest for our qualitative discussion and is therefore ignored]. The spin wave function for the triplet state ($M_S = 0$) is $S = (\alpha_1\beta_2 + \beta_1\alpha_2)$, where the plus sign signifies that the two spins are *in phase* and couple to generate a net spin of 1. Thus, we will let the singlet state (S) be represented by the symbols $\alpha\beta - \beta\alpha$, and the triplet state (T_0, where the subscript 0 refers to the spin quantum number $M_S = 0$) will be represented by the symbols $\alpha\beta + \beta\alpha$. Both of these states have the same quantum number ($M_S = 0$) but possess different wave functions and correspond to different vectorial orientations of the α and β spins. We can think of the minus sign in the singlet function $\alpha\beta - \beta\alpha$ as representing the "out-of-phase" character of the two spin vectors, causing the spin angular momentum of the individual spin vectors to exactly cancel (Fig. 2.10a), and the plus sign in the triplet wave function as representing the "in-phase" character, causing the individual spin vectors to add together and reinforce each other (Fig. 2.10b).

There is no Pauli restriction with respect to the interchange of spin if we label the wave function for the triplet state for which $M_S = +1$ as $T_+(\alpha\alpha)$ or if we label the triplet state for which $M_S = -1$ as $T_-(\beta\beta)$. For the state with $M_S = +1$, the spin function $\alpha\alpha$ is acceptable because both electrons possess the same orientation, so they are indistinguishable upon exchange. The same holds for the $M_S = -1$ state, for which the spin function $\beta\beta$ is acceptable according to the Pauli exclusion principle.

2.28   The Uncertainty Principle and Cones of Possible Orientations for Electron Spin

So far, we have discussed the spin vectors in terms of a 2D representation relative to a z-axis. Now, we proceed to a more realistic model of the spin vector in 3D relative to a z-axis. In 3D, the directional angle of S in the x–y plane is called the azimuthal angle. According to the uncertainty principle of quantum mechanics,[10] *if the value of S_z on the z-axis is measured precisely, then the azimuthal angle's position in space will be completely indeterminable.* Thus, only S_z can be measured precisely if we accept that there is no information at all that can be obtained experimentally concerning the azimuthal angle of S. In terms of the vector model, this means that an α spin, which makes a specific angle of $55°$ with the z-axis and a value of $S_z = 1/2$, must project to an unknown azimuthal angle in the x–y plane for any measurement of S_z (Fig. 2.9). Thus, there is an infinite set of positions that the spin vector can assume in space making an angle of $55°$ with the z-axis, any one of which could correspond to the actual position of the spin vector. *This set of possible positions constitutes a cone such that, whatever the specific position the spin vector takes on the cone, the angle of the vector with the z-axis and the projection of the vector on the z-axis are always the same, namely, $55°$ and $+\hbar/2$; however, the x and y components of the vector are completely undetermined in any experiment that measures S_z precisely.* Such a cone is called the *cone of possible orientations of an α spin* ($M_S = +1/2$). Fig. 2.11 (left, top) shows the cone of possible orientations of an α spin with one possible arbitrary orientation being shown explicitly. Although we can be sure that an α spin will lie somewhere in this cone, we cannot specify at all where it is located in the cone. A corresponding cone of possible orientations (Fig. 2.11, left, middle) also exists for a β spin ($M_S = -1$), with one possible arbitrary orientation being shown explicitly.

In the absence of an interacting magnetic field, the vector representing the spin angular momentum is imagined as being stationary and "resting" somewhere in the cone of possible orientations. If a coupling magnetic field is applied (e.g., a static applied laboratory field, an oscillating applied laboratory field, or fields due to the magnetic moments generated by the motion of spins in the environment, etc.), the coupling will cause the spin vector to sweep around the cone with the direction of the coupling field serving as the z-axis. This sweeping motion about the cone is called *spin precession*. We discuss spin precession in the cone of possible orientations in Chapter 3 when we consider transitions between magnetic states. Here we only describe the "static" magnetic states and the energies associated with them.

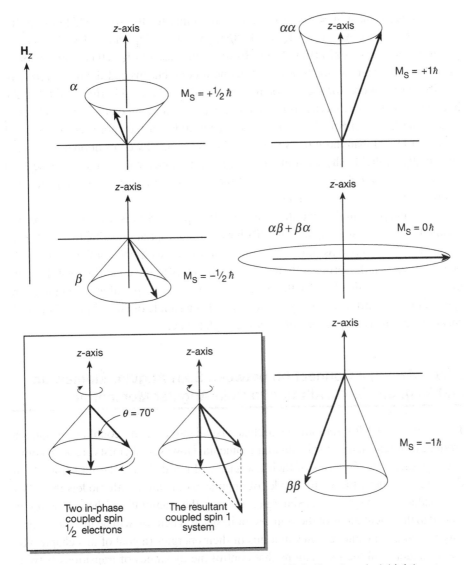

Figure 2.11   Cones of possible orientation for a spin 1/2 (left) and a spin 1 (right) system of angular momentum. An arbitrary position of the spin vectors is shown for each of the possible cones.

2.29   Cones of Possible Orientations for Two Coupled 1/2 Spins: Singlet and Triplet Cones of Orientation as a Basis for Visualizing the Interconversion of Spin States

Let us consider *R, which may be an excited singlet or a triplet state possessing an electronic configuration that is approximated by two singly occupied orbitals, such as an n,π^* or a π,π^* configuration. For *R, the two electron 1/2 spins can be viewed as being coupled to produce a net spin of 1, *R(T$_1$), or a net spin of 0, *R(S$_1$).

Now, consider the possible orientations of the spin vector **S** in the case of the triplet of **S** = 1 (Fig. 2.11, right). The situation with respect to a cone of possible orientations of the spin $= 1$ case is analogous to that for the spin $= 1/2$ case in the cases of $M_S = +1$ ($\alpha\alpha$) and $M_S = -1$ ($\beta\beta$), for which "up" and "down" cones of possible orientations exist (Fig. 2.11). However, the case of $M_S = 0$ (wavefunction $= \alpha\beta + \beta\alpha$), the spin vector $\mathbf{S}$ does not define a cone but instead defines a *circle* of possible orientations for the vector in the *x*–*y* plane (Fig. 2.11, right, middle). An interesting feature of this case is that the projection of the spin on the *z*-axis is 0, even though the length of $\mathbf{S}$ is $2^{1/2} = 1.4\hbar$. The magnitudes of the spin vectors for $S = 1$ are larger ($1.4\hbar$) than those for $S = 1/2 (0.87\hbar)$, although by convention the units are not usually explicitly shown in the figures. For the singlet state that also has $M_S = 0$ (wave function $= \alpha\beta - \beta\alpha$), the value of the spin vector is zero, so there is no cone of orientation for the singlet state.

Up to this point in the text, the term "parallel spins" ($\uparrow\uparrow$) is used to describe two coupled 1/2 spins in a triplet state. To have the appropriate resultant of $1.4\hbar$, two coupled spin 1/2 systems must make an angle of $70°$ with one another (Fig. 2.11). Now, we see that the term parallel spin actually refers to the condition that the spins make a relative angle of $70°$ with one another on a cone of possible orientation. The spins for the $\alpha\alpha$, $\beta\beta$, and $\alpha\beta + \beta\alpha$ states are all at an angle of $70°$. Thus, only loosely speaking are the two spins *parallel* for the triplet state.

2.30   Making a Connection between Spin Angular Momentum and Magnetic Moments Due to Spin Angular Momentum

The vector model for spin allows a clear visualization of the coupling of spins to produce different states of spin angular momenta. However, chemists are accustomed to discussing states in terms of their energies and not their angular momenta. Now, we need to connect the vector model of angular momentum with models that allow us to deduce the magnetic state energies and how the magnitude of these energies are related to the orientations of the spin vector in space. Then, we will be in a position not only to rank the magnetic states in terms of their energies (a goal of this chapter) but also to understand the interactions that control the dynamics of transitions between magnetic states (a goal of Chapter 3).

2.31   The Connection between Angular Momentum and Magnetic Moments: A Physical Model for an Electron with Angular Momentum

Now, we consider a physical model for electron spin that concentrates on the *magnetic properties* and the *magnetic energies* associated with the spin angular momentum.[8] Electrons and certain nuclei (e.g., ^{1}H and ^{13}C), which possess both electrostatic charge and spin angular momentum, also possess magnetic moments $\boldsymbol{\mu}$ (i.e., magnetic dipoles, which are vectors analogous to electric dipoles and spin angular momentum). These magnetic moments may interact with each other and with applied laboratory magnetic fields. Now, we examine the connection between spin angular momentum

(**S**) and the magnetic moments ($\boldsymbol{\mu}$) associated with **S**. First, we apply the vector model to a familiar physical model of a magnetic moment resulting from the orbital motion of an electron in a Bohr orbit. *This concrete physical model allows us to compute the magnetic moment of the orbiting electron in terms of its orbital angular momentum.* With a connection between electron *orbital* angular momentum and magnetic moments in hand, we shall see how this result can be used to deduce a connection between electron *spin* angular momentum and the magnetic moment due to spin.

In Section 2.7, we described how to qualitatively rank the relative energies of electronic orbitals and electronic states based on the filling of a set of energy-ranked electron orbitals following the aufbau principle and the Pauli exclusion principle. In addition, in Section 2.19, we learned how to qualitatively rank the relative energies of vibrational levels based on their quantum numbers and the number of nodes in their wave functions. It is now our goal to develop a model that allows us to rank the relative energies of *electronic spins and spin states.* The importance of magnetic moments is that the energy of their interactions with a magnetic field can be estimated and ranked in terms of a *magnetic energy diagram*, analogous to the electronic state energy diagram for orbitals (and the vibrational energy diagram for vibrations). Therefore, through an evaluation of the magnetic moments associated with electron spin in an applied magnetic field, we can construct magnetic energy diagrams ranking the energy of the various spin orientations corresponding to different states of angular momentum and thus to different magnetic moments. An important property of such magnetic energy diagrams is that, in contrast to the orbital and vibrational energy diagrams, *the separation between the energy levels depends on the strength of the coupled magnetic fields that interact with the electron spin's magnetic moment.*

2.32   The Magnetic Moment of an Electron in a Bohr Orbit

The Bohr model of an electron orbiting about a nucleus provides a concrete physical model (i.e., an electron as a negatively charged particle moving in a circular orbit) for showing how a magnetic moment ($\boldsymbol{\mu}$) is associated with the orbital angular momentum (**L**).[8] Through the model of the Bohr atom, we can connect the electron's **L** with the μ_L associated with orbital motion. After developing this model, we can use it as a basis for deducing the relationship between the quantum mechanical *spin angular momentum* (**S**) and the *magnetic moment* (μ_S) associated with spin motion.

To compute the magnetic motion associated with an electron in a circular Bohr orbit of radius r, first we suppose that an electron is a point of negative charge orbiting at constant velocity, **v**, in the plane of the circular orbit. Figure 2.12 presents a vectorial description of the relationship between μ_L and **L** (the figure is schematic to the extent that the sizes of the vectors are unitless and not to any particular scale). The circular motion of the electron generates orbital angular momentum, **L**, whose axis of rotation is perpendicular to the plane of the circular motion and passes through the nucleus (Fig. 2.12, left). The orbital motion of an electron of charge $-e$ generates a magnetic field (in fact, any charge moving in a circle generates a magnetic field). The magnetic field generated by the orbiting electron may be represented (Fig. 2.12,

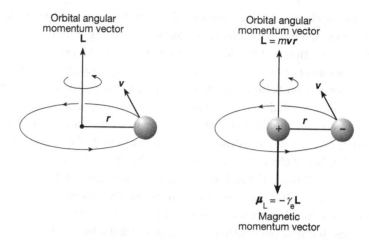

Figure 2.12   The vector model for the orbital angular momentum and the magnetic moment due to the circular motion of an electron in a Bohr orbit. The direction of the magnetic moment vector is opposite to that of the angular momentum vector for an electron. The units of **L** are $\hbar$ and the units of μ_L are joule per gauss (JG^{-1}).

right) as a magnetic dipole μ_L (a vector quantity analogous to an electric dipole) with a magnitude proportional to the angular momentum and a direction perpendicular to the plane of the orbit. For a constant circular velocity **v**, and fixed orbital angular momentum **L**, an orbiting Bohr electron generates a fixed magnetic moment, μ_L, that can be represented by a vector coinciding with the axis of rotation (Fig. 2.12). The direction of the head of the arrow representing the angular momentum vector follows the "right-hand-thumb rule" and points above the plane of rotation for the direction of motion in the orbit shown. This behavior is completely analogous to the classical picture of an electric current flowing in a circular wire and producing a magnetic moment positioned at the center of the motion and perpendicular to the plane of the wire.

From classical physics, the magnetic moment due to orbital motion of the electron, μ_L, is directly proportional to the magnitude of the orbital angular momentum, **L**. *This relationship is the important link connecting the value of angular momentum (in general) to the magnetic moment associated with the angular momentum.* From the model of the electron in the Bohr orbit, the proportionality constant between **L** and μ_L is $-e/2m$, the negative of one-half of the ratio of the unit of electric charge (e) to the electron's mass (m), as shown in Eq. 2.30a. Thus, the simple relationship expressed in Eq. 2.30b exists between the fundamental constants of an electron's charge and mass and the vectors μ_L and **L**.

$$\mu_L = -(e/2m)\mathbf{L} \tag{2.30a}$$

$$-\mu_L/\mathbf{L} = e/2m = \gamma_e \tag{2.30b}$$

The proportionality constant $(-e/2m)$ reflects a fundamental relationship between the magnetic moment μ_L and angular momentum $\mathbf{L}$ of a Bohr orbit electron and is therefore a fundamental quantity of quantum magnetism. This constant is called the *magnetogyric ratio* of the electron and is given the symbol γ_e, as shown in Eq. 2.30b. Thus, Eq. 2.30b may be rewritten in terms of γ_e, as shown in Eq. 2.31 (where γ_e is defined as a positive quantity).

$$\mu_L = -\gamma_e \mathbf{L} \tag{2.31}$$

If the electron possesses a single unit of *orbital* angular momentum $(\mathbf{L} = \hbar)$, the magnitude of its magnetic moment, μ_L is defined as exactly equal to $e/2m$. This fundamental unit of quantum magnetism is called the *Bohr magneton* and is given the symbol, μ_e. Its numerical value is 9.3×10^{-20} JG^{-1}. The symbol μ_e is the magnetic moment generated by an electron possessing an angular momentum of exactly $\hbar$.

The following important concepts can be deduced from Eqs. 2.30 and 2.31:

1. The vectors representing the magnetic moment (μ_L) and the orbital angular momentum $(\mathbf{L})$ are collinear (i.e., parallel in orientation).
2. The vector representing μ_L is *opposite (antiparallel) in direction* to that of $\mathbf{L}$ (the negative sign in Eq. 2.31 relating vectors means that the collinear vectors possess orientations 180° apart).
3. The proportionality factor γ_e reveals that the magnitude of the magnetic moment μ_L due to orbital motion is directly proportional to the charge of the electron (e) and inversely proportional to its mass (m).

Equations 2.30 and 2.31 are very important because they allow us to visualize both the electron's orbital angular momentum $(\mathbf{L})$ and magnetic moment (μ_L). We use these visualizations to understand the rankings of the energies of the magnetic states of electrons.

Now, we need to deduce how a spin magnetic moment (μ_S) is associated with the electron spin angular momentum $(\mathbf{S})$. We begin with the quantum mechanical results deduced previously from the interactions of $\mathbf{L}$ and μ_L. Then, we can transfer these ideas to electron spin, which can be modeled as resulting from a rotating sphere of negative charge. Finally, we can determine what quantum mechanical modifications of the model are necessary to apply our vector model to qualitatively visualize spin and its associated magnetic moment.

2.33   The Connection between Magnetic Moment and Electron Spin

The electron is a quantum particle, so it possesses many properties that are not understandable from observations of classical particles. Nonetheless, a clear and appealing visualization is possible if we start with a model that considers the electron as a classical particle of definite mass and of a spherical shape that possesses unit negative electric charge distributed uniformly over its surface. The classical properties of mass and charge are articulated in this simple physical model. It is natural to assume, therefore, that since the mass of the electron is fixed, and since its spin angular momentum is quantized (and therefore also a fixed quantity equal to $\hbar/2$), the spherical electron must spin about an axis with a fixed velocity, $\mathbf{v}$ (Fig. 2.13, left), in order to obey the fundamental law of the conservation of angular momentum. In other words, *because the electron's mass and charge are fixed, and since the electron possesses a fixed quantum unit of angular momentum (i.e., $\hbar/2$), the spinning velocity of the rotating electron must be constant for all electrons.* It is quite straightforward now to apply the results from the Bohr atom, discussed previously, to infer the relationship between the spin angular momentum ($\mathbf{S}$) of an electron to the magnetic moment ($\boldsymbol{\mu}_S$) due to its spin motion.

Analogous to a charged particle executing orbital motion, an electron executing spinning motion also generates a magnetic moment (also called a magnetic dipole) as shown in Fig. 2.13 (left). Accordingly, due to its spinning motion, the charged electron generates a magnetic moment ($\boldsymbol{\mu}_s$) in analogy to the magnetic moment ($\boldsymbol{\mu}_L$) generated by an electron in a circular Bohr orbit. The issue to be addressed is the relationship between $\boldsymbol{\mu}_s$ and $\mathbf{S}$.

It is tempting to start with a direct analogy to Eq. 2.31, the relationship between orbital angular momentum ($\mathbf{L}$) and the associated magnetic moment ($\boldsymbol{\mu}_L$), which would suggest that $\boldsymbol{\mu}_S = -\gamma_e \mathbf{S}$ (Eq. 2.32a). In other words, if the analogy of the magnetic moment due to orbital motion of a Bohr atom and the magnetic moment due to electron spin motion is valid, the magnetic moment due to spin ($\boldsymbol{\mu}_S$) should be directly proportional to the value of the spin ($\mathbf{S}$) and the proportionality constant should be γ_e. This simple analogy has the correct form qualitatively, but it has been shown experimentally that the expression is not quite correct quantitatively. Although a direct proportionality between $\boldsymbol{\mu}_s$ and $\mathbf{S}$ exists, for a free electron a "correction factor" g_e must be applied to relate $\boldsymbol{\mu}_s$ and $\mathbf{S}$ quantitatively. This quantitative relationship is thus given by Eq. 2.32b.

$$\boldsymbol{\mu}_s = -\gamma_e \mathbf{S} \tag{2.32a}$$

$$\boldsymbol{\mu}_s = -g_e \gamma_e \mathbf{S} \tag{2.32b}$$

In Eq. 2.32b, g_e is a dimensionless constant (a correction factor) called the g factor (or g value) of the free electron. The g factor has an experimental value close to 2 for a free electron in a vacuum. Thus, the simple model that attempts to transfer the properties of an orbiting electron to a spinning electron is correct qualitatively but is incorrect quantitatively by a factor very close to 2. This factor of 2 is fully justified by a more rigorous theory of the electron, which considers relativistic effects, but this rather subtle issue is of no concern for the qualitative features of spin that interest us in this text.

Fig. 2.13 uses the vector model to summarize the relationships between the spin angular momentum and the magnetic moment associated with electron spin. Fig. 2.13

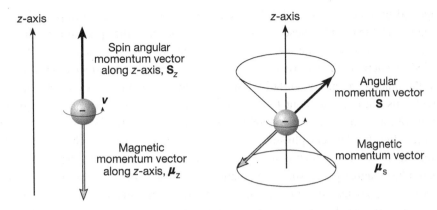

Figure 2.13   Vector representation of the spin angular momentum (**S**) and of the magnetic moment associated with spin (μ_S). The two vectors (**S** and μ_S) are collinear but antiparallel. The lengths shown are only schematic.

should be compared to the analogous form of the orbital angular momentum and its associated magnetic moment in Fig. 2.12.

Based on Eq. 2.32b and Fig. 2.13, the following important conclusions can be made concerning the relationship between electron spin and the magnetic moment due to spin:

1. Since electron spin is quantized and the spin vector and the magnetic moment vectors are directly related, *the magnetic moment associated with spin (μ_S), just as the angular momentum (**S**) from which it arises, is quantized in magnitude and orientation.*

2. Since the energy of a magnetic moment depends on its orientation in a magnetic field, *the energies of quantized spin states depend on the orientation of the spin vector in a magnetic field.*

3. In analogy to the relationship between the orbital angular momentum and the magnetic moment derived from orbital motion, the vectors μ_S and **S** are antiparallel (Fig. 2.13, left).

4. The vectors μ_S and **S** are both positioned in a cone of orientation that depends on the value of M_S [Fig. 2.13 (right) shows the case for **S** for $M_S = +1/2$].

2.34   Magnetic Energy Levels in an Applied Magnetic Field for a Classical Magnet

We have developed a model, specified in Eq. 2.32, that allows the association of a specific value of the magnetic moment (μ_S) with a specific value of spin angular momentum (**S**). Our next goal is to develop a model for the magnetic energy levels associated with different values (M_S) of spin for electrons in an applied magnetic field. First, let us consider what happens to the magnetic energy levels when a magnetic field is applied in the classical model. We examine the results for an exemplar classical magnet, and then apply these results to determine how the quantum mechanical magnet associated with the electron spin changes its energy when coupled to an applied magnetic field. The vector model not only provides an effective tool to deal with the qualitative and quantitative aspects of the magnetic energy levels but also provides us with an excellent tool for visualizing the qualitative and quantitative aspects of transitions between magnetic energy levels (Chapter 3).

According to classical physics, a laboratory magnetic field is characterized by its magnetic moment, which is a vector quantity we represent with the symbol $\mathbf{H}_z$. When a bar magnet possessing a magnetic moment μ is placed in a magnetic field $\mathbf{H}_z$, a torque operates on the magnetic moment of the bar magnet and twists it with a force that tries to align the direction of the moment with the direction of the field determined by $\mathbf{H}_z$. The precise equation describing the energy relationship of the bar magnet at various orientations in $\mathbf{H}_z$ is given by Eq. 2.33, where μ and $\mathbf{H}_z$ are the magnitudes of the magnetic moments of the bar magnet and the applied field in the z direction (defined as the direction from the south to north pole of the laboratory magnetic field), respectively. According to classical mechanics, any orientation of the bar magnet in the magnetic field is possible. However, the energy of a bar magnet depends on its orientation relative to the z-axis. Three important limiting orientations of the bar magnet relative to the magnetic field, namely, parallel $(0°)$, perpendicular $(90°)$, and antiparallel $(180°)$, are shown schematically in Fig. 2.14. The values for these three orientations are $-\mu\mathbf{H}_z$, 0, and $\mu\mathbf{H}_z$, respectively.

$$E_z(\text{magnetic}) = -\mu\mathbf{H}_z\cos\theta \qquad (2.33)$$

In Eq. 2.33, the negative sign before the quantity on the right side of the equation means that the system is *more stable* if the product $\mu\mathbf{H}_z\cos\theta$ is a positive quantity.

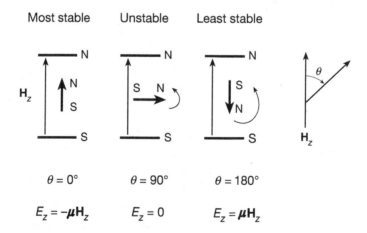

Figure 2.14   Energies of a classical bar magnet in a magnet field $\mathbf{H}_z$. The curved arrow indicates the force acting to rotate the magnet to align it with the direction of the field.

By definition, the magnitudes (lengths) of the vector quantities $\boldsymbol{\mu}$ and $\mathbf{H}_z$ are always positive, so whether the energy of the system is positive (less stable than the situation in zero field) or negative (more stable than the situation in zero field) depends on the sign of $\cos\theta$ (which is positive for θ between $0°$ and $90°$ and negative for θ between $90°$ and $180°$).

Consider the situations when $\boldsymbol{\mu}$ and $\mathbf{H}_z$ are two parallel vectors ($\theta = 0°$ or 0π), two perpendicular vectors ($\theta = 90°$ or $\pi/2$), or two antiparallel vectors ($\theta = 180°$ or π). For these cases, $\cos\theta = 1$, 0, and -1, respectively (Fig. 2.14). Thus, any orientation of the bar magnet with θ between 0 and just $< 90°$ is stabilizing (the energy E_z in Eq. 2.33 is negative) and any orientation of the bar magnet with θ between $90°$ and up to $180°$ is destabilizing (the energy E_z is positive). At an orientation of $90°$, the magnetic interaction between the applied magnetic field and the bar magnet is zero ($\cos 90° = 0$ in Eq. 2.33), that is, the same as in the absence of a field, when $\mathbf{H}_z = 0$. A more detailed discussion of the effect of the orientation of a bar magnet in a magnetic field is given in Section 2.39 for the interactions of two magnetic dipoles.

2.35   Quantum Magnets in the Absence of Coupling Magnetic Fields

In contrast to the classical magnet, which can assume any orientation in a magnetic field, the quantum magnet can only assume a set of orientations that depend on the value of the magnetic quantum number (M_S). We now review the notations and vector representations that are convenient for describing the most important spin situations that arise in organic photochemistry, namely, a single electron spin and two coupled electron spins. Table 2.4 summarizes the notations and vector representations for a single spin and two coupled spins; these conventions can be used to describe spin systems at both zero and high applied magnetic fields. We use the symbol D (doublet) to describe the state of a single spin, $S = 1/2$. For the case of $M_S = +1/2$ (α, or up spin pointing in the *positive* direction along the z-axis), we label the state D_+; for the case of $M_S = 1/2$ (β, or down spin pointing in the *negative* direction along the z-axis), we label the state D_-. In the case of two coupled spins, we use the symbol S to denote the singlet state (one α spin and one β spin; wave function $\mathbf{S} = \alpha\beta - \beta\alpha$). This spin state has the quantum number $M_S = 0$. For the triplet state we use the symbols T_+ (two α spins), T_0 (one α spin and one β spin; wave function $\mathbf{S} = \alpha\beta + \beta\alpha$), and T_- (two β spins) to label the states with quantum numbers $M_S = +1$, 0, and -1, respectively.

In the absence of magnetic interactions operating on the magnetic moments associated with the electron spins, all six of the states (D_+, D_-, S, T_-, T_0, and T_+) have exactly the same energy; that is, the magnetic energy levels of all these states are degenerate (of identical energy), since there are no magnetic interactions to split the energy levels ($\mathbf{H}_z = 0$ in Eq. 2.33). In the absence of any magnetic interactions, the vector model of electron spin views the spin vector representing the angular momentum and magnetic moment as stationary and possessing random arbitrary orientations

Table 2.4   Conventional Representations of the Singlet, Doublet, and Triplet States in an Applied Magnetic Field ($\mathbf{H}_z$)[a]

State	State Symbol	M_s	Magnetic Energy (E_z)	Spin Function	Vector Representation
Doublet	D_+	$+1/2$	$+(1/2)g\mu_e\mathbf{H}_z$	α	
Doublet	D_-	$-1/2$	$-(1/2)g\mu_e\mathbf{H}_z$	β	
Singlet	S	0	0	$\alpha\beta - \beta\alpha$	
Triplet	T_+	$+1$	$+(1)g\mu_e\mathbf{H}_z$	$\alpha\alpha$	
Triplet	T_0	0	0	$\alpha\beta + \beta\alpha$	
Triplet	T_-	-1	$-(1)g\mu_e\mathbf{H}_z$	$\beta\beta$	

a. The mathematical normalizing factor is not shown for the spin function.

in space. There is no pertinent M_S quantum number in zero field, because there is *no preferred* axis of orientation. However, even at zero field the total spin angular momenta for the singlet, doublet, and triplet ($S = 0$, 1/2, and 1) are well defined, but M_S is not. When $\mathbf{H}_z = 0$, therefore, the quantum number for total spin $\mathbf{S}$ is still a valid quantum number, but the quantum number M_S is not. In other words, a D or T state is magnetic and an S state is not, whether or not an applied magnetic field is present.

2.36 Quantum Mechanical Magnets in a Magnetic Field: Constructing a Magnetic State Energy Diagram for Spins in an Applied Magnetic Field

The magnetic energy resulting from the interaction of a magnetic moment μ and an applied field $\mathbf{H}_z$ (Fig. 2.14) is known as the *Zeeman energy* (E_z). Eq. 2.34 relates the energy of the quantum mechanical magnet to the strength of the applied field along the z-axis ($\mathbf{H}_z$), the magnetic quantum number for a given spin orientation (M_S), the g-value of the electron (g_e), and the magnetic moment of a free electron (μ_e).

$$E_z = M_S g \mu_e \mathbf{H}_z \tag{2.34}$$

The ranking of magnetic energies (relative to those in zero field) of the singlet, doublet, and triplet states are listed in Table 2.4.

2.37   Magnetic Energy Diagram for a Single Electron Spin and for Two Coupled Electron Spins

Fig. 2.15 displays the magnetic (Zeeman) energy-level state diagram[10] at zero magnetic field ($\mathbf{H}_z = 0$ G) and in the presence of a strong magnetic field ($\mathbf{H}_z \gg 0$ G) for the two fundamental cases: (1) a single electron spin, a doublet (D) state, and (2) two correlated electron spins, which may be either a triplet (T) or singlet (S) state. In zero field (for now we ignore the electron-exchange interaction (J) and only consider the magnetic interactions in the energy diagram), all of the magnetic energy levels are degenerate, because there is no preferred orientation of the angular momentum and therefore no preferred orientation of the magnetic moment due to spin. Although these states are all of the same energy, they are shown with levels slightly separated in Fig. 2.15 so that the number of degenerate states is clear. All of these states have identical energies in the absence of magnetic (and electron exchange) interactions.

The relative ordering of the magnetic states is always the same in an applied magnetic field ($\mathbf{H}_z$): $D_+ > D_-$ and $T_+ > T_0 > T_-$ (Fig. 2.15). The energy separation between adjacent levels depends, moreover, on the magnitude of the applied field ($\mathbf{H}_z$) (Eq. 2.34).

The zero-field situation is a benchmark for the calibration of the magnetic coupling energy in devising a magnetic energy diagram. The concept is the same as using the energy of a nonbonding p orbital as a benchmark of energy (set at $E = 0$) and then to treat bonding orbitals as lower in energy than a p orbital (i.e., negative energy relative to $E = 0$), and antibonding orbitals as higher in energy than a p-orbital (i.e., positive energy relative to $E = 0$). The magnitude of the exchange interaction, J (Section 2.12 and 2.13), is typically much larger (several to tens of kcal mol^{-1}) than

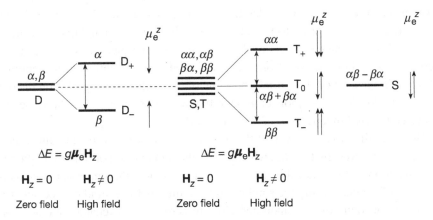

Figure 2.15   Magnetic energy diagram for a single electron spin and two correlated electron spins. The magnetic energy of T_0 and S are the same at $\mathbf{H}_z = 0$ or $\mathbf{H}_z \gg 0$. For these examples, $J = 0$, so the energy of T_0 and S are identical (i.e., there is no T_0 and S splitting when $J = 0$).

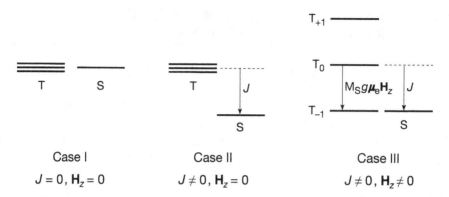

Figure 2.16   The effect of electron exchange on the magnetic energy levels of a triplet and singlet state.

that of magnetic interactions (much less than $1 \, kcal \, mol^{-1}$). However, the exchange interaction is Coulombic (i.e., an electrostatic interaction between charged particles) and is not magnetic. This finding allows us to consider the magnetic interactions first and independently, as shown in Fig. 2.15, and to later "turn on" the electrostatic exchange interactions J after considering the magnetic interactions (see Fig. 2.16).

2.38   Magnetic Energy Diagrams Including the Electron Exchange Interaction, J

The energy diagrams shown in Fig. 2.15 must be modified in the presence of electron exchange, J (Section 2.7). We saw in Section 2.14 that there is a significant difference between the magnitude of the singlet–triplet splittings for molecules (*R) and for reactive diradical intermediates, I(D), and that these differences are in large measure due to J. In the case discussed in Section 2.14, the singlet and triplet states being compared were S_1 and T_1, both derived from a HO–LU orbital configuration corresponding to *R. Even in the case of *R $=$ n,π*, the value of the exchange integral is on the order of several kilocalories per mole or greater. In these cases, $J \gg E_z$, so the value of E_z can be considered independent of the value of J.

However, the situation is very different when the value of J is on the same order as the value of E_z. Review Fig. 2.3, which shows the influence on orbital orientation and the separation of two radical centers. Let us now review the important conclusions in Section 2.14 concerning the singlet–triplet splittings for a diradical species, I(D), for which the value of J is small and close to zero. Weak exchange interactions in I(D) have two important consequences: (1) the magnitude of the exchange forces between electrons become on the same order of magnitude as the magnetic forces between electrons for I(D), and (2) T may no longer be lower in energy than S for some geometries of I(D), because bonding interactions become stronger than exchange interactions (Eq. 2.22). In these cases, the two orbitals containing the radicals begin to overlap significantly, and bonding effectively stabilizes the singlet state.

An important situation for photochemistry is the case where the radicals have their orbitals oriented for bond formation (Fig. 2.3, right). In this case, the S state ($\downarrow\uparrow$) becomes more stable than the T state ($\uparrow\uparrow$) as the radical centers approach each other. It is as if the energy of S decreases and the energy of T increases as the result of exchange (J). Thus, starting from a situation for which the S and T states are degenerate for a large separation of the radicals, the two electrons on each center exchange positions as the radical orbitals begin to overlap, and the system is stabilized by the covalent character of the exchange process.

There are three limiting situations (cases I, II, and III) for which the combination of exchange and magnetic field effects are important in determining the rate of intersystem crossing in the photochemical systems that are summarized in Fig. 2.16. (Transitions between spin states are covered in detail in Chapter 3. Here, the presentation is intended to describe how the magnetic energy levels are influenced by the exchange interaction, J.)

In case I (Fig. 2.16a), $J = 0$ and $H_z = 0$. Although S and T are distinct states, they are degenerate in energy because there are no magnetic interactions to split the magnetic levels and no exchange interactions to split the S and T states. Since the three sublevels of T (T_+, T_0, and T_-) are degenerate, they can be viewed as rapidly interconverting along a molecular axis (there is no z-axis available) as the molecule tumbles and rotates in solution. Case I is typical of situations for which there is no magnetic field present and the two spins of I(D) are separated by relatively large distances (> 3–5 Å) in space. Exemplars of this case are solvent-separated, spin-correlated geminate radical pairs and flexible biradicals whose odd electron centers are separated by > 3–5 Å. Intersystem crossing between the T and S states can be rapid for case I because of the degeneracy of all of the T states with S.

In case II (Fig. 2.16), $H_z = 0$ and J is finite, but small. Case II is typical of situations for which the two spins are close enough together so that their orbitals overlap slightly in space, as is the case for spin-correlated radical pairs and small biradicals. In case II, intersystem crossing between T and S can be slow because of the energy difference between T and S produced by J (for bonding situations such as that shown in Fig. 2.3b). In this case the effect of J is to drop S to a lower energy than T.

In case III (Fig. 2.16c), H_z is large relative to the magnetic interactions of the electron spins (so the T_+, T_0, and T_- states are split), and J values are 0 or are comparable to the Zeeman splitting. Figure 2.16 shows the case for which $J = 0$, so the S state is degenerate with T_0 (see the dotted line for S). Figure 2.16 also shows the case for which J is on the order of the energy splitting of T_0 and T_- (see the solid line for S).

As J increases from 0 to values on the order of the Zeeman energy and larger, the initial degeneracy of T_0 and S is broken. As S becomes more stabilized by increasing J, the T_- and S levels become degenerate. As J increases further, S drops much lower in energy than any of the T levels. The rates of intersystem crossing from T to S and from S to T is highly sensitive to whichever of these situations dominates for an actual I(D) system.

2.39   Interactions between Two Magnetic Dipoles: Orientation and Distance Dependence of the Energy of Magnetic Interactions

In Section 2.34, we learned that the magnetic energy of a bar magnet depends on the orientation of its magnetic moment (μ) in a magnetic field. A magnetic moment is a magnetic dipole; that is, the magnetic moment gives rise to a magnetic field in its vicinity (Fig. 2.17a). In this section we briefly investigate the mathematical form of the classical interactions of two magnetic dipoles in order to develop some intuition concerning the magnitude of the dipole–dipole interactions as a function of spin structure and the orientation of two interacting spins in space.

Insight into the nature of the magnetic dipole–dipole interaction as a function of the relative orientation of two dipoles in space is available from the mathematical formulation of the interaction and its interpretation in terms of the vector model (Eq. 2.35). The beauty of the mathematical formulation is that its representation provides an identical basis to consider all forms of dipole–dipole interactions. These may be due to electric dipoles interacting (two electric dipoles, an electric dipole and a nuclear dipole, or two nuclear dipoles) or to magnetic dipoles interacting (e.g., two electron spins, an electron spin and a nuclear spin, two nuclear spins, a spin and a magnetic field, a spin and an orbital magnetic dipole, etc.).

Classically, the dipole–dipole interaction energy depends on the relative orientation of the magnetic moments (e.g., two bar magnets).[11] To obtain some concrete insight into the dipolar interaction, consider the case in Fig. 2.17 for which the two magnetic dipoles, μ_1 and μ_2, are held parallel to one another. This is the case for two interacting magnetic dipoles, such as the magnetic dipoles associated with two electron spins in a strong magnetic field, which causes magnetic dipoles in the field to line up parallel with the field, $\mathbf{H}_z$, and therefore parallel to each other. The strength of the dipole–dipole interaction as a function of the orientation of the two dipoles in space is given by Eq. 2.35. In general, the strength of the interaction is proportional to the following factors: (1) the magnitudes of the individual interacting dipoles, μ_1 and μ_2; (2) the distance, r_{12}, separating the centers of the two interacting parallel dipoles; (3) the angle of orientation, θ, of the (parallel) dipoles relative to one another; and (4) the overlap integral of the energy states that satisfy the conservation of angular momentum and the conservation of energy. Strictly speaking, Eq. 2.35 refers to the interaction of two point dipoles (if r_{12}, the separation between the dipoles, is large relative to the dipole length, the dipole may be considered a point dipole).

$$E_{dd}(\text{dipole–dipole energy}) \propto [(\mu_1\mu_2)/r_{12}^3](3\cos^2\theta - 1) \qquad (2.35)$$

For dipole–dipole interactions in solution, the rate of processes involving dipole–dipole interactions is typically proportional to the *square* of the strength for the dipole–dipole interaction. Thus, the field strength falls off as $1/r_{12}^3$, but the rate of a process driven by dipolar interactions falls off as $1/r_{12}^6$. In Chapter 7, we show that this distance dependence is identical to a mechanism for electronic energy transfer, which can occur by the interaction of two electric dipoles.

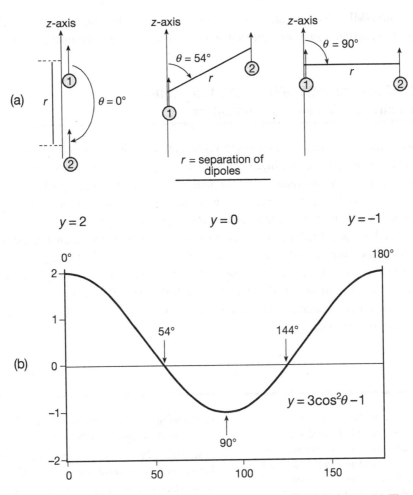

Figure 2.17   Dipole–dipole interactions of parallel magnetic moments. (a) The vector representation of dipoles interacting at a fixed separation, r, and various orientations relative to a z-axis. (b) The plot of the value of $3 \cos^2 \theta - 1$ (Eq. 2.35) as a function of θ.

The $3 \cos^2 \theta - 1$ term in Eq. 2.35, which is plotted in Fig. 2.17b for the same separation of the dipoles, r_{12}, is particularly important because of the following features: (1) for a fixed value of r_{12}, the separation of the interacting dipoles, the $3 \cos^2 \theta - 1$ term causes the interaction energy to be highly dependent on the angle θ that the vector $\mathbf{r}$ makes with the z-axis, and (2) the value of the $3 \cos^2 \theta - 1$ term averages to zero if all angles are represented, because the two dipoles are moving randomly in space (the average value of $\cos^2 \theta$ over all space is one-third, so $3 \cos^2 \theta - 1 = 0$ when averaged over all space). A plot of $E_{dd} = 3 \cos^2 \theta - 1$ for a fixed separation (r_{12}) is shown in the bottom one-half of Fig. 2.17b. The values of E_{dd} are symmetrical about $\theta = 90°$. For the values $\theta = 54°$ and $144°$, $E_{dd} = 0$ (i.e., for these particular angles of orientation, the dipolar interaction disappears even when the dipoles are close in space and large in magnitude). The values of $54°$ and $144°$ are the so-called magic angles employed to spin samples in the magnetic field of an NMR spectrometer for removing chemical shifts due to dipolar interactions in solid-state NMR. Certain values of y are positive (i.e., they increase the magnetic energy) and certain values of y are negative (i.e., they decrease the magnetic energy).

2.40 Summary: Structure and Energetics of Electrons, Vibrations, and Spins

In this chapter we have considered the visualization of the electronic, spin, and vibrational structure of the starting points for photochemical reactions, namely, R^* and diradical reactive intermediates, $I(D)$. In particular, a working paradigm has been developed for visualizing the structure of electrons in molecular orbitals, namely, the vibrations of a harmonic oscillator and electron spins as magnetic vectors in a magnetic field. For each structural representation, we can construct state energy diagrams that make it possible to understand the relative ranking of the electronic, vibrational, and magnetic states. Chapter 3 considers the visualization of the photophysical and photochemical transitions of R^*, and the way that classical and quantum mechanics provide a deep understanding of all transitions between states from a common conceptual framework.

References

1. (a) P. W. Atkins and R. Friedman, *Molecular Quantum Mechanics*, 5th ed., Oxford University Press, Oxford, UK, 2005. (b) W. Kautzmann, *Quantum Chemistry*, Academic Press, New York, 1957. (c) P. W. Atkins, *Quanta: A Handbook of Concepts*, 2nd ed., Oxford University Press, Oxford, UK, 1991. (d) M. Klessinger and J. Michl, *Excited States and Photochemistry of Organic Molecules*, VCH Publishers, New York, 1995.

2. N. J. Turro, *Angew. Chem. Int. Ed. Engl.* **25**, 882 (1986).

3. L. Salem and C. Rowland, *Angew. Chem. Int. Ed. Engl.* **11**, 92 (1971).

4. W. Kautzmann, *Quantum Chemistry*, Academic Press, New York, 1957, p. 200.

5. P. W. Atkins, *Quanta: A Handbook of Concepts*, 2nd ed., Oxford University Press, Oxford, UK, 1991, p. 153.

6. G. Herzberg, *Spectra of Diatomic Molecules*, Van Nostrand, Princeton, NJ, 1950, p. 91.

7. P. W. Atkins, *Physical Chemistry*, 3rd ed., Oxford University Press, Oxford, UK, 1982, p. 336.

8. K. A. McLauchlan, *Magnetic Resonance*, Oxford University Press, Oxford, UK, 1972, Chapter 1.

9. P. W. Atkins, *Quanta: A Handbook of Concepts*, 2nd ed., Oxford University Press, Oxford, UK, 1991, p. 368.

10. (a) A. Carrington, A. D. McLachlan, *Introduction to Magnetic Resonance*, Harper & Row, New York, 1967. (b) A. L. Buchachenko and V. L. Berdinsky, *Chem. Rev.* **102**, 603 (2002).

11. P. W. Atkins, *Quanta: A Handbook of Concepts*, 2nd ed., Oxford University Press, Oxford, UK, 1991, p. 183.

Transitions between States: Photophysical Processes

3.1 Transitions between States

The state energy diagram (Scheme 1.4) displays the *time-independent* energies for the electronic states of a molecule associated with a given "spatially frozen" nuclear geometry (i.e., the Born–Oppenheimer approximation, which allows us to focus on the *energetics* and *structures* of R and *R). In this chapter, we describe the *time-dependent photophysical transitions* between R and *R in which the energetics and structures change with time. Some of the transitions of interest to organic photophysics, shown in Scheme 3.1, are as follows: (a) radiative absorption of a photon by R to produce *R; (b) emission of a photon from *R to produce R; (c) radiationless transition from *R to produce R and heat; (d) radiationless transitions between electronically excited states, **R$_2$ (higher energy) and *R (lower energy); and (e) radiative transitions between *R$_2$ (higher energy) and *R$_1$ (lower energy). *Each of these transitions may involve singlet or triplet states*. Transitions of R and I(D) involving a change of electron spin are discussed in Section 3.12.

Exemplars of structure–reactivity and structure–efficiency relationships for the absorptive and emissive radiative transitions (e.g., R + $h\nu \rightarrow$ *R and *R $\rightarrow$ R + $h\nu$) are covered in Chapter 4. Exemplars of structure–reactivity and structure–efficiency relationships for the radiationless transitions (e.g., *R $\rightarrow$ R + heat and **R$_2 \rightarrow$ *R$_1$ + heat). Exemplars of the intersystem crossing in I(D) species are covered in Chapter 6. A structural and pictorial model for the primary *photochemical* transitions of Scheme 2.1 (i.e., *R $\rightarrow$ I and *R $\rightarrow$ F) are presented in Chapter 6.

According to the laws of quantum mechanics[1] (Section 2.2), the value of any observable property P$_1$ of a state may be computed from Eq. 3.1 if the wave function of the state Ψ_1 and the mathematical operator P_1 corresponding to the observable property are known. For example, if the electronic energy of a state E_1 is to be computed, the operator P_1 corresponds, in the Born–Oppenheimer approximation, to the classical *repulsive* Coulombic interactions (e^2/r) between two electrons in the field of the fixed

$$(a) \ R + h\nu \rightarrow {}^*R$$
$$(b) \ {}^*R \rightarrow R + h\nu$$
$$(c) \ {}^*R \rightarrow R + \text{heat}$$
$$(d) \ {}^{**}R_2 \rightarrow {}^*R_1 + \text{heat}$$
$$(e) \ {}^*R_2 \rightarrow {}^*R_1 + h\nu$$

Scheme 3.1   Important photophysical processes in molecular organic photochemistry.

positive nuclear framework. In the matrix element of Eq. 3.1, the two wave functions involved are identical. This means that the matrix element refers to the *property of a single state*, such as the energy of the state.

$$\text{Magnitude of observable property } P_1 = \ <\Psi_1|P_1|\Psi_1> \quad \text{Matrix element} \quad (3.1)$$

In Chapter 2, the most important observable properties P_1 of interest were the state energies (E_n) of the wave functions Ψ_n, where n is the quantum number for the state. In this chapter, we are interested in the rates of transitions between an *initial* state Ψ_1 (the initial state is given the subscript 1) and a second state Ψ_2 (the second state is given the subscript 2). By knowing the wave functions Ψ_1 and Ψ_2 and the laws of quantum mechanics, the rate k of a transition $\Psi_1 \rightarrow \Psi_2$ can be computed from the *square* of a matrix element corresponding to the transition (Eq. 3.2) if $P_{1\rightarrow 2}$, the operator that corresponds to the interaction that triggers the $\Psi_1 \rightarrow \Psi_2$ transition, is known. The rate of a transition that occurs in a single step (called an *elementary step*) is given the symbol k (i.e., the rate constant). The symbol "$\sim$" in an equation means that constants and unessential mathematical features have been omitted for simplicity. Notice that in the matrix element of Eq. 3.1, the two wave functions involved are different, which means that the matrix element refers to a *transition between two states*.

$$k \text{ for the } \Psi_1 \rightarrow \Psi_2 \text{ transition} \quad P_{1\rightarrow 2} \sim \ <\Psi_1|P_{1\rightarrow 2}|\Psi_2>^2 \quad\quad (3.2)$$

The rates for each of the transitions in Scheme 3.1 can be estimated by a form of the Eq. 3.2 matrix element. For example, the matrix element given by Eq. 3.3 corresponds to the probability of the transition $R + h\nu \rightarrow {}^*R$ in Scheme 3.1, where the wave functions are represented by the symbols for the initial and final states, $\Psi_1(R)$ and $\Psi_2({}^*R)$, respectively, and $P_{h\nu}$ is the appropriate operator corresponding to the interaction of the R electrons with a photon (or, more precisely, with the electromagnetic field). The use of the square of a matrix element to compute a rate for a transition is at the heart of *Fermi's golden rule* (Eq. 3.8) for transitions between weakly coupled states.

$$k \text{ for the } \Psi_1(R) + h\nu \rightarrow \Psi_2({}^*R) \text{ transition} \sim \ <\Psi_1(R)|P_{h\nu}|\Psi_2({}^*R)>^2 \quad (3.3)$$

In general, the interaction corresponding to the operator $P_{1\rightarrow 2}$ "distorts" the wave function Ψ_1. If this interaction makes Ψ_1 "look like" Ψ_2, a transition between Ψ_1 and Ψ_2 can be "triggered". In the language of wave mechanics, the interaction corresponding to $P_{1\rightarrow 2}$ causes the wave functions Ψ_1 and Ψ_2 to "mix" with one another. Effective

mixing of two waves occurs only under the very special condition of a *resonance* between the two waves Ψ_1 and Ψ_2. *We describe how this notion of resonance due to the
mixing of wave functions provides an excellent quantum intuition for the visualization of all of the transitions listed in Scheme 3.1.* For example, the visualization of
the resonance corresponding to Eq. 3.3 involves "picturing" how the electromagnetic
field of a light wave (a photon, $h\nu$) mixes $\Psi_1(R)$ with $\Psi_2(*R)$ and causes a resonance
between the two wave functions (when energy and momentum can be conserved). The
photon carries the energy and the interaction that is required to achieve the resonance
and cause the electronic transition $R + h\nu \rightarrow \,^*R$. Chapter 4 describes and visualizes
this resonance in detail with many experimental exemplars.

As with the process for a "zero-order" guess about the nature of the operators (P)
that correspond to the properties of states, the mathematical form of the operator $P_{1\rightarrow 2}$,
corresponding to the interaction (or perturbation) that causes the transition, is usually
made by appealing to a classical model for interactions corresponding to an operator
$P_{1\rightarrow 2}$ that can induce transitions $\Psi_1 \rightarrow \Psi_2$. The model for $P_{1\rightarrow 2}$ is then modified to
include the appropriate quantum and wave mechanical effects. Once this is done,
we can express the operator and wave function in pictorial terms and qualitatively
estimate the value of the mathematical integral or matrix element $< \Psi_1 | P_{1\rightarrow 2} | \Psi_2 >$ by
an equation of the form of Eq. 3.2. This qualitative evaluation of the matrix elements
provides useful *selection rules* for the transitions shown in Scheme 3.1. Selection
rules serve as a guide to the plausibility of a given transition and the probability of a
transition from a state when there are several plausible transitions. In general, $P_{1\rightarrow 2}$
represents the mathematical form of small interactions that can be considered as weak
first-order perturbations of the zero-order, or starting, approximation. This result will
always be the case when good zero-order electronic wave functions (Ψ_n) have been
selected.

3.2   A Starting Point for Modeling
Transitions between States

A selection rule is a statement of the plausibility that a state may undergo a specific
type of transition under a specific set of circumstances. For the photophysical transitions shown in Scheme 3.1, we seek to develop selection rules that will provide
quantum intuition as to whether the probability (or rate) of a transition is closer to the
hypothetically "strictly forbidden" (implausible) or "fully allowed" (plausible) limits. The pictorial process for transitions is an extension of the process of visualizing
states, except that the notion of time dependence (i.e., transitions between one state
and another) is introduced. We begin with a visualization of the wave functions corre-
sponding to the initial state (Ψ_1) and final state (Ψ_2) involved in a transition $\Psi_1 \rightarrow \Psi_2$.
With a picture of the wave function of the initial and final states in mind, we apply the
rules of quantum mechanics to estimate the (square of the) magnitude of the matrix
element (Section 2.5) that describes the qualitative rate of the transition (Eq. 3.2).

To qualitatively estimate the magnitude for the matrix element of Eq. 3.2, we need a
picture not only of the wave functions (i.e., the structures) for the initial and final states
but also of the operator, $P_{1\rightarrow 2}$. The operator represents the interactions or forces that

most effectively distort the initial state Ψ_1 and make Ψ_1 look like the wave function of the final state Ψ_2. It is natural to accept a physical picture for which transitions occur fastest between two states *when the two states are identical in energy and when the two states involved in the transition "look alike" or can be easily made to look alike as the result of a perturbation ($P_{1\to2}$).* This procedure follows the principle of "minimum quantum mechanical reorganization of wave functions" for the fastest transitions. The reorganization includes the energy required to change the molecular structure, motion (phase), and energy required to make Ψ_1 look like the molecular structure, motion (phase), and energy of Ψ_2. By "look alike" we mean "look alike in all respects."

When two classical waves are similar in energy and look very much alike, they are in excellent condition to go into *resonance* and *mix* with one another. Quantum mechanics picks up on this classical idea of the property of waves and states that the wave functions of the two states must have the same energy and look alike in order for *resonance* to occur and for transitions to occur between the two states. If the initial state Ψ_1 goes into a state of resonance with the final state Ψ_2, there is a certain probability that a transition $\Psi_1 \to \Psi_2$ will occur as the result of the resonance. In a schematic way, the transition $\Psi_1 \to \Psi_2$ can be viewed as occurring as a "reaction" between the wave function Ψ_1 as a substrate and a perturbation $P_{1\to2}$ as a "reagent." The systems interact and go into a "transition state" for which Ψ_2 is mixed into Ψ_1. The transition state contains a "mixture" of Ψ_1 and Ψ_2 and may be described as a "mixed" wave function, $\Psi_1 \pm \Psi_2$. The transition state has a certain probability of collapsing back to Ψ_1 or to Ψ_2. When this happens, a complete transition has occurred.

As was the case for the visualization of state properties, the "true" wave function can be approximated by the product of an electronic wave function Ψ_0, a vibrational wave function χ, and a spin wave function $\mathbf{S}$, and transitions between the electronic, vibrational, and spin portions of a transition can be pictured independently.

3.3 Classical Chemical Dynamics: Some Preliminary Comments

We can obtain *classical intuition* about the dynamics of molecular transitions through the concepts of classical mechanical dynamics,[2] which are based on the conservation laws (i.e., the conservation of energy and the conservation of momentum) and Newton's laws of interactions between particles. In particular, Newton's first and third laws for particles are the usual classical starting points when analyzing transitions between electronic states:

1. *The change in the motion of a system is proportional to the forces (interactions) acting on the system.* A central problem in understanding dynamic processes, such as transitions between states, is the identification of the interactions (the operators, $P_{1\to2}$, corresponding to the *forces* or *interactions*) involved in changing the motion and the energies of the particles in the initial state Ψ_1 and converting it to the final state Ψ_2. In general, these forces (interactions) are electric or magnetic. Typically, the most important forces involved in causing

transitions are electrostatic forces, such as electron–electron (electronic) interactions. Other interactions due to vibrations and spin are usually much weaker. Electronic and vibrational motions can often be visualized as associated oscillating harmonic motions along a conveniently selected axis or a molecular framework. Magnetic motions due to electron spins are associated with circular or rotational motions. A *torque* plays the same role in rotational motion that *force* does in linear motion. More precisely, for linear motion, force is equal to the rate of change of the linear momentum (the back-and-forth motion), whereas for rotational motion, torque is equal to the rate of change of the angular momentum (the twisting circular motion). The concepts of torque and rotational motion are key to understanding and visualizing spin and spin interconversions.

2. *To every action there is always an opposed and equal reaction.* Interactions that cause transitions result from interactions and occur reciprocally. Typically, if we can identify an interaction in one direction of a transition, we can deduce the nature of the interaction in the reverse direction.

The challenge in determining the plausibility of transitions (i.e., $\Psi_1 \rightarrow \Psi_2$) is to identify the *energies* (energy must be conserved) and the *interactions* (forces must be available to change the motions and the structure of the initial system) that are plausible in a particular system, and which of the possible available interactions that conserve energy make the transition $\Psi_1 \rightarrow \Psi_2$ plausible.

3.4   Quantum Dynamics: Transitions between States

In this chapter, we seek to estimate the relative rates of transitions between states by visualizing matrix elements of the form depicted in Eq. 3.2. Recall from Chapter 2 that we described pictorial models for the molecular wave function (Ψ_0) in terms of the approximate electronic (ψ), vibrational (χ), and spin (S) wave functions. Our main task in this chapter is to visualize the operators $P_{1 \rightarrow 2}$ in Eq. 3.2 and deduce how they operate on the wave functions ψ, χ, and S to produce a final value of the matrix element corresponding to the $\Psi_1 \rightarrow \Psi_2$ transition.

3.5   Perturbation Theory[1,3]

Mathematical methods were developed to generate approximate wave functions for complicated organic molecules from simpler systems for which the wave function is known more precisely and which resemble the molecular system of interest as closely as possible. These simpler, approximate wave functions are then "distorted" by a mathematical perturbation (P'), which provides solutions to the wave equation that are closer to the solutions of the true wave function and the true electronic energies. If the exact system resembles the approximate system closely, the distortion required is small and can be considered to be a small "perturbation" of the approximate wave function. *Perturbation theory* is a mathematical method that provides a recipe for using weak perturbations for mixing wave functions of the approximate system in an appropriate manner so as to achieve better and better approximations of the true system.

Suppose, for example, that we start with an approximate electronic wave function (Ψ_0). The solution of the wave equation (Eq. 2.1) provides the electronic energy (E_0) of Ψ_0. Such approximate wave functions and energies are called zero-order wave functions and zero-order electronic energies, respectively. If a zero-order wave function (Ψ_0) is a reasonable approximation to the true wave function (Ψ), perturbation theory can be employed to "distort" Ψ_0 and its E_0 in the direction of Ψ and the true state energies E_n. Mathematically, the approximate wave function Ψ_0 is said to be *perturbed* (or *corrected*) to look more like the true wave function (Ψ). The key to the successful use of perturbation theory is the judicious selection of zero-order wave functions (Ψ_0), and *the correct physical perturbation to serve as an operator (P') that mixes wave functions.*

A weak perturbation is defined as one that does not significantly change the energies associated with the zero-order wave function. Often, *weak perturbations* are responsible for triggering the transitions shown in Scheme 3.1. A weak perturbation, P', only slightly distorts the zero-order electronic (or vibrational or spin) wave function. This distortion can be interpreted as a "mixing" of the wave functions of the initial state (Ψ_1) and the final state (Ψ_2) as the result of a perturbation whose operator is $P_{1\rightarrow 2}$. As a result of the perturbation-induced mixing, Ψ_1 now contains a certain amount of Ψ_2. That is, there is a resonance between Ψ_1 and Ψ_2. This resonance can be expressed in terms of Eq. 3.4, where λ is a measure of the amount of Ψ_2 that is mixed into Ψ_1 as the result of the perturbation $P_{1\rightarrow 2}$. The value of λ can vary from 0 to 1.

$$\underbrace{\Psi_1 + P_{1\rightarrow 2}}_{\substack{\text{Initial state} \\ \text{+ interaction}}} \quad \rightarrow \quad \underbrace{\Psi_1 \pm \lambda\Psi_2}_{\text{Resonance}} \quad \rightarrow \quad \underbrace{\Psi_2}_{\substack{\text{Transition} \\ \text{to final state}}} \qquad (3.4)$$

The basic idea of perturbation theory is that there is a finite probability that after the weak perturbation $P_{1\rightarrow 2}$ has been applied to Ψ_1, the system will be able to achieve resonance so that a certain amount of Ψ_2 will be mixed into Ψ_1. The *mixing coefficient*, λ, is a measure of the extent of distortion of Ψ_1 toward Ψ_2 produced by the perturbation. Before the resonance, the system is approximated as "purely" Ψ_1, and after the resonance and relaxation, the system is approximated as "purely" Ψ_2. In Eq. 3.5, we show how λ is computed.

The modification of the approximate initial wave function Ψ_1 to make it look like the final state Ψ_2 is thus achieved by mixing into it other wave functions of the zero-order system in appropriate proportions through the interactions represented by the operator $P_{1\rightarrow 2}$. If the correct operator (interaction) $P_{1\rightarrow 2}$ has been selected and the proper conditions are present (i.e., the conservation laws are obeyed), the mixing makes Ψ_2 "look like" Ψ_1. The more the mixing makes Ψ_2 look like Ψ_1, the larger the mixing coefficient (λ) becomes, and the faster and more probable is the $\Psi_1 \rightarrow \Psi_2$ transition.

According to perturbation theory, the first-order correction of a wave function Ψ_1 is given by the mixing coefficient λ (Eq. 3.4), which is directly proportional to the strength of the perturbation P', and is inversely proportional to the separation of the energy between the interacting states (ΔE_{12}) being mixed (Eq. 3.5). The first-order wave function is obtained by multiplying Ψ_2 by λ and adding the result to Ψ_1 (Eq. 3.6).

$$\lambda = (\text{strength of the perturbation } P')/(\text{energy of separation of } \Psi_1 \text{ and } \Psi_2) \quad (3.5a)$$

$$\lambda = <\Psi_1|P'|\Psi_2> /\Delta E_{12} \quad (3.5b)$$

$$\Psi_1' \text{ (first-order wave function)} = \Psi_1 + \lambda\Psi_2 \text{ (zero-order wave function)}. \quad (3.6)$$

Based on Eqs. 3.5 and 3.6, there are two general rules of perturbation theory that provide us with very important quantum intuition for understanding the role of interactions in promoting effective and fast transitions between electronic states: *(1) the stronger the perturbation P', the stronger the mixing and distortion of the initial wave function Ψ_1, and (2) the smaller the energy separation ΔE between the two interacting wave functions, the stronger the mixing.* For two states (Ψ_1 and Ψ_2) that are widely separated in energy relative to the perturbation, therefore, the system is generally expected to be weakly responsive to any perturbation, and mixing is implausible (in other words, it will be relatively difficult to make Ψ_1 and Ψ_2 look alike, even for a strong perturbation). When a transition involves two states that are very close in energy, on the other hand, the initial system is very sensitive to perturbations and may be strongly perturbed, even by weak perturbations. When Ψ_1 and Ψ_2 have very similar energies, the two systems become "easy to mix" if the correct perturbation is available to operate on the system. The two states (Ψ_1 and Ψ_2) easily transform one into the other even through weak perturbations. This ease of mixing for states that are close in energy is a characteristic *resonance* feature of waves. Classically, it is easier to distort a weak spring (Fig. 2.4, right), but more difficult to distort a strong spring (Fig. 2.4, left) with external perturbations. The same situation holds for a quantum mechanical spring (e.g., vibrating electrons or atoms), because in quantum mechanics a stiff spring has widely separated energy levels (Fig. 2.5, left) and is difficult to perturb (i.e., its wave functions are relatively difficult to mix) compared to a soft spring (Fig. 2.5, right), which has closely spaced energy levels and is easier to perturb (i.e., its wave functions are relatively easy to mix).

The rates of "fully allowed" transitions between electronic states are limited only by the zero-point electronic motion change involved in the transition, provided the nuclear and spin configurations remain constant. Recall from Chapter 1 that we used the time scale for the completion of an orbit by a Bohr electron as a benchmark for the fastest rate of electronic motion. An electron completes its Bohr orbits at a rate of $\sim 10^{15}$–10^{16} s^{-1}, so this sets an approximate upper limit to the zero-point motion of an electronic system. However, if the nuclear and/or spin configurations change during a "fully allowed" electronic transition, the transition will be "rate limited" by the time it takes to change the nuclear or spin configuration, not by the time it takes the electron to make a zero-point motion. In other words, the electronic part of Ψ_1 may have a rate of 10^{15}–10^{16} s^{-1} in "looking like" Ψ_2, but the rate of the $\Psi_1 \rightarrow \Psi_2$ transition may be limited by the time it takes to make the vibrations or spins in the final state (Ψ_2) "look like" those in the initial state (Ψ_1). Such a view of transition rates provides benchmarks for the maximal rates of various "allowed" transitions. When a rate is slower than the maximal rate, electronic shape and motion, vibrational shape and motion, or spin configuration and motion may serve as kinetic "bottlenecks" in determining transition rates.

To understand molecular kinetics, we must obtain the rate constants (k) for the transitions between the electronic, vibrational, and spin states. It is convenient to consider the observed rate constant for the transition between two states (k_{obs}) in terms of the maximum *possible* rate constant (k_{max}^0) (the zero-point motions determine the rate constant) and the product of the prohibition factors (f) for the electronic, vibrational, and spin aspects of the transitions. In Eq. 3.7, for a given transition from Ψ_1 to Ψ_2, f_e is the prohibition factor associated with the electronic change (the orbital configuration change), f_v is the prohibition factor associated with the nuclear configuration change (usually described as a vibrational change in position or motion), and f_S is the prohibition factor associated with a spin configuration change ($f_S = 1$ for transitions for which there are no spin changes).

$$
\begin{array}{ccccc}
\text{Observed} & & \text{Zero-Point Motion-} & & \\
\text{Rate Constant} & & \text{Limited Rate Constant} & \text{"Fully Allowed Rate"} & \qquad (3.7)\\[4pt]
\underbrace{k_{obs}}_{\substack{\text{Prohibition to maximal}\\\text{caused by "selection rules"}}} & = & k_{max}^0 & \times & \underbrace{f_e \times f_v \times f_S}_{\substack{\text{Prohibition factors due to changes in}\\\text{electronic, nuclear, or spin configuration}}}
\end{array}
$$

In most cases k_{obs} is much smaller than k_{max}^0. When *weak* interactions between zero-order states trigger the transition, the rate of $\Psi_1 \rightarrow \Psi_2$ transitions is given by Fermi's golden rule[4] (Eq. 3.8), where ρ is the number of states of Ψ_2 that are of the same energy as Ψ_1 and are capable of being in resonance with Ψ_1 through the perturbation $P'_{1\rightarrow2}$. The term ρ is referred to as the *density of states that are capable of effectively mixing Ψ_1 with Ψ_2*. These are accessible states for which Ψ_1 and Ψ_2 can achieve the same energy during the time scale of the interaction that mixes the states. Pictorially, the higher the density of states for a transition that is capable of responding to a perturbation P', the more statistically probable that a transition will be triggered by any interaction. Fermi's golden rule applies to electronic, vibrational, and spin transitions that are triggered by interactions that are *weak* relative to the energy separations of the states involved. Consequently, the form of Eq. 3.8 shows up a number of times when discussing the rates of transitions that are induced by weak electronic, vibrational, or spin interactions. These weak perturbations include the interaction of the electromagnetic field with the electrons (responsible for the absorption and emission of light) and the interactions between the HOs and LUs of molecules that lead to chemical reaction and energy transfer.

$$k_{obs} \sim \rho[<\Psi_1|P'_{1\rightarrow2}|\Psi_2>]^2 \quad \text{Fermi's golden rule} \qquad (3.8)$$

For example, the distortion of the electron cloud of a molecule caused by the (weak) initial interaction of an electromagnetic field can be interpreted quantum mechanically as a mixing of the wave function of the ground state (R) with the wave function of one or more excited states (*R) to produce a perturbed wave function of R. Since the perturbed wave function of R has wave functions of excited states mixed into it, there is a finite probability of finding the system in an excited state, *R. In the case of an interaction with the electromagnetic field, the matrix element corresponds to the transition dipole moment (Chapter 4). This matrix element may be visualized as being a measure of the extent of the oscillating movement of negative electrical charge along the positive nuclear framework of the molecule as the result of the molecule's interaction with the electric component of the oscillating electromagnetic field. If the extent of the oscillation is great (i.e., if a large transition dipole is generated by the interaction), then the molecule's electrons and the electromagnetic field interact strongly (i.e., the magnitude of the matrix element is large) and the transition rate is high.

Fermi's golden rule provides a basis for the transitions of "selection rules" that are triggered by weak interactions; namely, if the value of the matrix element of Eq. 3.8 is zero (at a defined level of approximation), then the transition is zero and the transition is "forbidden."

In Section 2.3 (Eq. 2.4), we described how to use the Born–Oppenheimer approximation to approximate the true (but mathematically unattainable) wave function Ψ into a product of an electronic (ψ), vibrational (χ), and spin ($\mathbf{S}$) wave function. For transitions that do not involve a change in spin (i.e., $\mathbf{S_1} = \mathbf{S_2}$), electronic spin does not provide any prohibition on k_{obs}. In this case, *the rate of transition between Ψ_1 and Ψ_2 is limited by either the time it takes to make the electronic wave function ψ_1 look like ψ_2 or the time it takes for the vibrational wave function χ_1 to look like χ_2.*

For organic molecules, the most important perturbation for "mixing" electronic wave functions that initially do not look alike is vibrational nuclear motion that is coupled to the orbital motion of the electrons (i.e., vibronic coupling). Let the operator corresponding to vibronic coupling be called P_{vib}; the matrix element for the perturbation that vibrationally mixes ψ_1 and ψ_2 is then given by $< \psi_1|P_{vib}|\psi_2 >$. It is most important that the electronic wave function of ψ_1 be distorted into a shape that looks like ψ_2 by some vibration that couples the two states. It is this distortion, caused by molecular vibrations, that makes the two wave functions "look alike" and that will allow the transition to occur. When this is true, we need only consider the magnitude of the overlap integral of the vibrational wave functions, $< \chi_1|\chi_2 >$. The square of the vibrational overlap, $< \chi_1|\chi_2 >^2$, in Eq. 3.9 is called the *Franck–Condon (FC) factor*. The Franck–Condon factor is a measure of the overlap of the vibrational wave functions of the initial and final states and is mathematically similar to the electron orbital overlap integral (Section 2.14, Eq. 2.20). We show how to obtain a qualitative visualization of the FC factor in Sections 3.10 and 3.11. To summarize, using Fermi's golden rule (Eq. 3.8), k_{obs} is proportional to the square of the product of the vibronic coupling matrix element and the vibrational overlap matrix elements associated with the $\psi_1 \rightarrow \psi_2$ transition, as shown in Eq. 3.9.

$$k_{obs} = \underbrace{\left[\frac{k_{max}^0 <\psi_1|P_{vib}|\psi_2>^2}{\Delta E_{12}^2}\right]}_{\text{Vibrational coupling}} \times \underbrace{\left[<\chi_1|\chi_2>^2\right]}_{\substack{\text{Vibrational overlap} \\ \text{Franck-Condon factors}}} \qquad (3.9)$$

When the (radiationless or radiative) transition $\Psi_1 \rightarrow \Psi_2$ involves a change in the spin ($S_1 \neq S_2$), which perturbation is most likely to couple states of different spin? In molecular organic photochemistry, the most important transitions involving a change in spin are radiative or radiationless singlet–triplet or triplet–singlet transitions. For organic molecules, the most important perturbation available to make a pair of parallel triplet spins ($\uparrow\uparrow$) look like a pair of antiparallel singlet spins ($\uparrow\downarrow$) is the coupling of the electron spin motion with the electron orbital motion (termed *spin–orbit coupling*), which takes one of the parallel electron spins of a singlet state ($\uparrow\uparrow$) and twists it or flips it, making the spins antiparallel ($\uparrow\downarrow$). The terms "parallel" and "antiparallel" are approximations used here for simplicity; recall from Section 2.28 that the 3D representation of spin vectors is a more accurate description and is required when we analyze spin transitions in Section 3.12.

Let us label the operator that induced spin–orbit coupling as P_{so}, and the matrix element for the transition as $< \psi_1 | P_{so} | \psi_2 >$. This matrix element is a measure of the strength (or the energy) of the spin–orbit interactions. For simplicity, the spin wave functions, S_1 and S_2, need not be considered explicitly: recall from Table 2.1 that we use the symbols α and β to represent the wave functions of a spin up ($\uparrow$) and spin down ($\downarrow$), respectively. For transitions that involve a change in spin, we can modify Eq. 3.9 to produce Eq. 3.10, which includes the spin change prohibition.

$$k_{obs} = \underbrace{\left[\frac{k_{max}^0 < \psi_1 | P_{vib} | \psi_2 >^2}{\Delta E_{12}^2} \right]}_{\text{Vibrational coupling}} \times \underbrace{\left[\frac{< \psi_1 | P_{so} | \psi_2 >^2}{\Delta E_{12}^2} \right]}_{\text{Spin–orbital coupling}} \times \underbrace{\left[< \chi_1 | \chi_2 >^2 \right]}_{\substack{\text{Vibrational overlap} \\ \text{Franck-Condon factors}}}$$

(3.10)

3.6 The Spirit of Selection Rules for Transition Probabilities

As a starting approximation, a $\Psi_1 \rightarrow \Psi_2$ transition is "forbidden" (implausible) if the value of the matrix element equals zero, and the transition is "allowed" if the value of the matrix element is finite. The matrix element for a transition probability for the $\Psi_1 \rightarrow \Psi_2$ transition, $< \Psi_1 | P_{1\rightarrow 2} | \Psi_2 >$, may be calculated for a certain set of zero-order assumptions that assign an initial idealized symmetry for the wave functions for the electrons, nuclei, and spins (ψ, χ, and S) and a selected operator ($P_{1\rightarrow 2}$) that is assumed to trigger the transition. If the computed matrix element for the transition probability equals zero, the transition is strictly forbidden in the zero-order approximation. The plausibility of the transition in first order will depend on whether an interaction (e.g., vibronic or spin–orbit) exists that can overcome the forbidden character of the zero-order approximation.

When the approximate wave functions Ψ_1 and Ψ_2 possess a more realistic non-ideal symmetry, or when previously ignored forces and a different operator have been included, a *new* calculation of the matrix element may yield a first-order correction, and the value of the matrix element generally will be nonzero. If the transition probability corresponding to the matrix element is still small (e.g., < 1% of the maximum transition probability), then the process is "weakly allowed" (or *implausible*) in the sense that the rate of the process is not expected to compete with other fast transitions from the initial state. If the matrix element computed by the new calculation for the transition is large (e.g., close to the maximum transition rate, f, in Eq. 3.7), then the transition can be classified as "strongly allowed" (or *probable*) in the sense that its rate is expected to be among the fastest of the plausible transitions. Such qualitative descriptions can provide only a rough feeling for transition probabilities. Indeed, sometimes the breakdown of selection rules is so severe that the magnitude of the "forbidden" transition probability approaches that of the "allowed" transition probability. When this occurs, we have selected a poor zero-order starting point (the wave function Ψ_0 or the operator P) for our evaluation of the transition probability.

3.7   Nuclear Vibrational Motion As a Trigger for Electronic Transitions. Vibronic Coupling and Vibronic States: The Effect of Nuclear Motion on Electronic Energy and Electronic Structure[5]

For *spin-allowed electronic* transitions between Ψ_1 and Ψ_2, we need to devise a paradigm for evaluating the matrix elements for vibrational coupling of the electronic states of Ψ_1 and Ψ_2 (i.e., vibronic coupling). The goal is to estimate how the vibrational wave functions (Section 2.19) for spin-allowed transitions influence the rate of both radiative and radiationless electronic transitions (Scheme 3.1). The FC factors, $< \chi_1 | \chi_2 >^2$, are a measure of the similarity of the vibrational wave functions of Ψ_1 and Ψ_2 and are critical in determining whether a transition is allowed or forbidden in first order.

The Born–Oppenheimer approximation (Section 2.3, Eq. 2.4) allows the generation of a zero-order description of the electronic structure and electronic energy of a molecule, based on an assumed frozen, *nonvibrating* nuclear geometry. We must consider the effect of nuclear vibrational motion on the electronic structure and electronic energy of a molecule and how this vibrational motion can serve as a perturbation that mixes electronic wave functions and how this mixing can induce transitions between electronic states. Our goal is to replace the pure, classical "vibrationless" molecule with a vibrating molecule and to be able to visualize how this motion will modify our zero-order model. We call the states of a vibrating molecule "vibronic" states rather than pure "electronic" states, because *vibrations are constantly serving as a source of mixing of electronic states, especially electronic states that are of similar or identical energy in zero order.* The basic concept is that the vibrations of a molecule will distort the zero-order electronic wave function only slightly, therefore serving as a weak perturbation on the approximate wave function, but that certain vibrations will distort the approximate function so that it looks like the wave function of other electronic states to which transitions may occur. Again assuming only weak interactions, the *energy* E_V of a vibronic perturbation (from perturbation theory) is given by Eq. 3.11a. From Fermi's golden rule for weak interactions (Eq. 3.8), the *rate* of the transition from $\Psi_1 \rightarrow \Psi_2$ is proportional to the square of the matrix element (where $P'_{1 \rightarrow 2}$ is replaced by P_{vib}) for the transition multiplied by the density of states (ρ) capable of being mixed by P_{vib} during the time scale of the perturbation (Eq. 3.11b).

$$E_V = \; <\Psi_1|P_{vib}|\Psi_2>^2 / \Delta E_{12} \tag{3.11a}$$

$$k_{obs} \sim \rho \; <\Psi_1|P_{vib}|\Psi_2>^2 \tag{3.11b}$$

In Eq. 3.11a, Ψ_1 and Ψ_2 are two pure zero-order electronic states that are "mixed" by P_{vib} and ΔE_{12} is the energy difference between the zero-order electronic states that are being mixed by the vibronic interaction.

We have described interactions that allow the use of Fermi's golden rule as "weak" perturbations. At this point, we provide some numerical benchmarks for the values of the energy separations (ΔE_{12}) that are involved in the vibronic mixing of electronic states. What do we mean by a small or large value of ΔE_{12}? We know from perturbation theory (Eqs. 3.5 and 3.6) that if ΔE_{12} is large, the mixing (the value of λ) of the states will be small, and if ΔE_{12} is small, the mixing of the two states will be large. Intuitively, what we mean by small or large has to do with how the vibronic mixing energy (E_V) compares to the energy separation (ΔE_{12}) of the electronic states that are being mixed by vibronic interactions. If E_V is small (i.e., a few percent) compared to ΔE_{12}, the mixing of the states is expected to be small. In general, for organic molecules, the value of E_V is on the order of vibrational energies (Section 2.19), which range from ~ 10 kcal mol^{-1} for X—H stretching vibrations to ~ 1 kcal mol^{-1} for C—C—C bending vibrations. As a result, X—H vibrations can be very effective in mixing electronic states.

Vibronic interactions do not significantly mix electronic states whose energy separation (ΔE_{12}) is 50 kcal mol^{-1} or greater. The energy gaps between the ground state (R) and lowest excited states (*R) of most organic molecules are > 50 kcal mol^{-1}, so the vibronic mixing of R and *R is weak. This weak vibronic coupling is the reason that the Born–Oppenheimer approximation works so well for ground-state molecules; that is, vibrations in the ground state do not mix electronically excited states very effectively because the electronic energy gap between R and *R is much larger than the vibrational energies.

However, vibronic interactions are much more likely to be significant in mixing zero-order electronically *excited* states (*R$_2$ and *R$_1$), since ΔE_{12} between excited states is often on the order of only several kilocalories per mole or less, so excited states are typically packed together with much smaller energy gaps than the energy gaps that separate *R from R. *When electronic states are separated by small energy gaps, the electronic energy and electronic structure of the states may vary considerably during a vibration.* Thus, vibrational motion of the appropriate type (e.g., a motion that couples two electronic states that are close in energy) can be very effective in mixing excited states. Such effects are of great importance in triggering and, therefore, determining the rates of transitions from excited states.

The importance of similar energy states in determining the rates of transitions is apparent in Fermi's golden rule, where k_{obs} depends on the density of states (ρ) that have the same energy in both Ψ_1 and Ψ_2 (Eqs. 3.8 and 3.11b). We also conclude that

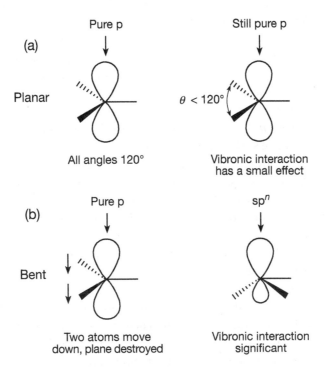

Figure 3.1   The effect of vibronic motion on the hybridization of a p orbital.

electronic transitions from any *R, both radiative and radiationless, depend on the ability of certain kinds of vibrations to couple the electronic wave function of *R with the wave functions of other states, particularly with other excited states.

As a simple exemplar (Fig. 3.1) of the effect of vibrational motion on the electronic orbital energy, consider the vibrations of a carbon atom that is bound to three other atoms (e.g., a methyl group as a radical, anion, or carbonium ion). When the system is planar and the angles between the atoms are $120°$, the carbon atom is a pure sp^2 hybrid and possesses a p orbital that serves as a "free valence" orbital. What happens to the shape and energy of this free valence orbital as the molecule vibrates? If the vibrations do not destroy the planar geometry (i.e., the H—C—H angle changes, but the system remains planar), the hybridization remains sp^2 and the spatial distribution of the free valence orbital above and below the plane must be *identical* because of the symmetry plane that contains all four atoms (Fig. 3.1a). In other words, if we put electrons into the free valence orbital, the electron density would have to be the same above and below the symmetry plane, since all conceivable interactions on one side of the plane are identical to those on the other side. In effect, *the p orbital remains essentially "pure p" during the in-plane bending vibration. Furthermore, since the system remains planar during the vibration, the energy of the orbital is not expected to change significantly as it executes the in-plane vibration.* We say that there is weak vibronic coupling of electronic (p orbital) and vibrational (in-plane) motions so that distortion of the p orbital induced by vibrations is small.

Now, consider a bending (umbrella-flipping) vibration that *breaks* the planar symmetry of the molecule and causes a change in the hybridization of the carbon atom (Fig. 3.1b). Intuitively, we expect the "pure p" orbital to change its shape in response to the fact that more electron density (due to the electrons in the bonds) is on one side of the plane. We say that a *rehybridization* of the carbon atom occurs, and that the "pure p" orbital begins to take on some s character; that is, *the out-of-plane vibration converts the pure p orbital into an* sp^n *orbital*, where n is a measure of the "p character" remaining. Since an s orbital is considerably lower in energy than a pure p orbital, an sp^n orbital will be lower in energy than a p orbital because the sp^n orbital has acquired some s character. Thus, the mixing due to out-of-plane vibrational motion can change the energy of the free valence orbital significantly.

In the extreme situation for sp^n, where $n = 3$, the out-of-plane vibration causes a continual oscillating electronic change, p (planar) $\leftrightarrow$ sp^3 (pyramidal), as pyramidal shapes interconvert through the planar shape. A significant vibronic coupling of the electronic and nuclear motion occurs due to this vibration, causing the value of n to oscillate between 2 and 3. Now, if the initial state (Ψ_1) is a pure p wave function and the final state (Ψ_2) is a pure sp^3 state, the out-of-plane vibrational motion makes Ψ_1 "look like" Ψ_2, but the in-plane vibrational motion does not, because the in-plane bending vibration does not introduce any s character into the p orbital. In other words, the out-of-plane vibrational motion "mixes" the hybridization of the free valence orbital, but the in-plane vibrational motion does not. In a convenient shorthand, we can write Ψ_1(p, planar) $\leftrightarrow$ Ψ_2(sp^3, pyramidal). If the operator that describes the in-plane (ip) vibronic interaction is called P_{ip} and the operator that describes the out-of-plane (op) vibronic interaction is called P_{op}, then the matrix element for in-plane vibronic mixing, $< \Psi_1|P_{ip}|\Psi_2 >$, equals 0, and the matrix element for out-of-plane vibronic mixing, $< \Psi_1|P_{op}|\Psi_2 >$, is finite in this case.

To summarize, some, but not all, vibrations are capable of perturbing the electronic wave functions and the electronic energy of zero-order electronic states. The energy difference of the zero-order electronic levels and vibronic levels may be small relative to the total electronic energy, yet the matrix element $< \Psi_1|P_{vib}|\Psi_2 >$ may "provide a first-order mechanism" for the transition from one vibronic state to another, even though the electronic transition is strictly forbidden (i.e., $< \Psi_1|P|\Psi_2 >= 0$) in the zero-order approximation.

3.8   The Effect of Vibrations on Transitions between Electronic States: The Franck–Condon Principle

The rates of transitions between electronic states ($\Psi_1 \rightarrow \Psi_2$) can be limited by either the rate at which the electrons in Ψ_1 can adjust to the nuclear geometry of Ψ_2, or the rate at which the nuclear geometry of Ψ_1 can adjust to the nuclear geometry of Ψ_2.

The Born–Oppenheimer approximation (Section 2.3) assumes that electron motion is so much faster than nuclear motion that the electrons "instantly" adjust to any change in the position of the nuclei in space. Since an electron jump between orbitals

(Section 1.13) generally takes $\sim 10^{-15}$–10^{-16} s to occur, whereas nuclear vibrations take $\sim 10^{-13}$–10^{-14} s to occur, the electron jump is generally much faster and will not be rate determining for transitions between two electronic states, $\Psi_1 \to \Psi_2$. Thus, the transition rate between electronic states (of the same spin) is limited by the ability of the system to adjust to the nuclear configuration and motion *after* the change in the electronic distribution of Ψ_1 to that of Ψ_2. *The rate of transitions induced by vibrations (nuclear motion) depends not only on how much the electronic distributions of the initial and final states look alike but also on how much the nuclear configuration and motion in the initial and final states look alike.*

Expressed in classical terms, the FC principle states that *because nuclei are much more massive than electrons (the mass of a proton is ~ 1000 times the mass of an electron), an electronic transition from one orbital to another takes place while the massive, higher-inertia nuclei are essentially stationary.* This means that, at the instant that a radiationless or radiative transition takes place between Ψ_1 and Ψ_2 (e.g., for any of the transitions shown in Scheme 3.1), the nuclear geometry of the massive nuclei momentarily remains fixed while the new electron configuration readjusts from that Ψ_1 to that of Ψ_2. After completion of the electronic transition, the nuclei experience the new electronic negative force field of Ψ_2 and begin to move and swing back and forth from the geometry of Ψ_1 until they adjust their nuclear geometry to that of Ψ_2. From the FC principle, we conclude that the conversion of electronic energy into vibrational energy is likely to be the rate-determining step in an electronic transition between states of significantly different nuclear geometry (but of the same spin).

Expressed in quantum mechanical terms, the FC principle states that *the most probable transitions between electronic states occur when the wave function of the initial vibrational state (χ_1) most closely resembles the wave function of the final vibrational state (χ_2).* In analogy to the orbital overlap integral $< \psi_1 | \psi_2 >$ (Section 2.14), which defines the extent of the mathematical orbital overlap for a pair of electronic wave functions or a set of orbitals, we define the vibrational overlap integral in terms of the extent of overlap for a pair of vibrational wave functions (χ_1 and χ_2) and use the symbol $< \chi_1 | \chi_2 >$ to indicate the degree of the overlap integral of the two vibrational wave functions, χ_1 and χ_2. Since two wave functions generally have a greater resemblance (i.e., look more alike) when the vibrational overlap integral $< \chi_1 | \chi_2 >$ is closer to 1 (the maximum value for complete overlap), the larger the value of the integral, the more probable the vibronic transition. From Eq. 3.9, the rate constant (k_{obs}) for the $\Psi_1 \to \Psi_2$ transition is proportional to $< \chi_1 | \chi_2 >^2$. We can now understand why $< \chi_1 | \chi_2 >^2$ in Eq. 3.9 is called the "Franck–Condon" factor.

In the following sections, we demonstrate that the FC principle provides a useful visualization of both radiative and radiationless electronic transitions. For radiative transitions, the *motions and geometries* of nuclei do not change during the time it takes for a photon to "interact with" and to be "absorbed," thus causing an electron to jump from one orbital to another. For radiationless transitions, nuclear *motions and geometries* do not change during the time it takes an electron to jump from one orbital to another.

3.9   A Classical and Semiclassical Harmonic Oscillator Model of the Franck–Condon Principle for Radiative Transitions ($R + h\nu \rightarrow {}^*R$ and ${}^*R \rightarrow R + h\nu$)

In the classical harmonic oscillator approximation (Section 2.16), the energies of the vibrations of diatomic molecules were discussed in terms of a parabola in which the potential energy (PE) of the system was displayed as a function of the displacement (Δr) from the equilibrium separation of the atoms (Eq. 2.24 and Figs. 2.3 and 2.4). The harmonic oscillator approximation for molecular vibrations applies to both ground states (R) and excited states (*R) and can be used as a starting point for both radiationless and radiative photophysical transitions. First, let us consider how the FC principle and FC factors apply to a radiative transition between two states in terms of the harmonic oscillator model.[6]

Figure 3.2 shows classical PE curves for a diatomic molecule (X—Y) that behaves as a harmonic oscillator. The top half of Fig. 3.2 is a representation of a classical harmonic oscillator for which one of the vibrating masses (X) is very large (X is attached to the left of the spring), and the other vibrating mass (Y) is much lighter (Y is attached to the right of the spring). This diatomic molecule can be viewed as a vibrating ball attached to a spring, which is affixed to a wall. This would be analogous to a light atom (the ball) bonded to a much heavier atom (the wall), for example, a C—H vibration where the carbon atom is analogous to the massive wall and the hydrogen atom is analogous to the light ball. Most of the motion of the two atoms is due to the movement in space of the lighter particle (the H atom).

Three PE curves are shown in Fig. 3.2 for three different situations with respect to an initial nuclear geometry of an R state relative to that of an *R state. In Fig. 3.2a, the equilibrium nuclear separation (r_{XY}) of R is essentially identical to the equilibrium nuclear separation (${}^*r_{XY}$) of the electronically excited *R molecule. In Fig. 3.2b, r_{XY} of R is slightly different from ${}^*r_{XY}$ of the *R molecule, with ${}^*r_{XY}$ being slightly longer because of (an assumed) slightly weaker bond resulting from electronic excitation and the placement of an electron in an antibonding orbital. In Fig. 3.2c, r_{XY} of R is considerably different from ${}^*r_{XY}$ of the *R molecule, with ${}^*r_{XY}$ being considerably longer because of the much weaker bond resulting from electronic excitation to put an electron into an antibonding orbital. The difference in excess vibrational energy (ΔE_{vib}) increases as the difference ($\Delta r = |{}^*r_{XY} - r_{XY}|$) in the equilibrium separations of R and *R increases. It is zero for the case in Fig. 3.2a, small for the case in Fig. 3.2b, and large for the case in Fig. 3.2c.

For each case in Fig. 3.2, a line is drawn vertically from the initial R state and intersects the upper PE curve at the point that will be the turning point in the *R state. This line represents a *vertical* electronic transition from R to *R. Radiative transitions are called vertical transitions with respect to nuclear geometry, since the nuclear geometry (r_{XY}, the horizontal axis) is fixed during the electronic transition. The length of the line representing the vertical electronic transition corresponds to the difference in energy between R and *R that is absorbed in the transition, that is, the energy of the absorbed photon is $|E_R - E_{*R}| = \Delta E = h\nu$.

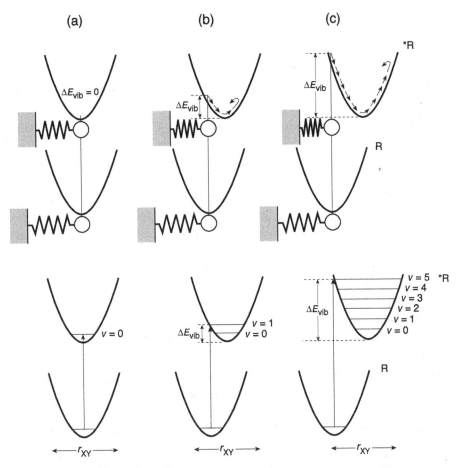

Figure 3.2   A mechanical representation of the Franck–Condon principle for radiative transitions of a diatomic molecule, XY. The motion of a point representing the vibrational motion of two atoms is shown by a sequence of arrows along the PE curve for the vibration in the top set of curves.

Now, let us consider how the FC principle influences a radiative $HO + h\nu \rightarrow LU$ orbital transition that takes R to *R. The time scale for photon absorption is on the order of 10^{-15}–10^{-16} s. According to the FC principle, the nuclear geometry (i.e., the separation of the two atoms) does not change during the time scale of an electronic transition or orbital jump; that is, immediately after the electronic transition, $r_{XY} = {^*r_{XY}}$. Thus, the geometry produced at the instance of the electronic transition on the upper surface by a radiative transition from a ground R state to an *R state is governed by the relative positions of the PE surfaces controlling the vibrational motion of R and *R.

If, for simplicity, we assume that the PE curves have similar shapes, and that the minimum of one curve lies directly over the minimum of the other (Fig. 3.2a), the Franck–Condon principle states that the most probable radiative electronic transitions would be from an initial state that has a separation of r_{XY} in R that is identical to the separation (r_{XY}) of the excited state *R. Since the two curves are assumed to lie exactly over one another, the most favored Franck–Condon transition will occur from the minimum of the ground surface to the minimum of the excited surface, that is, $R(v = 0) + hv \rightarrow {}^*R(v = 0)$. This situation is typical of the absorption of light to induce a $\pi \rightarrow \pi^*$ transition of aromatic hydrocarbons that have many bonding π electrons, so that the excitation of one π electron to a π^* orbital does not significantly change the structure of *R compared to R (Chapter 4). To a good approximation, therefore, we may regard the *absorption* of a photon as occurring from the *most probable* nuclear configuration of the ground state (R), which is the static, equilibrium arrangement of the nuclei in the classical model and is characterized by a separation r_{XY}. Based on Fig. 3.2, there is no excess vibrational energy produced by the transition.

By using the exemplar of Fig. 3.2b, let us consider the absorption of light from the HO of formaldehyde (the n_O orbital) to its LU (the π^* orbital). At the instance of completion of the electronic transition is complete, the nuclei are still in the same ground–state equilibrium (planar) geometry that they were before the transition, because the electronic jump occurs much faster than the nuclear vibrations. However, as the result of the orbital transition and the occupation of a π^* orbital, the electron density of *R about the nuclei is different from the electron density of R about the nuclei. Therefore, *R relaxes to a new geometry (that turns out to be a pyramidally shaped $H_2C=O$).

Figure 3.2c is an exemplar of a system that undergoes a very large structural change upon going from R to *R. The $\pi \rightarrow \pi^*$ excitation of ethylene is one such example. Although the equilibrium geometry of the ground state for ethylene (R) is planar, the equilibrium geometry of the excited state for ethylene (*R) is strongly twisted, leading to a large change in the equilibrium geometry of *R compared to R.

Now, let us consider the effect of the Franck–Condon principle on the excess of vibrational energy that is produced in *R upon absorption of a photon. In Fig. 3.2a, where the initial and final geometries of R and *R are assumed to be identical, there is no significant change in vibrational properties resulting from electronic excitation $(R + hv \rightarrow {}^*R)$, so *R is produced with no excess vibrational energy. However, in Fig. 3.2b and c, the electronic transition initially produces an *R state that is both a *vibrationally excited and an electronically excited species* as the result of the new force field experienced by the originally stationary nuclei of R. A few femtoseconds after the $R \rightarrow {}^*R$ transition, the atoms in *R will suddenly burst into a new vibrational motion in response to the new electronic force field of *R.

In the case of the n,π^* state of $H_2C=O$, an electron has been promoted into a π^* orbital, which will tend to make the C—O bond begin to vibrate and to stretch and become longer. This new force, provided by the sudden perturbation of the removal of an n electron and of the creation of a π^* electron, will induce a vibration along the C—O bond. The new vibrational motion of the molecule in *R(n,π^*) may be described in terms of a *representative point*, which represents the value of the internuclear separation and is constrained to follow the PE curve and execute harmonic oscillation.

The excess vibrational motion produced by absorption of a photon is indicated by the set of arrows on the PE surface in Fig. 3.2b and c. The maximum velocity of the motion of the representative point on the PE surface depends on the excess vibrational kinetic energy produced upon electronic excitation. The greater the excess vibrational motion produced by Franck–Condon excitation, the greater the velocity of vibrational motion produced immediately after electronic excitation. In the case of the $\pi \rightarrow \pi^*$ transition of ethylene, the loss of a π electron and the creation of a π^* electron strongly reduces the C=C bonding and essentially breaks the π double bond and creates a C—C single bond in *R. The new electronic distribution in the π,π^* state favors a twisting about the essentially C—C single bond and an equilibrium geometry that favors the two CH_2 groups perpendicular to each other rather than in the same plane (this situation is discussed in detail in Chapter 6).

For the classical case of Fig. 3.2b and c, *it follows that the original nuclear geometry of the ground state is a turning point of the new vibrational motion in the excited state, and that vibrational energy is stored by the molecule in the excited state.* Since the total energy of a harmonic oscillation is constant in the absence of friction, any PE lost as the spring decompresses is turned into the kinetic energy (KE) of the two masses attached to the spring, which sets the representative point into harmonic oscillation. Therefore, the PE at the turning points, E_{vib}, determines the energy at all displacements for that mode of oscillation. The greater the amount of vibrational energy that is produced in *R upon photoexcitation, the greater the amplitude of the vibration of *R.

Now, let us examine a "semiclassical" model (Fig. 3.2, bottom) that considers the effect of quantization of the vibrational levels of the harmonic oscillator and of zero-point motion on the classical model for a radiative electronic transition (we will consider the wave character of vibrations in Section 3.10). In Section 2.18, we learned that one of the effects of quantization on the harmonic oscillator is that only certain vibrational energies are allowed. In a semiclassical model, therefore, the classical PE curves must be replaced by PE curves displaying the quantized vibrational levels, each with a vibrational quantum number, $v = 0$. For example, Fig. 3.2a (bottom) shows the ground-state PE curve with a horizontal level corresponding to the $v = 0$ vibrational level. This level corresponds to a small range of nuclear geometries, determined by the zero-point vibrational motion, with the classical equilibrium geometry at the center of the vibration. *Radiative transitions from $v = 0$ will therefore not be initiated from a single geometry but will be initiated from a range of geometries that are explored during the zero-point motion of the vibration.* In Fig. 3.2a (bottom), the most probable transition is from the $v = 0$ level of R to the $v = 0$ level of *R. In Fig. 3.2b, the most probable transition is from the $v = 0$ level of R to the $v = 1$ level of *R. In Fig. 3.2c, the most probable transition is from the $v = 0$ of R to the $v = 5$ level of *R. As we go from Fig. 3.2a to b to c, the amount of excess vibrational excitation produced in *R by the electronic transition increases.

The final step in our visualization of the FC principle and radiative transitions is to determine how to picture the wave functions corresponding to the vibrational levels of R and *R. From this picture, we will see that the mathematical form of the vibrational wave functions of R and *R controls the probability of both radiative and radiationless electronic transitions between vibrational levels.

3.10   A Quantum Mechanical Interpretation of the Franck–Condon Principle and Radiative Transitions[7]

Recall from Section 2.19 that according to quantum mechanics, the classical concept of the precise position of nuclei in space and associated vibrational motion is replaced by the concept of a *vibrational wave function*, χ, which describes the nuclear configuration and nuclear vibrational momentum during a vibration. In the language of classical mechanics, the FC principle states that the most probable electronic transitions will occur between those states possessing a similar nuclear configuration and vibrational momentum at the instant of an electronic transition. In the language of quantum mechanics, the FC principle states that the most probable electronic transitions are those that possess vibrational wave functions that look most alike in the initial (χ_1) and final (χ_2) states at the instant of the electronic transition.

A measure of how much two states undergoing a transition "look alike" is given by the overlap integral (called the FC integral) of the vibrational wave functions of the two states, $< \chi_1 | \chi_2 >$. A net mathematical positive overlap of vibrational wave functions means that the initial and final vibrational states possess similar nuclear configurations and momentum in some region of space. Eq. 3.11b shows that the matrix element for any electronic transition is directly related to *the square* of the vibrational overlap integral (i.e., the FC factor, $< \chi_1 | \chi_2 >^2$, Eq. 3.9).

The larger the FC factor, the greater the net constructive overlap of the vibrational wave functions, the more similar χ_1 is to χ_2, and the more probable the transition. Thus, an understanding of the factors controlling the magnitude of $< \chi_1 | \chi_2 >^2$ is crucial for an understanding of the probabilities of radiative and radiationless transitions between electronic states. The FC factor may be considered a sort of nuclear "reorganization energy," similar to entropy, that is required for an electronic transition to occur. Recall that high organization implies a small degree of entropy and low organization implies a large degree of entropy. The greater the reorganization energy, the smaller the FC factor and the slower the electronic transition. The larger the FC factor, the smaller the reorganization energy and the more probable the electronic transition.

The FC principle provides a selection rule for the *relative probability of vibronic transitions*. Quantitatively, for *radiative* transitions of absorption or emission the FC factor ($< \chi_1 | \chi_2 >^2$) *governs* the relative intensities of vibrational bands in electronic absorption and emission spectra. The Franck–Condon factor is also important in determining the rates of *radiationless* transitions between electronic states. Since the value of $< \chi_1 | \chi_2 >^2$ parallels that of $< \chi_1 | \chi_2 >$, we need only consider the FC integral itself, rather than its square, for qualitative discussions of transition probabilities. We can obtain considerable quantum intuition simply by noting that the larger the difference in the vibrational quantum numbers for χ_1 compared to χ_2, the more likely it is that the equilibrium shape and/or momentum of the initial and the final states are different, and the more difficult and slower and less probable the transition $\chi_1 \rightarrow \chi_2$ will be. Indeed, this is exactly the result anticipated from the classical FC principle. In other words, the magnitude of the integral $< \chi_1 | \chi_2 >$ is related to the probability that an initial state χ_1 will have the same shape and momentum as χ_2. If this probability is high, the transition rate will be high also.

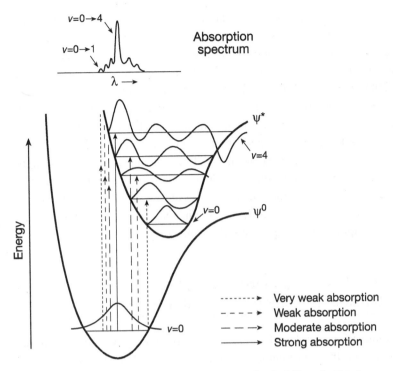

Figure 3.3   Representation of the quantum mechanical Franck–Condon interpretation of the absorption of light.

Consider Fig. 3.3, a schematic representation of the quantum mechanical basis of the FC principle for a radiative transition from an initial ground electronic state ψ_1 (i.e., R) to a final electronic excited state $^*\psi_2$ (i.e., *R). Absorption of a photon is assumed to start from the lowest-energy, $v = 0$ level of ψ_1, since this state is usually the most populated vibrational level in the ground state of R. According to the FC principle, the most likely radiative transition from $v = 0$ of ψ_1 to a vibrational level of $^*\psi_2$ will correspond to a vertical transition for which the overlap integral $< \chi_1 | \chi_2 >$ is maximal. By inspection of Fig. 3.3, the overlap integral ($< \chi_1 | \chi_2 >$) is maximal for the $v = 0 \rightarrow v = 4$ transition (χ_1 is positive everywhere and χ_2 is strongly positive vertically above the maximum of χ_1). Transitions from $v = 0$ to other vibrational levels of $^*\psi_2$ (e.g., from $v = 0$ of ψ_1 to $v = 3$ and $v = 5$ of $^*\psi_2$) may occur, but with lower probability because of the smaller overlap of χ_1 and χ_2 for these vertical transitions. A possible resulting absorption spectrum showing schematically how the intensities for an experimental absorption spectrum would vary is shown in Fig. 3.3, above the PE curves for ψ_1 and $^*\psi_2$. The intensities of the transitions are proportional to the values of the FC overlap integrals, with the $v = 0 \rightarrow v = 4$ transition being maximal.

The same general ideas of the FC principle apply to emission, except now *the important overlap is between the χ corresponding to $v = 0$ of $^*\psi_2$ (the equilibrium position of the excited state) and the various vibrational levels, χ_i, of ψ_1.* Chapter 4 discusses experimental examples of the FC principle for radiative transitions.

In Section 3.13, we shall seek quantum intuition concerning transitions involving electron spins in electronic radiative and radiationless processes. In this case, in addition to the Franck–Condon factors for vibrations, we are concerned with the overlap of two types of spin wave functions (**S**), the wave functions for singlets and triplets corresponding to *R. We have seen from the vector model discussed in Section 2.27 that the wave functions for singlets (with antiparallel spins ↑↓) and triplets (with parallel spins ↑↑) do not look alike at all! When two spin wave functions do not look alike, a *magnetic* perturbation, such as spin–orbit coupling, is required to distort the initial spin state to make it look like the final spin state and thereby induce a radiationless transition or a spin-forbidden radiative transition. The discussion of FC factors on radiative and radiationless transitions described in Section 3.11 are the same for transitions involving a spin change, but in addition to a good FC factor, transitions involving a spin change require simultaneously significant spin–orbit coupling.

3.11 The Franck–Condon Principle and Radiationless Transitions (*R → R + heat)[8]

The original FC principle stated that there is a preference for "vertical" jumps between PE curves for the representative point of a molecular system during a *radiative* transition. The classical and quantum mechanical ideas behind the FC principle for *radiative transitions* can be extended to *radiationless* transitions. The basic idea is the same for radiative and radiationless transitions; namely, (1) a small change in the initial and final nuclear structure and momentum is favored, and (2) energy must be conserved during the transition. For radiative transitions, energy is conserved during a transition by the absorption or emission of a photon ($h\nu$), which corresponds exactly to the energy difference between the initial and final states. For a radiationless transition, *the initial and final electronic states must have the same energy and the same nuclear geometry*. In other words, the initial and final states must look alike energetically and structurally.

In contrast to the situation for radiative transitions, *vertical jumps between* PE *curves separated by a large energy gap are improbable because of the need to conserve energy during a radiationless transition*. It is easiest to conserve energy for radiationless transitions *at points for which curves cross or come close together*, since at the crossing points the wave functions [e.g., ψ_1(*R) and ψ_2(R)] have exactly the same energy. Now, we can connect the quantum mechanical interpretation of radiationless transitions in terms of the FC factor ($< \chi_1 | \chi_2 >$) to the motion of the representative point on a PE surface.

Suppose a molecule starts off on an excited PE curve corresponding to the electronically excited state whose electronic wave function is ψ_1(*R) and undergoes a LU → HO electronic transition to ψ_2(R). Fig. 3.4 depicts just such a situation. On the left, the molecule begins in the ψ_1(*R) state, and on the right, the molecule has just been converted to the ψ_2(R) state. The representative point of ψ_1(*R), during its zero-point motion, makes a relatively small-amplitude oscillating trajectory between points A and B on the excited surface (Fig. 3.4a). But after the transition to ψ_2(R)

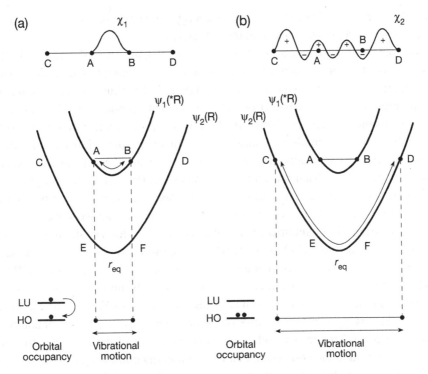

Figure 3.4   Visualization of the quantum mechanical basis for a slow rate of radiationless transitions due to low positive overlap of the vibrational wave functions.

the representative point makes a relatively large-amplitude oscillating trajectory between points C and D (Fig. 3.4b) because the electronic energy of $\psi_1(*R)$ has been converted into the vibrational energy of $\psi_2(R)$. For a radiationless transition from the $\psi_1(*R)$ curve to the ground-state $\psi_2(R)$ curve to be *possible*, energy and momentum must be conserved.

What happens when, in the limiting classical cases, a *horizontal* jump with conservation of PE occurs (Fig. 3.4a, A → C or B → D), or a *vertical* jump with conservation of geometry (Fig. 3.4b A → E or B → F) occurs? A classical horizontal "jump" from $\psi_1(*R)$ to $\psi_2(R)$ that conserves energy requires an unlikely abrupt change in nuclear geometry. The representative point, initially in $v = 0$ of $\psi_1(*R)$, will be undergoing a vibration of small amplitude between A and B before and after the horizontal transition to C or D [at points C and D, the atoms on $\psi_2(R)$ are momentarily moving slowly since they are at turning points of a vibration]. However, because the initial and final states have very different geometries, they do not "look alike" and therefore the horizontal jump is improbable. A classical *vertical* jump that conserves the initial geometry from $\psi_1(*R)$ to $\psi_2(R)$ will start an abrupt transformation of the representative point from a very small, mild-amplitude vibration with little KE and momentum between A and B to a very large-amplitude, high kinetic-energy oscillation between C and D. The motions of the representative points in $\psi_1(*R)$ to $\psi_2(R)$ do not look alike, and therefore the transition is improbable. The vertical jumps A → E or B → F are unlikely because, in order to conserve energy, some external energy sink must be available to suddenly absorb a great deal of energy.

The net result of either a horizontal or vertical jump from $\psi_1(*R)$ to $\psi_2(R)$ is that the vibration of the molecule will cause either an abrupt change in geometry (Fig. 3.4a) or an abrupt change in momentum (Fig. 3.4b). Thus, we can conclude that radiationless transitions between two PE curves that do not come close in energy are implausible. An abrupt change in the positional or momentum characteristics of a vibration corresponds to a large change in the organizational energy of the vibrating system. Classically, transitions requiring such large structural or dynamic reorganizations are resisted, and therefore implausible. Put in simpler terms, the initial and final state do not look very much alike in terms of their initial and final kinetic energies or of their initial and final structures, so a transition between them is slow.

Let us see how the quantum mechanical wave functions for the vibrations deal with the two transitions shown in Fig. 3.4 for the transition $\psi_1(*R) \to \psi_2(R)$. Suppose the wave functions χ_1 and χ_2 correspond to the $v = 0$ and $v = 6$ vibrations of $\psi_1(*R)$ and $\psi_2(R)$, respectively. The vibrational wave functions for the initial (i.e., χ_1, $v = 0$) and the final state (i.e., χ_2, $v = 6$), which are shown at the top of Fig. 3.4, do not look at all alike. That is, χ_1 is always positive (no nodes), whereas χ_2 oscillates a number of times ($v = 6$, so there are six nodes). The dissimilarities of the two wave functions immediately lead to the conclusion that the overlap integral ($< \chi_1 | \chi_2 >$) will have a value close to zero because of mathematical cancellation of the two functions. The poor net overlap is shown schematically in Fig. 3.5 (left).

Now, we consider the situation for which two PE curves come very close in energy (i.e., they actually intersect). Consider the mathematical form of the vibrational wave functions χ_1 and χ_2 for the initial and final vibrations shown in Fig. 3.5 for a radiationless transition, similar to that of Fig. 3.4, in which the two surfaces are far apart for all values of r [Fig. 3.5 (left), and a radiationless transition in which the two surfaces come close together and intersect or cross at a specific value of r (Fig. 3.5, right)]. Recall that for a radiative or radiationless transition to be probable according to the FC principle, there must be net positive overlap between these wave functions.

By inspection of the curves in Fig. 3.5 (left), namely, the case for which the two surfaces involved in the radiationless transition are far apart for all values of r, the vibrational wave function χ_1 (positive everywhere, no node) associated with $\psi_1(*R)$, plotted above the classical curve representing the excited state, is drastically different in form from the vibrational wave function χ_2 (highly oscillating between positive and negative values) associated with $\psi_2(R)$ at the energy where the transition occurs. Therefore, the mathematical overlap integral of χ_2 and χ_1 (i.e., $< \chi_1 | \chi_2 >$), the net overlap, will be zero or close to zero, because the initial function (χ_1) is positive everywhere about the point r_{eq} (the equilibrium separation) but the final function (χ_2) oscillates between positive and negative values about r_{eq}. The result is an effective cancellation of the mathematical overlap integral (Fig. 3.5, bottom left). Quantum intuition tells us, as does the FC principle, that if the overlap integral $< \chi_1 | \chi_2 >$ is very small, the probability of the radiationless transition from $\psi_1(*R)$ to $\psi_2(R)$ will also be very small. More simply stated, the wave functions χ_1 and χ_2 do not look very much alike and will be difficult to make look alike through electronic couplings. In

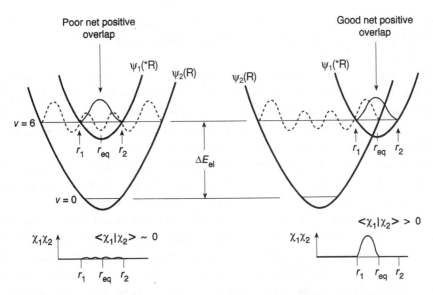

Figure 3.5   Schematic representation of situations for poor (left) and good (right) net positive overlap of vibrational wave functions. The value of the integral $< \chi_1 | \chi_2 >$ as a function of r is shown at the bottom of the figure.

terms of a selection rule, the transition as shown in Fig. 3.5 is implausible and will occur at a slow rate.

In Fig. 3.5 (right), there is a specific value of r where PE *curve-crossing* occurs between the wave functions $\psi_1(*R)$ and $\psi_2(R)$. How can visualization of the overlap of vibrational wave functions provide some quantum insight into the operation of the FC principle when two PE curves intersect one another? The poor overlap of the vibrational wave functions χ_1 and χ_2 for a molecule in the lowest vibrational level of $*\psi_1(*R)$ for the noncrossing situation (Fig. 3.5, left) contrasts with the significant overlap for the curve-crossing situation (Fig. 3.5, right). In both cases, χ_1 corresponds to the $v = 0$ level of $*\psi_1(*R)$ and χ_2 corresponds to the $v = 6$ level (six nodes in the wave function) of $\psi_2(R)$.

The amount of electronic energy (ΔE_{12}) that must be converted into vibrational energy and the vibrational quantum number (v) of the state produced by the transition are the same for both transitions shown in Fig. 3.5. The vibrational overlap integrals $< \chi_1 | \chi_2 >$ for the crossing and noncrossing situations are shown at the bottom of Fig. 3.5. The net overlap for the situation on the left is much less than that for the situation on the right. The wave function $\chi_2(v = 6)$ undergoes oscillations from mathematical plus to minus in the regions of space where the wave function $\chi_1(v = 0)$ is positive, thus causing a poor vibrational overlap integral. Consistent with the classical FC principle, therefore, quantum intuition states that radiationless transition for the surface-crossing situation (Fig. 3.5, right) will occur much faster than the nonsurface crossing, radiationless transition (Fig. 3.5, left) because the vibrational overlap integral $(< \chi_1 | \chi_2 >)$ is larger for the situation on the right. In terms of a selection rule, a radiationless transition when there is no surface is *Franck–Condon forbidden* (i.e., the FC factor $< \chi_i | \chi_j >^2 \sim 0$), whereas a radiationless transition at the surface-crossing (Fig. 3.5, right) is *Franck–Condon allowed* (i.e., the FC factor $< \chi_i | \chi_j >^2 \neq 0$).

For some vibrations that distort the energy surface of *R, there are those vibrations that may be better represented by Fig. 3.5 (left) and other vibrations that may be better represented by Fig. 3.5 (right). In other words, certain vibronic interactions, Fig. 3.5 (right), can lead to radiationless transitions between states by causing energy surfaces to intersect, and thereby enhance the FC factors for transition.

In summary, *radiationless transitions are most probable when two* PE *curves for vibration cross (or come very close to one another), because when this happens, it is easiest to conserve the energy, motion, and phase of the nuclei during the transition in the region of the crossing.* In other words, in the regions of curve crossings, the wave functions of *R and R look alike structurally, energetically, and dynamically. It has been assumed that the vibrational transition, rather than the electronic transition, is rate determining. This means that the curve crossing shown in Fig. 3.5 actually would not occur because the electronic states are mixed by the vibration in the region where the crossing occurs. This case is usually for radiationless electronic transitions involving no change in spin. Such crossings are more complicated in polyatomic molecules and are discussed in Chapter 6.

3.12   Radiationless and Radiative Transitions between Spin States of Different Multiplicity[8]

The spirit of the paradigm for the selection rules for spin transitions is similar to that for electronic and vibronic transitions. As before, we postulate that for all radiative or radiationless transitions, energy and momentum must be conserved and that transitions between states of different spin are allowed or plausible only when the initial and final states look alike in terms of structure and motion.

Some of the most important examples of spin transitions in organic photochemical reactions are listed in Scheme 3.2. These transitions are analogous to the general transitions of Scheme 3.1, except they are for transitions that involve a change of spin, specifically for the conversion of a singlet state to a triplet state or vice versa.

In Scheme 3.2, we will develop a model that will make it possible to understand and visualize the mechanisms for all of the spin transitions. The pictorial model uses the vector model of spin developed in Chapter 2. A precessing vector representing the spin wave function (**S**), analogous to the spirit of the pictorial model developed for vibrating nuclei or orbiting electrons, is employed. Consider the spin wave function of an initial spin state ($\mathbf{S}_1$) and a final spin state ($\mathbf{S}_2$). Analogous to the electronic overlap integral ($< \psi_1 | \psi_2 >$) and the vibrational overlap integral ($< \chi_1 | \chi_2 >$), there is a spin overlap integral $< \mathbf{S}_1 | \mathbf{S}_2 >$. When there is no spin change during the transition, then $< \mathbf{S}_1 | \mathbf{S}_2 > = 1$ (e.g., singlet–singlet, triplet–triplet, or doublet–doublet transitions). In this case, the initial and final spin states look alike in all respects and there is no spin prohibition on the electronic transition. However, when there is a spin change during

(a) $R(S_0) + h\nu \rightarrow {}^*R(T_1)$

(b) ${}^*R(T_1) \rightarrow R(S_0) + h\nu$

(c) ${}^*R(T_1) \rightarrow R(S_0) + \text{heat}$

(d) ${}^*R(S_1) \rightarrow {}^*R(T_1) + \text{heat}$

(e) ${}^3I(D) \rightarrow {}^1I(D)$ and ${}^1I(D) \rightarrow {}^3I(D)$

Scheme 3.2   Some important transitions involving intersystem crossing.

the transition, $< S_1 | S_2 > \neq 0$ (e.g., singlet–triplet transitions) and in the zero-order approximation, the transition is strictly forbidden.

In first order, transition between singlets and triplets becomes allowed only if an interaction for mixing spin states is available. In contrast to the mixing of electronic states, the mixing of spin states requires *magnetic interactions, not electrostatic interactions*. At a fundamental level, an electronic transition that involves a change of spin angular momentum requires some interaction (coupling) with another source of angular momentum that can both trigger the transition and allow conservation of the total angular momentum of the two interacting systems during the transition and provide for the conservation of *magnetic* energy. For organic molecules, the most important interaction that couples two spin states and that provides a means of conserving the total angular momentum of the system is the coupling of the electron spin with the orbital angular momentum (i.e., spin–orbit coupling).

From Scheme 3.2, the most important radiative transitions involving a spin change are the spin-forbidden absorption, $R(S_0) + h\nu \rightarrow {}^*R(T_1)$, and the spin-forbidden emission (phosphorescence), ${}^*R(T_1) \rightarrow R(S_0) + h\nu$. The most important radiationless transitions for ${}^*R$ involving a spin change are ${}^*R(T_1) \rightarrow R(S_0) + \text{heat}$ and ${}^*R(S_1) \rightarrow {}^*R(T_1) + \text{heat}$. Both are intersystem crossings (ISC). Primary photochemical processes, such as ${}^*R \rightarrow I(D)$, may be considered as elementary chemical steps for which a change of spin is forbidden by the spin selection rules. Thus, when the primary photochemical process is an ${}^*R(T_1) \rightarrow {}^3I(D)$ transition, the ISC process ${}^3I(D) \rightarrow {}^1I(D)$ shown in Scheme 3.2e must occur before products (P) can be formed from ${}^1I(D)$. This important ${}^3I(D) \rightarrow {}^1I(D)$ ISC step is discussed in Section 3.24 and in Chapter 6.

3.13   Spin Dynamics: Classical Precession of the Angular Momentum Vector[9]

In Section 2.24, we developed a model representing the electron spin angular momentum (S) as a vector. The spin vector S possesses an associated magnetic moment, μ_S (Eq. 2.32). In the absence of any other magnetic fields (i.e., other magnetic moments), both the S and μ_S vectors lie motionless in space and possess a magnetic energy (E). The situation is quite different when there are magnetic fields (i.e., magnetic moments), H_z, that couple with the magnetic moment of the electron spin, μ_S. The result of such a coupling is that the electron spin (and its associated magnetic moment, μ_S) orients itself either in a direction aligned with the coupling field (H_z) or in a direction

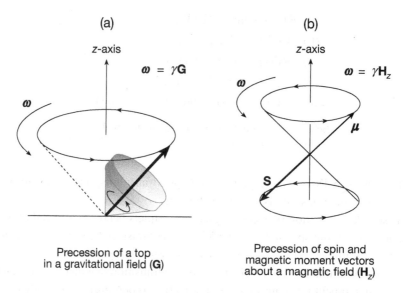

Figure 3.6   A vector diagram comparing the precessional motion of a spinning top (a) in a gravitational field (**G**) to the precessional motion of the spin angular momentum vector (**S**) and the precessional motion of the magnetic moment (μ_S), and (b) in the presence of an applied magnetic field (**H**$_z$).

opposed to **H**$_z$, and makes either a clockwise or counterclockwise precessional motion about the axis for the magnetic field (**H**$_z$), depending on the orientation of the vector μ_S. Precession is defined as the motion of the axis swept by a spinning body, such as a gyroscope or spinning top (Fig. 3.6). The precession of the axis for a spinning body sweeps out a conical surface, and the tip of the precessing axis sweeps out a circle (Fig. 3.6).

Now, we consider a model of the precessional motion for the vector representing the magnetic moment of a classical bar magnet and then adapt the model to include the precession of electron spin (i.e., a quantum mechanical magnet). The classical bar magnet in a magnetic field has already been employed in Section 2.34 as an exemplar of the energy states derived from an electron spin coupled to a magnetic field, similar to that of the spinning top shown in Fig. 3.6a. If the bar magnet in a magnetic field (**H**$_z$) possessed angular momentum by spinning about an axis, it would execute a precessional motion about the axis defined by the applied magnetic field. A classical analogue of the expected motion of a magnet possessing angular momentum in a magnetic field is available from the precessional motion of a rotating or spinning body, such as a toy top or a gyroscope (Fig. 3.6a). The angular momentum vector of a spinning top sweeps out a cone in space as it makes a *precessional* motion about the axis of rotation, and the tip of the vector sweeps out a circle. The cone of precession of a spinning top possesses a geometric form identical to the cone of orientations (Section 2.28) that are possible for the quantum spin vector, so the classical model can serve as a basis for understanding the quantum mechanical model.

A remarkable and nonintuitive feature of a spinning top is that it appears to defy gravity as it "stands up" and precesses after release, whereas a nonspinning top falls down upon release. The cause of the precessional motion and the spinning top's stability toward falling is attributed to the operation of the force of gravity, which although it applies a downward *force*, nevertheless exerts a nonintuitive *sideways (rather than downward) torque* on the angular momentum vector. A torque plays the same role in rotational (or circular) motion that force does in linear motion. More precisely, for linear motion, force is equal to the rate of change of linear momentum, whereas for rotational motion, torque is equal to the rate of change of angular momentum. The torque associated with a spinning top pulls the tip of the vector representing the angular momentum in a circular path and produces the nonintuitive result of precession induced by the force of gravity. *By analogy, in the presence of an applied field $\mathbf{H}_z$, the coupling of a spin's magnetic moment $\boldsymbol{\mu}_S$ with the magnetic field $\mathbf{H}_z$ produces a torque that, like gravity for the top, "grabs" the magnetic moment vector ($\boldsymbol{\mu}_S$) and causes it to precess about the field direction in the cone of possible orientations of the vector.* Since the spin's magnetic moment and angular momentum vectors are collinear (but opposite in direction, Fig. 2.12), the angular momentum vector ($\mathbf{S}$) also precesses at the same rate about the field direction (Fig. 3.6b).

Now, let us consider some of the precessional features of a gyroscope or top in a gravitational field as an exemplar and then adapt these features to generate a model of electron spin to obtain a quantitative relationship between the strength of the magnetic field and the rate of precession. A top precesses in a gravitational field ($\mathbf{G}$) with an angular frequency ($\boldsymbol{\omega}$). The angular frequency ($\boldsymbol{\omega}$) and strength of the gravitational field ($\mathbf{G}$) are both vector quantities that possess both a direction and a magnitude. When the symbol $\boldsymbol{\omega}$ is used, we are referring to a vector, the velocity of precession, which has both a magnitude and a direction (i.e., clockwise or counterclockwise) of precession. When the symbol ω is used, we are referring to a scalar quantity, the rate or speed of precession, which is independent of the sense or direction of precessional motion.

The two factors that determine the velocity of precession ($\boldsymbol{\omega}$) of a gyroscope are *the strength of the gravitational field ($\mathbf{G}$) and the magnitude of the spin angular momentum ($\mathbf{S}$) of the spinning gyroscope.* If $\mathbf{S}$, which is determined by the angular velocity of the spin, and the mass of the top is constant, then the rate of precession ($\boldsymbol{\omega}$) is determined only by the force of gravity ($\mathbf{G}$). Thus, when the spin angular momentum is constant (as is the case for an electron spin system), there is a direct proportionality between the velocity of precession ($\boldsymbol{\omega}$) and the force of gravity ($\mathbf{G}$), as shown in Eq. 3.12, where γ (cf. the magnetogyric ratio in Eq. 2.32) is a scalar proportionality constant between the precessional velocity and the force of gravity.

$$\boldsymbol{\omega} = \gamma \mathbf{G} \quad \text{Precessional velocity of a gyroscope in a gravitational field} \quad (3.12)$$

Now, let us apply the vector model of the precessing top to visualize the precessional motion of an electron spin. As a concrete exemplar, let us consider the specific case for which the amount of spin angular momentum is constant and equal to $\hbar$, the fundamental unit of angular momentum in the quantum world (i.e., the value of the angular momentum for a spin 1 system: $\mathbf{S} = \hbar$). By analogy to Eq. 3.12, the spin angular momentum vector for this exemplar and the magnetic moment vector due to

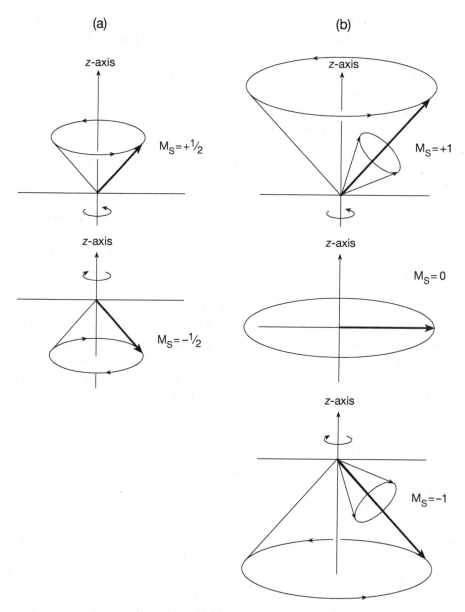

Figure 3.7   Vector model of vectors precessing in their possible cones of orientation. Only the spin angular momentum vectors are shown.

spin precess about the field direction with a characteristic angular rate (ω), given by Eq. 3.13, where γ_e is the magnetogyric ratio of the electron and $|\mathbf{H}_z|$ is the absolute value of the strength of the magnetic field. Note that it is the *rate* of precession (ω, a scalar quantity), not the *direction* that is being discussed. Examples of spin precession are shown in Fig. 3.7.

$$\omega = \gamma_e|\mathbf{H}_z| \qquad \text{Precession of an electron spin in a magnetic field} \qquad (3.13)$$

For the special case corresponding to one unit of angular momentum ($\hbar$), the rate of precession (ω) of the spin and magnetic moment vectors about the magnetic field ($\mathbf{H}_z$) *depends only on the magnitude of* γ_e *and* $\mathbf{H}_z$.

Recall an important feature from Section 2.35 that differentiates the classical magnet from the quantum mechanical magnet; namely, a classical magnet, may assume any arbitrary position in an applied field (although the energy will be different for different positions), but the quantum mechanical magnet can only achieve a finite set of orientations defining a *cone of possible orientations* with respect to an arbitrary axis. These allowed orientations in space are determined by the value of the magnetic quantum number M_S, which also determines the magnetic energy in a field $\mathbf{H}_z$. In the absence of a coupling magnetic field, the spin vector lies somewhere in the cone of possible orientations but does not precess because there is no torque from another magnetic field acting on it. However, upon application of a *constant* magnetic field $\mathbf{H}_z$ that couples to the electron spin, the coupling field imposes a constant torque on the magnetic moment of the electron spin, just as gravity imposes a constant torque on the angular momentum vector of a spinning top (Fig. 3.6, but replace the top with an electron spin). The spin vector then begins to precess about the magnetic field axis (we choose the z-axis for convenience, since the selection of the quantization axis is arbitrary). This precessional motion of the spin vector sweeps out one of the cones of possible orientations (Section 2.29).

Thus, we postulate that in the presence of a magnetic field, for each allowed orientation of the electron spin given by M_S, there exists a *cone of possible precession for the angular momentum and magnetic moment vectors associated with the electron's spin,* and that *the rate of precession (ω) of the spin vector in these cones is proportional to the strength of the coupling magnetic field* (Eq. 3.13). This conclusion has far-reaching consequences in molecular organic photochemistry for processes involving a change of spin (multiplicity) in photophysical and photochemical processes.

3.14   Precession of a Quantum Mechanical Magnet in the Cones of Possible Orientations

According to the principles of quantum mechanics, the positions of individual spin magnetic vectors in (a mathematical vectorial) space are confined to cones that are oriented along an arbitrary quantization axis (Fig. 3.6). Furthermore, the various cones of possible orientations (as shown in Fig. 2.10 for each value of the spin orientation quantum number, M_S) of the angular momentum vector also represent cones of precession when the spin vector is subject to a torque from some coupling magnetic field along the z-axis ($\mathbf{H}_z$). When a magnetic field is applied along the z-axis, *the spin states with different values of* M_S *will assume different orientations in the magnetic field* $\mathbf{H}_z$ *and will possess different energies because of the different orientations of their magnetic moments (μ_S) in the field.* (Recall from Fig. 3.6 that the direction of the magnetic moment vector is collinear with that of the angular momentum, so that if the angular momentum possesses different orientations so will the magnetic moment.)

These different energies correspond, in turn, to different angular rates of precession (ω) about the cones of orientations. In analogy to Eq. 2.32, we can formulate Eq. 3.14, indicating that the Zeeman magnetic energy (E_z) of a specific orientation of the angular momentum in a magnetic field is directly proportional to the magnitudes of quantities g, μ_e, and $\mathbf{H}_z$, and the magnitude and sign of M_S (see Eq. 2.34).

$$E_z = \hbar\omega_S = M_S g \mu_e \mathbf{H}_z \tag{3.14}$$

Eq. 3.14 may be rearranged algebraically to Eq. 3.15, which yields the value of the spin-vector precessional frequency, $\boldsymbol{\omega}_S$.

$$\boldsymbol{\omega}_S = [M_S g \mu_e \mathbf{H}_z]/\hbar \tag{3.15}$$

From Eq. 3.15, the values of $\boldsymbol{\omega}_S$ are directly related to the same factors as the magnetic energy associated with coupling to the field, namely, the g factor, the amount of angular momentum along the axis (M_S), the electron's magnetic moment (μ_e), and the magnetic field strength and direction ($\mathbf{H}_z$). For two states with the same absolute value of M_S, but different signs of M_S, the precessional frequencies ($\boldsymbol{\omega}_S$) are *identical in magnitude, but opposite in the sense (phase) of precession*. For example, if the tips of the vectors in Fig. 3.6 are viewed from above, they either trace out a circle via a clockwise motion for the lower-energy $M_S = -1$ state or trace out a circle via a counterclockwise motion for the higher-energy $M_S = +1$ state. These senses of rotation correspond to differing orientations of the magnetic moment vector in a magnetic field, and therefore different magnetic energies (E_z).

From classical physics, the rate of precession of a magnetic moment about an axis is proportional to the strength of coupling of the magnetic momentum to that axis (in analogy to Eq. 3.13). The same ideas hold if the coupling is due to sources other than an applied field ($\mathbf{H}_z$), that is, coupling with other forms of angular momentum and its associated magnetic moment. For example, if coupling of the spin angular momentum ($\mathbf{S}$) with the orbital angular momentum($\mathbf{L}$) (i.e., spin–orbit coupling) is strong, then the vectors $\mathbf{S}$ and $\mathbf{L}$ *precess rapidly about their resultant and are strongly coupled, behaving as a single vector and not as two independent vectors*. The idea is similar to the strong coupling of two electron spins (two doublets, D) of angular momentum 1/2, to form a new spin system, a triplet (T) of spin 1 or a singlet (S) of spin 0. When the coupling between the angular momenta is strong, the precessional motion about the resultant is fast, and it is difficult for other magnetic torques to break up the resultant coupled motion. However, if the coupling is weak, the precession is slow about the coupling axis, and relatively weak magnetic forces can break up the coupling. These ideas are of great importance in photochemistry because they provide insight into the coupling mechanisms of *radiative and radiationless transitions involving a change of spin* (Scheme 3.2).

Since the magnetic moment vector and the spin vector are collinear (but opposite in direction, as shown in Fig. 3.6b), *these two vectors will faithfully follow each other's precession, so that we do not need to draw each vector since we can deduce the characteristic motion of one vector from the characteristic motion of the other.* Unless specified, the vectors shown in the text will refer to the spin vector (S) and not the magnetic vector (μ_S). Remember, however, that the angular momentum vector has the units of angular momentum ($\hbar$), whereas the magnetic moment vector has the units of magnetic moment (Joule-Gauss = JG).

Fig. 3.7 (the precessional version for the static cones of Fig. 2.10) shows the vector model for the different types of precession for spin 1/2 states (Fig. 3.7a) and spin 1 states (Fig. 3.7b) in their possible cones of orientation. These five spin state exemplars (for $M_S = +1/2$, $-1/2$, $+1$, 0, and -1, respectively) are of great importance for the understanding of spin transitions for both radiative and radiationless transitions and are discussed in detail in the remainder of this chapter.

3.15   Important Characteristics of Spin Precession

In this section we summarize some important points concerning spin transitions. The process of rotation of the quantum mechanical magnet's vector about an axis of an applied magnetic field is called precession, and the specific precessional frequency of a given state under the influence of a specific field (H_z) is commonly called the "Larmor precessional frequency" (ω_z). The following characteristics of Larmor precession can be deduced from Eq. 3.15:

1. The value of the precessional frequency (ω_z) is proportional to the magnitude of the four quantities M_S, g, μ, and H_z.

2. For a given orientation and magnetic moment, the value of ω_S decreases with decreasing field, and in the limit of $H_z = 0$, $\omega_S = 0$, so the model requires precession to cease and the vector to lie motionless at some indeterminate position in space. For the $M_S = 0$ state of the triplet, there is no motion at any field strength (Eq. 3.14) and the spin vector can assume any orientation in a circle defined by a plane perpendicular to the z-axis (not a *cone* of possible orientations).

3. For a given field strength H_z and for a spin system with several values of M_S (multiplicity ≥ 1), the spin vector precesses fastest for the largest absolute values of M_S (e.g., T_+ and T_{T_-}) and is zero for states with $M_S = 0$ (e.g., T_0 and S).

4. For a given field strength H_z and for the same absolute value of M_S, different signs of M_S correspond to different directions (clockwise and counterclockwise) of precession, which have identical rates of precession but different magnetic energies (e.g., $M_S = +1$ or -1, which correspond to T_+ and T_).

5. A high precessional frequency (ω_z) and associated magnetic energy (E_z) is proportional to the strength of the coupling field (H_z).

6. For the same field strength H_z and quantum state M_S, the precessional frequency (ω_z) is proportional to the value of the g factor and the inherent magnetic moment (μ_e) of the electron spin.

3.16   Some Quantitative Benchmark Relationships between the Strength of a Coupled Magnetic Field and Precessional Rates

From Eq. 3.15 and the experimental value of γ_e (1.7×10^7 rad s^{-1}), we can obtain a quantitative relationship between the precessional rate (ω) of the vector representing a magnetic moment of a free electron and the coupling to an arbitrary magnetic field ($\mathbf{H}_z$), as shown in Eq. 3.16, where $\mathbf{H}_z$ is in units of gauss (G). For simplicity, we consider rates rather than frequencies, and only the absolute value for $\mathbf{H}_z$ needs to be considered when computing rates. In this case the direction of precession is not determined but the rate is the same for either clockwise or counterclockwise precession.

$$\omega \text{ (precessional rate)} = 1.7 \times 10^7 \text{ rad s}^{-1} \, \mathbf{H}_z \qquad (3.16)$$

Now, we can make a connection between the energy of a photon of frequency ν, and the precessional rate (ω) of an electron spin. The relationship of the energy of a photon or an energy gap to a frequency of a light wave (ν) is given by $\Delta E = h\nu$ (Eq. 1.1).

The circular precessional frequency of the spin vector tip in a field $\mathbf{H}_z$ may be related to the frequency at which the vector makes a complete cycle, called the "oscillation frequency." The relationship between the precessional rate (ω, in rad s^{-1}) and the oscillation frequency (ν, in s^{-1}) is $\nu = \omega/2\pi$ or $\omega = 2\pi\nu$ (where 1 rad $= 2\pi$). Since one complete cycle about a circle is equivalent to 2π rad, the value of ω for a fixed field in rad s^{-1} is always a larger number than the value of ν in s^{-1} for the same value of $\mathbf{H}_z$. Substituting this relationship between ω and ν into Eq. 3.16 leads to Eq. 3.17:

$$\nu \text{ (precessional frequency)} = 2.8 \times 10^6 \text{ s}^{-1} \, \mathbf{H}_z \qquad (3.17)$$

As an exemplar of the magnitudes of ν and ω for a given value of $\mathbf{H}_z$, consider a coupling field of 1 G (roughly the magnitude of the earth's magnetic field). The resulting precession rate of an electron spin is $\nu \cong 2.8 \times 10^6$ s^{-1} (2.8 MHz) or $\omega \cong 17 \times 10^6$ rad s^{-1}, consistent with $2\pi\nu = \omega$ or $\nu = \omega/2\pi$. Eq. 3.18 provides the relationship among the precessional frequency (ω), the resonance frequency of radiative transitions (ν), and the energy gap between the states undergoing transitions (ΔE), where $\hbar = h/2\pi$. We use both ν and ω throughout the following discussions. Pictorially, ν can be viewed as an oscillating linear vibrational motion back and forth with a certain period of time needed to return to a certain point along the vibration; ω, on the other hand, can be viewed as an oscillating circular motion with a certain period of time needed to return to a certain point in the circle. Both pictures are useful, depending on the situation under discussion.

$$\Delta E = h\nu = (h\omega/2\pi) = \hbar\omega \qquad (3.18)$$

From Eqs. 3.16 and 3.17, we can now compute numerical benchmarks for the magnitudes of precessional frequencies (ω), for various values of $\mathbf{H}_z$, and compare the

Table 3.1   Relationship between the Rate of Precession of an
Electron Spin as a Function of Magnetic Field Strength

H_z (G)	ω (rad s^{-1})	ν (s^{-1})	$\Delta E = h\nu$ (kcal mol^{-1})[a]
1	1.7×10^7	2.8×10^6	2.7×10^{-7}
10	1.7×10^8	2.8×10^7	2.7×10^{-6}
100	1.7×10^9	2.8×10^8	2.7×10^{-5}
1,000	1.7×10^{10}	2.8×10^9	2.7×10^{-4}
10,000	1.7×10^{11}	2.8×10^{10}	2.7×10^{-3}
100,000	1.7×10^{12}	2.8×10^{11}	2.7×10^{-2}
1,000,000	1.7×10^{13}	2.8×10^{12}	2.7×10^{-1}

a. Conversion factor: $h = 9.5 \times 10^{-14}$ kcal mol^{-1} s.

results to the values of the frequencies ν previously discussed for electronic motions in orbits and the vibrational motions of atoms in bonds (Table 3.1). Practically speaking, applied laboratory magnetic fields, whose strengths (H_z) can be conveniently varied from 0 to a maximum of $\sim 100,000$ G, are readily achievable. From Eq. 3.17, therefore, precessional rates as high as $\nu \sim 2.8 \times 10^{11}$ s^{-1} ($\omega \sim 1.7 \times 10^{12}$ rad s^{-1}) are achievable by applying very strong laboratory magnetic fields of 100,000 G to a free electron spin. Internal magnetic fields resulting from interactions with other electron spins or nuclear spins typically correspond to magnetic fields in the range from a few gauss to several thousand gauss, corresponding to precessional rates of $\nu \sim 10^7 - 10^9$ s^{-1}. In special cases systems involving molecules possessing "heavy nuclei" (Section 3.21) orbital motion can produce an exceptionally strong associated magnetic field on the order 1,000,000 G or larger (Section 3.21). In these special cases, spin–orbit coupling can produce precessional rates of $\sim 10^{12}$ s^{-1}. Included in Table 3.1 are the energy gaps of photons corresponding to the frequencies for precession in magnetic fields of varying strength. An important point is that even for an enormous magnetic field of 1,000,000 G, the value of ΔE is still < 1 kcal mol^{-1}, which is less than the order of many vibrational energies.

Now, we can compare the range of frequencies for precessional motion (ν and ω) to the rates of electronic orbital and nuclear vibrational motion. Electronic and vibrational frequencies can be visualized as "back-and-forth" linear periodic motion (i.e., electrons vibrating back and forth between atoms and atoms vibrating back and forth in bonds). With such a picture, it is natural to think of the rate of electronic or vibrational motion in terms of the frequency of vibration, ν, which has units of reciprocal seconds, s^{-1}. Thus, the more natural comparison with circular precessional motion is with the frequency of oscillation, ν, which also has units of s^{-1}. Electronic "vibrations" between atoms occur at frequencies that are typically on the order of $\nu = 10^{15}$–10^{16} s^{-1}, and the frequencies of vibrational motions of the functional groups of organic molecules are typically on the order of $\nu = 10^{12}$–10^{14} s^{-1}. Electron spins, even in a very strong laboratory magnetic field of 100,000 G, precess at a frequency of only $\nu \sim 10^{12}$ s^{-1} (Table 3.1). Thus, the precession of electron spins (except possibly in exceptionally strong magnetic fields produced by orbital motion of electrons in the field of heavy nuclei) is typically slower than electronic or vibrational motion. Therefore, in zero order, the assumption that electronic and vibrational motions are much faster than spin motion is justified. This assumption allows us to propose a FC principle for spin transitions. During the time period for an electronic transition or a vibration, the precessing spin is "frozen" in its precessional cone. Like all approximations, this one is subject to breakdown in special situations (e.g., exceptionally strong spin–orbit coupling). However, for organic molecules that contain only atoms of the first full row of the periodic table, this approximation holds well. We discuss the conditions for the breakdown of the approximation (e.g., the "heavy atom" effect for inducing ISC) in Chapter 5. Now, we consider some examples of magnetic coupling that are required for ISC to occur.

3.17 Transitions between Spin States: Magnetic Energies and Interactions

Transitions between the spin magnetic energy levels can be visualized as occurring through the result of magnetic *torques* exerted on the magnetic moment vectors (μ) of an electron spin, or equivalently as the result of spin angular momentum coupling to another form of angular momentum, in particular, *orbital* angular momentum (recall that a torque is a force that induces circular motion to the tip of a vector representing the angular momentum). The magnetic coupling causing the transition is due to interactions of the electron spin with the magnetic moment due to electron orbital motion, with the magnetic moment due to an external magnetic field, or with the magnetic moment due to a nuclear spin. The same vector model allows us to make an analogous visualization of the transitions resulting from the coupling of an electron spin to any magnetic moment. These features of the vector model provide a powerful and general tool for visualizing transitions involving a change of spin. In addition, the same features of the vector model provide a single common conceptual framework for understanding electron spin resonance (ESR) and nuclear magnetic resonance (NMR). For any spin transition to be plausible, though, the electronic and magnetic energies of the states undergoing transition (e.g., the T and S states) must be degenerate. Since the electron exchange interaction J (Section 2.38) can cause the electronic energies of the T and S states to split in energy, we also need to consider the role of this energy splitting on the rate of transitions involving a change of electron spin.

3.18 The Role of Electron Exchange (J) in Coupling Electron Spins

Electron exchange (Pauli exclusion principle) is a nonclassical, quantum effect resulting in a splitting of singlet and triplet states (Section 2.10). The same electron exchange is also responsible for a coupling of electron spin vectors, since upon exchange all properties of the electron must be preserved, including the phase of the relative orientations of the electron spin. When the exchange interaction (J) is strong between two coupled spins, the energy separation of the T and S states is very large and the two spins are strongly coupled; as a result of the strong coupling through exchange, the two electron spins must remain either in phase as a spin 1 system (triplet) or out of phase as a spin 0 system (singlet). The mathematical form of J in a spin Hamiltonian H_{EX} (the operator that indicates the interaction energies) is given by Eq. 3.19. The singlet–triplet splitting energy (ΔE_{ST}) is defined as $2J$ (i.e., an equal splitting of J above and below the energy corresponding to no exchange, Section 2.13).

$$H_{EX} = J\mathbf{S}_1 \cdot \mathbf{S}_2 \tag{3.19}$$

The energy associated with electron exchange refers to a correction to the energy of electron–electron repulsions and is therefore an electrostatic interaction of charges. It is not a magnetic interaction per se. When the exchange interaction causes a splitting of a singlet and triplet state that is much larger than available magnetic energy, singlet–triplet interconversions are said to be inhibited (or "quenched") by the exchange interaction J, because the S and T states have different energies. In this case, the two spin vectors precess about each other to produce a resultant net spin of 1 or 0, and it is more appropriate to think of the spins as a single spin state whose net spin is 1 or 0 than to think of the individual spins as two component spins of 1/2.

The magnitude of the exchange interaction J depends on the extent of overlap of orbitals containing the electron spins (Section 2.12). The exchange energy is approximated as an exponential function of the separation of the two orbitals (e.g., Eq. 3.20, where J_0 is a parameter that depends on the orbitals and r is the separation of the orbitals in space). It is commonly assumed that J will decrease in value exponentially as r increases, since orbital overlap decreases exponentially as the separation of the orbitals increases. This approximation, in turn, leads to an estimate that the value of J will be in the range of magnetic interactions for separations on the order of 5–10 Å or greater. This approximation means that for two radical centers that are closer than 5 Å on average, the exchange energy will be significant.

$$J = J_0 \exp\text{-}r \tag{3.20}$$

The role of J in determining the rate of transitions involving a spin change in electronically excited molecules, *R is somewhat different from its role in determining the rate of transitions in diradical reactive intermediates, I(D). For *R, there is usually a significant overlap between the half-filled HO and LU; for the *same* $(HO)^1(LU)^1$ electronic configuration, *the *R(T) state is always of lower energy than the *R(S) state*. Transitions between *R(S) and *R(T) occur from excited vibrational levels of *R(T_1) for which *R(S) and *R(T) have the same energy.

In the case of $^3I(D) \rightarrow {}^1I(D)$ and $^1I(D) \rightarrow {}^3I(D)$ processes, nonbonding orbitals are involved, so the value of J is generally smaller than the gap between vibrational levels and is a strong function of the orientation of the half-filled nonbonded orbitals (Fig. 2.3). In this case, as the orbitals begin to overlap (depending on the orbital separation and orientation), the value of J may vary considerably from values very close to 0 to relatively large values and either S or T may be lower in energy. We discuss the orbital orientational dependence of transitions involving a spin change in greater detail in Section 3.22.

3.19   Couplings of a Spin with a Magnetic Field: Visualization of Spin Transitions and Intersystem Crossing[9]

From the vector model of the triplet state (Section 2.28), we now deduce that for processes such as $*R(S_1) \rightarrow *R(T_1)$ and $^1I(D) \rightarrow {}^3I(D)$ *there are three distinct ISC transitions possible starting from a singlet state (S)*, depending on whether the final state is T_+, T_0, or T_-. These three transitions can be written as $S \rightarrow T_+$, $S \rightarrow T_0$, or $S \rightarrow T_-$, respectively. Likewise, for the important $R(T_1) \rightarrow R(S_0)$ and $^3I(D) \rightarrow {}^1I(D)$ processes, there are three important ISC processes that are possible, namely, $T_+ \rightarrow S$, $T_0 \rightarrow S$, and $T_- \rightarrow S$.

As an exemplar of how the vector model for electron spin provides a general visualization of a radiationless transition involving a spin change in a $^3I(D)$ species, let us consider the ISC of a T_0 state of $^3I(D)$ to a degenerate S state of $^1I(D)$ or vice versa (Fig. 3.8, top) and the ISC of a T_+ state of $^3I(D)$ to a degenerate S state of $^1I(D)$ or vice versa (Fig. 3.8, bottom). On the left and right sides of Fig. 3.8, the two electron spins, S_1 and S_2, are represented as "bonded" or "tightly" coupled to each other (by electron exchange, J) *by showing their precessional cones precessing about a common point connecting the two cones of precession*. These representations indicate that the two spins are strongly coupled. On Fig. 3.8 (left), the two spins are tightly coupled as a triplet (T_0 top or T_+ bottom), and on Fig. 3.8 (right), the two spins are tightly coupled as a singlet (S top and bottom). Intersystem crossing in such strong coupling situations is difficult to achieve because the value of J that couples the two spins is relatively large compared to available magnetic interactions. In other words, compared to the situation for $J = 0$, the singlet and triplet states are split apart from one another by $\Delta E_{ST} = 2J$, requiring a large amount of *magnetic* energy to match the energies of the singlet–triplet gap; exact energy matchings are difficult to find in the surroundings (in the language of magnetic resonance, the spectral density of any particular magnetic energy in the surroundings is small).

For ISC to occur, the electron exchange coupling J must decrease to a very small value that is comparable to that of magnetic torques available in the system and there also must be some magnetic interaction that can act selectively on one of the spin vectors (i.e., S_2 in Fig. 3.8, top, and Fig. 3.8, bottom). There is a decrease in the value of J (Eq. 3.20) that occurs when the electrons and their spins separate in space. This decrease in coupling is indicated in Fig. 3.8 by showing *the precessional cones separated by a dotted line to represent a "weakening" of the exchange interaction as the spins separate.* This weakening of the coupling is commonly produced when the two spins can separate from one another in space, thereby reducing the orbital overall and consequently coupling the spins and reducing the value of J. The latter is proportional to the orbital overlap and falls off exponentially (Eq. 3.20) as the electrons separate in space.

From the vector model for electron spin (shown in Fig. 3.8), we expect two distinct mechanisms for ISC: (1) the magnetic interaction with S_2 may occur along the z-axis, causing a "rephasing" of the spin vector S_1 relative to S_2 (the $T_0 \rightarrow S$ transition in Fig. 3.8, top); and (2) the magnetic interaction with S_2 may occur along the x- or y-axis, causing a spatial reorientation (along the z-axis) of the S_2 vector relative to the S_1 vector, called a "spin flip" (the $T_+ \rightarrow S$ transition in Fig. 3.8, bottom). To visualize

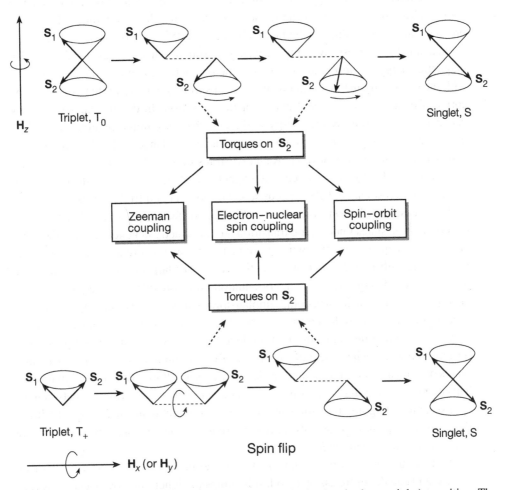

Figure 3.8   Magnetic couplings that cause a triplet-to-singlet (or singlet to triplet) transition. The torque on spin S_2 may result from the coupling of the magnetic moment of S_2 with magnetic moments due to any one of a number of sources, as indicated in the figure.

the $T_- \rightarrow S$ transition (not shown), start with two coupled down spins and perform an operation analogous to that in Fig. 3.8, bottom. These two ISC mechanisms are the same as the longitudinal T_1 relaxation in NMR (i.e., the transfer of spin energy to the environment or lattice leading to the Boltzmann distribution along the z-axis) and the transverse T_2 relaxation (i.e., the randomization of spins in the x–y plane) of magnetic nuclei in a magnetic field, respectively.

What magnetic forces are generally available to interact and provide torques on the magnetic moments of S_1 and S_2? Three of the most common sources of magnetic torques (Fig. 3.8, middle) on the spins S_1 or S_2 are (1) coupling with the magnetic moment of an applied laboratory magnetic field (called "Zeeman coupling"), (2) coupling with the magnetic moment of a nuclear spin (called "hyperfine coupling"), and (3) coupling with the magnetic moment due to orbital motion (called "spin–orbit coupling").

For *R, generally only the spin–orbit interactions are strong enough to induce ISC. For I(D), however, spin–orbit interactions, Zeeman interactions, and electron spin–nuclear spin interactions can induce ISC. Now, we show how the vector model provides an understanding of the factors influencing ISC in both *R and I(D).

In a radiative process involving the change in spin orientation (Scheme 2.2a and b), magnetic energy is conserved by coupling the magnetic moment of the electron spin with the oscillating magnetic moment of the electro*magnetic* field that possesses the correct frequency (energy) and phase for the transition. In detail, magnetic energy is conserved precisely by the energy of the absorbed or emitted photon (i.e., $\Delta E = h\nu$, where ν is the frequency required to achieve resonance and ΔE is the energy gap between the energy levels). Radiative transitions induced between magnetic sublevels (T_+, T_0, and T_-) studied in a magnetic field are known as ESR (or electron paramagnetic resonance, EPR) spectroscopy of doublet and triplet states.

In the case of radiationless transitions between different spin states (Scheme 3.2c, d, and e), if the states undergoing transition are not exactly degenerate, then the magnetic energy gap between the two states undergoing the transition must be conserved by coupling with a third source of magnetic energy. The larger the energy gap between the two states undergoing the spin reorientation, the more difficult effective coupling becomes, and the transition becomes implausible or very slow. The energy-conserving process for the radiationless transition may be accomplished by coupling the transition to the oscillating magnetic field produced by the motion of the components of the environment, such as the molecules of a solvent (which possess the magnetic moments of nuclear spins and oscillating electric fields resulting from dispersion forces). This energy-conserving process is viewed as a magnetic energy transfer between the spin system undergoing transition and some magnetic moments that are oscillating at the correct frequency in the solvent (the oscillating magnetic species in the solvent environment are called the *lattice*).

To maintain energy conservation, the lattice may either provide or absorb magnetic energy, and can therefore assist in both spin jumps to higher- and lower-energy levels. With this description for the lattice, we can imagine that the oscillating magnetic moments behave just like the oscillating magnetic field of electromagnetic radiation. Instead of photons, the lattice provides "phonons" or quanta of magnetic energy to the spin system. The lattice thus behaves analogously to a lamp emitting magnetic phonons or an absorber of magnetic phonons. The overall process is called "spin–lattice magnetic relaxation" and is simply magnetic energy transfer between spin states. The most important interaction that couples the electron spin to the lattice is usually a dipolar magnetic interaction (Eq. 2.35).

3.20   Vector Model for Transitions between Magnetic States[9]

Selection rules for the plausibility of transitions involving a spin change in organic molecules are based on the conservation of energy and the conservation of angular momentum. The spin selection rule states that during an electronic transition the two states undergoing transition must have the same energy and the electron spin must either remain unchanged or *change by one unit of angular momentum (i.e., by ℏ).* This selection rule can be satisfied only when a spin change is exactly compensated by an equal and opposite change in angular momentum that occurs from some other (coupled) interaction with another source of angular momentum. Any form of coupled quantum mechanical angular momentum equal to $ℏ$, not necessarily just electron spin angular momentum, will conserve angular momentum. For example, a photon possesses an angular momentum of $ℏ$, so it can couple to an electron spin and in principle induce any of the plausible spin transitions between two spin states for which the spin changes by exactly one unit. Thus, the conservation of angular momentum is the basis for the rule that the change in spin must be exactly $ℏ$ for radiative transitions. As another example, a proton or a ^{13}C nucleus possesses a nuclear spin angular momentum $\mathbf{I}$ of $ℏ/2$, so hyperfine coupling with these nuclei can induce any of the plausible spin transitions if the change in the nuclear spin angular momentum (e.g., $+ℏ/2 \rightarrow -ℏ/2$) is exactly the same as the change in the electron spin angular momentum $\mathbf{I}$ (e.g., $-ℏ/2 \rightarrow +ℏ/2$). Finally, if an electron jumps from an s orbital ($l = 0$) to a p orbital ($l = \pm 1$), or vice versa, the electron orbital angular momentum ($\mathbf{L}$) changes by $ℏ$; if the orbital jump is coupled with a spin change of $ℏ$, then spin–orbit coupling can induce the spin change. With this brief review of the conservation laws, we can now describe how spin–orbit coupling operates to induce the spin change in molecules where J is large and then how it operates in diradicals where J is small.

The preceding examples of angular momentum conservation refer to systems for which exactly one unit of angular momentum is transferred between interacting species. This situation is the case when the angular momentum can be characterized by a "good quantum number," such as 0, 1/2, 1, and so on. This situation holds pretty well for atoms that possess spherical symmetry and diatomic molecules that possess axial symmetry, but not for other molecules. However, angular momentum still must be maintained for any transitions in molecules. In general, appealing to the atomic model for which angular momentum is well defined can be useful as a starting point to determine the structural mechanism by which angular momentum is conserved.

3.21   Spin–Orbit Coupling:[8,10] A Dominant Mechanism for Inducing Spin Changes in Organic Molecules

According to perturbation theory, if a molecule, such as an electronically excited state *R, is prepared in a pure spin state, an interaction that either rephases or flips electron spins will cause the pure initial state to *evolve in time* to an oscillating mixture of singlet and triplet states. We have seen a number of times that the oscillation of two coupled quantum states corresponds to resonance, which is a characteristic feature of interacting waves. In addition to obeying the conservation laws of energy and momentum, there are two other fundamental requirements for an interaction to be effective in causing a resonance that will lead to a transition between two electronic states. First, there must be an interaction that causes a finite mixing of the electronic wave functions, and second, the magnitude of the energy separating the states must be on the order of the magnitude of the energy of the interaction (or smaller). The coupling of the angular momenta associated with electron spin and with orbital motion (spin–orbital coupling) is an important interaction for mixing different spin states for both *R and I(D).

The selection rules for spin changes induced by spin–orbital coupling may be deduced from the value of a matrix element for the orbitals involved in the transitions. In zero order, the matrix element for the coupling energy has the form $< \psi_1|H_{so}|\psi_2 >$, where H_{so} is the operator for spin–orbit coupling and ψ_1 and ψ_2 are the initial and final orbitals involved, respectively. It is convenient to visualize spin–orbit coupling as the interaction between the magnetic moment (μ_S) due to the electron's spin angular momentum (S) with the magnetic moment (μ_L) due to the electron's orbital angular momentum (L). The strength of the magnetic coupling depends on the orientation of the magnetic moments, as well as their magnitudes and separation in space (Section 2.39). The operator H_{so} represents the interaction of the spin–orbit coupling and has the form of Eq. 3.21, where ζ_{so} is the *spin–orbit coupling constant* and is related to the nuclear charge that the electron "sees" as it orbits key atoms involved in the ISC step. The magnitude of spin–orbit coupling (E_{so}) is given by a matrix element that has the form of Eq. 3.22. The magnitude of the spin–orbit coupling constant ζ_{so} for organic molecules containing only "light" atoms (e.g., H, C, O, N, and F) is typically much smaller (~ 0.01–0.1 kcal mol^{-1}) than vibrational energies (~ 5–0.5 kcal mol^{-1}), but the magnitude of ζ_{so} for "heavy" atoms (e.g., Br or Pb) can be large and approach and even exceed the values of vibrational energies.

$$\mathbf{H}_{so} = \zeta_{so}\mathbf{S}\mathbf{L} \sim \zeta_{so}\mu_S\mu_L \tag{3.21}$$

$$E_{so} = < \psi_1|H_{so}|\psi_2 > = < \psi_1|\zeta_{so}\mathbf{S}\mathbf{L}|\psi_2 > \sim < \psi_1|\zeta_{so}\mu_S\mu_L|\psi_2 > \tag{3.22}$$

We can select a specific, simple physical exemplar of a single electron atom (of nuclear charge Z) to understand how H_{so} operates on *atomic* wave functions and how this operation has a major impact on the strength of spin–orbit coupling. For atoms, the dependence on spin–orbit coupling is proportional to Z^4. In molecules, spin–orbit coupling is a local effect that is most effective when the electron is located near a heavy atom, so an atomic exemplar is expected to be appropriate, but the effect of the heavy atom needs to be scaled to reflect the probability that the electron of interest will "see" the heavy atom. In particular, let us examine the spin–orbit coupling associated with an electron in a p orbital, which is typically an important atomic orbital for both *R and I(D).

An electron in a p orbital possesses one unit of orbital angular momentum ($\hbar$), exactly the amount of angular momentum required to change an electron's spin orientation from $\alpha(+1/2) \rightarrow \beta(-1/2)$ or from $\beta(-1/2) \rightarrow \alpha(+1/2)$. *For angular momentum to be conserved during a change in spin orientation on a unit of angular momentum, the p orbital that is coupled to the electron spin must change its orientation by exactly one unit of angular momentum.* The p orbital can produce one unit of orbital angular momentum by undergoing a rotation of 90° about an arbitrary z-axis, a move that is equivalent to rotating and overlapping with an adjacent p orbital

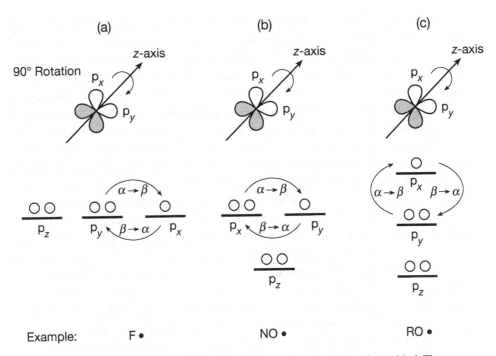

Figure 3.9   Schematic description of the effect of H_{so} on the orientation of a p orbital. The parameter H_{so} "twists" the p orbital 90°. See Figure 3.8 for the corresponding spin change.

(Fig. 3.9, top). It is this requirement to rotate about an axis in order to conserve angular momentum that is reflected in the form of the operator H_{so}. We can translate this quantum mechanical view of spin–orbit coupling into a physical picture by imagining the electron undergoing a rotation about the z-axis. The rotation is favored when the energy of the p_x and p_y orbitals are identical (Fig. 3.9a); when the energies differ, rotation is more difficult, because the electron is "trapped" in the p orbital of lower energy (Fig. 3.9c). We say that rotation is inhibited or that the angular momentum of the electron is "quenched."

Now, let us visualize the influence of H_{so} on p orbitals of the same principal quantum number (e.g., the three 2p orbitals) for the three exemplar orbital energy situations shown in Fig. 3.9. The exemplar in Fig. 3.9a corresponds to the situation for an atom (e.g., a fluorine atom, F•) that has spherical symmetry. In this case, all three p orbitals (p_x, p_y, and p_z) are energetically degenerate. The exemplar in Fig. 3.9b corresponds to the situation for a diatomic molecule (e.g., nitric oxide, NO•) that has cylindrical symmetry. In this case, two of the orbitals (defined as p_x and p_y, the two orbitals in the in the x–y plane) are degenerate but have a different energy from the p_z orbital along the z-axis, which is the bond axis. The p_z orbital is the one involved in bonding between the two atoms of the diatomic molecule, and therefore is shown as somewhat lower in energy than the p_x and p_y orbitals. The exemplar in Fig. 3.9c corresponds to the most common situation, namely, the one in which all three p orbitals have different energies (e.g., an alkoxy radical, RO•). In this exemplar the half-filled p_x orbital is highest in energy.

For the exemplars in Fig. 3.9a and b, the $p_x \rightarrow p_y$ jump can occur between orbitals of exactly the same energy, but for the exemplar in Fig. 3.9c, the $p_x \rightarrow p_y$ jump must occur between orbitals of different energies. From perturbation theory, the mixing of states with similar or identical energies is much stronger than the mixing of states with different energies (Eq. 3.5). The mixing of spin states corresponds to the generation of *orbital angular momentum* due to a $p_x \rightarrow p_y$ jump, which is required to couple with the spin angular momentum change with the orbital angular momentum change. From Fig. 3.9, we can visualize the circular orbital motion "twisting" about an axis, which is the essence of the generation of angular momentum. The larger the orbital angular momentum generated, the larger the magnetic moment generated by orbital motion (Eq. 2.31), and the stronger the spin–orbit coupling. Thus, the closer the system is to the case in Fig. 3.9a or b, the stronger the spin–orbit coupling is, and the faster the rate of ISC is, all other factors being similar. When *R possesses an electronic distribution for which there is a significant contribution of a half-filled orbital on a single atom (e.g., the situation for the n orbital of an n,π^* state), strong spin–orbit coupling is plausible. If, on the other hand, *R does *not* possess an electronic distribution for which there is a significant contribution of a half-filled orbital on a single atom (e.g., the situation for π,π^* states), strong spin–orbit coupling is *im*plausible.

Let us consider the magnetic interactions that occur when the p orbital rotates about the z-axis for the exemplar in Fig. 3.9a for an atom. The rotation corresponds physically to the jump of an electron from a p_x orbital to an empty p_y orbital (the p_z orbital is oriented parallel to or along the z-axis), and this rotation generates one unit of angular momentum (i.e., $\hbar$) about the z-axis. This orbital angular momentum possesses an associated magnetic moment (μ_L, Eq. 2.31) that can couple with the spin's magnetic moment (μ_S) and induce spin reorientation or spin rephasing (Fig. 3.8). The coupling of the spin and orbit magnetic moments means that μ_S serves as a torque that tends to twist the p orbital and make it rotate $90°$ around the z-axis of rotation. Reciprocally, μ_L serves as a torque that tends to twist the spin vector from an ↑ to an ↓ (or an ↓ to an ↑) orientation. The orbitally produced torque is effective in triggering spin changes in transitions when there is a significant matrix element coupling the initial and final electronic states.

Consider an initial electronic state ψ_1 and a final electronic state ψ_2 *that differ in spin* and that are coupled by a spin–orbit coupling operator H_{so}. The matrix element $< \psi_1 | H_{so} | \psi_2 >$ of Eq. 3.22 is a measure of the magnitude (energy) of spin–orbit coupling. We saw that the magnitude of matrix elements is large when the wave functions for the initial and final states "look alike." A remarkable feature of the operator H_{so} is that it makes ψ_1 look like ψ_2 by mathematically *rotating* the wave function ψ_1 by $90°$! Let us attempt to visualize the effect of H_{so} on the transition of a p_x to a p_y orbital; that is, let us attempt to visualize the matrix elements as shown in Fig. 3.10. Since the mathematical operation of H_{so} on p_y (the $< H_{so} | p_y >$ part of the matrix element) is to rotate the p_y orbital by $90°$, the mathematical operation H_{so} converts the p_y orbital into the p_x orbital (and vice versa) as shown in Fig. 3.10a and b.

The strength of the spin–orbit interaction depends on the net mathematical overlap of the orbitals when the matrix element $< p_x | H_{so} | p_y >$ is computed. In other words,

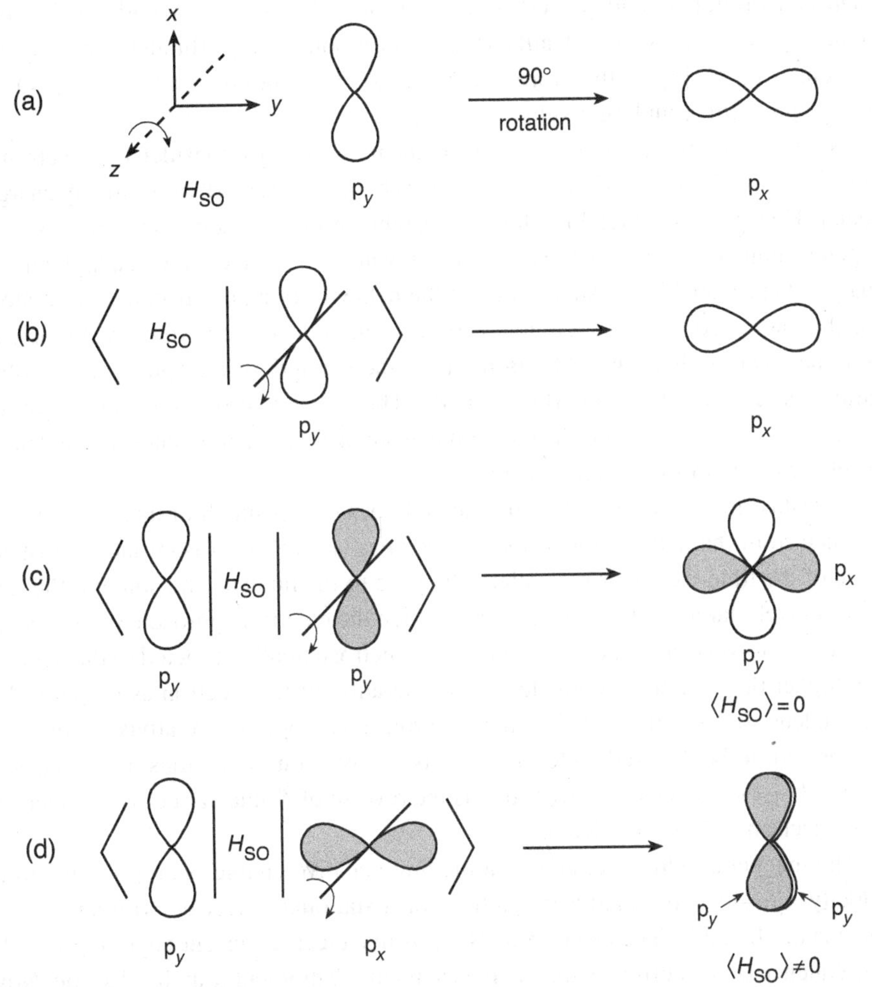

Figure 3.10   Visualization of the matrix element for spin–orbit coupling.

we seek to estimate the magnitude of the orbital overlap integral that results after the mathematical operation, $< |H_{so}|p_y >$, of Fig. 3.10b is performed. The orientation of the p_x and p_y orbitals relative to the x-, y-, and z-axes is shown in Fig. 3.10a. The visualization of the electronic version of the operation $< |H_{so}|p_y >= p_x$ is shown in Fig. 3.10b. Thus, the operation of H_{so} on the p_y orbital rotates it into a p_x orbital. If two p_y orbitals are involved in the spin–orbit coupling, the matrix element is $< p_y|H_{so}|p_y >$ (Fig. 3.10c) and the operation $< |H_{so}|p_y >$ produces a p_x orbital. For this case, the magnitude of the matrix element for spin–orbit coupling is proportional to the overlap integral $< p_y|p_x >$. The orbital overlap of a p_x orbital with a p_y orbital on the same atom is given by $< p_x|p_y >= 0$. Thus, the magnitude of the spin–orbit coupling matrix element $< p_y|H_{so}|p_y >$ is exactly zero! On the other hand, let us now determine what happens when a p_x orbital and a p_y orbital are involved in the coupling. In this case, the matrix element is $< p_y|H_{so}|p_x >$ (Fig. 3.10d), and the operation of H_{so} on p_x

rotates the orbital $90°$ and converts it into a p_y orbital. Now, the two orbitals involved in the interaction look exactly alike (Fig. 3.10d); both are p_y orbitals! For the matrix element $< p_y|H_{so}|p_x >$, therefore, there is a good overlap integral $(< p_y|p_y >)$ and strong spin–orbit coupling results.

For electrons in molecules, a more realistic view is to consider the electron as moving in an orbital, that is, moving in space about a framework of positively charged nuclei. However, considerable physical insight into the reason spin–orbit coupling depends dramatically on the presence of heavy atoms (i.e., atoms with a high nuclear charge, Z) is available via an analysis of the behavior of an electron in a Bohr atom, which is assumed to be in a circular orbit moving at a constant velocity. In the case of a one-electron, hydrogen-like Bohr atom, the strength of the spin–orbit coupling parameter (Eq. 3.21) is proportional to Z^4. Thus, a very strong dependence on the nuclear charge is expected if the electron is in an orbit that approaches the nucleus as the electron can in a one-electron atom.

In reality, the electron is in an orbital and does not spend all of its time at a fixed distance from the nucleus (as it does in the circular orbit of a Bohr atom). Instead, some of the time the electron is relatively close to the nucleus and some of the time it is relatively far away from the nucleus. When the electron approaches a nucleus, it accelerates to very high speeds because it is electrostatically attracted to the nucleus. The higher the Z value, the greater the acceleration of the electron as it approaches the nucleus. Indeed, the speed of the electron must approach relativistic velocities in order not to be "sucked" into the nucleus (on very rare occasions, the electron is captured by the nucleus, a process that is the reverse of β-nuclear decay, the ejection of an electron from the nucleus).

The magnetic field generated by a moving negative charge is proportional to its velocity. Consider the relativistic picture for a stationary electron with an orbiting nucleus of charge Z. The electron feels a positive electric current engulfing it and a consequence of the current is that there is a magnetic moment μ at the electron. Now, suppose the electron is moving in the vicinity of the nucleus at relativistic velocities. The electron's associated orbital magnetic moment (μ_L) may then be huge, on the order of 1 million G. Thus, the coupling of the spin (μ_S) and orbital (μ_L) magnetic moments will be a maximum when the electron is accelerating near Z. The larger the charge Z on the nucleus, the larger the acceleration the electron must achieve to avoid being sucked into the nucleus. As a result, there is a heavy-atom effect (more accurately, a nuclear-charge effect, Z) on the rate of spin–orbit induced transitions. The important conclusion is that as the Z felt by an orbiting electron increases, the degree of spin–orbit coupling increases, if the electron is in an orbital that allows a close approach to the nucleus to occur. Although this description is for a one-electron atom, a strong dependence of the extent of spin–orbit coupling on the value of Z is expected (for electrons in orbitals that can "see" the positive charge of the nucleus) and the strength or energy (E_{so}) of spin–orbit coupling is proportional to the spin–orbit proportionality constant (ζ_{so}) of Eqs. 3.21 and 3.22.

From this simplified, but useful model of spin–orbit coupling, the following generalizations can be made for spin–orbit coupling in molecules:

1. The strength, or energy (E_{so}), of spin–orbit coupling is directly proportional to the magnitude of the magnetic moment due to electron orbital motion (μ_L, a variable quantity depending on the orbit) and the electron spin (μ_S, a fixed quantity).

2. The magnitude of E_{so} will increase, for a given orbit, as the atomic number (Z; i.e., the charge on the nucleus) increases, since both the accelerating force attracting the electron and the spin–orbit coupling constant (ζ_{so}) are proportional to Z.

3. To maximize the effect of the nuclear charge, the electron must be in an orbital that approaches the nucleus closely (i.e., an orbital with some s character), since s orbitals have a finite probability of being located near or even in the nucleus.

4. Regardless of the magnitude of E_{so} for spin–orbit coupling, in order to induce a transition between states of different spin, the total angular momentum of the system (orbit plus spin) must be conserved. For example, a transition from an α-spin orientation to a β-spin orientation (angular momentum change of one unit) may be completely compensated for by a transition from a p orbital of orbital angular momentum 1 to a p orbital of angular momentum 0 (e.g., a $p_x \rightarrow p_y$ transition).

These generalizations lead to the following selection rules for *effective* spin–orbit coupling-induced ISC in organic molecules:

Rule 1.   The orbitals involved in the $p_x \rightarrow p_y$ transition must be similar in energy (Fig. 3.9). For a large energy difference between these orbitals, orbital angular momentum and therefore spin–orbit coupling through orbital angular momentum is "quenched."

Rule 2.   Spin–orbit coupling in organic molecules will be effective in inducing transitions between different spin states if a "$p_x \rightarrow p_y$" orbital transition *on a single atom* is involved because such an orbital transition provides both a means of conserving total angular momentum as well as a means of generating orbital angular momentum that can be employed in spin–orbit coupling (Fig. 3.10).

Rule 3.   Spin–orbit coupling in organic molecules will be effective in inducing transitions between different spin states if one (or both) of the electrons involved approaches a "heavy" atom nucleus that is capable of causing the electron to accelerate and thereby create a strong magnetic moment as the result of its orbital motion (in Eq. 3.21 for a one-electron atom, $\zeta_{so} \sim Z^4$).

Some representative values of ζ_{so} are listed in Table 3.2. Conventionally, for spin–orbit considerations, elements from $Z = 1$ (hydrogen) to $Z = 10$ (neon) are considered to be "light" atoms and elements with $Z > 10$ are considered to be "heavy" atoms.

We conclude the discussion of spin–orbit coupling with two concrete exemplars, namely, the n,π^* state of acetone and the π,π^* state of bromobenzene (Scheme 3.3). From Fig. 3.10, we have seen that a transition between p orbitals provides one of the best mechanisms for coupling spin and orbital angular momentum. Some insight into

Table 3.2   Approximate Spin–Orbit Coupling Parameters for Selected Atoms

Element	Atomic Number (Z)	ζ_{so} (kcal mol^{-1})
Hydrogen	1	<0.1
Carbon	6	0.1
Nitrogen	7	0.2
Oxygen	8	0.4
Fluorine	9	0.8
Chlorine	17	2
Bromine	35	7
Iodine	53	14

the probability of such a mechanism can be obtained by inspecting Lewis structures for *R. In the case of acetone, there are two resonance structures (Scheme 3.3, left) that approximate the n,π^* state. The single "dot" represents an electron in a p orbital on the oxygen atom and the single "x" represents an electron in an antibonding orbital. In both structures **1a** and **1b**, there is an odd electron in a p orbital that is on the oxygen atom. An odd electron on an oxygen atom corresponds to Fig. 3.9c, the alkoxy radical, which has an available p$_x$ → p$_y$ transition that can serve to induce spin–orbit coupling.

In the case of bromobenzene, there are a number of resonance structures that approximate the π,π^* state. Structure **2a** corresponds to the π,π^* state with excitation on the benzene group ("x" represents a π or π^* electron) and structure **2b** corresponds to the π,π^* state with excitation on the bromine (Br) atom (the "dot" on the bromine atom). To the extent that structure **2b** contributes to the π,π^* state, there will be a certain probability that one of the electrons in a half-filled orbital will be located for a period of time on the Br atom. During this time the electron will both experience the strong positive charge of the Br nucleus ($Z = +35$) and have an opportunity to undergo a p$_x$ → p$_y$ orbital jump, both of which favor spin–orbit coupling.

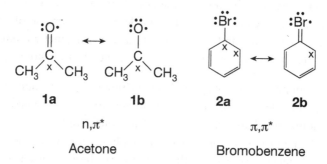

Scheme 3.3   Lewis structure representations of the n,π^* state of acetone and the π,π^* state of bromobenzene.

3.22 Coupling of Two Spins with a Third Spin: $T_+ \to S$ and $T_- \to S$ Transitions[11]

Figure 3.11 provides a vector model visualization of the important case of singlet–triplet and triplet–singlet ISC that is "catalyzed" by a third spin. In Fig. 3.11a, two electron spins, S_1 and S_2, are shown as coupled in a T_+ ($\alpha\alpha$) state (the coupling is indicated by showing the resultant vector produced by coupling and precession of S_1 and S_2 about the resultant). Now, suppose that a third spin, either an electron spin or a nuclear spin (represented as H_i in Fig. 3.11a), is capable of coupling specifically to the spin S_2 (Fig. 3.11a, middle). When S_2 and H_i are coupled, they precess about their resultant vector (Fig. 3.11a, right). The coupling of S_2 to H_i causes S_2 to precess about the x- or y-axis and to oscillate back and forth between the α and β orientations (Fig. 3.11b). According to the vector diagram, this oscillation produced by the coupling of S_2 and H_i causes triplet (T_+) to singlet (S) ISC to interconvert in an oscillating manner, as shown schematically in Fig. 3.11b. Thus, the third spin "catalyzes" the $T_+ \to S$ ISC and the $S \to T_+$ ISC.

At zero field, the singlet state (S) and the three triplet sublevels (T) are degenerate. Fig. 3.11c shows the situation for $J = 0$ at zero field ($H_z = 0$), and Fig. 3.11d shows the intersystem crossing between T and S for $J = 0$ at high field ($H_z \gg 0$). In both cases, the singlet–triplet gap between T_0 and S is zero; however, when a strong field is present, the T_+ and T_- states are split apart in energy from S and T_0 (Fig. 3.11d). In the latter case, radiationless transitions between S and T_+ and T_- are inhibited because they are not degenerate with S. However, transitions between T_0 and S are not inhibited by the strong field because the two states continue to be degenerate since J is assumed to be 0, even in the presence of a strong field. On the other hand, the splitting of the triplet sublevels allows for radiative transitions to occur between adjacent sublevels.

There are no possible radiative transitions between T and S at zero field because there is no energy gap between the states, and an energy gap is required for a radiative transition. If a photon is absorbed, there must be a transition between two states of different energy to accept the photon's energy if energy is to be conserved. Transitions between the three triplet sublevels (T_0, T_+, and T_-) are spin allowed and can occur in a radiationless or a radiative manner. The former is called electron spin sublevel relaxation.

At high field, the $T_+ \to S$ and $T_- \to S$ ISC transitions require some source of magnetic energy conservation by coupling with the lattice. This coupling is generally inefficient because there are so few states in the lattice (the environment's spectral density) with the correct magnetic energy and coupling. The plausibility of a radiative $T_+ \to S$ (and $T_- \to S$) ISC transition depends on the relative coupling of the electron spins to one another and to the radiative field. If the value of J is very small, the individual spins behave more or less independently, so that independent radiative transitions of each spin ("doublet" transitions) become plausible.

The vector diagram for the $T_- \to S$ transition can be constructed by comparing the symmetry relationships of the T_- vector representation to that of the T_+ vector representation. The conclusion from Fig. 3.11c and d is that when $J = 0$ and $H_z = 0$,

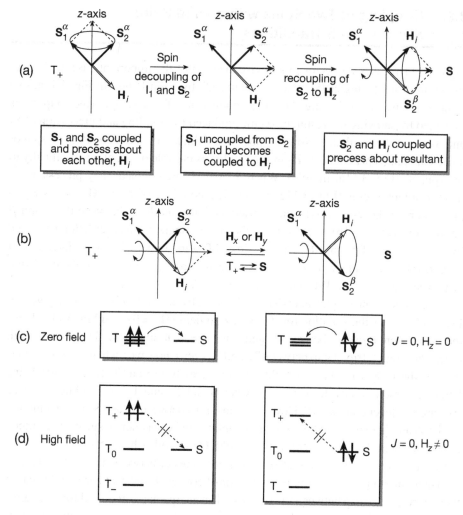

Figure 3.11   Schematic vector representation of two coupled spins coupled to a third spin along the x- or z-axis.

ISC is plausible between all three triplet sublevels and between each of the three triplet sublevels (T_+, T_0, and T_-) and S but that, at high field, direct ISC between T_+ (and T_-) and S is implausible and therefore slow.

3.23   Coupling Involving Two Correlated Spins: $T_0 \rightarrow$ S Transitions

It is also possible for $\mathbf{H}_i$ to operate on coupled electron spins along the z-axis and to cause an ISC from $T_0 \rightarrow$ S (Fig. 3.12). This situation is distinctly different from that shown in Fig. 3.11a for the $T_+ \rightarrow$ S ISC. As an exemplar, consider starting from an initial T_0 state ($\mathbf{S}_1 = \alpha$ and $\mathbf{S}_2 = \beta$, Fig. 3.12a). Under the condition that $J = 0$, rephasing along the z-axis occurs if $\mathbf{H}_i$ is coupled selectively to one of the electron

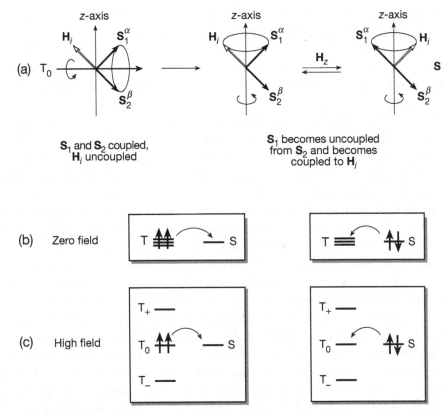

Figure 3.12 Vector representation of two correlated spins in a T_0 state coupled to a third spin along the z-axis that produces intersystem crossing to S.

spins (e.g., S_1). This *rephasing* causes the $T_0 \rightarrow$ S ISC at *both* low field and high field if $J = 0$. Although all three triplet sublevels can undergo ISC to S at zero field (or very low field), direct ISC between T_+ (and T_-) and S at high field is implausible and therefore is slow.

3.24   Intersystem Crossing in Diradicals, I(D): Radical Pairs, I(RP), and Biradicals, I(BR)[7,10]

Now, we consider the factors involved in the ISC processes of diradicals, I(D), which are produced in the common $*R \rightarrow$ I(D) primary photochemical processes. Recall that I(D) represents either a radical pair, I(RP) or a biradical I(BR). If the primary photochemical process involves an $S_1 \rightarrow {}^1I(D)$ process, there is no spin prohibition to the subsequent elementary step, $^1I \rightarrow P$, which produces the observed product(s). Only *singlet* radical pairs, $^1I(RP)$ or *singlet* biradicals $^1I(BR)$ are produced directly as reactive intermediates from S_1. Such species are expected to undergo one of two extremely rapid reactions that are plausible for reactions involving two radical centers, namely, recombination or disproportionation. The rates of these radical–radical reactions for ^{1}I are generally faster than diffusional separation of the fragments of the radical pair from the radical cage or often faster than the stereochemical change of conformation due to rotation about C—C single bonds. Thus, even though S_1 may produce a radical pair or a biradical, ^{1}I(D), the reactions of these species may be highly stereospecific.

However, if the primary photochemical reaction involves a $T_1 \rightarrow {}^3$I(D) step, there is a spin prohibition to the elementary step to the ^{3}I(D) $\rightarrow$ P step. Thus, there will be a time delay imposed on product formation until a mechanism for ISC, ^{3}I(D) $\rightarrow {}^1$I(D), is found somewhere along the reaction coordinate. If the rate of ISC for the triplet radical pair is slow relative to diffusional separation of the radical centers, I(D) will occur with high efficiency. For I(RP), efficient free radical formation will result; that is, ^{3}I(RP) $\rightarrow$ free radicals (FR). In the latter case, all of the products formed will proceed through free radicals and the pathway $T_1 \rightarrow {}^3$I(RP) $\rightarrow$ FR $\rightarrow {}^1$I(RP) $\rightarrow$ P. For ^{3}I(BR), efficient separation of the radical centers will occur, and if the rate of ISC for ^{3}I(BR) is slow relative to the rates of rotation about C—C bonds, loss of any initial stereochemistry will occur even for intramolecular reactions of the biradical. The latter is the typical situation for biradicals produced from $T_1 \rightarrow {}^3$I(BR) processes.

In summary, at some point or another a ^{3}I(D) reactive intermediate, in the process of forming the isolated singlet product P, must undergo intersystem crossing to the singlet surface. Understanding the geometries and orbital orientations of ^{3}I(D) at which such crossings occur is very important and may determine the structure of the products eventually formed from ^{3}I(D). In Section 3.25, we consider some rules relating orbital orientation to the effectiveness of intersystem crossing in ^{3}I(D).

3.25 Spin–Orbit Coupling in I(D): The Role of Relative Orbital Orientation[10]

Spin–orbit coupling, the most significant mechanism for inducing ISC in electronically excited molecules (*R), is also expected to be an important mechanism for ISC in radical pairs and biradicals, I(D). Salem[10a] formulated a set of simple rules for determining whether spin–orbit coupling will be a plausible mechanism for ISC in I(D) based on orbital separation and orientations of the interacting half-filled nonbonding orbitals of I(D). For situations when the orbital orientations and separation do not provide appropriate conditions for effective spin–orbit coupling, new mechanisms for producing magnetic interactions involving nuclear spins and external magnetic fields can operate to induce ISC. These situations lead to magnetic isotope and magnetic field effects on photochemical reactions.

According to *Salem's rules* for ISC in diradicals, ISC is fastest in biradicals and radical pairs when

1. The value of the exchange interaction (J) between the two radical centers is less than the strongest available magnetic coupling mechanism.

2. The nonbonding orbitals of the diradical are in an orbital orientation that can interact to some extent and can create orbital angular momentum that couples with the spin angular momentum during the ISC step.

3. The degree of electron pairing character in the singlet can become significant during the ISC step (Fig. 2.3).

These simple ideas are derived from the rules for the mixing of singlet and triplet states, for the distance dependence of mixing, and for the orientation dependence of mixing of the two half-filled NB orbitals of I(D) that were discussed in Section 2.15. According to rule (1), the electron-exchange rule, the stronger J is, the slower the rate of ISC. In order to mix the S and T states of I(D) effectively, the two states must have essentially the same energy; that is, they must be very close to degenerate. The magnitude of magnetic energies (Table 3.1) is tiny compared to that of electronic interactions. Since J causes the energy of the S and T state to "split," the value of J must approach zero if the energies of the S and the T states of I(D) are to become degenerate and mix effectively. Since the magnitude of J is a function of orbital overlap, and since orbital overlap decreases as an exponential function of distance (Eq. 3.20), the value of J decreases as an exponential function of the distance of separation of the NB orbitals. As a rule of thumb, the nonbonding orbitals of the diradical need to be separated by 5–10 Å for the value of J to be on the order of the value of the magnetic interactions or smaller. The value of the spin–orbit coupling also falls off with orbital overlap, roughly in an exponential manner (Section 3.21), *so as J decreases, and overlap decreases, the effectiveness of the spin–orbit interaction decreases.* As a result, *when the diradical structure possesses radical centers that are far apart (separated by 5–10 Å), spin–orbit coupling is no longer an effective mechanism for inducing ISC, opening the door for very weak interactions of the magnetic field and nuclear–electron hyperfine coupling to serve as the major mechanisms for inducing ISC.*

Next, we consider Salem's rule (2), the "orientation rule" (Fig. 3.13). In order to generate angular momentum, we have seen in Figs. 3.9 and 3.10 that an orbital jump of the "$p_z \rightarrow p_x$" type is required. The best orbital orientation for spin–orbit mixing is when the two nonbonded orbitals of the diradical are at $90°$ with respect to one another (Fig. 3.13a). This orbital orientation for strong spin–orbit coupling is not intuitive, since it is contrary to the orbital orientations that are most favorable for electronic mixing, a more familiar process. The perpendicular orientation in Fig. 3.13 is also favorable for a small value of J, since in the limit of completely orthogonal orbitals, $J = 0$, exactly.

Consider two other orientations (Fig. 3.13b) that are favorable for π bonding (i.e., two p orbitals overlapping and parallel to one another) and σ bonding (i.e., two p orbitals overlapping and "head to head" with one another). These two orientations are very poor for generating spin–orbit coupling, because there is significant orbital overlap (large J) and poor orbital orientation to generate orbital angular momentum. For I(RP) and I(BR), a wide range of relative orientations of the NB orbitals is possible due to the diffusion and rotation of I(D). The three possibilities shown in Fig. 3.13 are limiting cases.

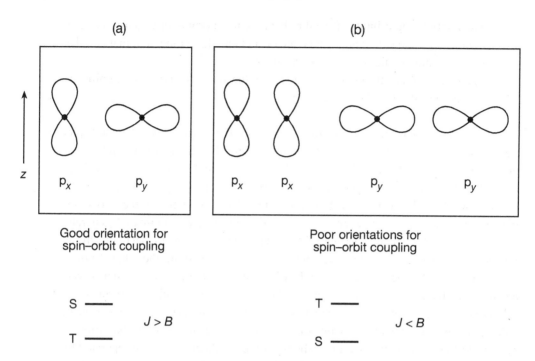

Figure 3.13   Limiting orbital orientations and orientational dependence of spin–orbit coupling. The term *B* refers to the bonding energy due to positive overlap of the two interacting orbitals.

Finally, to better understand Salem's rule (3), the singlet "ionic character" rule (Fig. 3.14), consider the orientation that is most favorable for generating orbital angular momentum (i.e., the $90°$ orientation in Fig. 3.14, left). In order to effectively generate orbital angular momentum, the electron must jump from one orbital to the other, a half-occupied orbital that is at a $90°$ orientation, as the singlet is created. The resulting situation consists of two electrons in a nonbonding orbital to a "zwitterionic" structure we can symbolize as $^1I(Z)$. The parameter $I(Z)$ *must* be spin paired because the two electrons are in the same orbital. Thus, for the most effective creation of angular momentum, the singlet must possess a certain amount of spin-paired character (in Chapter 6, we discuss in detail $^1I(Z)$ states that have spin-paired, zwitterionic character). However, if the $^1I(Z)$ structure possesses spin-paired character for which there is substantial bonding, this state may mix with a bonding singlet ground state that makes the S and T states of the I(D) very different in energy (Section 2.15). This energy splitting of S and T inhibits ISC. In Fig. 3.14, the $p_y \rightarrow p_x$ orbital jump is shown to occur simultaneously with the spin flip from $^3I(D)$ to $I(Z)$. Thus, *a balance of orbital interactions, orbital orientations, and orbital separations contrive to determine the effectiveness of ISC in diradicals.*

For example, imagine starting with a $^3I(D)$ that possesses radical centers exactly in the perpendicular orientation (Fig. 3.14, left). This geometry is very favorable for creating orbital angular momentum but is very poor for generating orbital overlap because the p_y and p_x orbitals are orthogonal. Imagine that the p_x orbital of the radical

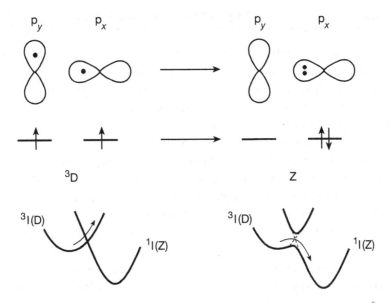

Figure 3.14   Representation of the coupling of the requirements of a 90°
twist and zwitterionic character for effective spin–orbit coupling.

center undergoes a certain amount of rehybridization to an sp^n orbital as the result of an
out-of-plane vibration (Fig. 3.1b). The s character that is added to the orbital provides
some singlet "ionic character" to the overall wave function of $^3I(D)$ and triggers the
$p_y(\uparrow)p_x(\uparrow) \to p_y()p_x(\uparrow\downarrow)$ orbital jump. The combination of orbital distortion through
vibrations can provide a mechanism for mixing $^3I(D)$ and I(Z), and triggering the ISC.

Although the factors responsible for effective ISC by spin–orbit coupling in diradicals are clear from Salem's rules, *we must consider several factors simultaneously* in
order to determine whether ISC is plausible in any actual example. Indeed, the values
of the spin–orbit coupling matrix element are expected to be very dependent on the
instantaneous diradical geometry, because J and the effectiveness of spin–orbit coupling vary with orientation and distance. A small thermal barrier to rotation between
different conformers of a biradical or the relative orientations of a radical pair in a
solvent cage may be required to achieve the optimum geometry for ISC.

In summary, geometries and orbital orientations that are expected to be favorable
for ISC are those for which both the covalent interactions (electron pairing) leading to
singlet bonding and for which the strongest spin–orbital coupling ($p_y \to p_x$ jump) can
occur simultaneously, that is, at geometries for which the localized singly occupied
orbitals overlap sufficiently before, as well as after a 90° rotation of one of the
p orbitals. After the jump to an essentially pure singlet surface from such geometries,
the representative point follows a path on the strongly binding covalent S_0 surface to
products.

In Section 3.26, we consider the effects of geometry on ISC for the concrete
exemplar of a flexible biradical.

3.26   Intersystem Crossing in Flexible Biradicals[10]

The requirement for a ^{3}I(BR) → ^{1}I(BR) → P path to products has a major effect on the kinetics (lifetimes) and product ratios produced from biradicals. The rate of ISC in a flexible ^{3}I(BR) is expected to be proportional to the rate of end-to-end encounters of the radical centers (Fig. 3.15) and the amount of singlet character that the biradicals develop during these encounters. Typically, the amount of singlet character is determined, in turn, by the spin–orbit coupling that occurs during the conformational dynamics of ^{3}I(BR). In some special cases, other couplings such as electron-nuclear hyperfine coupling may be important. If the ^{3}I(BR) → ^{1}I(BR) process occurs when the radical centers are within a few angstroms of separation, the ^{1}I(BR) → P reaction will be very efficient and not rate determining; that is, the ^{3}I(BR) → ^{1}I(BR) will determine the rate of reaction and the lifetime of the biradicals.

Figure 3.15 shows schematically some of the important processes for a flexible ^{3}I(BR) produced, say, by a Type I α-cleavage (Fig. 1.1) T_1 → ^{3}I(BR) primary process of a cyclic ketone. After formation of ^{3}I(BR), two processes occur: (1) a singlet biradical ^{1}(BR) is produced in a conformation that is favorable for ISC, and (2) internal rotation about bonds gives rise to chain dynamics, which changes the end-to-end distance and relative orientation of the NB orbitals continuously during the lifetime of ^{3}I(BR). Since ISC depends on the distance and the orientation of orbitals, the *efficiency of the ^{3}I(BR) → ^{1}I(BR) ISC step is different in each conformation because the distances of the separations and the orientations of the NB orbitals are different.* The structural dynamics of a flexible BR leads to a complicated, but understandable, dependence of the rate of ISC on conformer population, chain dynamics, and the distance and orientation dependence of mechanisms for ISC. Note that the disproportionation and combination products formed from ^{1}I(BR) can only occur from a small subset of conformers having a short end-to-end separation (Fig. 3.15b).

Now, let us continue the discussion of ISC in a flexible BR by considering the role of spin–spin interactions. Electronic spin–spin interactions can be conveniently divided into two types: interactions that affect only the gap between the singlet and triplet state (ΔE_{ST}), and interactions that can cause ISC. There are two interactions that influence (ΔE_{ST}): (1) the very weak *magnetic* interaction of the electron spins with an external field $\mathbf{H}_z$ (Zeeman splitting), which splits T_+ (to higher energy) and T_- (to lower energy) symmetrically about T_0 by the amount $g\beta H$ (Eq. 2.34), and (2) the *electrostatic* electron exchange interaction whose magnitude depends mainly on the biradical conformation and the end-to-end separation of the radical centers.

Figure 3.16 shows two of the key features that will determine the value of J and the influence of spin–orbit coupling for a flexible biradical: the separation of the radical centers and the orientation of the orbitals at the radical centers. When the two radical centers are close (2–3 Å or less), the value of J is large and ISC is slow because there is a large energy gap between ^{3}I(BR) and ^{1}I(BR). When the two radical centers are at an orientation of $90°$ to one another, the situation is most favorable for ISC. In general, from these qualitative considerations, it is expected that ISC will be fastest

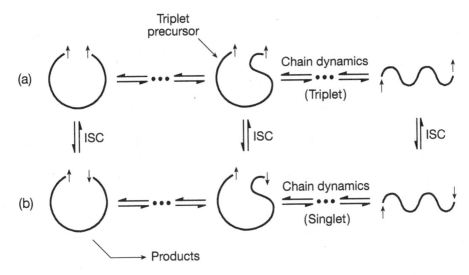

Figure 3.15   Schematic representation of the conformational dynamics of a flexible biradical with ISC occurring at closed (left), intermediate (center), and extended (right) end-to-end distances. The triplet biradical is represented by the structures in (a) and the singlet biradical is represented by the structures in (b).

when ^{3}I(BR) and ^{1}I(BR) are separated by > 5 Å or so for orbital orientations of $\sim 90°$ to one another.

The magnitude of $g\beta H$ for a magnetic field of 10,000 G is on the order of 3×10^{-3} kcal mol^{-1}; the magnitude of J can be close to 0 kcal mol^{-1} for end-to-end separations on the order of 5 Å or greater, but can be many kilocalories per mole for

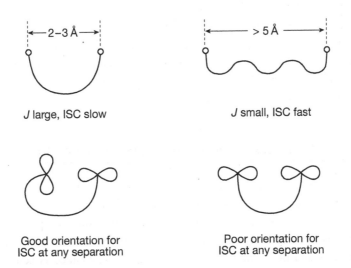

Figure 3.16   Schematic representation of the effect of the magnitude of J and orbital orientation on ISC of a biradical.

end-to-end separations of 3 Å or less (as bonding becomes important and stabilizes the S state). An important point is that the magnitude of the magnetic Zeeman splitting is independent of the separation of the end-to-end separation of the biradical, but the value of J drops off rapidly (roughly exponentially as the end-to-end distance between the radical centers increases, Eq. 3.20).

The two most important interactions that influence the strength of the magnetic forces that cause ISC are (1) the *electron spin–nuclear spin hyperfine interaction (HFI),* and (2) *the electron spin–orbit coupling interaction.* We have seen how spin–orbit coupling is influenced by the conformation of ^{3}I(BR) and ^{1}I(BR). The magnitude of the hyperfine coupling is on the order of 10^{-5} kcal mol^{-1} and *is a constant that is independent of the end-to-end separation.* This contrasts with the magnitude of the spin–orbit coupling, which like J, falls off rapidly as the end-to-end separation of the biradical increases, since it depends on orbital overlap and the magnitude of J (Eq. 3.20).

Thus, the rate of ISC is a function of two distance-independent interactions (the Zeeman and the hyperfine interactions), and two distance-dependent interactions (the electron exchange and the spin–orbit interactions). In zero or very low magnetic fields, the hyperfine interaction is the main mechanism for ISC for large separations for which both J and SOC are small; however, at small end-to-end distances only SOC has a significant magnitude and is the major mechanism for ISC. These considerations lead to the conclusion that for "small biradicals" SOC is likely to be the dominant mechanism for ISC and for "large biradicals" HFI is likely to be the dominant mechanism for ISC.

Depending on the biradical structure, either ISC (triggered by SOC or HFI) or chain dynamics (which determine the rate at which the radical ends come close to one another to achieve a conformation that is favorable for radical–radical reaction) can determine the lifetime and products from a biradical. If the chain dynamics are fast compared to ISC, then the slower latter process will be rate determining; if the chain dynamics are slow relative to ISC, then the former slower process will be rate limiting. For example, at high temperature, chain dynamics are fast and ISC is usually rate determining; at low temperature or in viscous solvents chain dynamics are slow and are usually rate determining. Examples of the control of biradical lifetimes by both chain dynamics or ISC are well established.[11]

3.27   What All Transitions between States Have in Common

The transitions between two electronic states in Scheme 3.1 all have in common the requirement that the total molecular wave functions corresponding to the initial (Ψ_1) and final (Ψ_2) states look alike at the instant of transition. The process of getting two wave functions to look alike requires an interaction that can couple the two wave functions and drive them into *resonance.* The resonance is achieved only if energy and momentum are strictly conserved during the transition. When the conservation laws are obeyed, transitions are induced by perturbations of the molecular wave functions through some appropriate coupling interaction, such as electron–electron interactions, electron–vibration-induced transitions between different vibrational levels, or electron–spin-induced ISC through spin–orbit coupling. The Franck–Condon principle determines the relative probability of radiative and radiationless transitions between electronic states. As a consequence of the principle, the most probable radiative transitions occur "vertically" between electronic states of identical nuclear configuration, and radiationless transitions occur "horizontally" near crossings of PE curves. Intersystem crossing is most probable for organic molecules when the transition involves a p_x to p_y orbital of comparable energy and is capable of undergoing a transition on a single atomic center. We describe some of the details of photophysical radiative transitions in Chapter 4 and of photophysical radiationless transitions in Chapter 5.

References

1. (a) W. Kautzmann, *Quantum Chemistry*, Academic Press, New York, 1957. (b) P. W. Atkins and R. Friedman, *Molecular Quantum Mechanics*, 5th ed., Oxford University Press, Oxford, UK, 2005.

2. For a more detailed discussion of classical mechanics, the reader is referred to any elementary physics textbook. For example, D. Halliday and R. Resnick, *Physics*, John Wiley & Sons, Inc., New York, 1967.

3. W. Kautzmann, *Quantum Chemistry*, Academic Press, New York, 1957, p. 524.

4. P. W. Atkins, *Molecular Quantum Mechanics*, Oxford University Press, Oxford, UK, 1983, Chapter 8.

5. S. P. McGlynn, F. J. Smith, and G. Cilento, *Photochem. Photobio.* **3**, 269 (1964).

6. G. Herzberg, *Spectra of Diatomic Molecules*, van Nostrand, Princeton, NJ, 1950.

7. (a) P. W. Atkins, *Quanta: A Handbook of Concepts*, 2nd ed., Oxford University Press, Oxford, UK, 1991. (b) M. Klessinger and J. Michl, *Excited States and Photochemistry of Organic Molecules*, VCH Publishers, New York, 1995. (c) J. Michl and V. Bonacic-Koutecky, *Electronic Aspects of Organic Photochemistry*, John Wiley & Sons, Inc., New York, 1990.

8. S. P. McGlynn, T. Azumi, and M. Kinoshita, *Molecular Spectroscopy of the Triplet State*, Prentice Hall, Englewood Cliffs, NJ, 1969, p. 183.

9. See the following references for excellent discussions of the vector model of spin and transitions between spin states. (a) P. W. Atkins and R. Friedman, *Molecular Quantum Mechanics*, 5th ed., Oxford University Press, Oxford, UK, 2005. (b) P. W. Atkins, *Quanta: A Handbook of Concepts*, 2nd ed., Oxford University Press, New York, 1991. (c) K. Salikov, Y. Molin, R. Sagdeev, and A. Buchachenko, *Spin Polarization and Magnetic Effects in Radical Reactions*, Elsevier, Amsterdam, The Netherlands, 1984.

10. (a) L. Salem and C. Rowland, *Angew. Chem. Intern. Ed. Engl.* **11**, 92 (1972). (b) C. E. Doubleday, Jr., N. J. Turro, and J. F. Wang, *Acc. Chem. Res.* **22**, 199 (1989).

11. A. Buchachenko and V. L. Berdinsky, *Chem. Rev.* **102**, 603 (2002).

Radiative Transitions
between Electronic States

4.1 The Absorption and Emission of Light
by Organic Molecules

Schemes 1.1–1.3 are key working paradigms used to initiate analysis of organic photophysics or organic photochemistry. In this chapter we are concerned with the portion of the working paradigm of molecular organic photochemistry that involves (1) the absorption (R + $h\nu$ → *R) of light by an organic molecule (R) to produce an electronically excited state (*R), and (2) the emission (*R → R + $h\nu$) of light from an electronically excited state (*R) to produce a ground state (R) (Scheme 4.1). In this chapter both spin-allowed and spin-forbidden radiative transitions will be discussed in detail. Of particular interest are the radiative transitions of light absorption and emission when *R is the lowest singlet (S_1) and triplet (T_1) state of an organic molecule (shown in the rectangle of Scheme 4.1).

4.2 The Nature of Light: A Series of Paradigm Shifts

The accepted paradigm that described the nature of light and its interaction with matter has changed dramatically three times since the 1700s. Each paradigm tried to answer the same questions: What is the nature of light, and What is the nature of the interaction of light with matter? Each paradigm answered the questions in a new and radically different way from that of the previous one, and each new one constituted a true *paradigm shift* for science. With each new paradigm shift, light was positioned as an ever more fundamental entity in the scientific universe; eventually, through the theory of relativity, light was placed at a level of significance equivalent to that of matter.

Historically, there was relatively little documented scientific study and discussion of the nature of light until the 1700s, when Newton proposed a paradigm that light was composed of a stream of *particles*. Newton employed concepts of the motion and

Radiative Processes

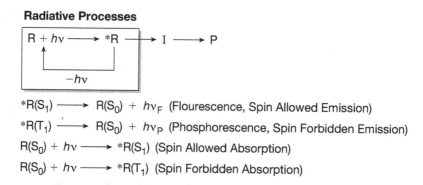

$*R(S_1) \longrightarrow R(S_0) + h\nu_F$ (Flourescence, Spin Allowed Emission)

$*R(T_1) \longrightarrow R(S_0) + h\nu_P$ (Phosphorescence, Spin Forbidden Emission)

$R(S_0) + h\nu \longrightarrow *R(S_1)$ (Spin Allowed Absorption)

$R(S_0) + h\nu \longrightarrow *R(T_1)$ (Spin Forbidden Absorption)

Scheme 4.1   Radiative transitions of organic molecules of greatest importance in organic photochemistry.

energy of point particles as a foundation for the hugely successful structure of classical mechanics. It was therefore natural for Newton to explain the properties of light by postulating that light also consisted of tiny particles emitted from light-producing objects such as the sun or a flame. These particles were imagined to move at great speeds through empty space or through transparent media. Newton's paradigm was supported by the action of a prism, which "decomposed" a ray of white light into its component "particles," possessing the diverse visible colors of the rainbow. Thus, each of the different particles of light could be associated with a color. The sensation of sight was interpreted as the result of the eye being excited by particles of light as they struck the eye. In large part (possibly because of Newton's towering reputation rather than the demonstration of a convincing array of experimental evidence), during the 1700s, the ruling paradigm of the nature of light was that it consisted of particles.

During the early 1800s, new experiments demonstrated aspects of light that Newton's particle theory completely failed to explain. In particular, Newton's theory could not explain the phenomena of *interference* by which two light rays can interact with one another *constructively* to make a more intense light ray or *destructively* to make a light ray completely disappear. On the other hand, interference was a well-known property of waves. For example, the interference of waves is commonly observed when one produces waves by disturbing the surface of a still sample of water. The theory of matter waves readily explained the phenomenon of interference. It seemed obvious that particles could commingle with one another, create a sort of constructive interference, and amplify each other's effect; on the other hand, there was no known way for particles to "cancel each other" and explain the phenomenon of destructive interference. The easily demonstrated and reproducible observation that light rays could interfere with one another was the beginning of the end for the paradigm of light as particles. A paradigm shift was about to occur.

In the mid-1800s, Maxwell put forward a new paradigm for the nature of light, proposing that light is composed of a *force field of oscillating electric charges that have the characteristics of waves, not particles.* If electric charges are oscillating, they generate not only an oscillating electric field but also an associated oscillating magnetic field. In this paradigm, the wave nature of light was contained in an elegant set of mathematical equations (Maxwell's equations), which described light as a wave, driven by an oscillating electromagnetic field surrounding oscillating charged particles (later identified as negatively charged electrons and positively charged nuclei). Maxwell's paradigm for the nature of light also provided a beautiful, previously unrecognized synthesis of the electrical and magnetic forces in both light and matter. Maxwell's equations quantitatively explained the phenomena of interference, scattering, reflection, and refraction. Maxwell's paradigm probably possessed a special appeal to many scientists because it was formulated in an elegant mathematical language and because it integrated electrical and magnetic phenomena. By the end of the 1800s, Maxwell's paradigm for light as a form of electromagnetic waves was universally accepted by the scientific community and was widely considered a universal and unshakable paradigm of physics. Physicists refer to Maxwell's paradigm as the *classical paradigm of light as electromagnetic waves.*

In spite of its mathematical elegance and ability to integrate electricity and magnetism, Maxwell's paradigm for the wave nature of light ran into serious difficulties because of certain experiments that were unexplainable if light was fundamentally an electromagnetic wave. Toward the end of the 1800s, the validity of the classical paradigm for the electromagnetic wave nature of light was called into question because of its inability to explain two very simple experiments, one involving the *emission of light* and the other involving the *absorption of light*: (1) the first experiment dealt with the measurement of the wavelength dependence of the energy distribution for light *emitted* by a hot object, such as a heated metal bar (so-called *black-body* radiation, Fig. 4.1); and (2) the second experiment dealt with the measurement of the

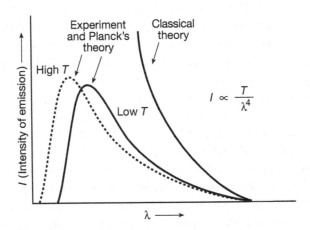

Figure 4.1   The UV catastrophe. The classical theory of light predicted that the intensity (I) of light emitted by a metal bar should be proportional to the temperature (T) of the bar and the inverse fourth power of the wavelength (λ) of the emitted light.

wavelength (or frequency) dependence of the kinetic energy (KE) for electrons emitted when light is *absorbed* by a metal (the so-called *photoelectric effect*, Fig. 4.2). At the beginning of the 1900s, the issue of the nature of light and its interaction with matter to produce the absorption and emission of light was a source of great debate among physicists.

4.3   Black-Body Radiation and the "Ultraviolet Catastrophe" and Planck's Quantization of Light Energy: The Energy Quantum Is Postulated

The energy distribution of light emitted from a heated metal bar is found experimentally (Fig. 4.1) to depend on temperature (T). At lower temperatures the heated bar emits at wavelengths, λ (or related frequencies, ν), that maximize in the infrared (IR) range; as the temperature of the metal is increased, the bar begins to glow red, green, then white, as all wavelengths in the visible (vis) region of the spectrum are emitted. A typical heated metal bar emits an energy distribution that possesses a maximum in the visible or ultraviolet (UV) region of the electromagnetic spectrum. The distribution of wavelengths emitted by the heated metal was modeled by a "black body," or a sample of matter that absorbs all of the light energy that strikes it. Black-body radiation is simply the emitted radiation of the electromagnetic field (light) that is in equilibrium with the body (matter) at a given temperature. The classical theory of light predicts that the electromagnetic field associated with the black body will possess a certain distribution of wavelengths (frequencies) characteristic of the temperature of the black body (Fig. 4.1).

In quantitative terms, the classical theory predicted (Fig. 4.1, right) that the intensity I (energy per unit time) of emitted light by a metal bar at a given temperature (T) is proportional to T/λ^4. Thus, according to classical theory, as $\lambda \to 0$, *the intensity of the emitted light should become infinite*, a preposterous prediction! If the intensity I indeed was proportional to $1/\lambda^4$, then a firefly, when it flashed its light, would release sufficient energy to cause the entire universe to be destroyed! The prediction of unlimited energy in the high-energy (UV) region of the spectrum was named (possibly by a perplexed physicist) the "ultraviolet catastrophe."

The beginning of a new paradigm shift for the nature of light was triggered by Planck's explanation of how to avoid the ultraviolet catastrophe and to theoretically fit the experimental data. Planck's radical paradigm-shifting contribution was to show *mathematically* that the energy distribution of a heated metal would have a maximum, in excellent agreement with experiment (Fig. 4.1), if the *energy of a light wave was quantized* and *if the energy of light was directly related to frequency* by a remarkably simple and now familiar relationship (Eq. 4.1). From this equation, the energy of a light is assumed to be directly proportional to its frequency (ν) through a *proportionality constant* (h), which *required* a value of 6.6×10^{-34} J s in order to fit the experimental data. We now know that this constant, h, is a fundamental quantity in quantum chemistry and its appearance in any unit indicates the quantity obeys the laws of quantum mechanics. In honor of his contribution, h is now known as Planck's constant and the bit of energy, E (Eq. 4.1), that corresponds to the frequency ν is known as a *quantum*. When we say that the energy of the electromagnetic field is quantized we understand that this means light of a given frequency ν can only be absorbed (or emitted) in energy steps equal to $h\nu$ [i.e., light cannot be absorbed (or emitted) continuously]. However, there is no limitation to the value that ν can take.

$$E = h\nu \tag{4.1}$$

Recall that the classical theory viewed light as an electromagnetic field that was created by oscillating electric charges. Using the classical harmonic oscillator (Chapters 2 and 3) as a guide, it was assumed that the charges could be stationary or they could oscillate. Charges that were not oscillating were considered to be in a "ground state" and incapable of emitting light; charges that were oscillating were considered to be in an "excited state" and to be capable of emitting light energy on return to the ground state. Planck's basic idea, which solved the paradox of the ultraviolet catastrophe, was that at a given temperature the *energy* ($h\nu$) associated with a quantum of very short-wavelength light (a very high-frequency oscillator) was so high that the electric oscillators contained in a light wave that are associated with very short wavelengths are not excited; the energy available to the black body was insufficient to excite very short-wavelength (high-frequency) oscillators. It was a very simple idea: As the wavelength, λ, of emitted light decreases, the energy, $E = h\nu$, associated with the quantum required to excite the oscillator increases. Thus it become less and less likely that the high-energy oscillator will be excited at a given temperature, since energy must be conserved. Since the high-energy oscillator is not excited, it cannot emit light (oscillators in their ground state cannot emit light according to the classical theory of a harmonic oscillator). The quantization of energy effectively discriminates against the population of the short-wavelength, high-frequency oscillators, and thereby mathematically eliminates the ultraviolet catastrophe by preventing the intensity of emitted light to approach infinity as the wavelength decreases. Of course, Planck's mathematical "trick," while impressive, was completely nonintuitive based on classical physics and completely at odds with the paradigm of the classical theory of light, which viewed the absorption or emission of light as being continuous, and therefore able to be associated with any arbitrary energy.

4.4   The "Photoelectric Effect" and Einstein's Quantization of Light—The Quantum of Light: Photons

For proponents of the classical paradigm of light, another experimental observation called the "photoelectric effect" (Fig. 4.2) was just as baffling as the ultraviolet catastrophe. The *photoelectric effect* was the name given to the following phenomenon: When light of a certain wavelength λ (or associated frequency, ν) is *absorbed* by a metal surface, electrons are emitted from the metal (the phenomenon is the basis for the common and familiar "electric eyes" that open and close "automatic" doors). The ejected electrons possess a certain amount of KE. The *maximum* KE of the ejected

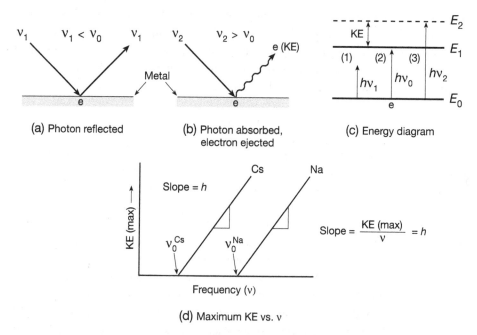

(a) Photon reflected

(b) Photon absorbed,
electron ejected

(c) Energy diagram

(d) Maximum KE vs. ν

Figure 4.2   The photoelectric effect. (a) Light of frequency $\nu_1 < \nu_0$ is reflected from the surface of a metal. (b) Light of frequency $\nu_1 > \nu_0$ is absorbed by the metal and ejects an electron, which possesses a certain measurable KE. (c) Energy diagram showing the relationship of the frequency of light striking the metal, the energy (E_1) required to eject an electron from the surface of the metal, and the excess KE possessed by the ejected electron. For $h\nu_1$ there is insufficient energy to eject an electron; for $h\nu_0$ there is just enough energy for the light to be absorbed; for $h\nu_2$ the absorbed light contains sufficient energy to cause electrons to be ejected with excess KE. (d) Plots of the excess KE as a function of frequency of absorbed light for Cs (left) and Na (right).

electrons can be readily measured experimentally as a function of ν (or λ) for the absorbed light. However, the frequency of the light, ν, capable of emitting electrons had to be greater than a certain minimum value, ν_0.

Einstein showed that the simple relationships of Eqs. 4.2 and 4.3 provide a quantitative relationship between the maximum KE of the emitted electrons and the experimentally measured frequencies ν and ν_0.

$$\text{Maximum KE of emitted electrons} = h\nu - h\nu_0 \tag{4.2}$$

$$\text{Maximum KE of emitted electrons} = h(\nu - \nu_0) \tag{4.3}$$

According to Maxwell's classical theory, the energy of a light wave should be absorbed continuously by the metal (as more and more electron oscillators are excited) and, in addition, *the energy absorbed should depend only on the square of the amplitude, A, of the light wave, not on the frequency, ν, of the light wave.* However, the experimental measurements of the photoelectric effect demonstrated a number of surprises (Fig. 4.2) that could not be explained by classical theory:

1. Electrons are emitted from the metal surface only if the frequency of the light (v) striking the metal is *larger* than a threshold value (v_0).
2. The value of v_0 is characteristic of the metal and differs for different metals (e.g., the value of v_0 is smaller for Cs than for Na).
3. If the frequency of light, v_1, striking the surface of the metal is less than the threshold value ($v_0 > v_1$), the light is completely reflected and is not absorbed (Fig. 4.2a) even at high intensities.
4. If the frequency of light, v_2, striking the metal is greater than the threshold frequency ($v_2 > v_0$), electrons with a certain amount of KE are emitted "instantaneously" (Fig. 4.2b).
5. The *maximum* KE of the ejected electrons depends *linearly* on the frequency of the light once the threshold frequency, v_0, is exceeded (Eq. 4.2c).
6. The slopes of plots of the maximum KE as a function of v are *identical* for all metals (Fig. 4.2d). Most remarkable, the slope of such plots (KE_{max}/v) are equal to the value of h (6.6×10^{-34}J s), Planck's constant!

Observations (1) and (3) taken together imply that energy is transferred from the light to the metal surface but that the energy transferred does not accumulate, as expected for a wave hitting a surface. Einstein interpreted these results in terms of the instantaneous absorption of energy of the light striking the metal surface, reminiscent of the instantaneous exchange of energy between two colliding particles. The colliding particles were viewed as photons of the light striking the electrons on the surface of the metal. Thus, Einstein concluded that not only is the energy of light quantized as *quanta*, as described by Planck, but light itself is quantized and consists of energy-carrying particles, termed "photons." Since a photon is a quantized particle, its energy can only be transferred to the electrons of the metal as an all-or-nothing event. In other words, when a photon strikes a metal surface, it can only eject an electron by transmitting *none (reflection of the photon) or all (absorption of the photon)* of its energy to the metal.

Furthermore, the energy imparted to the emitted electrons is directly related, by Eq. 4.1, to the frequency of the absorbed light (v_2) after the threshold value, v_0, has been achieved. Observations (2) and (5) imply that the threshold energy, $E = hv_0$, needed to remove an electron from a metal depends on the tendency of the metal to hold on to its electrons. Cesium (Cs) does not hold on to its electrons as effectively as sodium (Na), as expected from the lower position of Cs in the first column of the periodic table. Thus, the energy required to remove an electron from Cs metal is less (Fig. 4.2d) than the energy required to remove an electron from Na metal (the frequency of light required for electron ejection, v_0, is lower for Cs than for Na). Observation (4) implies that a certain amount of energy (a threshold quantum, $E = hv_0$) is required to remove an electron from the surface of the metal, and that if the quantum of energy provided by the absorbed photon exceeds that required to do the work needed just to remove the electron, this excess light energy shows up as the KE of the electron (in order to obey the law of conservation of energy). Observation (6) is the most striking since it demonstrates that *there is a universal relationship of the energy, E, imparted to electrons by the absorption of light of a given frequency, v; the proportionality constant between E and v is Planck's constant, h.*

Thus, only five years after Planck's proposal of the quantization of the energy of light, Einstein connected Planck's ad hoc explanation of the quantization of light energy and the existence of the quantum with a proposal that light itself (and, by inference, all of electromagnetic radiation) was quantized and consisted of particles that possess discrete "bundles" of energy that he called photons. As a result, two crucial intellectual building blocks of a new paradigm of the nature for light and the beginning of quantum mechanics were put in place: quanta of energy and photons of light. (Most chemists do not use the word "quantized" to refer to matter, but in reality, matter had long been accepted by chemists as quantized in the form of atoms and molecules.) Einstein's brilliant idea was more than a speculation, since it could be used to "prove" the validity of Planck's Eq. 4.1 *quantitatively*.

4.5   If Light Waves Have the Properties of Particles, Do Particles Have the Properties of Waves? —de Broglie Integrates Matter and Light

If Planck and Einstein were correct and light consisted of particles (photons) with quantized energy (quanta) that was proportional to the light's frequency, v, and if *at the same time* Maxwell was correct and light consists of waves, then there must be a correspondence between these two apparently incompatible views, however paradoxical the compatibility may seem at first. de Broglie postulated a fusion of the idea of particle and wave by postulating that *every* particle possesses some wavelike characteristics and that *every* wave possesses some particle characteristics. He hypothesized that the conditions of measurement determined whether the wave or particle characteristics were dominant in a given observation. In particular, de Broglie proposed that the wavelength (λ) of a particle is related to Planck's constant (h), and also to the mass (m) and velocity (v) of the particle through Eq. 4.4, termed the de Broglie equation. Since the product of mass and velocity (mv), is equal to the linear momentum of the particle, then the wavelength, λ, of a particle is inversely related to its linear momentum. The de Broglie equation elegantly connects the nature of particles and waves through h as a proportionality constant. Whenever h appears in an expression, we know that we are dealing with a quantum phenomenon.

$$\text{Wave property}   \lambda = h/mv   \text{Particle property} \qquad (4.4)$$

The relationship between a photon's energy and its associated wavelength is given by connecting Eqs. 4.1 and 4.4 to yield Eq. 4.5:

$$E = hv = h(c/\lambda) \qquad (4.5)$$

Algebraically manipulating Eqs. 4.4 and 4.5, the energy (E) can be related to the particle's momentum (mv), producing relationships among energy, frequency, wavelength, the speed of light, and momentum in Eq. 4.6:

$$E = hv = h(c/\lambda) = mvc \qquad (4.6)$$

If we accept that the velocity (v) of a particle can be replaced with the velocity of light for a photon in Eq. 4.6 (i.e., $v = c$), then we can derive an expression (Eq. 4.7), that is, the remarkable and well-known Einstein equation. This equation couples the *paradigm of relativity to the paradigm of light as photons* and also demonstrates the equivalence of photons (light) and mass.

$$\text{Photons} \quad E = h\nu = mvc = mc^2 \quad \text{Relativity} \tag{4.7}$$

Note that the rather drastic paradigm shifts concerning the nature of light (from Newton's particles to Maxwell's oscillating electromagnetic waves to the wave particle of Planck–Einstein–de Broglie) should be a warning that *no matter how powerful a guiding paradigm may appear to be to a community, in the face of new results and new concepts, all current paradigms must be considered useful, but tentative and conditional—and subject to eventual replacement by more powerful or more universally governing paradigms.*

Finally, the classical electromagnetic wave theory of light as an oscillating electromagnetic field resulting from oscillating electrons in matter continues to be a useful quantitative paradigm to explain certain phenomena involving light, such as interference effects and the ability to assign light a wavelength (interference and wavelength both being signature properties of waves). On the other hand, the quantum mechanical theory of light, as bundles of photons possessing energy and momentum (both signature properties of particles), best explains phenomena such as the intensity of black-body emission of light by a heated metal and the KE of ejected electrons after light absorption by a metal in the photoelectric effect. The classical theory of light is at its worst in trying to explain phenomena associated with light absorption (the photoelectric effect) and emission (the ultraviolet catastrophe). In a qualitative manner, when light is unperturbed by strong interactions with matter, it is well characterized as a wave and displays wave properties that are completely explained by Maxwell's equations. In this model, light is part of the electromagnetic field that is spread out and fills the entire universe; the wave can be considered to propagate through the universe "at the speed of light." When light is absorbed by a molecule, the "spread out" wave is suddenly "localized" in the small space occupied by the molecule. The wave function of the photon can be viewed as "collapsing" from one that is very diffuse and spread out (infinite dimensions) and "wavelike" to one that is highly localized (molecular dimensions) and "particlelike." So a photon behaves more like a wave when it is not strongly interacting with matter and more like a particle when it is strongly interacting with matter (absorption and emission). We can view the initial, weak interactions of light and matter in terms of an electromagnetic *wave* weakly coupling with the electrons of a molecule; this weak coupling leads to scattering of light, which is well understood in terms of classical wave theory. On the other hand, as the strength of the interaction increases, we view the interactions in terms of a *photon* strongly interacting with the electrons of a molecule leading to absorption of a photon (emission is the reverse of the absorption process).

In closing this historical introduction to the nature of light, note that the *uncertainty principle* provides the ultimate quantum explanation of the apparent wave–particle duality of light: For certain types of experiments, the measurement of a property that pins down light as a photon (absorption and emission) will cause complete lack of knowledge of all of its wave properties, and the measurement of a property that pins down light as a wave (interference) will cause complete lack of knowledge of all of its particle properties. Since this chapter is concerned with the absorption and emission of light (Scheme 4.1) by the electrons of molecules, from the above discussion it might seem that the photon model of light will be most useful. While this is true, we will see that describing the *initial interaction* of light with the electrons of a molecule is best modeled by considering light as an electromagnetic field that oscillates like a wave and interacts with electrons that can be driven into oscillation by the absorption of light.

4.6   Absorption and Emission Spectra of Organic Molecules: The State Energy Diagram as a Paradigm for Molecular Photophysics

The general paradigm of Scheme 4.1 can readily be integrated into the state energy diagram of Scheme 1.4, which is a very useful starting point for discussing radiative transitions. Electronic absorption and emission spectra provide important information concerning the structure, energetics, and dynamics of electronically excited states *R, in particular on the following parameters of *R: structures, energies, lifetimes, electron configurations, and quantum yields. For example, from knowledge of the $S_0 + h\nu \rightarrow S_1$ and $S_0 + h\nu \rightarrow T_1$ absorption processes, and of the $S_1 \rightarrow S_0 + h\nu$ and $T_1 \rightarrow S_0 + h\nu$ emission processes, one can often construct a fairly complete state energy diagram (Scheme 1.4), which includes the electron configurations of S_1 and T_1 and the energies of these two excited states relative to S_0. From measurements of the lifetimes of S_1 and T_1 and of the quantum efficiencies of emission Φ, we can deduce the rate constants (k) of the radiative and radiationless photophysical pathways available to S_1 and T_1. These energies and rates will set the stage against which photochemical processes must compete if they are to occur with significant efficiency.

4.7   Some Examples of Experimental Absorption and Emission Spectra of Organic Molecules: Benchmarks

In Chapter 1, we learned that for organic molecules the energy required to excite an electron from an occupied *valence* orbital (σ, π, or n) to an unoccupied antibonding orbital (π^* or σ^*) corresponds to light whose wavelength is typically in the range of 200 nm (UV light, 143 kcal mol^{-1}) to 700 nm (red light, 41 kcal mol^{-1}). In initiating a photochemical study of an organic molecule, the photochemist starts by measuring the electronic absorption and emission spectra of the starting materials (solutes, solvents, and reaction vessels). Saturated organic compounds (alkanes) are generally "transparent" to light in the region ~ 200–700 nm (Table 4.1). The lowest-energy absorption corresponds to a HO → LU (highest occupied → lowest unoccupied) orbital jump of an electron; for saturated hydrocarbons this jump corresponds to a σ (HO) → σ^*(LU) orbital transition. The energy gap between σ and σ^* orbitals for saturated hydrocarbons corresponds to energies greater than that of a 200-nm photon ($\sim 143 \, kcal \, mol^{-1}$).

On the other hand, unsaturated organic molecules (ketones, olefins, conjugated polyenes, enones, aromatic hydrocarbons, etc.) possess several absorption bands in the conventional "photochemical" region of the electromagnetic spectrum, 250 – −700 nm. Absorption of light in this region corresponds to π (HO) → π^*(LU) transitions for olefins and aromatic compounds that possess a π electron in the HO, or to n(HO) → π^*(LU) transitions for compounds, such as ketones, that possess an n electron in the HO.

The shorter wavelength limit of the photochemical region is set by the absorption of light by reaction vessels and solvents through which the light must pass (quartz and common organic solvents absorb strongly at 200-nm and shorter wavelengths). The longer wavelength limit is set by considerations of the minimum energy required to excite electrons (electronic excitation of organic molecules usually requires light of wavelengths < 700 nm). Wavelengths in the range of 700–10,000 nm correspond to near-IR and IR radiation. The energy of photons corresponding to these wavelengths is generally too small to excite electrons of organic molecules from a HO to a LU. However, such light excites fundamental vibrations or overtones of fundamental vibrations when absorbed.

A *chromophore* ("color bearer") is defined as an atom or group of atoms that behave as a unit in light absorption. A *lumophore* ("light bearer") is an atom or group of atoms that behave as a unit in light emission (fluorescence or phosphorescence, Schema 4.1). Typical organic chromophores and lumophores are the common organic functional groups, such as ketones (C=O), olefins (C=C), conjugated polyenes (C=C–C=C), conjugated enones (C=C–C=O), and aromatic compounds (benzene ring and condensed benzene rings). In this chapter and in Chapter 5, we concentrate on these common chromophores as exemplars for the photophysical properties of organic molecules.

Table 4.1 lists some numerical benchmarks for the maximum of the longest-wavelength absorption bands (λ_{max}) and the extinction coefficient at maximum absorption (ε_{max}) of some common organic chromophores, and assigns an electronic orbital transition to the band. The transitions listed generally correspond to the lowest-energy (longest-wavelength) electronic HO → LU orbital transition of the chromophore. The magnitude of ε_{max} determines the "absorption strength" of a chromophore; for organic molecules the value of ε_{max} for spin-allowed absorption may vary over several orders of magnitude. The usual units of ε_{max} are $cm^{-1} \, M^{-1}$ (a reciprocal length per mole). Since $M^{-1} = cm^3 \, mol^{-1}$, then an equivalent unit for ε is $cm^2 \, mol^{-1}$ (area per mol). Thus, the units of ε are the same as a *surface area per mole of chromophore molecules*. These units suggest that we may interpret ε as the "cross-sectional area" that a mole of chromophores present to passing photons of a given wavelength, λ (as discussed in Section 4.15). The data in Table 4.1 show that the wavelength (λ) or corresponding frequency (ν) of absorption maxima vary greatly with the chromophore structure, as does the strength of the absorption as measured

Table 4.1   Long-Wavelength Absorption Bands (Corresponding to HO → LU Transitions) of Some Typical Organic Chromophores

Chromophore	λ_{max}(nm)	ε_{max}	Transition Type
C—C	<180	1000	σ,σ^*
C—H	<180	1000	σ,σ^*
C=C	180	10,000	π,π^*
C=C—C=C	220	20,000	π,π^*
Benzene	260	200	π,π^*
Napththalene	310	200	π,π^*
Anthracene	380	10,000	π,π^*
C=O	280	20	n,π^*
N=N	350	100	n,π^*
N=O	660	200	n,π^*
C=C—C=O	350	30	n,π^*
C=C—C=O	220	20,000	π,π^*

by ε_{max}. We shall postpone discussion of emission parameters until Section 4.16; we simply point out that both absorption and emission parameters vary widely as a function of molecular structure. *Note that the absorptions in Table 4.1 correspond to spin-allowed singlet–singlet transitions.* The values of ε_{max} for spin-forbidden, singlet–triplet transitions, while finite, are usually $\ll 1\,\mathrm{cm}^{-1}\,\mathrm{M}^{-1}$, so that a sample is effectively "transparent" at the wavelengths corresponding to the transition. For essentially all purposes, when an organic photochemist discusses absorption spectra, it is understood that spin-allowed singlet–singlet absorption is responsible for the absorption.

Inspection of Table 4.1 poses a number of interesting and important questions concerning $R + h\nu \rightarrow {}^*R$ (or ${}^*R \rightarrow R + h\nu$ transitions) that will be answered in this chapter:

1. Why is there such a wide variation in electronic absorption (and emission) parameters of λ_{max} and ε_{max} for the different chromophores?
2. Why is the value of ε_{max} smaller for some aromatic molecules (e.g., benzene, naphthalene) and larger for others (e.g., anthracene)?
3. How is the orbital configuration (HO → LU) for the electronic transition $R + h\nu \rightarrow {}^*R$ related to and assigned to a given absorption (or emission) band?
4. How are experimental absorption parameters related to theoretical quantities, such as quantum mechanical matrix elements?
5. How are the processes of electronic absorption and electronic emission related mechanistically?
6. How do vibrations influence electronic transition in absorption and emission?
7. What can we say about the interactions that provide a mechanism for "spin-forbidden" transitions in absorption ($S_0 + h\nu \rightarrow T_n$) and emission ($T_1 \rightarrow S_0 + h\nu$)?

In order to answer these and other related questions, we will construct a pictorial model of the radiative processes of Scheme 4.1 that relates the molecular structure of R and *R (electronic, nuclear, and spin configurations) to interactions with the electromagnetic field and to spectroscopic parameters. We will start by developing a simple paradigm for the structure of light and for the interactions of light with the electrons of molecules that are responsible for electronic absorption and emission.

4.8   The Nature of Light: From Particles to Waves to Wave Particles[1,2]

As discussed in Sections 4.2–4.5, the classical wave theory of light as a wave has been shown to be inadequate to explain the details of absorption and emission of light by molecules. Nevertheless, the classical theory of light is still a useful starting point for producing a qualitative pictorial representation of the initial, weak interaction between light as the oscillating electromagnetic wave and the electrons of a molecule. We start with Maxwell's model of light as an oscillating *electrical force field* resulting from oscillating electric charges (electrons) in molecules. This oscillating force field is used as a basis for constructing a quantum mechanical operator for the calculation of matrix elements for the absorption and emission of *photons* by organic molecules. A justification for using the classical theory of light as a starting point for the absorption and emission of light is that it provides a concrete pictorial explanation (and therefore an intuitive physical explanation) for the initial interaction of light and molecules, if not the overall process of absorption (or emission). This pictorial intuitive representation is generally much easier for the organic chemist to grasp than the highly mathematical quantum theory.

4.9   A Pictorial Representation of the Absorption of Light

The basic idea for visualizing the interaction of light and the electrons of a molecule is borrowed directly from the classical theory of light as a wave.[3] Photons are viewed as particles that allow the exchange of energy between the (electric portion of the) electromagnetic field and the electrons of a molecule under the rules of quantum mechanics. The most important interaction between the electromagnetic field and the electrons of a molecule can be modeled as the interaction of two oscillating electric dipole systems: *the oscillating electromagnetic field that fills the entire universe* and *the oscillating electrons that are fixed to the nuclear framework of a molecule in matter.* These two oscillating electric systems, when coupled to one another, behave as a reciprocally interacting and coupled system of potential energy (PE) donor and acceptor attempting to participate in a common resonance if a common frequency (ν) can be found. The electromagnetic field is visualized as a field of electric dipoles that pervade the universe and oscillate at a range of frequencies (ν). If the electrons of a molecule possess a "natural" (resonance) oscillation frequency (ν) that corresponds to a "natural" (resonance) oscillation frequency (ν) of one of the oscillating dipoles (photons) in the available electromagnetic field, and if the electrons and the field are coupled by a significant dipole–dipole interaction, the electromagnetic field can interact with the electrons and exchange energy with the electrons by driving the electrons into oscillation and result in absorption of photons from the electromagnetic field.

The interactions are completely analogous to that of two interacting antennae, one an energy transmitter and the other an energy acceptor. The dipole–dipole interaction can cause a coupling of the two antennae (i.e., a *resonance results between the two antennae*). This resonance is most efficient when there exists a frequency (ν) that is common to both the electromagnetic field and the electronic transition for the photon ($E = h\nu$) and for the energy gap for the transition of the electron from one state to another ($\Delta E = h\nu$). In the classical harmonic oscillator model for the electromagnetic field, the field contains energy by virtue of its oscillating electrons (oscillating electrons are excited, and therefore possess energy that can be transferred). Transfer of energy from the field (absorption of a photon by a molecule, R) reduces the oscillations and energy of the electromagnetic field, whereas transfer of energy to the field (emission of a photon by an electronically excited molecule, *R) increases the oscillations (energy) of the field. *In the ground state (R) the electrons of a molecule are considered to be at rest, and in the excited state (*R) the electrons of a molecule are considered to be oscillating along the molecular framework in some manner.*

At the special resonance frequency (ν) corresponding to $\Delta E = h\nu$, the electrons of a molecule (R) can absorb energy from the electromagnetic field (by absorbing a photon). The electromagnetic field is then impoverished by one photon, and an excited oscillating electron of a molecule takes all of the energy of the photon and becomes an electronically excited molecular (*R). Emission is viewed as the reverse process, in which an excited oscillating electron interacts through dipole–dipole coupling with the electromagnetic field and the electromagnetic field becomes excited by the photon emitted by the molecule (*R). The photon is transferred from the oscillating electron of the excited molecule to the electromagnetic field; the field's energy is increased by the energy of one more photon and an electron of the molecule returns from an excited state (*R) to its ground state (R). From the above qualitative, intuitive classical description, it is now relatively easy to make a simple quantum modification of the classical picture of energy transfer from the electromagnetic field to the electrons of a molecule, and vice versa, as follows: *The absorption of energy from the electromagnetic field corresponds to the removal of a photon from the electromagnetic field, and the emission of energy from a molecule corresponds to the addition of a photon to the electromagnetic field.* Both pictures involve the coupling and resonance of two oscillating electric fields.

4.10   The Interaction of Electrons with the Electric and Magnetic Forces of Light

The absorption and emission of energy from the *electric* portion of the *electro*magnetic field by electrons are described by the field of *electronic molecular spectroscopy*; the absorption and emission of the electric portion of electromagnetic radiation by vibrating nuclei are the basis for the field of *vibrational molecular spectroscopy*. The absorption and emission of energy from the *magnetic* portion of the electro*magnetic* field by the electrons are the basis for the field of *magnetic resonance spectroscopy*. All three forms of spectroscopy have relevance for an understanding of molecular organic photophysics and photochemistry.

Now, we analyze more deeply the resonance condition that is required for the absorption or emission of light by molecules. The absorption or emission of light, as for all transitions of molecules, requires the conservation of energy. When two electronic states separated by an energy, $\Delta E = E_1 - E_2$, are coupled by some interaction, the electron density (square of the wave function) appears to oscillate and vary with time. From the Einstein resonance relation (Eq. 4.8a), an energy separation of $E_1 - E_2$ corresponds to a frequency of oscillation of $\nu = (E_1 - E_2)/h$. This oscillation of electric charge, or electric dipole moment (at the frequency ν) interacts with the oscillating dipolar electric field of electromagnetic radiation at this frequency and corresponds to the resonance condition for the absorption or emission of light (Eq. 4.8b).

$$\Delta E = E_1 - E_2 = h\nu \qquad (4.8a)$$

$$\nu = (E_1 - E_2)/h \qquad (4.8b)$$

Now, we describe a concrete classical picture (Fig. 4.3) of an electromagnetic wave and analyze the electric and magnetic features of the *initial* interactions of the electric and magnetic fields of light with the electrons of a molecule.

An electromagnetic wave exerts both electric and magnetic forces on charged particles (e.g., electrons and nuclei) and on magnetic dipoles (e.g., the magnetic

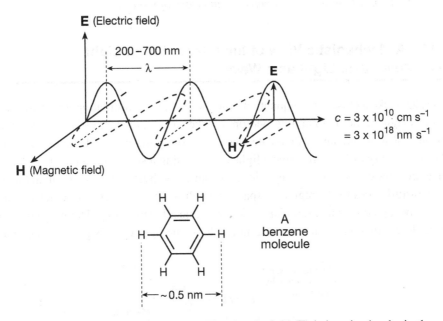

Figure 4.3   An electromagnetic wave. The electric field (**E**) is imagined to be in the plane of the page, and the magnetic field (**H**) is imagined to be perpendicular to the plane of the page.

moments associated with electron and nuclear spins). We can view light as mapping oscillating dipolar electric and magnetic force fields into the neighborhood of space about its direction of propagation (Fig. 4.3). In the volume of this space, two vectors can be drawn: an electric vector (**E**) that represents the source of the *electrostatic force* of the light wave, and a magnetic vector (**H**) that represents the source of the *magnetic force* of the light wave. The magnitudes of **E** and **H** at any point in space vary as a function of time and oscillate from mathematically positive (attractive) to mathematically negative (repulsive) values. A stationary spectator measuring the magnitude of **E** (or **H**), as the wave passes, would thus record oscillating values of **E** (or **H**) as a function of time. A test electric charge in space that can be coupled to **E** and that possesses the frequency ν can be set into oscillation by the oscillating values of **E**. To both the spectator and the test charge, the light wave appears to have the characteristics of a *harmonically oscillating electric and magnetic dipole*. A characteristic of this harmonic motion is the back-and-forth *linear oscillation* of the electric force field of the light wave and the electron cloud of the molecule.

A key idea in understanding the interaction of light with organic molecules is that electrons can be set into resonant oscillation (resonance) by the oscillating dipolar electric field of light *only* when Eq. 4.8 is obeyed. Under the condition of resonance, an electron (of R) may absorb energy from the electromagnetic field set up by the light wave, or an electron of *R may emit a photon as electromagnetic radiation. Thus, we can visualize the interaction of light by molecules as a process in which energy is exchanged *by resonance* between a collection of oscillating dipoles (electrons) that are coupled to a radiation field (an oscillating electric field that pervades the universe). *In more concrete chemical terms, the oscillation of the dipoles corresponds to the movements of electrons in bonds relative to positively charged nuclei in matter; that is, electrons oscillate about the nuclear framework of molecules.*

4.11   A Mechanistic View of the Interaction of Light with Molecules: Light as a Wave

Now, let's take a closer, more quantitative view on how the oscillating light wave (the electromagnetic field containing photons) makes a ground-state electronic configuration of R look like an excited-state electronic configuration of *R according to the classical theory of light. Imagine a light wave passing a stationary molecule.[3] As we have seen above, the electromagnetic wave causes both periodic electrical and magnetic disturbances in the region of space through which it passes, particularly in the region occupied by our exemplar stationary molecule (Fig. 4.4). The magnitude of force (**F**) exerted on an electron in a molecule by the light wave is given by Eq. 4.9:

Total force exerted on an electron by a light wave

$$\mathbf{F} = e\mathbf{E} + \frac{e[\mathbf{H}\nu]}{c}$$

Electrical force

Magnetic force

(4.9)

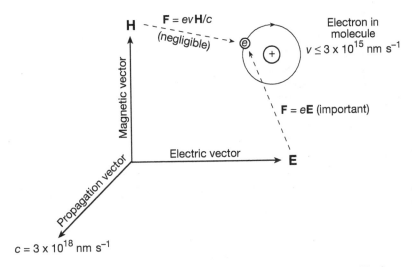

Figure 4.4 Interaction of the electric field **E** and the magnetic field **H** of an electromagnetic wave on an electron in an orbit about a nucleus.

where e is the charge of an electron, **E** is the electric field strength, **H** is the magnetic field strength, v is the velocity of an electron, and c is the speed of light. Since the speed of light (3×10^{10} cm s^{-1} = 3×10^{17} nm s^{-1}) is much greater than the possible speed of an orbiting electron ($v_{max} \sim 10^{8}$ cm s^{-1} = 10^{15} nm s^{-1}, from the Bohr atom model), the magnitude of e**E**, in general, *will be considerably greater* than the value of the magnetic force (e/c)[**H**v]; that is, e**E** $\gg$ (e/c)[**H**v]. Therefore, we conclude that the electric force of the oscillating light wave operating on electrons is much larger than the magnetic force operating on the electrons. Therefore, we can ignore the magnetic force to a good first approximation when we are dealing with electronic excitation by the electromagnetic field. Thus, because e**E** $\gg$ (e/c)[**H**v], Eq. 4.9 is approximated by Eq. 4.10, if the magnetic term of Eq. 4.9 is ignored.

$$\begin{array}{c} \text{Force on electron} \\ \text{ignoring magnetic interaction} \\ \mathbf{F} \cong e\mathbf{E} \end{array} \qquad (4.10)$$

Thus, although there is a simultaneous interaction of the electrons with the oscillating magnetic field (**H**), this interaction is negligible compared to the electrical interaction. However, the oscillating magnetic field of electromagnetic radiation interacts strongly with the *magnetic* dipoles of electron and nuclear spins and is the basis of magnetic resonance spectroscopy.

4.12 An Exemplar of the Interaction of Light with Matter: The Hydrogen Atom

Let us consider the simple exemplar of an electron in a Bohr orbit of a hydrogen atom interacting with the oscillating electromagnetic field (Fig. 4.5a). Here, we assume that

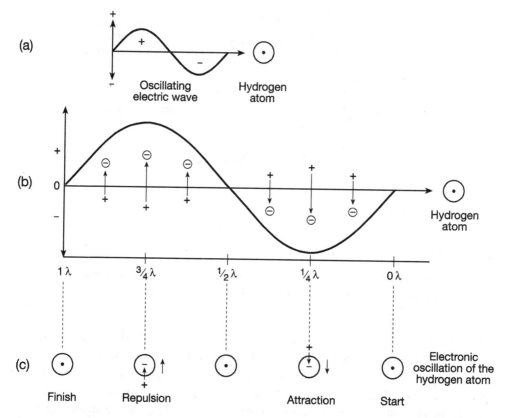

Figure 4.5   Schematic of the electric portion of the light wave interacting with a hydrogen atom.

the massive nucleus holds the atom fixed in space as the light wave zips by, but the electron of the atom can interact with and follow to some extent the oscillating electric field of the passing light wave. The maximum interaction between **E** and the electron will occur if the electron possesses a natural resonance frequency of oscillation (v) that is equal to the frequency of oscillation of the passing light wave. Suppose that the frequency v of the passing light wave corresponds to that of some natural frequency of the hydrogen atom; this frequency corresponds to an energy gap ($\Delta E = E_1 - E_2$) through Eq. 4.8b.

At the start of the interaction (Fig. 4.5b, right, 0 λ), for the sake of argument, let us imagine that the electric field **E** has a value of zero; that is, at this particular point of its period, the light wave does not exert an attractive or a repulsive force on the electrons of the hydrogen atom and the shape of the atom is spherical (Fig. 4.5c, right). One-quarter of a wavelength ($\lambda/4$) later, we imagine that the value of **E** has decreased to a negative maximum (defined as a repulsive electrical force on the electron) value and that the light wave exerts the same force on the electron of the hydrogen atom as an electric dipole with its negative end closest to the electron. This repulsion between negative charges continues until one-half of a period ($\lambda/2$) is completed. Thus, during the first half-cycle of the passing light wave, the hydrogen atom's electron is repelled away from the passing light wave with the maximum force occurring after $\lambda/4$ of the wave has passed. When exactly one-half of a wavelength ($\lambda/2$) has passed the atom, the force on the electrical force on the electron has dropped to 0. After $3\lambda/4$ has passed, we imagine that the value of **E** has increased to a positive (defined as an attractive force on the electron) maximum value and that the light wave exerts the same force on the electron of the hydrogen atom as an electric dipole with its positive end closest to the electron. During the second half-cycle of the passing light wave, the hydrogen atom's electron is attracted toward the passing light wave, with the maximum force occurring when $3\lambda/4$ has passed. When one full wavelength λ of the light wave has passed, the value of **E** is back to 0, and the electromagnetic field exerts no force on the electron.

What is the positive nucleus doing during the passing of the light wave? After all, the nucleus is a charged electrical particle just like the electron. Surely the electromagnetic field was also exerting a force on the nucleus as the wave passed. Indeed, this is correct: The nucleus does feel the force of the passing electromagnetic wave. However, because the nucleus is so massive compared to the electron that its electrical interactions with electromagnetic radiation occur at much lower frequencies ($\nu \sim 10^{13}$–10^{14} s^{-1}) than those that set electrons into oscillation ($\nu \sim 10^{15}$–10^{16} s^{-1}). This conclusion is easily deduced from the harmonic oscillator model (Eq. 2.25), for which the frequency of oscillation is inversely proportional to the mass of the particle undergoing oscillation. Thus, *the nucleus is not set into resonance by frequencies that set the electrons to resonance*. In fact, the nucleus is set into vibrational resonance at frequencies in the IR portion of the electromagnetic field and is the basis of vibrational spectroscopy.

The effect of **E** on the electrons of a hydrogen atom may be compared to the effect on an electron cloud of a hydrogen atom that is "fixed" in space between two charged plates. If one plate is charged positive and the other is charged negative, an induced dipole is produced in the hydrogen atom, with the negative end of the electric dipole pointing toward the positively charged plate. The important picture that emerges is that as far as the electron of an atom is concerned, the oscillating **E** of the light wave is an oscillating dipole that can interact with the electrons of a molecule; if the electron can oscillate at the correct frequency (ν), a resonance occurs and energy moves back and forth from the electromagnetic field to the electron's motion about the nucleus (and back and forth with the electromagnetic field). In the Bohr model, the electron in resonance with the electromagnetic field would be viewed as making harmonic oscillations back and forth between two Bohr orbits during the resonance period. We say that the interaction between **E** and the electron produces a transitory (or *transition*) *dipole moment* in the hydrogen atom as it oscillates between the two Bohr orbits. The greater the strength of the interaction of the charged plates with the hydrogen atom, the greater the size of the transition dipole. The greater the ease with which the electron can oscillate between two orbits, the more "polarizable" the electron and the larger the transition dipole moment. We shall see how this classical idea of a transition dipole carries over to help us understand the probability of absorption and emission of light by molecules.

4.13   From the Classical Representation to a Quantum Mechanical Representation of Light Absorption by a Hydrogen Atom and a Hydrogen Molecule[4,5]

Now, let us modify the classical picture of the interaction of light and introduce the required quantum mechanical features of wave functions so that we can obtain some quantum intuition and can deduce the basis for the important selection rules for the absorption and emission of light. Instead of an electron in a Bohr orbit (Fig. 4.5), now we consider the electron of the hydrogen atom to be in a 1 s orbital. The wave function of a 1s state is spherically symmetric about the nucleus, and therefore does not possess a net dipole moment. Figure 4.6a shows schematically how the oscillating $\mathbf{E}$ force will alternately cause the 1s electron cloud of the hydrogen atom to move toward and away from the passing light wave (in an analogous manner to that in which the oscillating $\mathbf{E}$ force causes a distortion of an electron in a Bohr orbit in Fig. 4.5). As a result, the light wave "reshapes" the electron distribution from one that is spherically symmetric about the nucleus to one that is alternatively more concentrated on one side of the nucleus, closer to the light wave, and then is more concentrated on the other side of the nucleus, farther away from the light wave (Fig. 4.6a). The oscillation of negative charge back and forth from one side of the atom to the other has the appearance of a transitory oscillating dipole. The *time average* of the oscillating electron "looks like" a p orbital (Fig. 4.6a, right), which possesses an electron distribution above and below a nodal plane containing the nucleus. This picture provides the intuition that the resonance interaction of an electromagnetic wave with a hydrogen atom in a spherically shaped 1s orbital will change the shape of the orbital and make it "look like" a 2p orbital. The intuition gained is that there must be a selection rule saying that this sort of shape change in the electron cloud corresponds to an "allowed" absorption.

In the above pictorial representation, the electron *vibrates* back and forth, just like a *harmonic oscillator according to the classical theory of light!* In the 1s state, the electron corresponds to a classical oscillator *that is not vibrating*. Translating this classical picture into a quantum mechanical one, we say that, in the 1s orbital, the electron has zero orbital angular momentum ($l = 0$) and that the interaction with the light wave causes the electron to pick up exactly one unit of orbital angular momentum as it is excited to a 2p orbital ($l = 1$). At this point, it is important to recognize that a photon possesses one unit of spin angular momentum. The excitation process requires both the conservation of energy ($\Delta E = 0$, because $E = h\nu =$ energy gap of 1s $\rightarrow$ 2p transition, exactly) and the conservation of angular momentum ($\Delta l = 0$, because angular momentum change increase for the 1s $\rightarrow$ 2p transition must equal the loss of one unit of angular momentum of the absorbed photon). Experimentally, the wavelength of the 1s $\rightarrow$ 2p transition in the hydrogen atom is 122 nm (deep UV) and corresponds to an energy gap of 234 kcal mol^{-1}.

From the above description of examining the interaction of a light with the hydrogen atom's 1s wave function, we deduce (Fig. 4.6a) that a nodal plane of the 2p orbital is produced at right angles to the oscillating electric vector ($\mathbf{E}$). Thus, we conclude that the interaction of $\mathbf{E}$ sets the electron into a harmonic oscillation (vibration) selectively along the direction of motion of the electric vector. This idea, in turn, provides

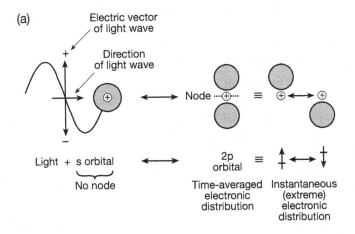

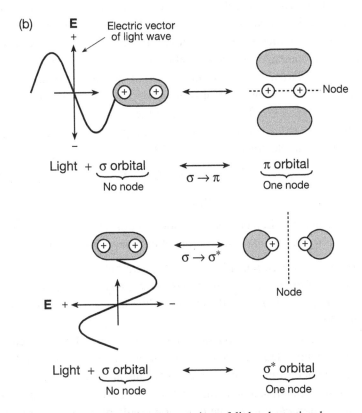

Figure 4.6 (a) Pictorial representation of light absorption by a hydrogen atom. The light wave and the 1s state of hydrogen are viewed as being in resonance with the 2p state. The latter may be represented as a time-averaged dumbbell shaped electron cloud or as an oscillating dipole. The transition is found experimentally at 122 nm. (b) Pictorial representation of light absorption by a hydrogen molecule. The light wave and the 1σ state have two possible resonances: a $\sigma \rightarrow \pi$ transition (at 90 nm) and a $\sigma \rightarrow \sigma^*$ transition (at 100 nm). One interaction of the electric field (upper drawing) drives the electron into oscillations perpendicular to the bond axis; the other interaction (lower drawing) drives the electron into oscillations along the bond axis.

a pictorial representation of the absorption of *polarized* light: The electron of the hydrogen atom can be set in motion selectively along one of the three Cartesian axes (x, y, or z) corresponding to production of one of the three possible 2p orbitals (p_x, p_y, or p_z). The production of a single node in a wave function corresponds to the change of angular momentum by one unit, $\hbar$ (this symbol characterizes angular momentum that has the units of Planck's constant h divided by 2π, i.e., $\hbar = h/2\pi$). An increase in the number of nodes (e.g., 1s → 2p transition) during the absorption of a photon is an essential feature of the absorption process. Conversely, a decrease in the number of nodes (e.g., 2p → 1s transition) is an essential feature of the emission of a photon from an excited hydrogen atom.

When the resonance condition is satisfied, the strength of the interaction between an electron and **E** is related to the ability of the electron to couple with and "follow" the electric force of the light wave, and to the magnitude of the maximal charge separation, Δr, effected by the interaction of **E** and the charge on the electron, e. Classically, the magnitude of development of charge separation as one proceeds from an s to a p orbital is related to α, the polarizability of the electron cloud, which is defined as the transitory (or transition) dipole moment (μ_i) that is induced in the electron cloud by an applied electric field (**E**) (Eq. 4.11). The magnitude of the transition dipole moment (μ_i) in turn is given by the extent of displacement of the positive and negative centers of charge (**r**) times a unit electric charge (Eq. 4.12).

$$\alpha = \mu_i/\mathbf{E} \tag{4.11}$$

$$\mu_i = e\mathbf{r} \tag{4.12}$$

The fundamental requirement for absorption or emission of light by an atom or a molecule may now be summarized in terms of the simple models discussed above:

1. The *energy conservation rule* (Eq. 4.8): There must be an exact matching of the energy difference that corresponds to the energy required for the transition (ΔE) between orbitals and the energy of the photon (hν); that is, ΔE must exactly equal hν (Eq. 4.8).

2. The *momentum conservation rule*: There must be an exact matching of the angular momentum gained (or lost) during the transition and the angular momentum of the photon; in quantum mechanical terms, the transition between orbitals must generate a node (absorption) or destroy a node (emission).

3. The *finite interaction rule*: The transition dipole moment (μ_i) created by the interaction of the electron with the electromagnetic field must be finite. The larger the value of μ_i, the more probable the absorption of light by R, and conversely the more probable (the faster) the emission of light from *R.

4. The *frequency matching (resonance) rule*: There must be a matching of a frequency (ν) of the oscillating light wave and a frequency that corresponds to the formation of a transition dipole moment. Since $\Delta E = h\nu$ and $\nu = \Delta E/h$, this rule is related to rule 1 above, the energy conservation rule. A matching of energy, when there is an energy gap of ΔE, is equivalent to a matching of frequency, which is the resonance condition.

Now, let us extend the ideas we have developed for a hydrogen atom and apply them to the simplest molecular system, the H_2 molecule (Fig. 4.6b). In going from atoms to diatomic molecules, the atomic s and p orbitals are replaced with molecular σ and π orbitals. In the simple case of H_2, the absorption of a photon promotes an electron in the ground-state HO(σ) orbital, which does not possess a node, into one of two low-energy unoccupied orbitals containing a single node (for convenience we refer to both of these orbitals as LU orbitals): a π orbital or a σ^* orbital. The bonding π LU orbital (not an antibonding π^* orbital!) possesses a single node along the bond axis, and the σ^* LU orbital possesses a single node perpendicular to the bond axis (Fig. 4.6b).

In the ground state, the hydrogen molecule possesses two electrons in the σ HO orbital, which is cylindrically symmetric about the internuclear axis. Compared to an atom, the nuclear axis imbues the molecule with an inherent axial asymmetry, and one can imagine electronic oscillations of two types: one oscillation that is parallel and one oscillation that is perpendicular to the bond axis.

Notice that both the $\sigma \rightarrow \pi$ and the $\sigma \rightarrow \sigma^*$ transitions are similar to the 1s $\rightarrow$ 2p transformation of atoms, respectively, in that a node is produced when the HO $\rightarrow$ LU transition occurs for each. The major difference is that two symmetry-distinct electronic oscillations are possible for the molecule: One oscillation (corresponding to the transition dipole of the $\sigma \rightarrow \sigma^*$ transition) is in a plane perpendicular to the bond axis and between the two nuclei (Fig. 4.6b, lower drawing) and the other oscillation (corresponding to the transition dipole of the $\sigma \rightarrow \pi$ transition) is in a plane containing the bond axis and the nuclei (Fig. 4.6b, upper drawing). Experimentally, the wavelength of the $\sigma \rightarrow \pi$ transition in the hydrogen molecule is 100 nm (very deep UV), which corresponds to an energy gap of 286 kcal mol^{-1}, and the wavelength of the $\sigma \rightarrow \sigma^*$ is 110 nm (deep UV), which corresponds to an energy gap of 264 kcal mol^{-1}.

Although an organic molecule is much more complex than a hydrogen atom or a hydrogen molecule, the two systems provide exemplars such that the basic ideas we have developed for these two simple species shall be sufficient as a starting point to develop a working paradigm for the absorption and emission of light by organic molecules.

In summary, an atom (or molecule) and the electromagnetic field of light are viewed as analogous to two coupled harmonic oscillators, such as two coupled pendulums. The electrons of the molecule possess a set of "natural transition frequencies," which correspond to the values of $v_i = \Delta E_i / h$, where ΔE_i correspond to the energy gap between two allowed electronic energy levels. We imagine that by changing the frequencies of the incident light passing by a molecule we can achieve values of v for the light that correspond to the values v_i of the electrons of the molecule. When a matching of the value of v of the light and the value of v_i of the electrons of the molecules occur, the electrons of the molecule behave like two coupled pendulums and energy is transferred from the light field (absorption of a photon by the molecule) or is transferred from an excited electron to the light field (emission of a photon by the molecule).

4.14 Photons as Massless Reagents

In spite of its shortcomings in describing the details of the *weak initial* interaction of light with electrons, the concept of a photon has the concreteness that is associated with the concept of a particle and provides a powerful intuition in dealing with quantum phenomena. The concept of a photon works well when the interaction of light with electrons is strong and absorption occurs. Indeed, the concept of a photon as a particle allows the organic chemist to consider the photon as a "massless reagent." A photon reagent may "collide" with molecules and "react" with them (i.e., be absorbed). Since *R is clearly a completely different species than R in terms of its energy, its electronic distribution, and its chemical reactivity, the photon reagent has certainly caused a chemical reaction, $R + hv \rightarrow {}^*R$, to occur! Like the molecules of an organic reagent, which can be counted, we can also count photons. A source of light of frequency v can be regarded as being composed of N photons, each of which possesses the energy hv. Each photon of wavelength $\lambda = c/v$ carries an energy hv and a linear momentum hv/c. *Low-frequency (long-wavelength) photons carry little energy and momentum; high-frequency (short-wavelength) photons carry a great deal of both energy and momentum.* Eq. 4.13 provides a quantitative connection between the energy (E) for Avogadro's number (N_0) of photons with a light source of frequency v (wavelength λ).

$$E = N_0 hv = N_0(c/\lambda) \tag{4.13}$$

From Eq. 4.13, Table 4.2 can be constructed to describe the relationship between the number of photons (N) that correspond to 100 kcal of energy (selected as an arbitrary numerical benchmark energy for photons of varying frequency and wavelength over the entire electromagnetic spectrum). From Table 4.2, it can be seen that 100 kcal corresponds to about an einstein (a mole of photons) for the special case of 1 mol of photons whose $\lambda = 286$ nm ($v = 1 \times 10^{15}$ s^{-1}). However, for light of $\lambda = 1000$ nm (IR), the same 100 kcal of energy corresponds to over 3 mol of photons, and for radiowaves ($\lambda = 10^{11}$ nm) 100 kcal of energy corresponds to 10^7 mol of photons! On the other hand, for X Rays, 100 kcal mol^{-1} of energy corresponds to < 1 millimole of photons, and for γ-rays 100 kcal mol^{-1} of energy corresponds to a few micromoles of photons! The concentration of photons can be computed from knowledge of the number of photons absorbed in a volume.

It is very important in photochemistry to understand the difference between the *energy* and *intensity* of photons. The intensity of a beam of monochromatic light of frequency v refers to the *number* of photons in the beam. The greater the intensity of a beam of monochromatic light, the greater the number of photons in the beam. No matter what the intensity of the beam, each photon carries the energy $E = hv$. The greater the frequency of a beam of monochromatic light, the greater the energy of the photons in the beam. Thus, a weak light beam of high frequency may possess sufficient energy to break strong bonds, whereas a strong beam of light of low frequency may not be able to break even weak bonds. This difference between energy and intensity was noted by Einstein in his explanation of the photoelectric effect (Fig. 4.2).

Analogous to organic reagents, photons may be viewed as *chiral* reagents in that they may possess a quality of "handedness" or "helical circularity" analogous to that of optically active molecules.[6] In the discussion of light absorption, we noted that a photon possesses one unit of angular momentum, $\hbar$. A photon's angular momentum is due to its inherent spin. *The existence of left- and right-circularly polarized light*

Table 4.2 Relationship of Number of Photons (einsteins) Corresponding to 100 kcal mol^{-1}

Spectra Region	λ (nm)	ν (s^{-1})	einsteins (N)
Gamma	0.001	1.0×10^{20}	3.5×10^{-6}
X Ray	0.1	1.0×10^{18}	3.5×10^{-4}
UV	300	1.0×10^{15}	1.1
Violet	400	7.5×10^{14}	1.5
Green	500	6.0×10^{14}	1.8
Red	700	4.3×10^{14}	2.5
NIR	1000	3.0×10^{14}	3.5
IR	5000	0.6×10^{14}	17.3
Microwave	10^7	3.0×10^{10}	3.3×10^4
Radiowave	10^{11}	3.0×10^6	3.3×10^7

is a manifestation of the spin angular momentum of a photon. A beam of circularly polarized light passing through a quartz crystal (which itself has right or left chirality) produces a torque that causes the crystal to acquire an angular momentum and turn to the left or to the right! As we have seen in the above description of light absorption, in any absorption or emission process, the photon and the electrons of the molecule "exchange" angular momentum, and the total system of photon-plus-molecule experiences no *net* change in angular momentum. The fact that a photon possesses an angular momentum is the basis of the selection rule that requires the creation or destruction of nodes in the electron cloud of a molecule as the result of absorption or emission of light (Section 4.13). Perhaps the most convincing evidence for the chirality of photons is the observation that racemic mixtures of organic molecules may be resolved if one enantiomer of a racemic pair absorbs circularly polarized light more efficiently than the other enantiomer, and if a reaction follows the act of absorption.[6]

Viewing the photon as a particle elicits a picture of absorption as an energy and momentum transfer between two colliding particles, the photon and the electron. It is possible to evaluate the "cross section," or size of the "target," that the electrons of a molecule present for being struck by the photon. By cross section, we mean the area of space around a molecule that is accessible to being struck by a passing photon. If we view the molecule as a target of a given diameter (d), then the experimental extinction coefficient (ε, in cm^{-1}M^{-1}) for absorption is given by Eq. 4.14.[5,7]

$$\varepsilon = 10^{20} d^2 \qquad (4.14)$$

In Eq. 4.14, d^2 is the area or cross section of the molecule in squared centimeters (cm^2). From the experimentally maximal value of ε_{max}, which is on the order of 10^5 cm^{-1} M^{-1} for organic molecules, we calculate (Eq. 4.15) d^2_{max} to be

$$d^2_{max} \sim 10^5/10^{20} = 10 \times 10^{-16} \text{cm}^2 \sim 10 \text{ Å}^2 \qquad (4.15)$$

According to this evaluation, we have a numerical benchmark for the largest cross section of an individual chromophore to be on the order of 10 Å^2, which corresponds to a diameter of 3.2 Å. The latter value is on the order of one or two bond lengths. This is on the correct order of the size of typical organic chromophores (Table 4.1).

4.15   Relationship of Experimental Spectroscopic Quantities to Theoretical Quantities[4]

Now, we use the qualitative classical model for the interaction of light with electrons as an intuitive basis for developing a quantum mechanical picture that will allow us to establish a relationship between experimental quantities and theoretical electronic wave functions (and matrix elements) for the radiative processes shown in Scheme 4.1. Let us consider the simplest case for radiative processes between two energy levels, one corresponding to the energy of R and one corresponding to the energy of *R. The fundamental experimental spectroscopic quantities related to absorption and emission of light between two energy levels are (1) the extinction coefficient for absorption (ε) for the $R + h\nu \rightarrow {}^*R$ process as a function of wavelength, λ; (2) the intensity of emission (I) for the ${}^*R \rightarrow R + h\nu$ process as a function of wavelength, λ; and (3) the inherent rate constant of decay of emission (k_e^0), the ${}^*R \rightarrow R + h\nu$ process (which is usually independent of the emission wavelength). From the full state energy diagram (Fig. 1.4), we understand that an important issue in understanding photophysical phenomena is the fact that there generally will be a competition between radiative ${}^*R \rightarrow R + h\nu$ processes and radiationless (physical and chemical) processes. However, for simplicity, we start our discussion by assuming that emission is the only pathway for excited-state deactivation of *R. This means that there are no competitive radiationless pathways from the emitting state; in this special case k_e^0 represents the rate constant for deactivation of the state by emission. By definition, the *lifetime* of a state is equal to the reciprocal of the rate for deactivation of the state; for a unimolecular decay $k = 1/\tau$. Thus, $k_e^0 = 1/\tau_e^0$, so that we immediately know the decay time of emission τ_e^0 if we know k_e^0, the rate of decay of emission, and vice versa. Now, we seek to answer the question, How do these experimentally measurable quantities ε and k_e^0 relate to theoretical quantum mechanical quantities (matrix elements) for absorption or emission of light?

According to quantum mechanics (Section 3.4), the rate of a measurable experimental transition, P_{12}, such as ε or k_e^0, may be computed in terms of the square of a theoretical quantity, a matrix element (i.e., for transitions between an initial state, Ψ_1, and a final state, Ψ_2), as shown in Eq. 4.16:

$$\begin{array}{ccc} \text{Experimental quantity:} & P_{12} \rightarrow <\Psi_1|P|\Psi_2>^2 & \text{Theoretical quantity:} \\ \text{transition rate } (\varepsilon \text{ or } k_e^0) & & <\text{matrix element}>^2 \end{array} \quad (4.16)$$

To qualitatively or pictorially determine the value of P_{12}, we need to answer the questions: In the matrix element of Eq. 4.16, to what electronic states of a molecule do Ψ_1 and Ψ_2 correspond, and what is the nature of the operator P_{12} that triggers the transition? If we can compute or approximate Ψ_1 and Ψ_2, and if we have identified an appropriate operator P_{12}, we can then proceed to evaluate the matrix element given in Eq. 4.16. Then, we may set up the calculation of a matrix element such that the result allows us to compute the probability of absorption or the rate of emission, ε or k_e^0, respectively. From the results of the classical theory of light absorption,[4,5] which considers electrons as negative charges that can oscillate in specified ways along a molecular framework,[8] the values of ε and k_e^0 can be related by way of Eq. 4.16.

4.16   The Oscillator Strength Concept[4,5]

In classical theory, light is viewed as a harmonically oscillating electromagnetic wave, and the electrons of a molecule are viewed as negatively charged harmonic oscillators that can be driven into oscillation by interactions with light of certain frequencies corresponding to the resonance condition. In the classical theory of light, the *oscillator strength* (f) is defined as a quantity that measures the intensity or probability of an electronic transition that is induced by the interaction of electrons in matter with the electromagnetic field of a light wave. In a simple model, f may be viewed as the ratio of the intensity or radiation absorbed or emitted by an actual molecule compared to a single electron behaving as a "perfect" harmonic oscillator bound to a molecule. For such an idealized electron, $f = 1$; that is, the electron is considered a "perfect" harmonic oscillator. The basic idea is that for an electron with $f = 1$, when light interacts with this electron, the probability that light is absorbed at the correct resonance frequency, v_i, will be very close to a maximal value (in the limit a probability of light absorption of a passing photon will be equal to 1 and every photon that interacts with the electron will be absorbed). In classical theory, electrons in the ground state of a molecule, R, are not oscillating, and are motionless (a situation that is forbidden by quantum mechanics!). The process of light absorption causes the excitation of an electron; the excited electron is viewed as an oscillating electron, but an unexcited electron is viewed as not oscillating. In the language of photons, the process of light absorption removes a photon from the electromagnetic field and causes an electron to oscillate; the process of light emission adds a photon to the electromagnetic field and causes an electron to stop oscillating.

As a concrete exemplar capable of relating the quantities ε and k_e^0 to the oscillator strength (f), an excited oscillating electron is approximated as a one-dimensional harmonic oscillator,[4c] that is, an oscillating dipole. For this simple case, the *theoretical quantity* of oscillator strength f in the classical theory of light absorption is related qualitatively to the experimental quantity ε by the expression given in Eq. 4.17:[4]

$$\text{Theoretical oscillator strength} \quad f \equiv 4.3 \times 10^{-9} \int \varepsilon d\bar{v} \quad \text{Experimental absorption} \quad (4.17)$$

In Eq. 4.17, ε is the experimental extinction coefficient and $\bar{v}$ is the energy (conventionally given in units of $1/\lambda$, typically in reciprocal centimeters cm^{-1}, termed wavenumbers) of the absorption in question.

Experimentally, the integral $\int \varepsilon d\bar{v}$ in Eq. 4.17 corresponds to the value of the area under a curve of a plot of the molecular extinction coefficient ε against wavenumber

$\bar{v}$ corresponding to a single electron oscillator. The rate constant (k_e^0) for emission (probability of emission of photons per unit time), according to classical theory, is related to the extinction coefficient for absorption[8,9] by Eq. 4.18.

$$\text{Radiative} \qquad k_e^0 = 3 \times 10^{-9} \bar{v}_0^2 \int \varepsilon d\bar{v} \cong \bar{v}_0^2 f \qquad (4.18)$$
$$\text{rate constant}$$

In Eq. 4.18, $\bar{v}_0$ is the wavenumber (energy in $1/\lambda$ units) corresponding to the maximum wavelength of absorption, and the integral $\int \varepsilon d\bar{v}$ is the same as that given in Eq. 4.17. We can see from Eq. 4.18 that the probability of light absorption as measured by f is directly related to the experimental extinction coefficient, ε, and the radiative rate, k_e^0, and depends on the *square* of the frequency of the absorption (since $\bar{v}$ is directly proportional to $1/\lambda$). Immediately, we deduce from Eq. 4.18 that all other factors being equal, the rate of emission of light will be faster for an emission at a shorter wavelength.

The perfect electron oscillator in a molecule is predicted to have an oscillator strength of 1, corresponding to the maximum values of f (and the related values of ε). However, from Table 4.4 (page 219), the experimental values of f, calculated from Eq. 4.17, are found to vary over an enormous range (from values near 1–10^{-10} cm^{-1} M^{-1}). One of the major failings of the classical theory of light absorption and emission was its inability to provide an adequate basis for understanding the wide variation in the observed values of f computed from Eq. 4.17. Nonetheless, the oscillator strength concept possesses the fundamentally correct form of the initial interaction of light with electrons, and upon reinterpretation in quantum mechanical terms, the wide variation in the value of f (and of experimental values of ε and k_e^0) can be explained in terms of the wave functions of the initial and final states involved in a radiative transition, as well as an operator representing the dipolar electric forces that an oscillating electromagnetic field imposes on an electron. Wave functions possess properties that must reflect the *electronic symmetries* of R and *R. In addition, they must reflect the *vibrational and spin properties* of these states. Since these important features were not considered at all in the classical theory of light absorption and emission, it is not surprising that the wide range of experimental values of f could not be predicted by classical theory and that the spread will be due at least in part to considerations of molecular symmetry, molecular vibrations, and electron spins.

4.17   The Relationship between the Classical Concept of Oscillator Strength and the Quantum Mechanical Transition Dipole Moment

The classical theory of light approximates the excited electron in a molecule as a linear harmonic oscillator or oscillating electric dipole. Let us consider some properties of electric dipoles in order to obtain some insight into the connection between f and the magnitude (strength) of an electric dipole. Then, we exploit this classical intuition to understand how the values of ε and k_e^0 can be related to the dipole strength of a radiative transition. If two equal and opposite electric charges (e) are separated by a (vectorial) distance (**r**), a dipole moment (μ) of magnitude equal to e**r** is created (Eq. 4.12). For an electronic transition of the type R + $h\nu \rightarrow$ *R, an oscillating dipole must be induced by the interaction of an electron with the electromagnetic field. According to classical theory,[4,5] the magnitude of f is proportional to the *square* of the induced (or transition) dipole moment (μ_i) produced by the action of a light wave on an electric dipole (Eq. 4.19):

$$\text{Oscillator strength } f \, \alpha \, \mu_i^2 = (e\mathbf{r})^2 \quad \text{Transition dipole moment} \quad (4.19)$$

In Eq. 4.19, μ_i is the induced *transition dipole moment* (or *dipole strength*) corresponding to the electronic transition (absorption or emission). The dipole strength of a transition may be set equal to e**r**, which can be viewed as the average size of the transition dipole, where **r** is the dipole length. By combining the classical oscillator strength with the quantization of the oscillation of electrons, we have an expression relating f and μ_i, which is given by

$$f = \left(\frac{8\pi m_e \bar{\nu}}{3he^2} \right) \mu_i^2 \cong 10^{-5} \bar{\nu} |e\mathbf{r}_i|^2 \quad (4.20)$$

Eq. 4.20, where m_e is the mass of the electron, $\bar{\nu}$ is the energy of the transition (in cm^{-1}), h is Planck's constant, and **r** is the length (in cm) of the transition dipole.

Now, we can identify μ_i with an observable quantity that can be computed as a matrix element, that is, $\mu_i = < \psi_1|P|\psi_2 >$, and produce Eq. 4.21:

$$\text{Classical} \rightarrow f = \left(\frac{8\pi m_e \bar{\nu}}{3he^2} \right) < P >^2 \leftarrow \text{Quantum mechanical} \quad (4.21)$$

Equation 4.21 connects the classical mechanical oscillator strength (f) to the quantum mechanical square of the matrix element, $< \Psi_1|P|\Psi_2 >^2$ of Eq. 4.16. We are now in a position to derive the relationships between the quantum mechanical quantity, $< \Psi_1|P|\Psi_2 >$, and experimental quantities, since both f and $< \Psi_1|P|\Psi_2 >$ may be directly related to experimental quantities, such as ε, k_e^0 ($= \tau_e^{0^{-1}}$) through Eq. 4.18.

4.18   Examples of the Relationships of ε, k_e^0, τ_e^0, $< \Psi_1|P|\Psi_2 >$, and f

Note that the expressions given above that relate theory and experiment are simplified and given only to provide quantum insight into the important factors controlling the nature of the radiative transitions R + $h\nu \rightarrow$ *R and *R $\rightarrow$ R + $h\nu$. Nevertheless, the use of these equations is expected to provide at least a *qualitative order-of-magnitude* agreement with experiment and to serve as calibration points and numerical benchmarks for expectations and comparison with experimental results. With these qualifications in mind, let us present some exemplars as numerical benchmarks in order to acquire a feel for the orders and limits of magnitude of quantities associated with various radiative transitions.

A molecular absorption spectrum corresponds to the absorption of light over a range of many wavelengths. Consequently, Eq. 4.18 must be integrated over all of the wavelengths at which absorption occurs. The integration of Eq. 4.18 is simplified by the assumption that the absorption spectrum is a symmetrical curve that can be approximated by an isosceles triangle.[9] With this assumption, we have Eq. 4.22:

$$\int \varepsilon d\bar{\nu} \sim \varepsilon_{max}\Delta\bar{\nu}_{1/2} \tag{4.22}$$

In Eq. 4.22, ε_{max} is the value of ε at the absorption maximum and $\Delta\bar{\nu}_{1/2}$ is the width of the absorption band at a value of $1/2\ \varepsilon_{max}$ in wavenumber (energy) units.

As an example, let us take the absorption spectrum of a molecule with $\varepsilon_{max} = 5 \times 10^4\ M^{-1}\ cm^{-1}$ at $\bar{\nu} = 20,000\ cm^{-1}$ (500 nm) and a half-width of $\Delta\bar{\nu}_{1/2} = 5000\ cm^{-1}$. Such a half-width for an absorption band is typical for organic molecules, and the extinction of maximum absorption in the example is close to the maximum found for typical organic molecules. In other words, *this example is an exemplar for a fully allowed electronic absorption* and would correspond closely to the classical case of an oscillator strength, $f = 1$.

Now, let us relate the experimental quantity ε to its theoretical counterparts, namely, f and $< \Psi_1|P|\Psi_2 >^2$ of Eq. 4.16. Guided by the classical model, the matrix element $< \Psi_1|P|\Psi_2 >$ is identified with a transition dipole moment, $\mu_i = er$. The value of the matrix element $< \Psi_1|P|\Psi_2 >$ is then directly related to the value of the transition dipole $e<r>$. Approximate expressions relating f and $<r>^2$ are given by Eqs. 4.23 and 4.24:

$$f \sim \frac{\varepsilon_{max}\Delta\bar{\nu}_{1/2}}{2.5 \times 10^8} \quad \text{(unitless)} \tag{4.23}$$

$$<r>^2 \sim \frac{\varepsilon_{max}\Delta\bar{\nu}_{1/2}}{2.5 \times 10^{19}\bar{\nu}} \quad \text{(units cm}^2) \tag{4.24}$$

Evaluating for $\varepsilon_{max} = 5 \times 10^4\ M^{-1}\ cm^{-1}$ at $20,000\ cm^{-1}$ with $\Delta\bar{\nu}_{1/2} = 5000\ cm^{-1}$, we find for f and $\mathbf{r}$:

$$f \sim \frac{(5 \times 10^4)(5 \times 10^3)}{2.5 \times 10^8} = 1.0 \tag{4.25}$$

$$<r>^2 \sim \frac{(5 \times 10^4)(5 \times 10^3)}{(2.5 \times 10^{19})(2 \times 10^4)} = 5 \times 10^{-16} \quad \text{(units cm}^2) \tag{4.26}$$

Thus, we see (Eq. 4.25) that indeed such a large value for ε_{max} corresponds to an *oscillator strength f of the order of* 1.0.

In classical theory, such an exemplar system would correspond to an ideal electron harmonic oscillator.[4c] The transition dipole length $\mathbf{r}$ for the exemplar under discussion is $2.2 \times 10^{-8}\ cm = 2.2\ Å$. Thus, according to quantum theory, our exemplar system would have a transition dipole moment length of 2.2 Å, corresponding to a transition dipole moment er of 2.2×10^{-8} cm $\times 4.8 \times 10^{-10} \sim 10$ D (the symbol D here stands for debye, the conventional unit for dipole moments). Thus, the strong electronic transition in the exemplar has associated with it a transition dipole moment of ~ 10 D. In other words, during the interaction of the light wave and the molecule, the electron cloud is distorted enough to produce a transition dipole moment of 10 D. For a numerical benchmark comparison, the permanent dipole moment of water is ~ 2 D.

Now, let us estimate the value of the radiative rate constant (k_e^0) associated with emission of light from the state producing the absorption spectrum we have just discussed. From our calculation of f and with the use of Eq. 4.18, we have Eq. 4.27:

$$k_e^0 \ (\equiv 1/\tau^0) \sim \bar{v}_0^2 f \sim (2 \times 10^4)^2 \sim 4 \times 10^8 \ \text{s}^{-1} \qquad (4.27)$$

In the calculation of k_e^0, we take the *theoretical* relationship of absorption to emission through the oscillator strength, f, and then make a prediction of the relationship between *experimental* quantities, that is, the integrated absorption spectrum and the inherent emission lifetime τ^0 (which is defined as $1/k_e^0$) for a corresponding radiative transition.

Now, consider a second exemplar of a molecule whose absorption spectrum is identical in shape and spectral position (same frequency) to our first case, but whose ε_{max} is only ~ 10. We find that

$$f = 2 \times 10^{-4} \quad \mathbf{r} = 0.3 \ \text{Å} \quad k_e^0 \sim 10^5 \ \text{s}^{-1} \qquad (4.28)$$

Note that the value of the oscillator strength f is orders of magnitude smaller than the maximum value of 1.0, as are the associated dipole strength and radiative rate, and that the size of the transition dipole is correspondingly also much smaller.

These order-of-magnitude calculations for exemplars allow us to generate numerical benchmark estimates of the most intense radiative absorptions (measured by ε_{max}) or the fastest radiative emissions (measured by k_e^0) that we expect to encounter experimentally for organic molecules. How large can the value of ε_{max} be for an organic molecule? If one uses arguments derived from the classical theory of light absorption, the largest value of ε_{max} is associated with an oscillator strength of 1.0. A limiting value of $\varepsilon_{max} \sim 100,000$ is predicted for absorptions occurring near 400 nm (20,000 cm^{-1}). The corresponding radiative rate is $\sim 10^9$ s^{-1}. Thus we have the values given in Eq. 4.29:

$$\text{Spin-allowed absorption} \quad \varepsilon_{max} \to 10^5 \ \text{cm}^2 \ \text{M}^{-1} \quad \text{(limit)} \qquad (4.29a)$$

$$\text{Spin-allowed emission} \quad k_e^0 \to 10^9 \ \text{s}^{-1} \qquad \text{(limit)} \qquad (4.29b)$$

The typical bandwidths ($\Delta \bar{v}_{1/2}$) of many absorption bands in the vis and near-UV regions are on average ~ 3000 cm^{-1} (at room temperature), so that from Eqs. 4.23

and 4.27 a convenient approximate relationship between the rate of emission, k_e^0, and ε_{max} is given by

$$k_e^0 \sim \varepsilon_{max} \Delta \bar{\nu}_{1/2} \sim 10^4 \varepsilon_{max} = 1/\tau^0 \qquad (4.30)$$

Note that lifetimes calculated in this way are *pure radiative* lifetimes (τ^0) that is, lifetimes that would be observed in the absence of all radiationless processes by which the excited molecule *R could return to the ground state. The values of $1/\tau^0$ in turn are associated with rate constants (k_e^0), which correspond to pure radiative processes from *R. The experimentally observed lifetimes, τ_{obs}, are almost always less than the calculated values because of competing radiationless processes, both photophysical (*R → R) and photochemical (*R → I, *R → F), which compete with the *R → R + $h\nu$ radiative process.

Now, let us seek a benchmark for the smallest value of ε_{max} (smallest value of k_e^0). The smallest values of the matrix element for ε are expected when the R + $h\nu$ → *R process is spin forbidden (classical theory is of no help here because it does not consider an electron's spin at all). Of course, the theoretical limit for a spin-forbidden process is precisely zero. However, finite perturbations due to spin-orbit coupling[10] will always be present in organic molecules, so that, although the value of ε for a spin-forbidden radiative transition will always be much smaller than the value for an analogous spin allowed transition, the value is finite. The experimental lower limit for ε for organic molecules is in the range of $\varepsilon_{max} \sim 10^{-4}$. This corresponds to a radiative rate $k_e^0 \sim 10^{-1}$–10^{-2} s^{-1}, which is the benchmark for spin-forbidden radiative transitions. When spin–orbit coupling is particularly strong, an upper limit of $\varepsilon_{max} \sim 10^0$ may be observed (molecules possessing heavy atoms).

4.19 Experimental Tests of the Quantitative Theory Relating Emission and Absorption to Spectroscopic Quantities

Experimental tests[9] of a slightly modified form of Eq. 4.18 have been made for singlet–singlet transitions; the results are given in Table 4.3. The agreement between the calculated and experimental values is excellent when the geometries of the ground and excited states are not significantly different and when the symmetry of the molecule is not too high. Large geometry changes between R and *R result in a breakdown in the assumptions of the above equations. High symmetry (e.g., benzene) may cause an "orbital" forbiddenness in a transition to cause a lower experimental value of ε than expected. The reason for this can be understood qualitatively as being due to the inability of the light wave to find a suitable molecular axis along which to generate a transition dipole by a HO → LU transition. This is the situation for the lowest-energy transition of benzene, naphthalene, and pyrene and other molecules possessing high symmetry. An important collateral property of these molecules is that they possess very long fluorescence lifetimes!

Although application of Eq. 4.18 to electronically forbidden singlet–singlet and spin–forbidden singlet–triplet radiative transitions is not theoretically justified, since

Table 4.3 Experimental and Calculated Radiative Lifetimes for Singlet–Singlet Transitions

Compound	$\tau^0 (\times 10^9)^a$	$\tau (\times 10^9)^b$
Rubrene	22	16
Anthracene	13	17
Perylene	4	5
9,10-Diphenylanthracene	9	9
9,10-Dichloroanthracene	11	14
Acridone	15	14
Fluorescein	5	4
9-Aminoacridine	15	14
Rhodamine B	6	6
Acetone	10,000	1,000
Benzene	140	60

a. Calculated
b. Experimental

the classical theory does not consider the influence of molecular symmetry or electron spin on radiative transitions,[10] nonetheless it appears that relative values of k_e^0 still may be approximately derived from absorption data.[11]

4.20   The Shapes of Absorption and Emission Spectra[12]

We will examine a number of electronic absorption and emission spectra of several exemplar organic molecules. Some of these spectra show relatively "broad" bands, while others show a number of "sharp" bands. Now, we provide a qualitative description of the shape of the bands in an absorption or emission spectrum and provide a structural interpretation of the basis for their shapes.

Let the energy of the ground state R be E and the energy of the excited state *R be *E. The resonance equation $\Delta E = h\nu$ then corresponds to transitions between two energy levels, E and *E, with an energy gap, $\Delta E = *E - E$. For the absorption (R + $h\nu \rightarrow$ *R), the energy change corresponds to $\Delta E = h\nu = |E - *E|$, and for the emission (*R $\rightarrow$ R + $h\nu$), the energy change corresponds to $\Delta E = h\nu = |*E - E|$. We might expect that absorption and emission spectra will be observed experimentally as "sharp lines" with respect to the frequency ν of the absorbed or emitted light. In fact, only the absorption and emission spectra of atoms are close to being "sharp lines" (see Fig. 4.7a).

This sharpness of an atomic spectrum appears because the energies E and *E of the electronic states of atoms can be accurately described by specifying the *electronic energies* of the electronic orbitals. At low pressures in the gas phase for an atom, there are no rotations, vibrations, or collisions that "broaden" the values of E and *E for the ground and excited states of R and *R for atoms, and the value of $\Delta E = |*E - E| = h\nu$

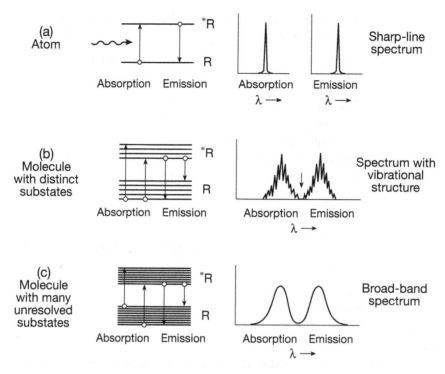

Figure 4.7   (a) Sharp-line absorption and emission spectrum typical of atoms at low pressure in the vapor phase. (b) Broad-band absorption and emission spectrum possessing vibrational typical of certain rigid molecules at low pressure in the vapor phase. (c) Structureless broad absorption and emission spectrum typical of molecules in solvents. Each absorption or emission corresponds to a single electronic transition.

has a sharply and precisely defined magnitude since E and $^*E$ are sharply defined. Thus, at low pressure in the gas phase, atomic absorption and atomic emission spectra show very sharp lines, since the values of E and $^*E$ are sharply defined and the value of $\Delta E = h\nu$ is therefore sharply defined. An atomic electronic transition from the ground state, R, to an excited state, *R, requires a quantum of well-defined energy; consequently, the absorption (R + $h\nu \rightarrow$ *R) or emission (*R $\rightarrow$ R + $h\nu$) spectrum of an atom in the gas phase is a very narrow band of wavelengths (Fig. 4.7a). For example, both the absorption and emission spectra of the gaseous H atom in the *visible* region of the spectrum (corresponding to transitions between the $n = 2$ (2s, 2p) and $n = 3$ (3s, 3p, 3d) states of the H atom) consists of four sharp lines at 410, 434, 486, and 656 nm, respectively. There is nearly an exact correspondence between positions of the absorption and emission, because the energy of orbitals involved for R and *R in the transitions do not change significantly for an atom.

For a molecule, even in the gas phase at low pressures, because of coupling between the electrons and vibrations, the transitions between R and *R are not "pure" electronic transitions, but rather are "vibronic" transitions that can possess a range of energies. In order to describe the electronic states of a molecule, one must consider not only the motions of the electrons, but also the motions of nuclei relative to one another and the molecules as a whole (e.g., vibrations and rotations). Thus, a molecular electronic transition between R and *R does not correspond to a well-defined single quantum of energy, because an instantaneous *ensemble* of different nuclear shapes may correspond to the initial or the final states (Chapter 3). As a result, the absorption (and emission) spectrum for a HO → LU transition of a molecule may involve many vibrational transitions over a range of energies corresponding to slightly different conformations of R and *R, even in the vapor phase at low pressures (Fig. 4.7b). The sharp "line" that characterized atomic transitions is replaced in molecular absorption by a set of closely spaced lines characterizing the molecular vibrations. Usually, these sets of closely spaced lines cannot be resolved and are termed an absorption or emission "band."

For organic molecules in solution, the situation becomes even more complex (Fig. 4.7c). In solution, R and *R are surrounded by solvent molecules that may be instantaneously intermolecularly oriented about R and *R in many different *supramolecular configurations*; in addition, the vibrations are coupled to within the molecule and to some extent to the solvent molecules. These supramolecular configurations of solvent molecules around R may be described by the supramolecular terminology R@solvent. Each one of these supramolecular configurations for the R@solvent + hv → *R@solvent transitions will have a slightly different energy gap for absorption or emission, so that compared to the gas phase, the molecular spectrum is greatly broadened and the molecular vibrational structure (Fig. 4.7b) is further broadened, or may be blurred out (Fig. 4.7c) or lost completely (termed a "featureless band with a single maximum). Similarly, for *R@solvent, the energy levels are broadened by the range of solvent orientations about *R so that the *R@solvent → R@solvent + hv emissions will be broadened, often leading to a single, featureless band.

In certain cases in solution, some molecular vibrational structure is still apparent in a band corresponding to a single electronic transition. This situation occurs when the couplings of the electronic transitions to the solvent are weak. The most prominent vibrational progression of an electronic absorption or emission band is often associated with a vibration whose equilibrium position is greatly changed by the radiative electronic transition.[13] Thus, a prominent vibrational progression reveals the most important nuclear distortion that occurs during a transition. For "weak" spin-allowed electronic transitions ($f < 10^{-2}$), often the vibration in question is the one that destroys the molecular symmetry to such an extent that a forbidden transition (in idealized perfect molecular symmetry) is caused to become partially allowed via electronic–vibrational interactions. When this is the case, the vibrational structure for absorption provides information on the geometry of the excited state, and the vibrational structure for emission provides information on the geometry of the ground state. [Looking ahead, an experimental example of vibrational structure for an aromatic hydrocarbon, pyrene, is shown in Fig. 4.15. The vibrational structure is a mix of C=C and C—H motions that are involved in the $\pi \rightarrow \pi^*$ transition in going from R → *R in the absorption spectrum, and those involved in the $^*\pi \rightarrow \pi$ transition in going from *R → R in the emission spectrum.]

In some cases, the absorption and emission possess a vibrational structure that can be assigned to a specific vibrational motion. For example, during the n,π^* radiative transitions of a ketone, in the $R(n^2) + h\nu \rightarrow {}^*R(n,\pi^*)$ transition, an n electron is removed from the HO and is placed in a π^*LU. Thus, in the light absorption process, a π^* electron is produced along the C=O axis so that the C=O bond is suddenly weakened and a C=O vibration is selectively excited. The structure of the C=O vibration of the excited state dominates the vibrational progression absorption spectrum. On the other hand, the ${}^*R(n,\pi^*) \rightarrow R(n^2) + h\nu$ process selectively removes an electron from a π^* orbital (LU) and places it in an n orbital (HO). Consequently, the C=O vibration of the ground state dominates the emission spectrum.

The absorption and emission spectra of benzophenone provide excellent exemplars of the concepts relating vibrational features to the electron configuration associated with radiative transitions. [Again looking ahead, Fig. 4.16 shows that some vibrational structure occurs in the bands of the n,π^* absorption of benzophenone in solution.] The vibrational structure of the n $\rightarrow \pi^*$ transitions of benzophenone corresponds to the two adjacent maxima in Fig. 4.16. The adjacent vibrational levels are separated by $\sim 1200 \text{ cm}^{-1}$, which corresponds to the C=O stretching vibration of *R. This value of 1200 cm^{-1} for the *R state is confirmed by direct time-resolved IR measurements of the C=O stretch in R and *R. Figure 4.19a shows the emission spectrum of benzophenone at 77 K. Under these conditions, the vibration bands in the (phosphorescence) emission spectrum of the $T_1(n,\pi^*)$ state of benzophenone are well resolved and show a separation of $\sim 1700 \text{ cm}^{-1}$ between adjacent bands. This separation is in good agreement with the energy of the C=O stretching vibration of the *ground-state* benzophenone (as measured by IR spectroscopy). Thus, the separation of the vibrational bands of the absorption spectrum reflects the vibrational structure of *R and the separation of the vibrational bands of the emission spectrum reflects the vibrational structure of R. In passing, we note that the smaller value of 1200 cm^{-1} for the *R state is consistent with weaker bonding in *R, due to the existence of an antibonding electron.

Now, we will see how vibration intensities in electronic absorption and emission spectra are controlled by the Franck–Condon principle.

4.21 The Franck–Condon Principle and Absorption Spectra of Organic Molecules[13]

The Franck–Condon principle (Section 3.10, especially Fig. 3.3) leads to the conclusion that there will be a difference in the probability of vibrational transitions between the wave functions ψ_0 and $^*\psi$ (corresponding to the R and *R states, respectively) of a diatomic molecule X—Y. Here, we present examples of the role of the Franck–Condon principle in terms of a *semi*classical model (the vibrations are quantized, but the vibrational wave functions are not considered explicitly). In Fig. 4.8, a situation is shown for which the two potential curves for ψ_0 and $^*\psi$ are similar and not displaced vertically; that is, the equilibrium separation $\mathbf{r}_{eq}$ is the same for ψ_0 and $^*\psi$. The situation in Fig. 4.8 corresponds to an electronic orbital jump for which the overall bonding is similar in both ψ_0 and $^*\psi$. As a result, the equilibrium geometries of R and *R should be similar. In such a situation, the Franck–Condon principle requires that, since absorption must occur vertically, a relatively strong $v = 0 \rightarrow v = 0$ transi-

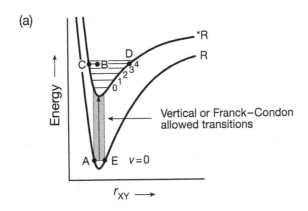

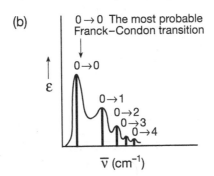

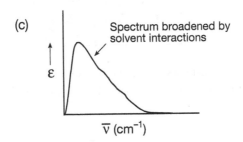

Figure 4.8   (a) Potential energy curves for similar minimum for R and *R showing the Franck–Condon allowed vertical transition for $R \rightarrow$ *R transitions. (b) Form of the observed absorption spectrum. For an experimental exemplar, see Fig. 4.9. (c) Effect of solvent broadening on the vibrational structure of the absorption spectrum.

tion (termed a $0 \rightarrow 0$ band) is observed for both absorption and emission (compared with the intensity to other vibrational bands) and the $0 \rightarrow 0$ band of the absorption and emission spectrum overlap significantly. The molecule in R undergoes zero-point motion ($v = 0$) between points A and E. A vertical transition to point B ($v = 4$) is Franck–Condon forbidden, as are transitions to points C and D ($v = 4$). As a result, the $0 \rightarrow 4$ transition is very weak.

Examples of a molecule with a small displacement of the minimum of the energy curves[14] are rigid aromatic hydrocarbons, such as anthracene (Fig. 4.9). In the case of

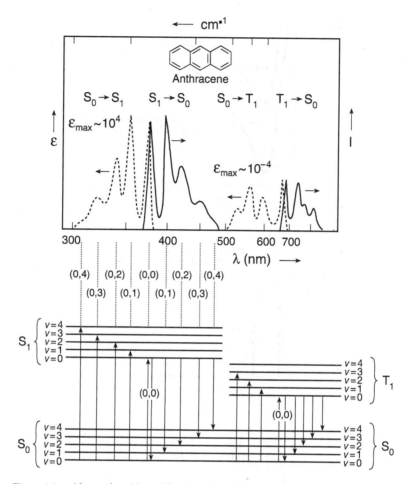

Figure 4.9   Absorption (dotted lines) and emission (solid lines) of anthracene in solution. The lower portion displays an energy level diagram as the basis for vibrational assignments.

anthracene, the excited-state curve minimum is slightly displaced from the ground-state curve minimum because the molecule in the excited state bends slightly to make a "v" shape symmetrical about the 9,10 position of the molecule. In this case, relatively strong $0 \rightarrow 0$ and $0 \rightarrow 1$ vibrational bands are observed in both absorption and emission, and the 0,0 band of absorption and emission overlap. Note that the vibrational patterns for the spin-allowed $S_0 \leftrightarrow S_1$ and spin-forbidden $S_0 \leftrightarrow T_1$ radiative transitions are slightly different. A possible reason is that the vibrations that best mix in triplet character are not exactly the same as those that have the best Franck–Condon factors.

Figures 4.10 and 4.11 represent situations for which the excited curve minima for $*\psi$ are significantly displaced relative to ψ_0 (r_{eq} is assumed to be larger in $*\psi$ than in ψ_0 because the antibonding electron weakens the bonds of $*R$, and therefore makes the equilibrium geometry, r_{eq}, larger in $*R$). In Fig. 4.10, the $0 \rightarrow 2$ and $0 \rightarrow 3$ vibrational bands are relatively intense and the $0 \rightarrow 0$ and $0 \rightarrow 1$ vibrational bands are relatively

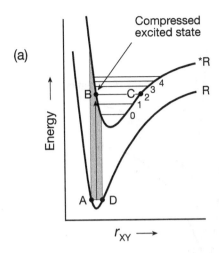

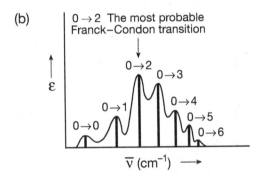

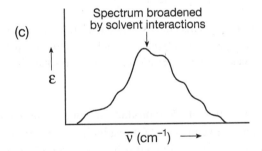

Figure 4.10   (a) Potential energy curves for significantly different minimum for R and *R showing the Franck–Condon allowed vertical transitions for R → *R transitions. (b) Form of the observed absorption spectrum (see Fig. 4.15 for an exemplar). (c) Effect of solvent broadening on the vibration structure of the absorption spectrum. A typical example is an aromatic ketone undergoing n,π^* absorption (see Fig. 4.15).

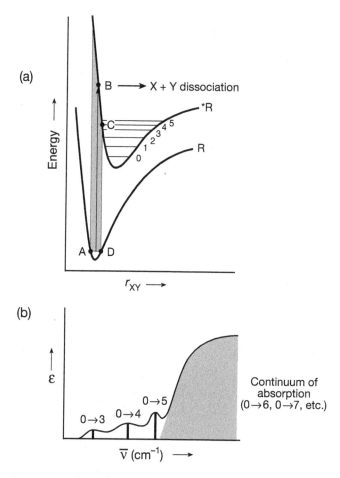

Figure 4.11   The effect of absorption to a dissociative state on the absorption spectrum. A typical example is a molecule possessing a very weak σ bond, for example, CH_3I or CH_3OOCH_3.

weak. In Fig. 4.11, excitation of $*\psi$ to produce geometries more contracted than point C in the figure (points between B and C) results in dissociation of the diatomic XY molecule into two atoms $X + Y$. When this happens, there is no vibrational structure in the absorption spectrum since the atoms dissociate immediately after absorbing a photon and do not undergo any vibrations.

4.22   The Franck–Condon Principle and Emission Spectra[13]

In condensed phases, the rate of vibrational and electronic energy relaxation among excited states is very rapid compared to the rate of emission (Chapter 3). As a result, emission will generally occur *only* from the $v = 0$ vibrational level of the lowest excited states of *R. Now, let us apply the Franck–Condon principle to the $*R \rightarrow R + h\nu$ emission (Fig. 4.12).

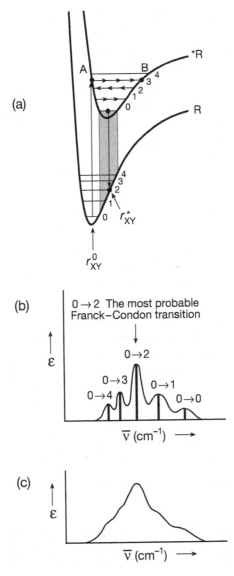

Figure 4.12   (a) Potential energy curves, (b) emission spectrum, and (c) solvent-broadened emission spectrum. As shown, the $0 \rightarrow 2$ emission is the most probable Franck–Condon transition. The most probable absorption is from r_{XY}, and the most probable emission is from r_{XY}^*. Emission from R* produces an elongated ground state.

In analogy to absorption, the most probable emissions will be those that occur "vertically" with the smallest change in the value of the equilibrium geometry of *R, which will generally be those emissions from the $v = 0$ level of *R. The equilibrium separation of the ground-state PE curve minimum is smaller than that of the excited-state curve (because *R possesses an antibonding electron, whereas R does not), so that the most probable vertical transitions from *R produce an elongated ground state (while absorption produces a compressed excited state) immediately after transition. It is important to note that the frequency (energy) of emission cannot be greater than the frequency (energy) of the $0 \rightarrow 0$ emission, since the emissions from $0 \rightarrow 1$, $0 \rightarrow 2$, and so on emissions correspond to smaller energies than the $0 \rightarrow 0$ emission. Since the energy of the $0 \rightarrow 0$ transition is the maximum energy that can be produced in the *R $\rightarrow$ R $+ h\nu$ transition, this energy is defined as the excitation energy of the state and is given the symbol $*E_S$ for the energy of *R(S$_1$) and the symbol $*E_T$ for the energy of *R(T$_1$). Figure 4.12 shows the emission spectrum expected from a molecule whose PE curves are similar to those in Fig. 4.10.

Consider the two PE curves of Fig. 4.12a. When absorption occurs vertically from the ground-state surface, the excited state is "born" near point A, a turning point for the compressed vibration in $v = 3$. The molecule will begin to vibrate in the $v = 3$ state and, in the absence of any external perturbation (say, in the gas phase at low pressure), the atoms XY would continue to persist in the $v = 3$ state. In solution, however, there are many perturbations induced by collisions, which can also remove energy. In addition, in a polyatomic molecule, vibrations in one part of the molecule may act as a perturbation to vibrations in another part of the same molecule, so that energy is rapidly transferred among vibrational modes (time scales on the order of a few ps or less). Thus, vibrational energy is generally removed very rapidly from upper vibrational levels, and transitions between vibrational levels seem to occur about as fast as vibrational energy can either be removed by the environment or redistribute itself within a molecule. In Fig. 4.12, this decrease in vibrational energy from $v = 3$ to 0 is shown as a sequence of arrows. The mechanisms of removal of vibrational energy from excited energy levels will be discussed in more detail in Chapter 5, which deals with radiationless transitions. This rapid relaxation of vibrational energy in electronically excited states is the fundamental basis for Kasha's rule.

4.23   The Effect of Orbital Configuration Mixing and Multiplicity Mixing on Radiative Transitions[14,15]

In Section 3.4, we saw that the zero-order approximation of electronic states (ignore electron–electron interaction and electron–vibrational interactions) is only a starting point for describing transitions between electronic states, and that we must consider electronic state mixing in order to understand the probability of radiative transitions between electronic states. A basic result of state mixing is to imbue a zero-order state, originally described in terms of a single electronic orbital configuration (or spin multiplicity), with characteristics of a second electronic orbital configuration (or spin multiplicity). We may, of course, consider the mixing of many orbital configurations. Often, however, mixing of only one other orbital configuration (usually the orbital that is closest in energy to the zero-order orbital of interest) is sufficient to interpret a great deal of experimental data. How does mixing affect zero-order predictions? For radiative (and radiationless) transitions, first order mixing is generally a significant mechanism "allowing" processes that are strictly forbidden in zero-order to occur with measurable probability. We can view the mixing as an interaction between wave functions that are close in energy. Typically, a wave function for which a transition is forbidden in zero-order borrows some of the character of a wave function for which the transition becomes allowed in second order.

As an exemplar, let us consider the radiative processes involving the absorptive transition of an electron from an n orbital to a π^* orbital (or the emissive transition from a π^* orbital to an n orbital). We start with the simple one-electron (zero-order) description of an n and a π^* orbital. This approximation does not include any electron–electron or electron–vibrational interactions. We can term the zero-order wave function that uses these orbitals as a "pure" state (Eq. 4.31a).

Consider the magnitude of f, the oscillator strength (Eq. 4.21), which is directly related to the probability of absorption (ε) and of emission (k_e^0). In zero-order, $f = 0$ for a $S_0 \rightarrow S_1$ (pure n,π^*) radiative transition, since the n and π^* orbitals undergoing transition are strictly orthogonal and therefore there is no orbital overlap. From Eq. 4.21 $< n|P|\pi^* > = 0$, since $<n|\pi^*> = 0$. In first order, we consider that vibrations or electron–electron interactions may "mix" n,π^*, and π,π^* configurations (e.g., compare to Fig. 3.1). The result is that the n and π^* orbitals are no longer orthogonal and the value of $<n|P|\pi^*> \neq 0$. We say that the transition that is forbidden in zero order is "partially" allowed in first order because of vibronic interactions. Including vibronic interactions, Eq. 4.31b is therefore a better description of S_1, which is a state of "mixed" configurations, and the coefficient λ (a mixing coefficient, not a wavelength) is a measure of the extent of mixing.

$$S_1 \text{ (pure)} = n,\pi^* \tag{4.31a}$$

$$S_1 \text{ (mixed)} = n,\pi^* + \lambda(\pi,\pi^*) \tag{4.31b}$$

Since the $S_0 \rightarrow \pi,\pi^*$ transition is generally "allowed" (i.e., in Eq. 4.21 $<P> \neq 0$), a transition $S_0 \rightarrow S_1$ (mixed) is plausible through the contribution of $\lambda(\pi,\pi^*)$ to the wave function describing S_0, although f is expected to be relatively weak compared to a fully allowed transition of the $S_0 \rightarrow \pi,\pi^*$ type. Now, we may estimate the value of f for Eq. 4.32.

$$S_0 \underset{-h\nu}{\overset{+h\nu}{\rightleftarrows}} S_1 \quad \text{(mixed)} \tag{4.32}$$

The expression for f becomes Eq. 4.33:

$$f(S_0 \rightleftarrows S_1) = \lambda^2 f(S_0 \rightleftarrows S_2) \tag{4.33}$$

In other words, the first-order $S_0 \rightarrow S_1$ (mixed) transitions are allowed only to the extent that $S_2(\pi,\pi)$ is mixed into $S_1(n,\pi^*)$. The amount of mixing is given by λ, the coefficient predicted from perturbation theory (Section 3.4). Thus, the observed value of $f(S_0 \rightarrow S_1)$ may be evaluated in terms of the theoretical value of f for the zero-order $S_0 \rightarrow S_2(\pi,\pi^*)$ transition.

According to perturbation theory, the value of λ is given by the value of the matrix element for mixing of S_0 and S_2 divided by the energy gap between S_0 and S_2 (Eq. 4.34).

$$\text{Mixing coefficient} \rightarrow \lambda = \left| \frac{\langle \psi_a | P | \psi_b \rangle}{E_a - E_b} \right| \quad \begin{array}{l} \leftarrow \text{Matrix element} \\ \leftarrow \text{Energy separation} \end{array} \quad (4.34)$$

If we replace λ in Eq. 4.33 by its equivalent from Eq. 4.34, we produce Eq. 4.35,

$$\underset{\substack{\text{Observed} \\ \text{mixed state}}}{f(S_0 \rightleftarrows S_1)} = \left| \frac{\overset{\text{Mixing}}{\underset{\text{coefficient}}{\langle n, \pi^* | P | \pi, \pi^* \rangle}}}{E_{\pi,\pi^*} - E_{n,\pi^*}} \right|^2 \overset{\text{Zero-order}}{f(S_0 \rightleftarrows \pi, \pi^*)} \quad S_2 = \pi, \pi^* \quad (4.35)$$

where n, π^* and π, π^* refer to the zero-order states, and P corresponds to the interaction that mixes S_1 and S_2.

From Eq. 4.35, we see that the measured value of $f(S_0 \rightarrow S_1)$ reflects both λ (Eq. 4.34) and $f(S_0 \rightarrow \pi, \pi^*)$. The magnitude of $f(S_0 \rightarrow S_1)$ depends on three factors:

1. The magnitude of the matrix element $<n, \pi^* | P | \pi, \pi^*>$.
2. The energy gap $^*E_{\pi,\pi^*} - E_{n,\pi^*}$.
3. The zero-order oscillator strength, $f(S_0 \rightarrow \pi, \pi^*)$, of the allowed transition.

In effect, theory predicts that $f[S_0 \rightarrow S_1 + \lambda^2 S_2(\pi, \pi^*)]$ will possess all of the characteristic properties of $S_0 \rightarrow \pi, \pi^*$ transitions except that the probability of the observed transitions will be decreased by the factor λ^2 (Eq. 4.33). For example, the electronic polarization (orientation of the absorbed light relative to the molecular framework) of the $S_0 \rightarrow S_1$ transitions will be derived from the polarization of the $S_0 \rightarrow S_2(\pi, \pi^*)$ transitions. For aromatic molecules, $S_0 \rightarrow S_2(\pi, \pi^*)$ transitions are in-plane polarized.[14] Thus, if S_1 is a mixture of n, π^* and π, π^* states, the $S_0 \rightarrow S_1$ transitions will be in-plane polarized.

Now, we may straightforwardly apply the same ideas of mixing to the qualitative evaluation of a *spin-forbidden* transition, that is, the value of $f(S_0 \rightarrow T_1)$. We assume (Section 3.21) that $^3(n, \pi^*)$ and $^1(\pi, \pi^*)$ mixing is dominant for a $S_0(n^2) \rightarrow T_1(n, \pi^*)$ transition and that spin–orbit coupling is the dominant interaction that mixes singlet and triplet states, and from perturbation theory we are led to Eq. 4.36:

$$f(S_0 \rightleftarrows T_1) = \left| \frac{\langle {}^3(n, \pi^*) | P_{SO} | {}^1(\pi, \pi^*) \rangle}{E_{\pi,\pi^*} - E_{n,\pi^*}} \right|^2 f(S_0 \rightleftarrows \pi, \pi^*) \quad (4.36)$$

We see that the spin-"forbidden" $S_0(n^2) \rightarrow T_1(n, \pi^*)$ transitions pick up finite oscillator strength via a mechanism that mixes the S and T states. The magnitude of f for $S_0 \rightarrow T_1(n, \pi^*)$ depends on the value of the matrix element and the oscillator strength of the "pure" spin-allowed $S_0 \rightarrow S_n(\pi, \pi^*)$ transitions. Both $S_0 \rightarrow T_1(n, \pi^*)$

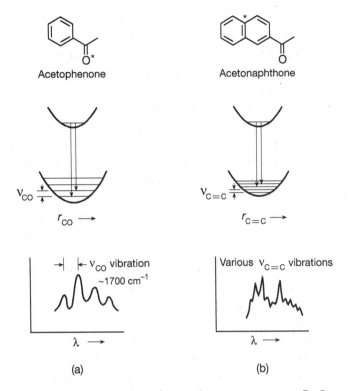

Figure 4.13   (a) Emission of acetophenone possesses a C=O vibrational pattern. (b) Emission of acetonaphthone possesses a C=C vibrational pattern.

absorption and $T_1(n,\pi^*) \rightarrow S_0$ emission are predicted to be in-plane polarized, because the oscillator strength of the "real" transition is due to the π,π^* state mixed into T_1 and because π,π^* states are generally in-plane polarized.

From the above discussion, it is clear that measurement of "forbidden" absorptions and emissions yield evidence of the identity of the mixing state that provides an interaction mechanism that makes a zero-order forbidden transition allowed in first order. A study of the vibrational structure of an absorption or emission also provides clues as to which molecular motions are most effective in mixing states. For example, like the situation for benzophenone, the vibrational structure of the $S_0 + h\nu \rightarrow S_1(n,\pi^*)$ absorptions of acetophenone shows a regular progression of vibrational bands separated by $\sim 1200 \text{ cm}^{-1}$.[16] This separation corresponds to the energy required to stretch the C=O bond in S_1. The same C=O stretching vibrations are important in mixing allowed π,π^* character into S_1, which is a nominally n,π^* state if it is strictly planar and not vibrating. As soon as the C=O stretch (and bend) is turned on, the S_1 state begins to mix with other states.

In contrast, again like benzophenone, the vibrational structure of $T_1 \rightarrow S_0 + h\nu$ phosphorescence spectrum of acetophenone shows a vibrational pattern with separations between adjacent bands of ~ 1700 cm^{-1}, a value that is characteristic of the C=O vibrational stretch in S_0 (shown schematically in Fig. 4.13a).[16] Such a vibrational pattern is expected for emission from an n,π^* state if the transition is localized on the C=O group; that is, the $\pi^* \rightarrow n$ electron jump leaves excess vibrational energy on the C and O atoms specifically because an antibonding node between these atoms disappears as a result of the orbital jump. On the other hand, the phosphorescence spectrum of acetonaphthone (Fig. 4.13b) does *not* show the characteristic C=O vibrational pattern. Instead, a complicated pattern of C=C vibrations characteristic of the aromatic ring in S_0 is observed. This is expected for emission from a π,π^* state, since the $\pi^* \rightarrow \pi$ electron jump leaves excess vibrational energy between certain C atoms of the aromatic ring as the result of the destruction of a node.

In summary, numerous spectroscopic criteria from absorption and emission exist for the assignment of the electronic configuration to S_1 and T_1. *Photochemical reactivity* can also serve to classify S_1 and T_1 in terms of their electronic configurations. Because of the two correlations, (a) spectroscopic parameters $\Leftrightarrow$ orbital configurations, and (b) orbital configuration $\Leftrightarrow$ photoreactivity, we expect (and will find) correlations between spectroscopic parameters and photoreactivity. The theoretical basis for this relationship will be made via frontier orbital methods and state correlation diagrams (Chapter 6).

4.24 Experimental Exemplars of the Absorption and Emission of Light by Organic Molecules[16]

Now, we will consider a number of experimental examples of the radiative transitions $R + h\nu \rightarrow {}^*R$ and ${}^*R \rightarrow R + h\nu$ (Scheme 4.1). We only consider the lowest excited singlet state (S_1) or the lowest triplet state (T_1) as likely candidates for the initiation of an emission. This generalization, which is based on *Kasha's rule*,[17] results from the experimental observation that the majority of photoreactions and photoemissions studied to date do not appear to involve higher-order electronic states (S_2, T_2, etc.), because rapid radiationless conversions (e.g., $S_n \rightarrow S_1$ and $T_n \rightarrow T_1$) compete favorably with emission from upper electronically excited states ($S_n, T_n, n > 1$). (We will discuss the reasons why this is generally the case in Chapter 5.) Thus, it is the spectroscopy of S_1 and T_1 that is of greatest interest to the organic photochemist, since these two states are the most likely starting point for initiating photophysical and photochemical processes. Accordingly, we consider the following radiative processes in detail:

1. $S_0 + h\nu \rightarrow S_1$ Spin-allowed absorption (singlet–singlet absorption)
2. $S_0 + h\nu \rightarrow T_1$ Spin-forbidden absorption (singlet–triplet absorption)
3. $S_1 \rightarrow S_0 + h\nu$ Spin-allowed emission (fluorescence)
4. $T_1 \rightarrow S_0 + h\nu$ Spin-forbidden emission (phosphorescence)

An exemplar that shows experimental examples of all four of these key radiative transitions has already been given for anthracene in Fig. 4.9. For the allowed radiative transition, the $S_0 \rightarrow S_1$ absorption occurs at the higher energies (~ 300 to ~ 380 nm), and $S_1 \rightarrow S_0$ fluorescence emission occurs at lower energies (~ 380 to ~ 480 nm). The bands that nearly overlap in $S_0 \rightarrow S_1$ absorption and in $S_1 \rightarrow S_0$ emission near 380 nm correspond to transitions between the 0 vibrational level of S_0 and S_1, that is, the 0,0 *bands for absorption and emission* (see the energy-level diagram below the spectrum of anthracene in Fig. 4.9). At still lower energies, the very weak $S_0 \rightarrow T_1$ absorption (~ 500 to ~ 700 nm) appears. In the spectrum, the signal for this transition is amplified many times in order to be visible. The value of ε_{max} for the $S_0 \rightarrow T_1$ absorption is $\sim 10^8$ *times smaller* than the value of ε_{max} for $S_0 \rightarrow S_1$ absorption. Finally, the $T_1 \rightarrow S_0$ phosphorescence emission is observed at lowest energies.

There are several exemplar chromophores that are representative of the types encountered in most organic photochemical systems. These are as follows:

1. The carbonyl chromophore
2. The ethylene chromophore and conjugated polyenes
3. Combinations of the carbonyl and ethylene chromophores (enones)
4. The benzene chromophore and aromatic chromophores and their derivatives

From these exemplar chromophores, we can examine and understand all of the important principles and features of radiative transitions. Building on the knowledge of the photophysics of this set of exemplar chromophores, we can infer the influence of substituted derivatives on the radiative properties of a wide range of organic molecules.

4.25   Absorption, Emission, and Excitation Spectra

The experimental measurement of an *electronic absorption spectrum* is based on two important principles: Lambert's and Beer's laws.[18] *Lambert's law* states that the proportion of light absorbed by a medium is independent of the initial intensity of the light, I_0. This law is a good approximation for ordinary light sources, such as lamps, but it breaks down when high-intensity lasers are employed. *Beer's law* states that the amount of light absorbed is proportional to the concentration of absorbing molecules in the light path. This law is a good approximation unless molecules begin to form aggregates at higher concentrations. The experimental quantity related to absorption that is conventionally measured is called the *optical density* (OD, Eq. 4.37a), where I_0 is the intensity of incident light falling on the sample and I_t is the intensity of transmitted light through the sample (usually understood to be 1 cm in depth). For example, an OD of 2.0 corresponds to $\sim 1\%$ transmission or $\sim 99\%$ absorption; an OD of 1.0 corresponds to $\sim 10\%$ transmission and 90% absorption; an OD of 0.01 corresponds to $\sim 98\%$ transmission or $\sim 2\%$ absorption. It is important to note that for a sample with an optical density > 2.0, most of the light is absorbed in a very small volume of the sample near to the place where the light impinges on the sample.

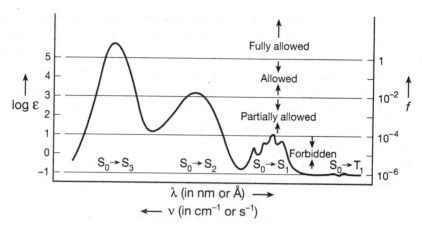

Figure 4.14   Schematic representation of the absorption spectrum of a simple organic molecule. Note the y-axis (left) plots $\log \varepsilon$ and the y-axis (right) plots f. Often the first ($S_0 \rightarrow S_1$) transition is relatively weak compared to $S_0 \rightarrow S_2$ and $S_0 \rightarrow S_3$ transitions. The $S_0 \rightarrow T_1$ transition, while always present, is often too weak to measure experimentally. Note that the ordinate is a log scale, each entry specifying a power of 10.

An absorption spectrum is completely described by a graph of optical density as the ordinate and the wavelength (λ units typically nm, or Å) of absorbed light as the abscissa (experimental example, Fig. 4.9; schematic examples, Fig. 4.14). In some cases, it is more informative to use energy units (reciprocal wavelength, cm^{-1}, or frequency, s^{-1}) instead of wavelength. Conventionally, the molar extinction coefficient (ε) is employed in such graphs rather than absorption intensity, and is given by Eq. 4.37b.

$$OD = \log (I_0/I_t) \tag{4.37a}$$

$$\varepsilon = [\log (I_0/I_t)]l/[A] \tag{4.37b}$$

In Eqs. 4.37a and b, I_0 and I_t are the intensity of the incident and transmitted light, respectively; l is the optical path length (usually 1 cm); and [A] is the concentration of absorbing material, A. The coefficient (ε) is a fundamental molecular property and is independent of concentration and path length if Lambert's and Beer's laws hold. Because of the wide variation of the values of ε, absorption spectra are sometimes plotted as $\log \varepsilon$ versus wavelength (λ, nm), as shown in Fig. 4.14 and as discussed in Section 4.14. The units of ε are $cm^{-1} M^{-1}$ (understood as the units of absorption). For comparison with ε, the oscillator strength, f, is plotted on the y-axis to the right of Fig. 4.14. As discussed in Section 4.14, it is interesting to note that ε has the dimensions of *area*/mol (i.e., $cm^{-1} M^{-1} = cm^{-1} mol^{-1} l = cm^2 mol^{-1}$).

An *emission spectrum* is a plot in nm (or Å) of emission intensity I_e (at a fixed excitation wavelength and constant exciting intensity I_0) as a function of wavelength (or energy) of exciting light. For a weakly absorbing solution (OD < 0.1) of a luminescent molecule A, I_e is given by Eq. 4.38.

$$I_e = 2.3 \, I_0 \varepsilon_A \, l \, \Phi^A [A] \tag{4.38}$$

In Eq. 4.38, ε_A is the extinction coefficient of the absorbing molecule, l is the optical path length, Φ^A is the quantum yield of emission of A (discussed in Section 4.27),

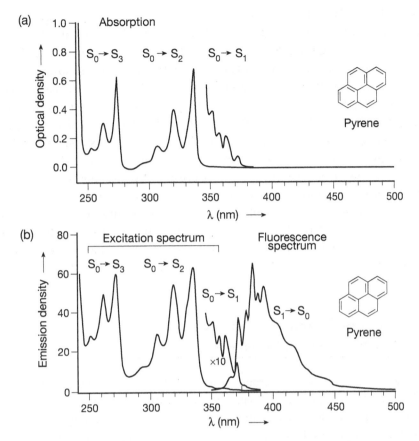

Figure 4.15   (a) Absorption spectrum of pyrene. (b) Excitation spectra of pyrene (left) and fluorescence spectrum of pyrene (right) All spectra are in cyclohexane solvent at room temperature. The insert of bands between 340–380 nm correspond to the $S_0 \rightarrow S_1$ absorption expanded by a factor of 10. This π, π^* transition is symmetry forbidden (Section 4.26) and possesses a much smaller extinction coefficient than the shorter wavelength $S_0 \rightarrow S_2$ and $S_0 \rightarrow S_3$ absorptive transitions.

and [A] is the concentration of A. The quantum yield of emission, Φ^A, is usually independent of exciting wavelength (Kasha's rule).[17] Thus, from Eq. 4.38, at fixed values of [A], I_0, and l, the intensity of emitted light from a sample, I_e, is directly proportional to the extinction coefficient (ε_A). A plot of I_e as a function of wavelength of exciting light that will vary as ε_A is termed an *excitation spectrum* and has the *same spectral shape and appearance as the absorption spectrum*. An advantage of an excitation spectrum over a standard absorption spectrum is the greater sensitivity of luminescence techniques, which often allows observation of an excitation spectra at [A] too low to be directly measured by absorption spectroscopy.

Figure 4.15 is an example of an absorption, fluorescence, and fluorescence excitation spectrum, with pyrene as an exemplar. The excitation spectrum is shown on the lower left portion of the figure together with the fluorescence spectrum to its right. The absorption spectrum of pyrene is shown for comparison directly above the fluorescence excitation spectrum. Notice, as expected from Eq. 4.38, the close correspondence of the absorption spectrum and the excitation spectrum. Note, also, the relationship of the vibrational spectra of the $S_0 \rightarrow S_1$ absorption and the $S_1 \rightarrow S_0$ fluorescence. Compare this relationship to that shown in Fig. 4.9 for anthracene. Both anthracene and pyrene are examples of aromatic hydrocarbons whose absorption spectra are "mirror images" of the fluorescence spectrum. However, there is an important difference in the $S_0 + h\nu \rightarrow S_1$ transitions: the transition is allowed for anthracene ($\varepsilon_{max} \sim 100,000$ for the $S_0 + h\nu \rightarrow S_1$ transition) and is partially forbidden for pyrene ($\varepsilon_{max} \sim 500$, $S_0 \rightarrow S_1$ transition). Because the $S_0 + h\nu \rightarrow S_1$ transition is partially forbidden for pyrene, the $S_1 \rightarrow S_0 + h\nu$ fluorescence transition is also partially forbidden and relatively long lived. The energy gap between energy of the fluorescence transition is easily perturbed by the polarity of solvents.[18c] This sensitivity to solvent polarity makes pyrene fluorescence an excellent probe of the polarity of supramolecular media.

We have seen (Eq. 4.18) that the value of ε_{max} is proportional to the radiative rate of emission. Consequently, the rate constant of fluorescence from anthracene ($\sim 10^8$ s^{-1}) is much faster than the rate constant for fluorescence of pyrene ($\sim 10^6$ s^{-1}).

4.26 Order of Magnitude Estimates of Radiative Transition Parameters

By the term "spin-allowed electronic radiative transition," we mean any radiative transition that does not involve a change in spin multiplicity. For organic molecules, spin-allowed electronic radiative transitions are of two types: singlet–singlet and triplet–triplet transitions. Examples are $S_0 \rightarrow S_n$ or $T_1 \rightarrow T_n$ absorptions and $S_1 \rightarrow S_0$ or $T_2 \rightarrow T_1$ emissions.

The probability of spin-"allowed" transitions ranges over four orders of magnitude (Fig. 4.14 and Table 4.4). Thus, we must accept a wide range of "allowedness" in radiative transitions, even for spin-allowed radiative transitions. *We must understand that the terms "allowed" and "forbidden" are relative and not absolute.* We need to think of these terms as the *relative* probability, or rate of one type of process compared to another. It is important in making such comparisons that the processes being compared have similar mechanistic features (analogous forces that induce the transitions). For example, it is not proper to compare the allowed (or forbidden) character of a radiative transition (e.g., $S_1 \rightarrow S_0 + h\nu$) to the allowed (or forbidden) character of a radiationless transition (e.g., $S_1 \rightarrow S_0 + heat$) since the former involves forces of interactions between the electrons of S_1 and the electromagnetic field and the latter involves forces of interactions between the electrons of S_1 and intramolecular and intermolecular vibrations. We can obtain some insight to the terms allowed and forbidden by considering the classical theory of the interaction of light with molecules, where the "allowedness" or strength of radiative transitions was measured by the oscillator strength (f). Order-of-magnitude estimations of f, ε, and k_e^0 can be made by use of the relationships given in Eq. 4.39.

$$f \, \alpha \int \varepsilon dv \sim \varepsilon_{max} \Delta \bar{\nu}_{1/2} \quad \text{and} \quad f \, \alpha \, k_e^0 (\bar{\nu}^2)^{-1} \qquad (4.39)$$

Table 4.4   Some Representative Examples of ε_{max} and f Values for Prototype Transitions[a]

	k_e^0 (s^{-1})	Example	Transition type	ε_{max}	f	ν_{max}(cm^{-1})
	10^9	p-Terphenyl	$S_1(\pi,\pi^*) \leftrightarrow S_0$	3×10^4	1	30,000
Spin	10^8	Perylene	$S_1(\pi,\pi^*) \leftrightarrow S_0$	4×10^4	10^{-1}	22,850
Allowed	10^7	1,4-Dimethylbenzene	$S_1(\pi,\pi^*) \leftrightarrow S_0$	7×10^2	10^{-2}	36,000
	10^6	Pyrene	$S_1(\pi,\pi^*) \leftrightarrow S_0$	5×10^2	10^{-3}	26,850
	10^5	Acetone	$S_1(n,\pi^*) \leftrightarrow S_0$	10	10^{-4}	~30,000
	10^4	Xanthone	$T_1(n,\pi^*) \leftrightarrow S_0$	1	10^{-5}	~15,000
Spin	10^3	Acetone	$T_1(n,\pi^*) \leftrightarrow S_0$	10^{-1}	10^{-6}	~27,000
Forbidden	10^2	1-Bromonaphthalene	$T_1(\pi,\pi^*) \leftrightarrow S_0$	10^{-2}	10^{-7}	20,000
	10	1-Chloronaphthalene	$T_1(\pi,\pi^*) \leftrightarrow S_0$	10^{-3}	10^{-8}	20,600
	10^{-1}	Naphthalene	$T_1(\pi,\pi^*) \leftrightarrow S_0$	10^{-4}	10^{-9}	21,300

a. These values represent orders of magnitude only.

For orders-of-magnitude estimates, we assume that $\Delta\bar{\nu}_{1/2}$, the one-half width of the absorption band (in units of cm^{-1}, called wavenumbers, $\bar{\nu}$, which are directly proportional to energy), is roughly constant for commonly encountered transitions, so that f is proportional to ε_{max} (Eq. 4.17). Thus, a qualitative relationship between the commonly measured experimental quantity ε_{max} and the theoretical quantity f is available. Table 4.4 lists some exemplars of this relationship. "Allowed" transitions ($f \sim 1$) correspond to values of ε_{max} on the order of 10^4–10^5 (exemplars, p-terphenyl and perylene). The strongest spin-"allowed" transitions possess $\varepsilon_{max} \sim 10$ (exemplar, acetone) and therefore correspond to $f \sim 10^{-4}$. Fig. 4.13 compares ε_{max} and f in terms of absorption spectra. Table 4.4 compares values of k_e^0, ε_{max}, and f. Importantly, the value of f is also related to the probability, or rate per unit time, of emission, k_e^0. The more probable the absorption (the larger the value of ε_{max}), the faster the related emission (the larger the value of k_e^0). For values of f close to 1, the fastest rates of emission are on the order of $10^9 \cdot$s^{-1} (exemplar, p-terphenyl).

For more quantitative comparisons of experiment with theory, we note that the relationship between f and k_e^0 (Eq. 4.27) depends on $\bar{\nu}^2$, the square of the *frequency* of the emission. Thus, the rate of emission depends not only on ε_{max} but also on the wavelength of emission, which determines the frequency of emission. For example (Table 4.4), while 1,4-dimethylbenzene and pyrene possess similar values of ε_{max} (~500–700), they absorb and emit at very different wavelengths (1,4-dimethylbenzene at 277 nm and pyrene at 372 nm). As a result, 1,4-dimethylbenzene and pyrene possess quite different oscillator strengths, f, and values of k_e^0 for fluorescence. Similarly, perylene possesses a slightly larger ε_{max} than p-terphenyl, but the latter possesses a larger f because it possesses a larger $\bar{\nu}_{max}$ of the two compounds.

An important conclusion from the data in Table 4.4 is that *even for spin-allowed transitions, there are factors that cause a certain degree of forbiddenness to absorption and emission.* If we think of a perfectly allowed transition as having an oscillator strength $f_{max} = 1.0$, then we may think of an observed measured f_{obs} value in terms of the product of the individual "forbiddenness factors" f_i, which *reduce* the value of f_{max} from that of the ideal system, Eq. 4.40:

$$f_{obs} = (f_e \times f_v \times f_s) f_{max} \qquad (4.40)$$

where f_e is the prohibition due to electronic factors, f_v is the prohibition due to vibrational (Franck–Condon) factors, and f_s is the prohibition due to spin factors. For a spin-allowed transition, $f_s = 1$, and for a spin-forbidden transition, f_s depends on the spin–orbit coupling available during the transition (typical values of f_s range from 10^{-6}–10^{-11}).

The electronic factor f_e may be subclassified in terms of different kinds of forbiddenness:

1. *Overlap* forbiddenness, which results from poor spatial overlap of the orbitals involved in the HO $\rightarrow$ LU electronic transition. An exemplar is the n,π^* transition in ketones (Section 2.11), for which the HO and LU are orthogonal to one another in zero-order and the overlap integral $<n|\pi^*>$ is close to zero.

2. *Orbital symmetry* forbiddenness, which results from orbital wave functions (involved in the transition) that overlap in space but have their overlap integral canceled because of the symmetry of the wave functions. Examples are the $S_0 + hv \rightarrow S_1(\pi,\pi^*)$ and $S_1(\pi,\pi^*) \rightarrow S_0 + hv$ transitions in benzene, naphthalene, and pyrene. Inspection of the details of the phases of HO and LU for these transitions beyond the scope of this text is required for an understanding of orbital forbiddenness. However, some insight is available from classical theory. In order to produce a transition electric dipole moment, the oscillating electric field of the light wave needs to drive an electron back and forth along a molecular axis; that is, a transition oscillating dipole must result from the interaction of the electromagnetic field and the electron. To a good approximation, we need consider only the HO $\rightarrow$ LU transition as determining the transition dipole along the molecular axis. For molecules possessing high symmetry, quite often for the HO $\rightarrow$ LU transition, there is no good axis along which a significant transition dipole can be generated. Benzene and pyrene, for example, are very symmetrical molecules for which f is $\sim 10^{-3}$. For *p*-terphenyl, however, a favorable axis exists along the 1,4-positions of the three benzene rings, and $f \sim 1$.

Generally, for *spin-allowed* transitions the electronic factor f_e is the major factor in the determination of the observed values of f. From Table 4.4, note that perylene and *p*-terphenyl possess "strong" $S_0 \rightarrow S_1$ absorptions ($f \sim 1$–10^{-1}, $\varepsilon_{max} \sim 10^5$–$10^4$). These absorptions essentially correspond to fully allowed ($\pi \rightarrow \pi^*$) transitions. For pyrene (and for benzene and naphthalene), the $S_0 \rightarrow S_1(\pi,\pi^*)$ transition is *orbital symmetry forbidden,* and an electronic "forbiddenness factor" of $f_e \sim 10^2$–10^{-3} results in a relatively weak ε_{max} of $\sim 10^2$. For acetone, the $S_0 \rightarrow S_1$ transition corresponds to an $n \rightarrow \pi^*$ transition. This transition is *forbidden by* both *orbital overlap and symmetry* and would be predicted to have $f \sim 0$ if the n orbital were a pure p orbital, and if the molecules were strictly planar. Experimentally, $\varepsilon_{max} \sim 10$ for this n $\to \pi^*$ transition as the result of vibronic mixing of the n and π^* orbitals.

Let us see how vibronic mixing operates. Out-of-plane vibrations (Fig. 3.1) allow the n orbital to pick up s "character." In the case of benzophenone, "mixing" of the n,π^* state with nearby π,π^* states makes S_1 a hybrid of these two transition types (Eq. 3.9). As a result, the $S_0 \to S_1$ transition, which is an overlap forbidden n,π^* transition in zero order, picks up finite oscillator strength because of the π,π^* character "mixed" into S_1 (Eqs. 4.31 and 4.35). In a manner of speaking, the S_1 state picks up absorption intensity from its acquired π,π^* character. In the case of acetone, S_1 is more nearly "pure" n,π^* because of the poorer mixing (*ΔE is much larger in the denominator of Eq. 4.35*), and the absorption intensity is correspondingly lower.

Because of the direct relationship between f and the fluorescence rate constant k_F^0, (Eq. 4.27, $k_e^0 = k_F^0$), the factors determining the magnitude of f automatically are proportional to those determining k_F^0. This approximation is a good one for allowed transitions if we can ignore f_v (i.e., rigid molecules) as a major factor determining the value of f or k_F^0. If the nuclear geometry of the equilibrated excited state is very different from that of the initial ground state, then the value of k_F^0 for the $S_1 \to S_0$ transition will be determined by f_e and by different Franck–Condon factors (see Section 4.22) that relate to f for the $S_0 \to S_1$ transition. In the special cases for which the equilibrium geometry and predominant vibrational progressions of S_0 and S_1 are similar, a "mirror-image" relationship is sometimes observed for the absorption and related emission spectra (see Figs. 4.9 and 4.15); that is, $S_0 \to S_1$ "mirrors" $S_1 \to S_0$ and $S_0 \to T_1$ "mirrors" $T_1 \to S_0$.

Based on information from absorption spectra, an orbital configuration may be assigned to the electronic transition responsible for an absorption band. For anthracene, benzene, pyrene, and other aromatic hydrocarbons, the entire π system can be assumed to behave approximately as a single chromophore since the molecular orbitals are delocalized over the molecular framework and only $\pi \to \pi^*$ transitions are energetically feasible in the 200–700-nm region. For benzophenone and aromatic carbonyl compounds, both n $\to \pi^*$ (longer wavelength) and $\pi \to \pi^*$ (shorter wavelength) transitions are possible. Empirically, a number of criteria have been developed that allow an orbital configuration change to be identified from characteristics of absorption and emission spectra; these criteria are presented in Table 4.5.

Consider Fig. 4.16a to be an example for the use of information in Table 4.5 to identify the orbitals involved in radiative transitions that display the spin-allowed absorption spectrum of benzophenone; this spectrum in cyclohexane consists of two major absorption bands, one maximizing at ~ 350 nm and the other maximizing at ~ 250 nm. Note that the n,π^* absorption of a rigid cyclanone (Fig. 4.16b) occurs in a similar wavelength region to that for benzophenone. The relatively low value of ε_{max} (~ 100) for the longer-wavelength band of benzophenone and the wavelength of the maximum allows assignment of an orbitally and spatially forbidden n $\to \pi^*$ transition to the long wavelength band (see Table 4.5). This assignment is consistent with the "blue shift" of the maximum upon going from cyclohexane to ethanol, since the energy gap between the n and π^* orbitals is expected to be larger (therefore absorption occurs at shorter wavelengths) in ethanol than in cyclohexane. The reason is because

Table 4.5 Empirical Criteria for the Assignment of Orbital Configurations of Ketones

Property	$n-\pi^*$		$\pi-\pi^*$	
	$S_0 \rightarrow S_1$	$S_0 \rightarrow T_1$	$S_0 \rightarrow S_1$	$S_0 \rightarrow T_1$
ε_{max}	<200	>10^{-2}	>1000	<10^{-3}
$k_e^0(s^{-1})$	10^5–10^6	10^3–10^2	10^7–10^8	1–10^{-1}
Solvent shift	Shorter wavelengths with increasing solvent polarity		Longer wavelengths with increasing solvent polarity	
Vibrational structure	Localized vibrations		Delocalized vibrations	
Heavy atom effect	None		Increases probability of all S → T transitions	
ΔE_{ST}	Small (< 10 kcal mol^{-1})		Large (> 20 kcal mol^{-1})	
Polarization of transition moment	Perpendicular to molecular plane	Parallel to molecular plane	Parallel to molecular plane	Perpendicular to molecular plane
$\Phi_e^{77 K}$	< 0.01	~0.5	1.0–0.05	< 0.5
E_T	< 75	< 65	Variable	

hydrogen bonding stabilizes the n orbital in S_0 and destabilizes the π^* orbital in S_1. The destabilization of the π^* orbital results from the Franck–Condon excitation of an n electron into a π^* orbital. The ethanol solvent molecules do not have time to reorient during the time scale of the electron orbital jump. As a result, the solvent dipoles are oriented about the carbonyl oxygen atom with their positive ends (hydrogen bonds) pointing toward the oxygen. Immediately after the electron jump occurs, the n orbital is half-filled and is much more electronegative than it was in the ground state. The solvent dipoles are in an unfavorable spatial distribution about the oxygen atom, and this corresponds to an energy increase in the energy of the π^* orbital and a higher energy required for the n $\rightarrow$ π^* transition in ethanol relative to cyclohexane.

To the photochemist, an idea of the limiting values of k_F^0 is important to calibrate the maximum time allowed for reaction in S_1, because if a reaction from S_1 is to occur efficiently, its rate must be competitive with k_F^0. Otherwise, fluorescence will dominate as a decay pathway for S_0. In terms of numerical benchmarks, note from Table 4.4 that, among organic molecules, a benchmark for the "world's record" for the largest k_F^0 is $\sim 10^9$ s^{-1} for p-terphenyl (a fully allowed $\pi \rightarrow \pi^*$ transition) and the benchmark for the "world's record" for the smallest k_F^0 is $\sim 10^5$ s^{-1} for acetone (a weakly allowed n $\rightarrow \pi^*$ transition). These calibration points put the limits on the range of lifetimes limited by the rate of fluorescence in the range of 10^{-9}–10^{-5} s and provide two useful rules for calibration: If an electronically excited organic molecule possesses a lifetime shorter than 10^{-9} s, its lifetime is limited not by fluorescence but by some photophysical or photochemical radiationless process. If an electronically excited organic molecule possesses a lifetime longer than 10^{-5} s, it cannot be a singlet state.

As benchmarks for spin–forbidden transitions, the largest value of the phosphorescence rate constant k_P^0 is on the order of 10^3 s^{-1} (lifetime 10^{-3} s), and the smallest

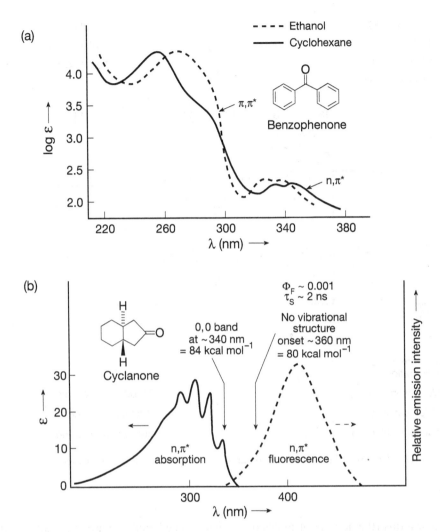

Figure 4.16 (a) Absorption spectrum of benzophenone in ethanol (dashed line) and cyclohexane (solid line). (b) Absorption and emission spectrum of a cyclanone.

value of k_P^0 is on the order of 10^{-1} s^{-1} (lifetime 10 s). Thus, $^*R(T_1)$ will persist many orders of magnitude longer than $^*R(S_1)$ before emitting a photon. This long-lifetime characteristic of triplets has important implications concerning their photochemistry.

4.27   Quantum Yields for Emission ($^*R \rightarrow R + h\nu$)

The rate constants of photophysical and photochemical processes from $^*R(S_1)$ and $^*R(T_1)$ determine the efficiencies of the processes that occur from these electronically excited states. The quantum yield (Φ) is an efficiency parameter measuring the fraction of absorbed photons that produces a specific sequence shown in the paradigm of organic photochemistry (Scheme 2.1). The parameter Φ may be expressed in

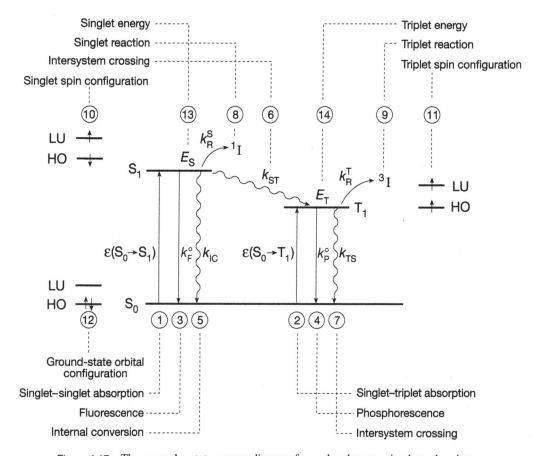

Figure 4.17   The exemplar state energy diagram for molecular organic photochemistry.

molar terms (the number of moles of *R that proceed along a particular pathway in Scheme 2.1 relative to the number of moles of photons absorbed by R) or in kinetic terms (the rate of a pathway of interest from *R compared to the sum of the rates of all pathways for decay of *R). The state energy diagram (Fig. 4.17) serves as a convenient paradigm for bookkeeping of the state electronic configurations, rates radiative and radiationless transitions, energies, and efficiencies of radiative and radiationless photophysical processes. Note that the state energy diagram refers to a nuclear geometry that is very close to the equilibrium geometry of the ground state. In Chapter 6, we discuss an extension of the state energy diagram, namely, the state correlation diagram showing the way the states of R and *R correlate with I and F during photochemical reactions.

The absolute quantum yield of emission for an organic molecule upon absorption of light (Φ_e) is an important experimental parameter containing useful information relating the structure and dynamics of electronically excited states. In addition, emission from organic molecules has become a valuable tool in analytical chemistry and modern photonics in the development of light-operated switches and sensors. Examples of the total (fluorescence and phosphorescence) emission spectra of different types of organic molecules are given in Figs. 4.18–4.21 for exemplar aromatic lumophores (π,π^* emission) and the ketone lumophore (n,π^* emission).

We noted a number of times that the paradigm for analyzing emission of organic molecules is Kasha's rule,[17] which states that, upon photoexcitation of an organic molecule, only fluorescence from an (thermally equilibrated) S_1 state or phosphorescence from a (thermally equilibrated) T_1 state is observed experimentally. Whether any emission is observed at all from S_1 or T_1 for an organic molecule is determined by the experimental quantum yield for emission (Φ_F for fluorescence and Φ_P for phosphorescence). The value of Φ is a direct and absolute measure of the efficiency of an emission process and is defined as photons out (emitted) versus photons in (absorbed). Although all excited states emit a finite number of photons *in principle*, it is found *in practice* that quantum yields of $< 10^{-5}$ are difficult to measure experimentally and are prone to experimental artifacts. Thus, for practical purposes a molecule is considered to be (measurably) fluorescent or (measurably) phosphorescent if the quantum yield for emission, Φ_e, is $> 10^{-5}$.

A general expression for a quantum yield of emission Φ_e from a specific state, *R(S_1) or *R(T_1), is given by Eq. 4.41:

$$\Phi_e = {}^*\Phi k_e^0 (k_e^0 + \Sigma k_i)^{-1} = {}^*\Phi k_e^0 \tau \qquad (4.41)$$

where $^*\Phi$ is the formation efficiency of the emitting state, k_e^0 is the rate constant (k_F^0 or k_P^0) for emission from the state, and Σk_i is the sum of all rate constants (unimolecular or pseudo-unimolecular) that radiationlessly deactivate the emitting state; that is, $\tau = (k_e^0 + \Sigma k_i)^{-1}$. The experimental lifetime (τ), and therefore the experimental quantum yield of emission, Φ_e, depend crucially on the magnitude of Σk_i relative to the value of k_e^0. The latter does not usually change very much with experimental conditions, but Σk_i can vary by many orders of magnitude depending on experimental conditions.

For example, in fluid solution at room temperature, bimolecular, diffusional quenching processes (oxygen is a ubiquitous and efficient quencher of electronically excited states), photophysical radiationless deactivations, and photochemical reactions may compete with radiative decay of an excited state (see the state-energy diagram, Fig. 4.17). Thus, Φ_e may be very small *even* if $^*\Phi$ is close to unity.

Thus, in order to observe an electronic emission spectrum routinely, it is usually necessary to minimize Σk_i. This is accomplished by cooling the sample to a very low temperature (77 K, the boiling point of liquid nitrogen, is an experimentally convenient temperature) and/or by making the sample a rigid solid, such as a polymer (all organic solvents are solids at 77 K). Numerous solvents form optically clear solid solutions at 77 K and are called *glasses* at this temperature. The low temperature and sample rigidity cause terms in Σk_i that correspond to the rate constants of processes activated by several kilocalories per mole or more to become small relative to k_e^0. The rigidity of the sample also eliminates terms in Σk_i that are due to bimolecular quenching processes, since diffusion is eliminated in solid solution. In addition to preventing diffusional quenching, a rigid solvent matrix may restrict certain molecular motions (e.g., twisting of C=C bonds or extensive stretching of C—C bonds), which

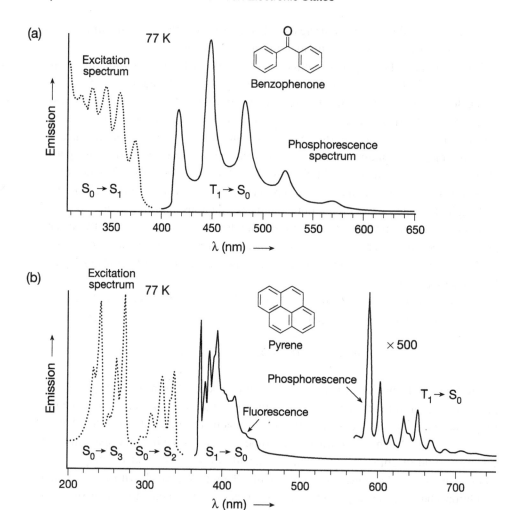

Figure 4.18 Emission spectrum of benzophenone and pyrene at 77 K of (a) benzophenone and (b) pyrene as exemplars for ketones and aromatic compounds. The spectrum of 1-chloronaphthalene (c) is shown as an exemplar of the heavy atom effect. (See opposite page for panel (c).)

are particularly effective at promoting physical and chemical radiationless transitions. Thus, taking an emission spectrum at 77 K usually "quenches" many processes that compete with emission from *R, allowing the quantum yield for emission to reach a value that can be measured experimentally.

Even when fluorescence and phosphorescence spectra of organic molecules are measured at 77 K in organic glasses, the total quantum yields of emission ($\Phi_F + \Phi_P$) are generally < 1.00. Evidently, some radiationless processes from S_1 occur even at 77 K in rigid glasses, as shown in Eq. 4.42.

$$\Phi_F + \Phi_P + \Sigma\Phi_R = 1 \qquad (4.42)$$

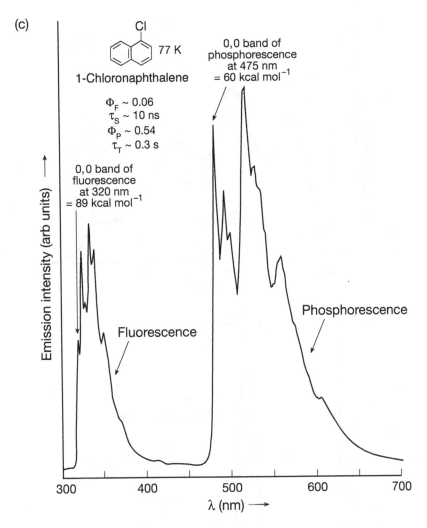

(c)

Cl

1-Chloronaphthalene

77 K

$\Phi_F \sim 0.06$
$\tau_S \sim 10$ ns
$\Phi_P \sim 0.54$
$\tau_T \sim 0.3$ s

0,0 band of
phosphorescence
at 475 nm
= 60 kcal mol^{-1}

0,0 band of
fluorescence
at 320 nm
= 89 kcal mol^{-1}

Fluorescence

Phosphorescence

Emission intensity (arb units)

300   400   500   600   700

λ (nm) →

Figure 4.18 *(continued)*

In Eq. 4.42, $\Sigma\Phi_R$ is the sum of quantum yields for photochemical and photophysical radiationless transitions from S_1 and T_1. Identification and evaluation of the radiationless photophysical processes of Φ_R will be discussed in Chapter 5, and the photochemical sources will be discussed in Chapter 6. For our purposes, here we note simply that radiationless processes can compete with k_F^0 for deactivation of S_1 and with k_P^0 for deactivation of T_1, even at 77 K.

Data derived from *fluorescence* spectra at 77 K are conveniently analyzed and interpreted in terms of Eq. 4.43, which is a specific form of Eq. 4.42 ($\Phi = 1.00$, since the emitting state is the absorbing state, $k_e^0 = k_F^0$ and $\Sigma k_i = k_{ST}$).

$$\Phi_F = k_F^0(k_F^0 + k_{ST})^{-1} = k_F^0\tau_s \qquad (4.43)$$

where τ_S is defined as $(k_F^0 + k_{ST})^{-1}$.

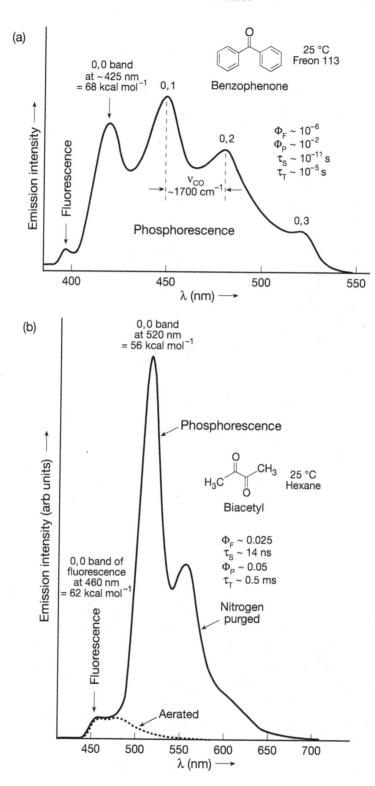

Figure 4.19   Emission spectra of (a) benzophenone in freon113 solvent, (b) biacetyl in hexane solvent, and (c) 1,4-dibromonaphthalene in acetonitrile solvent at room temperature. (See opposite page for panel (c).)

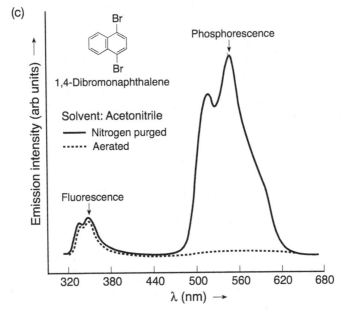

(c)

1,4-Dibromonaphthalene

Solvent: Acetonitrile
— Nitrogen purged
····· Aerated

Phosphorescence

Fluorescence

Figure 4.19 *(continued)*

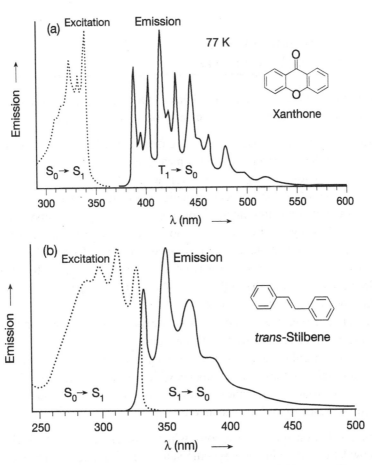

(a) Excitation Emission 77 K

Xanthone

$S_0 \rightarrow S_1$ $T_1 \rightarrow S_0$

(b) Excitation Emission

trans-Stilbene

$S_0 \rightarrow S_1$ $S_1 \rightarrow S_0$

Figure 4.20 Emission spectra (77 K) of (a) xanthone and (b) *trans*-stilbene
as exemplars for substituent effects.

Equation 4.43 has two limiting situations: (a) $k_F^0 \gg k_{ST}$, in which case $\Phi_F \sim 1.00$, and (b) $k_{ST} \gg k_F^0$, in which case $\Phi_F = k_F^0/k_{ST}$. In terms of these limits, note that Φ_F will $\to 1$ when k_F^0 is very large or k_{ST} is very small. Also, $\Phi_F \to 0$ when k_F^0 is very small or k_{ST} is very large. From the limits of ε_{max} for $S_0 \to S_1$ absorption, the limits of the rate constant of fluorescence for organic molecules are expected to be in the range (Section 4.26) given by Eq. 4.44.

$$10^9 \text{ s}^{-1} > k_F^0 > 10^5 \text{ s}^{-1} \tag{4.44}$$

The limits of k_{ST} for organic molecules from experiment (Section 4.31) turn out to be

$$10^{11} \text{ s}^{-1} > k_{ST} > 10^5 \text{ s}^{-1} \tag{4.45}$$

Thus, it is expected that simply upon consideration of the competition between fluorescence and intersystem crossing from S_1, the value Φ_F can vary over orders of magnitude. For example, in one limiting case the fluorescence is intense and easy to observe; in the other limiting case the fluorescence may be either intense or weak or nonobservable, depending on the values of *both* k_F^0 and k_{ST}.

4.28 Experimental Examples of Fluorescence Quantum Yields

Some data for fluorescence quantum yields (77 K, rigid organic glass) are given in Table 4.6. The following generalizations may be made from the data:

1. Most rigid aromatic hydrocarbons (benzene, naphthalene, anthracene, pyrene, etc.) and their derivatives possess measurable, but variable, fluorescence quantum yields ($1 > \Phi_F > 0.01$), even at 77 K.

2. Low values of Φ_F for nonrigid aromatic hydrocarbons are common and usually the result of competing internal conversion ($S_1 \to S_0$) or intersystem crossing ($S_1 \to T_1$) triggered by molecular motion. Spin-allowed and spin-forbidden radiationless transitions will be discussed in detail in Chapter 5.

3. Substitution of Cl, Br, or I, for H on an aromatic ring generally results in a decrease in Φ_F such that $\Phi_F^H > \Phi_F^F > \Phi_F^{Cl} > \Phi_F^{Br} > \Phi_F^I$ (cf. naphthalene with the halonaphthalenes in Table 4.6).

4. Substitution of C=O for H on an aromatic ring generally results in a substantial decrease in Φ_F (cf. benzene with benzophenone).

5. Molecular rigidity (due to structural or environmental constraints) enhances Φ_F (cf. rigid aromatics with stilbene, which possesses a flexible C=C bond).

6. For rigid aromatic molecules, internal conversion does not compete favorably with fluorescence or intersystem crossing.

Starting with aromatic hydrocarbons, which generally possess the highest values of Φ_F, now let us determine how specific values of k_F^0 and k_{ST} contrive to make the value of Φ_F so variable.

For benzene, naphthalene, and pyrene, the $S_0 \to S_1$ transitions are "orbital symmetry forbidden." These molecules are sufficiently symmetrical that the electric vector

Table 4.6   Some Examples of Fluorescence Quantum Yields and Other Emission Parameters[a]

Compound	Φ_F^a	ε_{max}	k_F^0	k_{ST}	Configuration of S_1
Benzene	~0.2	250	2×10^6	10^7	π,π^*
Naphthalene	~0.2	270	2×10^6	5×10^6	π,π^*
Anthracene	~0.4	8,500	5×10^7	$\sim5 \times 10^7$	π,π^*
Tetracene	~0.2	14,000	2×10^7	$<10^8$	π,π^*
9,10-Diphenylanthracene	~1.0	12,600	$\sim5 \times 10^8$	$<10^7$	π,π^*
Pyrene	~0.7	510	$\sim10^6$	$<10^5$	π,π^*
Triphenylene	~0.1	355	$\sim2 \times 10^6$	$\sim10^7$	π,π^*
Perylene	~1.0	39,500	$\sim10^8$	$<10^7$	π,π^*
Stilbene	~0.05	24,000	$\sim10^8$	$\sim10^9$	π,π^*
1-Chloronaphthalene	~0.05	~300	$\sim10^6$	5×10^8	π,π^*
1-Bromonaphthalene	~0.002	~300	$\sim10^6$	$\sim10^9$	π,π^*
1-Iodonaphthalene	~0.000	~300	$\sim10^6$	$\sim10^{10}$	π,π^*
Benzophenone	~0.000	~200	$\sim10^6$	$\sim10^{11}$	n,π^*
Biacetyl	~0.002	~20	$\sim10^5$	$\sim10^8$	n,π^*
Diaza[2.2.2]bicyclooctene	~1.0	~200	$\sim10^6$	$<10^5$	n,π^*
Acetone	~0.001	~20	$\sim10^5$	$\sim10^9$	n,π^*
Perfluoroacetone	~0.1	~20	$\sim10^5$	$\sim10^7$	n,π^*
3-Bromoperylene	~1.0	~40,000	$\sim10^8$	$<10^6$	π,π^*
Pyrene-3-carboxaldehyde	~0.25	~70,000	$\sim10^8$	$\sim10^8$	π,π^*
Cyclobutanone	~0.0001	~20	$\sim10^5$	$\sim10^9$	n,π^*
Diaza[2.2.1]bicycloheptene	~0.0001	400	$\sim10^6$	$\sim10^6$	n,π^*

a. ε_{max} in $M^{-1}cm^{-1}$; k_F^0 and k_{ST} in s^{-1}.

of the light cannot easily find an effective axis along which to oscillate an electron and generate a significant transition dipole. Therefore, these molecules have a relatively low oscillator strength, f, and a relatively low value of ε_{max}. Consequently, for these hydrocarbons, $\varepsilon_{max} \sim 10^2$ and $k_F^0 \sim 10^6$ s^{-1}. The latter value of k_F^0 is close to the smallest for organic molecules. However, the rate of intersystem crossing from $S_1 \rightarrow T_1$ for aromatic hydrocarbons is also on the order of $\sim 10^6$ s^{-1} for both benzene and naphthalene and is somewhat slower for pyrene. Thus, benzene, naphthalene, and pyrene all fluoresce with a moderate quantum yield $\Phi_F > 0.20$. The S_1 molecules that do not fluoresce generally undergo intersystem crossing to T_1 (i.e., $\Phi_{ST} \sim 1.0$).

For anthracene, the $S_0 \rightarrow S_1$ transition is symmetry allowed (the electric vector of light now easily recognizes the long or the short axis of anthracene as an excellent axis for induction of electron oscillation). As a result, $\varepsilon_{max} \sim 10^4$ and $k_F^0 \sim 10^8$ s^{-1}. In the case of diphenylanthracene, $\Phi_F \sim 1.0$; that is, essentially every excited singlet state formed fluoresces. In this case, a large value of k_F^0 and small values of k_{ST} (poor spin–orbit coupling) and k_{IC} (large energy gap between S_1 and S_0) contribute to produce a value of Φ_F that is close to unity.

Now, consider the decrease in Φ_F that generally accompanies the replacement of H with halogen or C=O functions. The small value of Φ_F (within the framework of our assumption of negligible internal conversion) means that intersystem crossing is much faster than fluorescence for these molecules; that is, $k_{ST} \gg k_F^0$. For halogenated naphthalenes, the substitution of halogen for hydrogen affects ε_{max} by a factor of only ~ 2, whereas Φ_F varies over several orders of magnitude. From the constancy of ε_{max}, we conclude that k_F^0 does not vary much in this series, so that k_{ST} must be the changing variable leading to the radical variation in Φ_F. The enhancement of probability of spin–forbidden transitions that result from the replacement of hydrogen by halogen is known as the "heavy-atom effect," where a "light" atom is defined as any atom in the first two rows of the periodic table (e.g., H, C, N, O, F) and a "heavy" atom is defined as any atom beyond the third row of the periodic table (e.g., Cl, Br, I). The theoretical basis of this effect is due to enhanced spin–orbit coupling induced by heavy atoms (discussed in Section 3.21).

Aromatic ketones (e.g., acetophenone and benzophenone) do not possess heavy atoms, yet generally possess a very small value of Φ_F, implying a relatively fast intersystem crossing compared to the rate of fluorescence. This low value of Φ_F ($\Phi_F \sim 0.01$–0.0001) is a general feature of the emission of ketones possessing $S_1(n,\pi^*)$ states (Figs. 4.18–4.20 and Table 4.6). Since the radiative rate of the orbital symmetry forbidden transition $S_1(n,\pi^*) \to S_0 + h\nu$ fluorescence is relatively slow ($\sim 10^5$ s^{-1} from Table 4.6), the magnitude of k_{ST} need not be much larger than it is for aromatic hydrocarbons. Strikingly however, values of k_{ST} may reach values of 10^{11} s^{-1} for certain ketones (e.g., benzophenone), implying a very enhanced value of k_{ST} relative to aromatic hydrocarbons (for an explanation of this effect see Section 4.31). Thus, a small value of k_F^0 and a large value of k_{ST} combine to make Φ_F very small for aromatic ketones.

In a few special cases (e.g., unstrained cyclic azoalkanes) a relatively small value of k_F^0 is accompanied by an even smaller value of k_{ST}, so that Φ_F is still ~ 1.0.

The general rule that Φ_F is small for halogenated compounds and carbonyl compounds has exceptions, which are informative to analyze. From Table 4.6, bromoperylene ($\Phi_F \sim 1.0$) and pyrene-3-aldehydes ($\Phi_F \sim 0.70$) are examples. A large energy gap between S_1 and any other T_n states produces a Franck–Condon inhibition to direct $S_1 \to T_n$ intersystem crossing (Section 4.30), so that the occurrence of a second triplet (usually T_2) that is of lower energy than S_1 is required for state mixing and fast intersystem crossing. In bromoperylene and pyrene-3-aldehyde, k_{ST} is exceptionally slow in spite of the attachment of a bromine atom or an aldehyde function to the aromatic ring, because T_2 lies well above S_1, and state mixing is inhibited.

Anthracene presents an interesting case for which T_2 has an energy that is very close to the energy of S_1. The energy is so close that T_2 may possess a higher energy than S_1 or a lower energy than S_1 depending on the solvent or substitutents. When $^*E(T_2) > E(S_1)$, intersystem crossing is slow and the fluorescence yield is high; when $^*E(T_2) < E(S_1)$, the intersystem crossing rate is fast and the fluorescence yield is low. The effect of the position of T_2 relative to S_1 will be discussed in Chapter 5.

In addition to a competing fast intersystem crossing, Φ_F may be small for molecules that undergo a very fast photochemical reaction in S_1. For example, cyclobutanone ($\Phi_F \sim 0.0001$) undergoes an efficient very fast cleavage of a CO–C bond in S_1 that competes effectively with *both* fluorescence and intersystem crossing.

The lesson to be learned from these examples is that the experimental value of Φ_F represents an *efficiency* that compares *relative* transition *probabilities* and does *not* relate directly to *rates or rate constants*; that is, $\Phi_F \sim 1.0$ for 9,10-diphenylanthracene, for which $k_F^0 \sim 5 \times 10^8 \text{ s}^{-1}$, and $\Phi_F \sim 1.0$ for diaza[2.2.2]bicyclooctane, for which $k_F^0 \sim 10^6 \text{ s}^{-1}$. In the latter case, Φ_F is high in spite of a low value of k_F^0 because both photophysical or photochemical processes from S_1 are much slower than k_F^0.

Note that bimolecular quenching processes (oxygen, impurities, solvent, etc.), which are not related to the molecular structure of the fluorescing molecule, may determine the observed value of Φ_F in fluid solution, especially for molecules possessing a long S_1 lifetime.

Saturated compounds[19a,b] and simple alkenes, such as ethylenes[19c] and polyenes,[20] generally do not fluoresce efficiently. For example, tetramethylethylene shows a very broad weak fluorescence ($\lambda_{max}^F \sim 265 \text{ nm}$), with $\Phi_F \sim 10^{-4}$ and $\tau_S \sim 10^{-11} \text{ s}$. Such short lifetimes and low emission efficiencies are typical of "flexible" molecules for which a rapid radiationless deactivation may occur via a stretching motion along a C—C (or C—H) bond or via a twisting motion about a C=C bond.[21] The role of stretching and twisting motions in determining the rates of photophysical radiationless deactivations will be discussed in detail in Chapter 5.

As exemplars of the role of stretching motions in determining Φ_F, consider[22] the aromatic hydrocarbons toluene (**1**) and *tert*-butyl benzene (**2**). The latter possesses a "looser" side-chain vibration and a lower value of Φ_F than the former. The dissipation of electronic energy through coupling to "loose" stretching vibrational modes has been termed the "loose-bolt effect" of radiationless transitions (see further discussion in Section 5.12).

Toluene (**1**)
$\Phi_F = 0.14$

tert-Butyl benzene (**2**)
$\Phi_F = 0.032$

For the role of twisting motions in determining Φ_F, consider the flexible stilbenes **3**[23a] and **4**[23b] and their rigid cyclic derivatives **5** and **6**[23c,d], which provide a nice exemplar of how either molecular structure or environmental rigidity may control the measured values of Φ_F:

3
trans-Stilbene

4
cis-Stilbene

5

6

Temperature	*trans*-Stilbene **3**	*cis*-Stilbene **4**	**5**	**6**
25 °C	0.05	0.00	1.0	1.0
77 K	0.75	0.75	1.0	1.0

Although *trans*-stilbene (**3**) is only weakly fluorescent ($\Phi_F = 0.05$) and *cis*-stilbene (**4**) is essentially nonfluorescent in fluid solution at room temperature, in a solid solution at 77 K, *both compounds are strongly fluorescent* ($\Phi_F \sim 0.75$). Both temperature and environmental rigidity contribute to this enhancement of fluorescence. For example,[23] the fluorescence efficiency of (**3**) increases by a factor of 3 in going from nonviscous organic solvents ($\Phi_F \sim 0.05$) to viscous glycerol ($\Phi_F \sim 0.15$). Presumably, twisting about the C=C bond induces the radiationless process that competes with fluorescence, and the ability to twist is inhibited in the more viscous solvent and entirely inhibited in rigid solvents. In contrast, the fluorescence yields[23c,d] of the structurally rigid stilbene analogues **5** and **6** are ~ 1.0 at both 25 °C and 77 K. Here, rigidity is built into the molecular structure, and inhibition of the twisting motion does not require assistance from the external environmental structure or temperature. The dissipation of electronic energy through coupling to "loose" twisting vibrational modes has been termed the "free rotor effect" of radiationless transitions (Section 5.12).

Most ethylenes and polyenes do not display fluorescence or phosphorescence even at 77 K. In several exceptional cases,[20] structured fluorescence emission has been reported from polyenes for which both loose stretching and twisting modes are inhibited by cyclic molecular structures. For example, the rigid 1,3-dienes **7a** and **7b** steroidal compounds sufficiently show structured fluorescence emission. Presumably, structural rigidity of the steroid framework enhances the efficiency of light by inhibiting radiationless processes (especially internal conversion) that compete with fluorescence, and by enhancing k_F^0 by preventing a large Franck–Condon geometry difference between S_1 and S_0 (see Fig. 3.2), as well as by inhibiting a free rotor effect about C=C bonds.

7a
$\lambda_F^{0,0} = 305$ nm
$E_S \sim 95$ kcal mol^{-1}

7b
$\lambda_F^{0,0} = 312$ nm
$E_S \sim 92$ kcal mol^{-1}

4.29   Determination of "State Energies" E_S and E_T from Emission Spectra

The electronic energy of *R is an important property since it can be used as free energy to drive photochemical reactions. For example, the higher the energy of *R, the stronger the bonds that can be broken in a primary photochemical process. The electronic energy of *R may be determined *directly* from its emission spectrum. The *highest*-energy (highest frequency, shortest wavelength) vibrational band in an emission spectrum corresponds to the 0,0 transition (Fig. 4.9). The energy gap corresponding to this 0,0 transition characterizes the energy of the excited state, *R, which is responsible for the emission. The energy gap of a 0,0 transition is the *maximum* energy derivable from the excited state if S_0 is regenerated by emission of a photon. The singlet-state energy (E_S) and the triplet-state energy (E_T) are defined as the 0,0 energy gap for fluorescence, S_1 ($v = 0$) $\rightarrow$ $S_0(v = 0)$, and phosphorescence, $T_1(v = 0)$ $\rightarrow$ $S_0(v = 0)$, respectively.

Sometimes an emission spectrum does not show sufficiently resolved fine structure for an accurate estimate of E_S of E_T to be made from emission spectra. In this case, the "onset" or the high-energy (short-wavelength portion) of the emission spectrum must be used to estimate the upper limit of E_S or E_T. If vibrational structure appears in the absorption spectrum, the 0,0 band of absorption serves as a safe guide for an upper limit to state energies, even if the emission spectrum is not available. In Figures 4.18–4.20, the 0,0 bands of fluorescence and/or phosphorescence are noted with an arrow, and the values of E_S or E_T derived from the 0,0 vibrational band are given in the associated energy diagrams.

4.30 Spin–Orbit Coupling and Spin-Forbidden Radiative Transitions

The radiative $S_0 \rightarrow T_1$ and $T_1 \rightarrow S_0$ processes are formally "spin-forbidden," but nonetheless are generally observed experimentally because of mixing of T_1 and excited singlet states or mixing of T_1 and S_0. The magnitude of the values of $\varepsilon(S_0 \rightarrow T_1)$ or of k_P^0 ($T_1 \rightarrow S_0$) is directly related to the degree of spin–orbit coupling that mixes S_0 and T. From Section 3.20, the degree of spin–orbit coupling was shown to depend strongly on (a) the ability of the electrons in the HO or LU of *R to approach a nucleus closely, (b) the magnitude of the positive charge (atomic number) of the nucleus that the HO or LU electrons approach and experience, (c) the availability of transitions between orthogonal (or nearly orthogonal) orbitals, and (d) the availability of a "one-atom center" $p_x \rightarrow p_z$ transition that can generate orbital angular momentum that could couple with spin angular momentum.

The degree of spin–orbit coupling between two states of an atom is related to ζ_{SO}, the spin(S)–orbit(L) coupling constant available from atomic spectra (Section 3.21, Eq. 3.21).[24]

$$\mathbf{H}_{SO} = \zeta_{SO}\mathbf{SL} \tag{4.46}$$

The correlation between the magnitude of spin–orbit coupling within a molecule (as judged from the magnitude of the spin–orbit coupling constant, ζ_{SO}, of atoms) and the magnitude of $\varepsilon(S_0 \rightarrow T_1)$ absorption and $k_P^0(T_1 \rightarrow S_0)$ emission played a decisive role in establishing the triplet state as an important entity in molecular photochemistry.[25] The oscillator strength of a spin–forbidden radiative transition is expected to be related to the magnitude of ζ_{SO}, since spin–orbit coupling is the major interaction responsible for mixing singlet and triplet states in molecules.[24] This means the value of $\varepsilon(S_0 \rightarrow T_1)$

Table 4.7   Spin–Orbit Coupling in Atoms[a,b]

Atom	Atomic number	ζ (kcal mol^{-1})	Atom	Atomic number	ζ (kcal mol^{-1})
C[c]	6	0.1	I	53	14.0
N[c]	7	0.2	Kr	36	15
O[c]	8	0.4	Xe	54	28
F[c]	9	0.7	Pb	82	21
Si[c]	14	0.4	Hg	80	18
P[c]	15	0.7	Na	11	0.1
S[c]	16	1.0	K	19	0.2
Cl[c]	17	1.7	Rb	37	1.0
Br	35	7.0	Cs	55	2.4

a. Values are only representative and depend on electron configuration; they are intended here to show only trends.

b. The values are in kilocalories per reciprocal moles. They are rounded off and, strictly speaking, apply to the radical part of ζ. In effect, the angular part of ζ is considered to be close to unity. Values of ζ are for the lowest-energy atomic configurations.

c. Because of substantial configuration interaction for these atoms, the values given are extrapolated from nearby atoms in the periodic table by assuming that ζ varies with atomic number Z^4.

and $k_P^0(T_1 \rightarrow S_0)$ will increase as ζ_{SO} increases, if factors involved in the transition are similar. However, the magnitude of ζ_{SO} depends on the orbital configurations of the states involved (Table 4.7). The important points to be derived from Table 4.7 are (a) there is a rapid increase in the magnitude or spin–orbit coupling parameter, ζ_{SO}, as the atomic number increases; (b) the magnitude of spin-orbit coupling is smaller than the energy of vibrational couplings ($\sim$ 1–5 kcal mol^{-1}) for first- and second-row atoms, such as H, C, N, and O; and (c) for very heavy atoms (Pb, Xe), the magnitude of spin–orbit coupling surpasses the value of vibrational energy levels and begins to approach the value of electronic energy gaps and strong electronic interactions ($\sim$ 20–30 kcal mol^{-1}). For the molecules containing very heavy atoms, spin inversion can occur on a time scale comparable to that of vibrational motions, and the zero-order distinction between singlet and triplet states begins to break down; that is, the mixing between spin states is very strong and the usual zero-order approximation is no longer adequate to describe singlet and triplet states as "pure."

Under certain circumstances, spin–orbit coupling can be very effective even in organic molecules containing only the "light" C, N, O, and H atoms. Effective spin–orbit coupling occurs when the two states involved in mixing are close in energy, so that effective mixing occurs because of the small energy gap, which brings the states close together and allows for effective mixing through resonance. However, in addition to a small energy gap, some other conditions involving the orbitals involved in the intersystem crossing are required for strong spin–orbit coupling involving only light atoms. We discuss these orbital factors in Section 4.31.

4.31   Radiative Transitions Involving a Change in Multiplicity: $S_0 \leftrightarrow T(n,\pi^*)$ and $S_0 \leftrightarrow T(\pi,\pi^*)$ Transitions as Exemplars

For spin-forbidden transitions in organic molecules, oscillator strengths are very small and fall in the range $f \sim 10^{-5}$–10^{-9}, whereas the oscillator strengths for spin-allowed transitions are generally much larger and in the range $f \sim 1$–10^{-3} (Table 4.3). This means that radiative $S_0 \leftrightarrow T_n$ transitions are strongly forbidden relative to spin-allowed $S_0 \leftrightarrow S_n$ transitions. We can picture spin forbiddenness as the result of the requirement that the light wave must "catch" the electrons of a molecule in a situation such that spin–orbit coupling is operating on the electron spins at the same time that the electric vector of the light wave is operating on the electron cloud. This situation is generally of low probability unless heavy atoms are involved or the transition involves two states that are strongly spin–orbit coupled because of a small energy gap. More precisely, from the quantum mechanical point of view, the wave functions of the singlet and triplet states of the initial and final states must be mixed by spin–orbit coupling, as the light wave interacts with the singlet portion of the wave function.

Experimentally, it is found that, for a spin-forbidden radiative transition, $S_0(n^2) \leftrightarrow T(n,\pi^*)$ transitions possess a much greater oscillator strength than $S_0(\pi^2) \leftrightarrow T(\pi,\pi^*)$ transitions, as indicated by Eq. 4.47.

$$f[S_0 \leftrightarrow T(n,\pi^*)] \gg f[S_0 \leftrightarrow T(\pi,\pi^*)] \tag{4.47}$$

This is exactly the opposite of the situation for $S_0 \leftrightarrow S_n$ transitions, for which $f(\pi,\pi^*) > f(n,\pi^*)$. Evidently this turnaround in the relative magnitude of the oscillator strength occurs because the spin–orbit force on an electron spin is much more effective when $n^2 \leftrightarrow n,\pi^*$ transitions occur than when $\pi^2 \leftrightarrow \pi,\pi^*$ transitions occur. Let us examine why this should be the case.

The situation may be viewed schematically as follows. The f values of the $S_0(n^2) \leftrightarrow T(n,\pi^*)$ and $S_0(\pi 2) \leftrightarrow T(\pi,\pi^*)$ transitions are composed of three parts (Eq. 4.40): the electronic (f_e), vibrational (f_v), and spin factors (f_s). We know that, in general, $f_e f_v(\pi,\pi^*) > f_e f_v(n,\pi^*)$ because $\varepsilon(\pi,\pi^*) > \varepsilon(n,\pi^*)$ for singlet–singlet transitions for which spin cannot be a contributing factor. This implies that for spin-forbidden radiative transitions, $f_s(n,\pi^*) \gg f_s(\pi,\pi^*)$.

We can obtain a pictorial understanding of the reason why $f_s(n,\pi^*) \gg f_s(\pi,\pi^*)$ from application of the rules for spin–orbit coupling to the $n^2 \leftrightarrow n,\pi^*$ and $\pi^2 \leftrightarrow \pi,\pi^*$ transitions. First, as exemplars, let us consider a radiative $S_0(n^2) + h\nu \rightarrow T(n,\pi^*)$ transition for formaldehyde (Fig. 4.21) and a $S_0(\pi^2) + h\nu \rightarrow T(\pi,\pi^*)$ transition for ethylene (Fig. 4.22). The spin change for formaldehyde is due to an $n \rightarrow \pi^*$ transition, which may be viewed as a jump of an electron from a p orbital localized on oxygen (say, p_x) in the plane of the molecule to a p orbital (say, p_y) perpendicular to the plane of the molecule (i.e., the atomic p orbital on oxygen that makes up one-half of the π^* orbital). The simultaneous $p_x \rightarrow p_y$ orbital jump is thus a one-center jump involving an orbital angular momentum change. *This type of situation is*

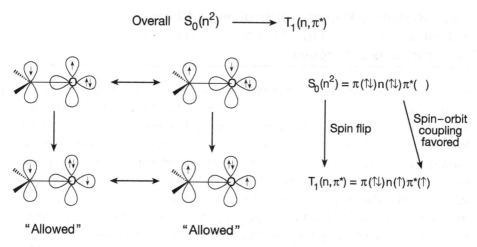

Figure 4.21   Orbital description of the spin–orbit selection rules for a radiative transition involving a spin flip. The $n^2 \rightarrow n,\pi^*$ transition involves an orbital angular momentum change that can be coupled with a spin momentum change on a single (oxygen) atom and is therefore spin–orbit "allowed."

precisely what is required for generating orbital angular momentum and favors strong spin–orbit coupling (Fig. 3.10); that is, the orbital momentum change associated with the $\alpha\beta \rightarrow \alpha\alpha$ (or $\alpha\beta \rightarrow \beta\beta$) spin flip can be coupled with a $p_x \rightarrow p_y$ orbital jump.

Now, let us compare this qualitative picture to the situation for a radiative $S_0(\pi^2) \rightarrow T(\pi,\pi^*)$ transition, for example, for ethylene (Fig. 4.22). It is immediately seen that for a planar ground state, there is no orbital of low energy in the molecular plane into which the π electron can jump (a σ^* orbital is available but has a very high energy, which strongly inhibits spin–orbital coupling); that is, the analogue of the low-energy $p_x \rightarrow p_y$ jump of ketones does not exist for ethylene. Consequently, there are no "one-center" spin–orbit interactions to help flip spins when a light wave interacts with the π electrons of the ethylene, spin–orbit coupling is inhibited, and the oscillator strength of the transition is small.

We conclude that the matrix element for spin–orbit coupling is much larger for $n^2 \rightarrow n,\pi^*$ transitions than for $\pi^2 \rightarrow \pi,\pi^*$ transitions. Since the oscillator strength of a spin-forbidden transition depends directly on the square of the matrix element (Eq. 4.36) corresponding to the perturbation (spin–orbit coupling) that mixes the states undergoing transition, we can conclude that Eq. 4.48 holds.

$$f[S_0(n^2) \leftrightarrow T(n,\pi^*)] \gg f[S_0(\pi^2) \rightarrow T(\pi,\pi^*)] \qquad (4.48)$$

Equation 4.48 (and its extension to other transitions) represents a general situation for organic molecules and is known as El-Sayed's rule.[24c] As seen in the case of ethylene, an aromatic hydrocarbon, such as benzene, cannot invoke an $n \rightarrow \pi^*$ mixing to generate a spin–orbital coupling mechanism.[14] The closest analogous $p_x \rightarrow p_y$ orbital mixing (in-plane to out-of-plane and vice versa) for benzene is of the $\sigma \rightarrow \pi^*$ or $\pi \rightarrow \sigma^*$ type; however, both of these mixings are ineffective because of the large

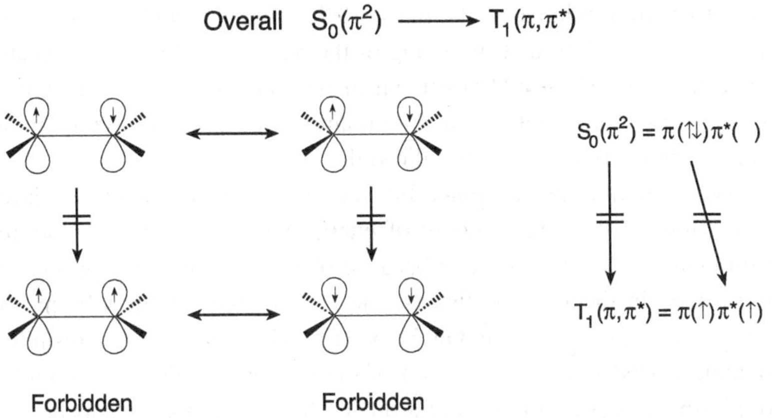

Overall $S_0(\pi^2) \longrightarrow T_1(\pi,\pi^*)$

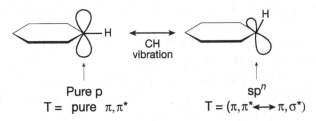

$S_0(\pi^2) = \pi(\uparrow\downarrow)\pi^*(\ )$

$T_1(\pi,\pi^*) = \pi(\uparrow)\pi^*(\uparrow)$

Forbidden   Forbidden

Figure 4.22   Orbital description of the spin–orbit selection rules for a radiative transition involving a spin flip. The $\pi^2 \to \pi,\pi^*$ transition does not involve an orbital angular momentum change and is spin–orbit "forbidden."

energy gap between the bonding and antibonding orbitals involved. We have seen how out-of-plane vibrations can induce vibronic mixing between an n,π^* and π,π^* state (Fig. 3.1) and can thereby add oscillator strength to a spin-allowed transition that is overlap forbidden in zero order. Now, let us try to picture an analogous theoretical connection among the out-of-plane vibrations, the observation of spin-forbidden phosphorescence emission, and the allowed transitions involving mixing with σ or σ^* orbitals. Consider Fig. 4.23, which shows (a) a planar benzene molecule and one of the p orbitals involved in the π system, and (b) another benzene molecule undergoing out-of-plane C—H vibrations on the carbon atom associated with the p orbital. As long as the molecule is planar, the π,π^* states and, say, the π,σ^* (or σ,π^*) states do not mix, because there is no net orbital overlap between orthogonal orbitals. An out-of-plane C—H vibration, however, destroys the planar symmetry of the molecule and allows mixing of the π,π^* and π,σ^* (or σ,π^*) states. In the extreme case, a substantial out-of-plane C—H vibration would cause a p orbital (originally symmetric above and below the molecular plane) to be transformed into an orbital that possesses an asymmetric

Pure p sp^n
T = pure π,π^* T = $(\pi,\pi^* \longleftrightarrow \pi,\sigma^*)$

Figure 4.23   Schematic of the effect of an out-of-plane C—H vibration on the hybridization of a carbon atom in benzene. The vibration induces "s character" in the carbon atom and provides a "weak" mechanism for spin–orbit coupling.

electronic distribution relative to the plane of the molecule and that acquires some "s character" as a sp^n hybrid. The mixing of the π,π^* and π,σ^* states means some of the character of the $\pi^* \to \sigma^*$ transition is mixed into the $\pi \to \pi^*$ transition so that the latter transitions can "pick up" a certain amount of spin–orbit coupling and the spin-forbidden transition is "weakly" allowed.

This mechanism of inducing spin–orbit coupling is not expected to be particularly effective because of the large amount of energy required to deform the aromatic π electron cloud in this manner and because of the large energy gap between the orbitals involved in the mixing; that is, the matrix element for vibronic mixing, $<\pi,\pi^*|P_{CH}|\pi,\sigma^*>$, is very small. On the other hand, no better mechanism for spin–orbit coupling exists! Indeed, as a result of the very weak spin–orbit coupling, the radiative phosphorescence lifetimes of aromatic hydrocarbons, such as benzene and naphthalene, are very long—on the order of 10 s ($f \sim 10^{-9}$)!

4.32 Experimental Exemplars of Spin-Forbidden Radiative Transitions: $S_0 \to T_1$ Absorption and $T_1 \to S_0$ Phosphorescence[25]

Some experimental data for radiative $S_0 \leftrightarrow T_1$ transitions were given in Tables 4.4 and 4.8. The following points may be derived from the data. The oscillator strengths of radiative spin-forbidden $S_0(\pi^2) \leftrightarrow T_1(\pi,\pi^*)$ transitions are very small ($\sim 10^{-7}$–10^{-9}). Indeed, the values of $\varepsilon_{max}(S_0 \to T_1)$ and k_P^0 for $S_0 \leftrightarrow T_1(\pi,\pi^*)$ transitions are among the smallest observed for organic molecules; that is, $\varepsilon_{max} \sim 10^{-5}$–$10^{-6}$ and $k_P^0 \sim 1$–10^{-1} s^{-1}. The largest values of $\varepsilon_{max}(S_0 \to T_1)$ and k_P^0 are found for $T_1(n,\pi^*)$ states or for $T_1(\pi,\pi^*)$ states that possess a heavy atom (e.g., Br or I) conjugated to the π systems or that possess a strong mixing of n,π^* and π,π^* states (heavy atom effect on spin-forbidden transitions). For these systems, $\varepsilon_{max} \sim 10^{-1}$–$10^{-2}$ and $k_P^0 \sim 10$–10^2 s^{-1}. For some organometallic compounds (e.g., tetraphenyl lead), ε_{max} values as high as ~ 10 have been reported.[26a] Generally, there is a wide variation found in the value of Φ_P in going from 77 K to 25 °C; this variation is usually the result of the onset of diffusional quenching of long-lived triplet lifetimes at the higher temperatures (in fluid solutions). The fact that triplet lifetimes at 25 °C in plastic films (a rigid medium that prevents diffusional quenching) are often comparable to those at 77 K is strong support for this conclusion. For example, the lifetime of triphenylene is 16 s at 77 K and is 12 s at 25 °C in a plastic film.[26c]

The substantial difference between the values of ε_{max} (and k_P^0) for π,π^* relative to n,π^* triplet states provides an experimental means of classifying molecules in terms of the orbital configuration of T_1. The rule is as follows. For non–heavy-atom-containing molecules possessing "pure" π,π^* configurations: The value of $\varepsilon_{max}(S_0 \to T_1)$ and $k_P^0(T_1 \to S_0)$ will be on the order of 10^{-5}–10^{-6} $cm^{-1}M^{-1}$ and 10^{-1}–10^{-2} s^{-1}, respectively. For molecules possessing "pure" n,π^* configurations: The value of $\varepsilon_{max}(S_0 \to T_1)$ and k_P^0 will be on the order of 10^2–10^1 $cm^{-1}M^{-1}$ and 10^2–10^{-1} s^{-1}, respectively.

Table 4.8   Quantum Yields for Phosphorescence and Other Triplet Emission Parameters.[a]

Compound	Φ_P 77 K	Φ_P 25 °C	Φ_{ST}	k_P^0	Configuration of T_1
Benzene	~0.2	(<10^{-4})	~0.7	~10^{-1}	π,π^*
Naphthalene	~0.05	(<10^{-4})	~0.7	~10^{-1}	π,π^*
1-Fluoronaphthalene	~0.05	(<10^{-4})		~0.3	π,π^*
1-Chloronaphthalene	~0.3	(<10^{-4})	~1.0	~2	π,π^*
1-Bromonaphthalene	~0.3	(<10^{-4})	~1.0	~30	π,π^*
1-Iodonaphthalene	~0.4		~1.0	~300	π,π^*
Triphenylene	~0.5	(<10^{-4})	~0.9	~10^{-1}	π,π^*
Benzophenone	~0.9	(~0.1)[b]	~1.0	~10^2	n,π^*
Biacetyl	~0.3	(~0.1)[c]	~1.0	~10^2	n,π^*
Acetone	~0.3	(~0.01)[c]	~1.0	~10^2	n,π^*
4-Phenylbenzophenone			~1.0	1.0	π,π^*
Acetophenone	~0.7	(~0.03)[b]	~1.0	~10^2	n,π^*
Cyclobutanone	0.0	0.0	0.0		n,π^*

a. Unless specified, the temperature is 77 K in an organic solvent.
b. In deaerated perfuloromethylcyclohexane at room temperature.
c. In acetonitrile at room temperature.

Some examples of the relation of k_P^0 and orbital configuration (Table 4.8) are available for aromatic ketones. For aromatic ketones, T_1 may be either "pure" n,π^*, pure π,π^*, or a hybrid mixture of the two configurations. An exemplar for a "pure" $T_1(n,\pi^*)$ state is acetone, for which $k_P^0 = 60$ s^{-1}; an exemplar for a "pure" $T_1(\pi,\pi^*)$ state is naphthalene, for which $k_P^0 = 0.1$ s^{-1}. Examination of Table 4.8 shows that aromatic ketones may be classified as acetone-like (k_P^0 within an order of magnitude of 60 s^{-1}), or naphthalene-like (k_P^0 within an order of magnitude of 0.1 s^{-1}). For example, we may assign a nearly "pure" n,π^* configuration to T_1 of benzophenone ($k_P^0 = 20$ s^{-1}) and a mixed $n,\pi^* \leftrightarrow \pi,\pi^*$ configuration to T_1 of 4-phenyl benzophenone ($k_P^0 \sim 1$ s^{-1}).

Note in passing that molecular oxygen can enhance the intensity of the $S_0 \rightarrow T_1$ transition of aromatic hydrocarbons.[27] The enhancement of the forbidden transition is explained as the result of charge-transfer interaction of the hydrocarbon with oxygen and mixing of the triplet of the oxygen with the hydrocarbon's singlet state.

As an example of the *internal* heavy-atom effect on a spin-forbidden radiative transition, consider 2-bromobenzene undergoing a $T_1 \rightarrow S_0$ transition. The triplet state may be represented as a set of resonance structures of diradical character (see resonance structures **8a** $\leftrightarrow$ **8b**), at least one of which will place the odd electron on the 1-carbon atom, which has a bromine atom attached. Since bromine is capable of expanding its valence octet, some delocalization of the odd electron onto the bromine atom is possible. This finite amount of localization on the "heavy atom" produces a good mechanism for spin inversion because of the strong spin–orbit coupling the π^*

electron experiences when it is on the bromine atom (Chapter 3). The odd electron that undergoes the spin inversion does so by simultaneously jumping from one orbital to another in order to satisfy the requirement of conservation of total angular momentum.

8a   **8b**

The heavy-atom effect on absorption spectra strongly enhances $\varepsilon(S_0 \rightarrow T_1)$, but not $\varepsilon(S_0 \rightarrow S_1)$. Because of the relationships between $\varepsilon(S_0 \rightarrow T_1)$ and k_P^0 and between $\varepsilon(S_0 \rightarrow S_1)$ and k_F^0 (Eq. 4.23), we expect that k_P^0, but not k_F^0, will be influenced by heavy-atom perturbation. For example, the shapes of the fluorescence and phosphorescence spectra of molecules **9a**, **9b**, and **10** are very similar in appearance:[28]

9a	**9b**	**10**
$k_F^0 = 1 \times 10^6 \text{ s}^{-1}$	$1 \times 10^6 \text{ s}^{-1}$	$1 \times 10^6 \text{ s}^{-1}$
$k_P^0 = 3 \times 10^{-2} \text{ s}^{-1}$	$500 \times 10^{-2} \text{ s}^{-1}$	$1000 \times 10^{-2} \text{ s}^{-1}$

On the other hand, the fluorescence and phosphorescence *quantum yields* for the molecule **9a**, which contains only "light atoms," are $\Phi_F \sim 0.5$ and $\Phi_P \sim 0.06$, and for **9b** and **10**, $\Phi_F \sim 10^{-3}$ and $\Phi_P \sim 0.6$. Although the values of k_F^0 are essentially constant in this series, the values of k_P^0 are greatly enhanced in the bromine-containing molecules. The much higher values of Φ_P reflect both a greater efficiency of population of T_1 (k_{ST} enhanced by the heavy-atom effect; see Section 5.11) and a greater efficiency of emission from T_1 (k_P^0 is enhanced *more* than k_{TS}).

An example of the *external* heavy-atom effect[29] is shown for 1-chloronaphthalene[30] (Fig. 4.24). When external heavy atoms are contained in the solvent (ethyl iodide as solvent or a high pressure of Xe gas),[31] there is a significant increase in absorption in the region from 350–500 nm. The vibrational structure of the absorption and its mirror-image relationship to the phosphorescence spectrum of 1-chloronaphthalene strongly suggests that the new absorption is due to the enhancement of the $S_0 \rightarrow T_1$ absorption for 1-chloronaphthalene. For the case of 1-chloronaphthalene (Fig. 4.24), the pure liquid exhibits a number of weak absorption bands near 470 nm. A solution of 1-chloronaphthalene in ethyl iodide shows that the weak bands are greatly enhanced in intensity.[30] The 0,0 band of the enhanced $S_0 \rightarrow T_1$ absorption (58 kcal mol^{-1}) occurs at nearly the same energy as the 0,0 band of normal $T \rightarrow S_0$ phosphorescence, confirming that in this case the effect is mainly on the oscillator strength of the spin-forbidden transitions and not on the energies of the states undergoing the transitions.

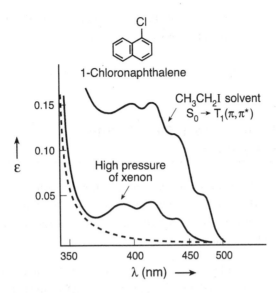

Figure 4.24   Heavy-atom perturbation of the $S_0 \rightarrow T_1$ absorption of 1-chloronaphthalene. The dashed line indicates the absorption spectrum in a "light-atom" solvent.

4.33   Quantum Yields of Phosphorescence, Φ_P: The $T_1 \rightarrow S_0 + h\nu$ Process

A general expression for the quantum yield of phosphorescence (Φ_P) is given by Eq. 4.49.

$$\Phi_P = \Phi_{ST} k_P^0 (k_P^0 + \Sigma k_d + \Sigma k_q [Q]^{-1}) = \Phi_{ST} k_P^0 \tau_T \qquad (4.49)$$

In Eq. 4.49, Φ_{ST} is the quantum yield for intersystem crossing, $S_1 \rightarrow T_1$; k_P^0 is the radiative rate of phosphorescence; Σk_d is the sum of the rate constants of all unimolecular radiationless deactivations of T_1 (including photochemical reactions); and $\Sigma k_q [Q]$ is the sum of all bimolecular deactivations of T_1 (including photochemical reactions). By definition, the experimental lifetime of T_1 is given by $\tau_T = (k_P^0 + \Sigma k_d + \Sigma k_q [Q])^{-1}$. From Eq. 4.49, we see that the quantum yield for phosphorescence is the product of a number of factors. Unless these factors can be experimentally identified and controlled, Φ_P is not a reliable parameter for characterizing T_1, although it may be a useful parameter in certain kinetic analyses.

Data for the Φ_P value of molecules at 77 K in rigid glasses (optically clear frozen solvents) is given in Table 4.8. Experimentally, a wide range of values of Φ_P are found for organic molecules. High values of $\Phi_P (\sim 1)$ require that $\Phi_{ST} \sim 1$ and $k_P^0 > (\Sigma k_d + \Sigma k_q [Q])$. At 77 K, it appears that all of the major deactivation processes of bimolecular diffusional quenching ($\Sigma k_q [Q]$ term) are inhibited, so that the main radiationless deactivation of T_1 is $T_1 \rightarrow S_0$ intersystem crossing. In this limiting case, the quantum yield of phosphorescence is simplified to Eq. 4.50; that is, the value of Φ_P depends only on the value of Φ_{ST} and the competition between the rates of phosphorescence emission and intersystem crossing.

$$\Phi_P = \Phi_{ST} k_P^0 (k_P^0 + k_{TS})^{-1} \quad \text{(at 77 K)} \tag{4.50}$$

4.34 Phosphorescence in Fluid Solution at Room Temperature[32]

The observation of phosphorescence in fluid solution at room temperature was once considered a rare and unusual phenomenon. Now, it is clear that if phosphorescence is observed at 77 K, it can also *generally* be observed at room temperature in fluid solution if two conditions are met:

1. Impurities (e.g., molecular oxygen) and other ground and excited molecules required in the system (R and *R) capable of deactivating triplets by diffusional quenching are rigorously minimized.
2. The triplet does not undergo an activated unimolecular deactivation (photophysical or photochemical) that possesses a rate of greater than $\sim 10^4 \, k_P^0$ at room temperature.

The routine experimental observation of measurable phosphorescence requires a value of $\Phi_P \sim 10^{-5}$. The value of Φ_P may be expressed in terms of the rate of phosphorescence and all processes that deactivate T_1. From Eq. 4.48, in the case for which triplet formation from S_1 is efficient, Φ_P is given by Eq. 4.51.

$$\Phi_P \sim \frac{k_P^0}{k_d + k_q[Q] + k_P^0} \sim \frac{k_P^0}{k_d + k_q[Q]} \quad \text{(in most fluid solutions)} \tag{4.51}$$

In Eq. 4.51, k_d represents the sum of *all* unimolecular deactivations of T_1, and $k_q[Q]$ represents the sum of *all* bimolecular deactivations of T_1.

As discussed above, a typical value of k_P^0 for a "pure" $T_1(n,\pi^*)$ is 10^2 s^{-1}, and a typical value of k_P^0 for $T_1(\pi,\pi^*)$ is 10^{-1} s^{-1}. For $\Phi_P \sim 10^{-4}$, we find Eqs. 4.52 and 4.53:

$$k_d + k_q[Q] \sim 10^6 \, \text{s}^{-1} \quad \text{for} \quad T_1(n,\pi^*) \tag{4.52}$$

$$k_d + k_q[Q] \sim 10^3 \, \text{s}^{-1} \quad \text{for} \quad T_1(\pi,\pi^*) \tag{4.53}$$

Let us calculate the maximum value for the concentration of a quencher, [Q], that is tolerable for observation of phosphorescence if Q is a diffusional quencher. For a nonviscous organic solvent, the *maximal* rate constant for diffusion is given by $k_{dif} \sim 10^{10}$ M^{-1} s^{-1}. Therefore,

$$\text{if} \quad k_{dif}[Q] < 10^6 \, \text{s}^{-1} \quad \text{then} \quad [Q] < 10^{-4} \, \text{M} \tag{4.54}$$

and

$$\text{if} \quad k_{dif}[Q] < 10^3 \, \text{s}^{-1} \quad \text{then} \quad [Q] < 10^{-7} \, \text{M} \tag{4.55}$$

The limit of 10^{-4} M for [Q] is relatively easily obtained, but the limit of 10^{-7} M is more difficult to achieve without very careful purification of solvents and degassing to remove oxygen. These qualitative considerations allow us to understand why compounds that phosphoresce from $T_1(n,\pi^*)$ states are commonly observed in fluid solutions, but phosphorescence from $T_1(\pi,\pi^*)$ is rarely observed, unless extraordinary care is taken to eliminate bimolecular quenching.

The value of k_P^0 may be increased for aromatic hydrocarbons by external or internal heavy-atom perturbation. In certain heavy-atom solvents, the value of k_P^0 is increased to values approaching 10–10^2 s^{-1} (e.g., bromonaphthalene). In these cases, phosphorescence is observed even for aromatic hydrocarbons if the heavy-atom solvent is not itself a triplet quencher.[32e] Examples of phosphorescence data in fluid solutions were given in Table 4.8 and Fig. 4.19. In very favorable or contrived circumstances, phosphorescence from aromatic hydrocarbons occurs even in the vapor phase.[32f] An interesting method to prevent triplet diffusional quenching is to replace a molecular solvent cage with a supramolecular cage (a supercage), such as the hydrophobic core of a micelle.

4.35   Absorption Spectra of Electronically Excited States[33]

We have seen that the electronically excited states $^*R(S_1)$ and $^*R(T_1)$ undergo fluorescence and phosphorescence emission, respectively. Since $^*R$ is a "real" molecular structure, it must also possess absorption spectra: a $^*R + h\nu \rightarrow {^{**}R}$ processes (where $^{**}R$ is $S_{n>1}$ or $T_{n>1}$). Indeed, $S_1 + h\nu \rightarrow S_{n>1}$ and $T_1 + h\nu \rightarrow T_n$ radiative transitions can be observed experimentally through very fast excitation and detection methods, termed pulse–probe flash spectroscopy.[33] Since 1950, the speed of detection has increased steadily from $\sim 10^{-3}$ s (ms) and has now approached a limit of $\sim 10^{-15}$ s (fs);[33c] as a result, it is possible under favorable circumstances to detect $^*R$ by pulse–probe spectroscopy even when the lifetime of $^*R$ is of the order of femtoseconds!

The idea behind flash spectroscopy is to deliver an intense *preparation pulse* (a laser flash) of photons to an absorbing sample to produce $^*R$ in as short a period as possible and then, with a weaker *pulse probe* of photons, to detect and characterize the transient species ($^*R$, I, P) that were produced after the preparation pulse *in real time*. These short, intense pulses are readily provided by pulsed lasers. Lasers have been developed that are capable of delivering intense pulses (10^{16}–10^{18} photons) in time periods as short as from 10^{-12} (a ps) to 10^{-15} (a fs) second! In order to gain an appreciation of the time scale implied by a picosecond, consider the following. A bullet traveling 1000 m s^{-1} takes 10^6 ps to travel 1 mm, and light travels 0.3 mm in 1 ps. In a femtosecond, an electron in a Bohr orbit can move only a few angstroms.

Figure 4.25 shows an example[34] of the $T_1 \rightarrow T_{n>1}$ absorption spectra of naphthalene along with the state energy diagram that schematically shows the transitions relative to S_0. Such absorption spectra may be used to characterize the configurations of the states involved in the transitions, but this is a difficult task for various technical and theoretical reasons. However, the use of excited-state absorption spectra in following the concentration of excited states is a very important tool for photochemical kinetics.

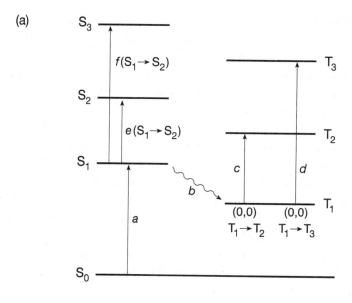

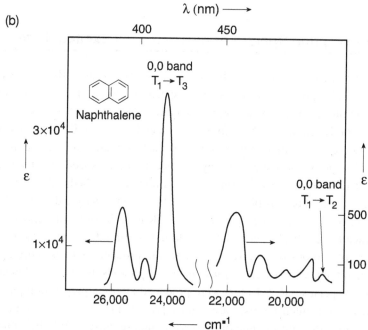

Figure 4.25   The triplet–triplet (T–T) absorption spectrum of naphthalene. (a) A state diagram showing the pathway leading to T–T absorption. Absorption (a) is followed by intersystem crossing (b) to populate T_1. The latter is capable of absorbing photons and undergoing $T_1 \rightarrow T_2$ and $T_1 \rightarrow T_3$ transitions. (b) The experimental T–T absorption spectrum of naphthalene.

4.36   Radiative Transitions Involving Two Molecules: Absorption Complexes and Exciplexes[35]

Thus far, we have considered absorption and emission processes that involve a single molecule (surrounded by "inert," noninteracting solvent molecules). In certain cases, two or more molecules may participate in *cooperative* absorption or emission; that is, the absorption or emission can only be understood as arising from ground- or excited-state *complexes* of a definite stoichiometry. Such complexes, formed by two or more molecules, are termed *supramolecular* complexes. Commonly, the stoichiometry of such complexes consists of two molecules, so we use this situation as an exemplar. When two molecules act cooperatively to absorb a photon, we say that an *absorption (supramolecular) complex* exists in a ground state and is responsible for the absorption. An excited molecular complex of definite stoichiometry, but which is dissociated in its ground state, is called an *exciplex*. Thus, if two molecules act cooperatively to emit a photon to a dissociative ground state, we say an *exciplex* exists; an exciplex can be detected directly and characterized by the emitted photon. The important conceptual distinction between an absorption complex and an exciplex is that the absorption complex possesses some stability in its ground state, whereas the exciplex does not.

Some significant experimental spectroscopic characteristics of absorption complexes and exciplexes are one or more of the following:

1. *Absorption Complex:* The observation of a new absorption band, usually occurring at longer wavelengths than the absorption of either molecular component, that is characteristic of the complex but not of either of the individual molecular components of the ground-state complex.

2. *Exciplex:* The observation of a new emission band, usually structureless and at longer wavelengths than the absorption for either of the molecular components, that is characteristic of the exciplex but not of the individual components of the exciplex.

3. *Absorption Complex and Exciplex:* A concentration dependence of the new absorption or emission intensity.

In the special case, where the molecular components of the exciplex are the same, the excited molecular complex is termed an "excimer," rather than the more general term "exciplex." Exciplex is reserved for excited-state complexes consisting of two different molecular components. The notion of a specific stoichiometry is included in our definitions of exciplex and excimer because we wish to distinguish these species from those solvated excited molecules for which an undefined number of unexcited solvent molecules provide an environment. Both are examples of *supramolecular assemblies*, which comprise the family of molecules that are held together by intermolecular forces.[36]

4.37   Examples of Ground-State Charge-Transfer Absorption Complexes[37]

Solutions of mixtures of molecules that possess a low ionization potential (electron donors, D) or a high electron affinity (electron acceptors, A) often exhibit absorption bands that are not shown separately by either component. Generally, the new band is due to an electron donor-acceptor (EDA) or charge-transfer (CT) complex in which D has donated an electron (charge) to a certain extent to A. Typical examples[38] of CT spectra are shown in Fig. 4.26. Generally, the absorption band of a CT complex is broad and devoid of vibrational structure. This breadth occurs because the rather small binding energies of EDA complexes allows many different structural configurations for the complex to exist in equilibrium with one another. The absorption energy for each configuration will differ and cause a broadening of the band. Since the bonding is weak and intermolecular, there are no characteristic vibrational bands that appear in the spectra. In addition, the excited complex may be very short lived and not have sufficient time to vibrate in *R.

An important experimental characteristic of an EDA absorption band is its sensitivity-to-solvent polarity. For example, the maxima of the EDA absorption band for enol ethers (donors) and tetracyanoethylene (acceptor) molecules shown in Fig. 4.26 vary substantially as solvent polarity is varied.[38] The energy required for

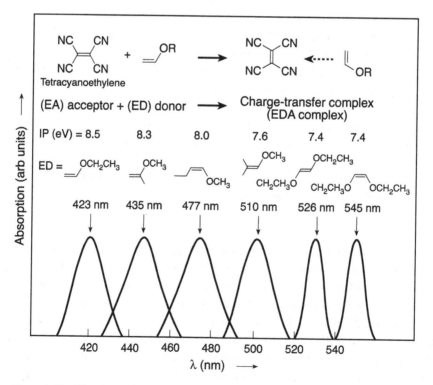

Figure 4.26   The absorption spectra of some EDA complexes of tetracyanoethylene and a variety of enol ethers.

absorption decreases as the solvent polarity increases. This effect is understood in terms of solvent-assisted mixing of the wave functions for the states involved in the EDA transition, which cause a solvent-dependent energy separation of the HO and LU responsible for the CT transition.

4.38 Excimers and Exciplexes[35]

Consider a pair of molecules R and N for which R absorbs a photon to form *R and then *R collides with N. What factors contribute to the stability of an excited-state complex, R-*-N (where excitation is *shared* to some extent by both molecular components) that are missing in the ground-state complex, R/N? An electronically excited-state *R possesses a much stronger electron affinity and a much lower ionization potential than the ground state, because of the occurrence of an electrophilic half-filled HO and a nucleophilic half-filled LU (Chapter 7). Consequently, these orbitals may participate in CT interactions with other polar or polarizable species they encounter intermolecularly. For example, a collision complex between an electronically excited molecular species (*R) with *any* polar or polarizable ground-state molecule, N, will generally be stabilized by some CT interaction involving the HO and the LU or *R with N. This energetic stabilization will in turn cause the R-*-N collision complex to possess a longer lifetime than the corresponding R/N (ground-state) collision complex. The R-*-N collision complex should possess observable spectroscopic and chemical properties that are distinct from those of R*. When this is the case, the R-*-N collision complex can be considered an electronically excited state that is distinct from *R alone; that is, the R-*-N collision complex is a *supramolecular* electronically excited species, held together by attractive intermolecular forces. As mentioned in Section 4.37, such a supramolecular excited complex is termed an *exciplex* if R and N are different molecules (Eq. 4.56a), whereas if R and N are the same molecule (Eq. 4.56b), then the excited complex R-*-R is termed an *excimer*:

$$*R + N \rightarrow R\text{-*-}N \quad \text{Exciplex} \qquad (4.56a)$$

$$*R + R \rightarrow R\text{-*-}R \quad \text{Excimer} \qquad (4.56b)$$

A simple theoretical basis for the enhanced stabilization of a R-*-N collision pair relative to a R/N ground-state collision pair is available from the simple theory of MO interactions (Fig. 4.27). If we consider R (or *R) and N colliding, then the major electronic interactions will involve their highest-energy filled (HO) and lowest-energy unfilled (LU) orbitals. According to the rules of perturbation theory, the HO of R will interact with the HO of N to yield two new HOs of the ground-state collision complex or exciplex. Similarly, the LU of R will interact with that of N to produce two new LUs of the collision complex or exciplex. The new HOs and LUs are split in energy relative to the original HOs and LUs of R and N as shown in Fig. 4.27. In both, the collision complex and exciplex, one of the new HOs is lower in energy than the original HOs and one is higher. Similarly, the LUs of the collision complex and exciplex split in energy above and below the original LUs.

Orbital interactions   States

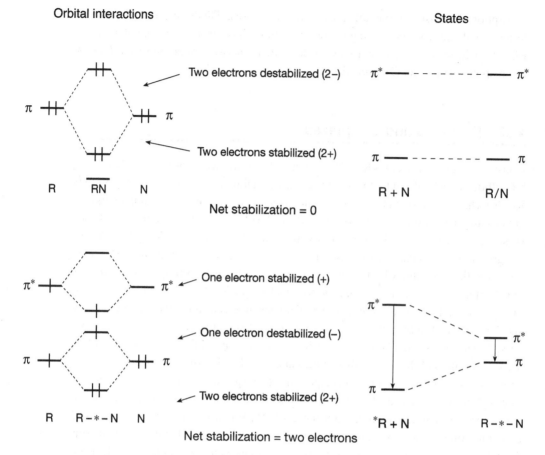

Figure 4.27   Orbital interactions of RN collision pairs and R-*-N exciplexes.

In the collision complex of the ground-state molecules R and N, the four electrons that occupied the HOs of R and N occupy the new set of HOs according to the aufbau principle and fill the lowest-energy orbital. Two electrons are stabilized in the HO and two electrons are destabilized in the LU (Fig. 4.27, top); *thus, no gain in energy is achieved by interaction of R and N during their collisions since the bonding and antibonding interactions are equal and cancel each other*. In the exciplex, however, since one of the partners (*R) is electronically excited, *three* electrons are stabilized (two in the lower-energy HO and one in the lower-energy LU). Only one electron is destabilized (in the higher-energy HO) as the electrons redistribute themselves from their original noninteracting orbitals to the new orbitals of the exciplex (Fig. 4.27, bottom). *Thus, a net gain in energy is always achieved by interaction of *R and N during their collision!* This analysis provides the remarkable conclusion that an electronically excited state has an inherent tendency to form a supramolecular complex with other molecules. The only issue is the strength of the excimer or exciplex binding.

Now, consider the states produced by the collisions corresponding to the collision complex and the exciplex. If there is only a very weak interaction between *R and N, the emission of the collision complex will look very much like that of the monomer

*R, and the energy of the emission will be close to that for the molecular *R → R + $h\nu$ process. If the orbital interactions between the excited molecule and the ground-state molecule are sufficiently strong, the collision complex between *R and N becomes an exciplex (R-*-N), and the energy of the latter *decreases* relative to that of the ground-state complex (R/N). The emission (R-*-N) → R/N + $h\nu$ produces the unstable, ground-state collision complex.

As distinct chemical species, exciplexes and excimers are expected to possess distinct and characteristic photophysical and photochemical properties. Perhaps the most general distinguishing characteristic of an electronically excited state is its emission to produce a ground state and a photon. Thus, if exciplexes exist, they should in principle exhibit fluorescence (singlet exciplexes) or phosphorescence (triplet exciplexes). The emission from R-*-N will in general be different from that of *R. Furthermore, since the ground-state collision complex R/N will generally be less bound than R-*-N, emission from the exciplex will usually occur to a weakly bound or dissociative ground state.

Figure 4.28 shows a PE surface description of excimer (or exciplex) formation and emission and relates the energy surface to absorption and emission of photons. The situation shown assumes that the ground-state complex R/N experiences relatively strong repulsions as the two molecules approach each other to close distances, whereas the exciplex (R-*-N) experiences significant stabilizing attractions as the excited ground- and excited-state molecules approach each other. Since exciplexes are typically stabilized by CT interactions, we can replace the symbols R and N with the labels D and A for donor and acceptor molecules, respectively. At large separations of D and A, the absorption spectrum of either component would be identical to that of each monomer; that is, neither component would influence the other. As D and A approach, the absorption spectrum remains constant. Eventually, D and A undergo collisions. If there are no substantial attractions between D and A in their ground states (lower surface), collisions will raise the energy of the system, and very few collision complexes will exist at any given time. Consequently, their concentrations will be quite low and *no new absorption due to the collision complex will be observed*. The "instability" of the ground-state complex is a somewhat arbitrary feature of the excimer and exciplex definitions. The essential idea is that the ground-state DA collision complexes are *unstable*, low-structured species, *not* that they lack a measurable absorption spectrum.

Consider the situation for the approach and collision of *D and A (or *A and D) on the excited surface. At a large separation of *D and A, the emission spectrum is that of the monomer, *D. As the two molecules approach, the bonding between them may increase due to CT or excitation–exchange interactions. These interactions will cause a minimum to occur in the PE curve; if this enthalpy decrease is not offset by an entropy decrease, an excited-state complex (an exciplex) will form. Emission from the exciplex will occur according to the Franck–Condon principle, that is, vertically from the excited-state minimum. If the separation of D and A in the excited-state minimum corresponds to a point on the repulsive part of the ground-state potential curve, Franck–Condon emission will lead exclusively to repulsive states on the ground surface. Immediately after formation, D and A will fly apart. This

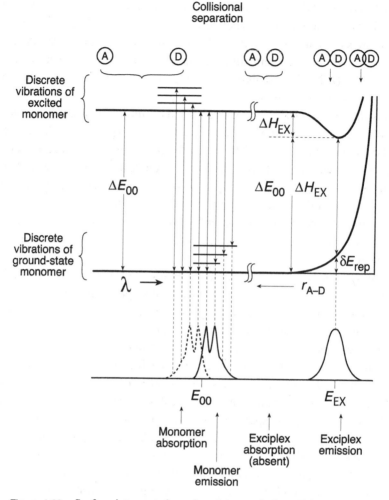

Figure 4.28   Surface interpretation of exciplex emission. Either D or A may be excited on the excited surface.

process is the emission analogue of predissociative or directly dissociative absorption (Fig. 4.11). The short lifetime and indefinite character for the "vibrations" of the final state (collision complex D/A) result in a total absence of *vibrational structure* in the emission spectra of excimers and exciplexes.

Now, we are in a position to appreciate the single most definitive kind of direct spectroscopic evidence for the formation of an excimer or exciplex: the observation of a concentration-dependent, vibrationally unstructured emission spectrum, which occurs on the red (lower-energy, longer-wavelength) of the absorption spectrum of both D and A but does not correspond to the monomer emission spectrum of either D or A.

From Fig. 4.28, we also notice how the important quantities ΔE_{00}, ΔE_{EX}, and ΔH^* are related. The parameter ΔE_{00} is either the excitation energy required to raise the monomer from $v = 0$ of the ground state to $v = 0$ or the excited state or the excitation energy released when the monomer excited state ($v = 0$) emits a photon and produces the ground state ($v = 0$).

Triplet excimers and exciplexes, while less often directly observed by their emission, are well-established excited-state species.[39] They tend to be more weakly bound and to possess different structures than singlet excimers and exciplexes, probably because of the generally decreased CT character of triplets.

Our theoretical analysis of exciplexes indicates that D-*-A should possess the typical properties of electronically excited states (e.g., emission, radiationless transitions, and photochemistry). The exciplex may be treated stoichiometrically as a supramolecular species and its reactions considered *unimolecular*, although its formation is bimolecular.

4.39   Exemplars of Excimers: Pyrene and Aromatic Compounds

The pyrene excimer serves as a classic exemplar of excimer formation and excimer emission.[35,40] Figure 4.29 shows the *fluorescence* of pyrene in methylcyclohexane as a function of pyrene concentration. At concentrations of $\sim 10^{-5}$ M or less, the fluorescence is concentration independent and is composed of pure pyrene monomer fluorescence (Fig. 4.15), which shows vibrational structure and occurs with a maximum at ~ 380 nm. As the pyrene concentration increases to values in the vicinity of $\sim 10^{-5}$ M, two effects are observed: (1) a new broad, structureless fluorescence emission, due to the pyrene excimer, appears on the longer wavelengths of the monomer emission; and (2) the relative amount of monomer emission-to-excimer emission decreases in intensity. As the concentration of pyrene is increased, the excimer intensity continues to increase relative to the monomer (in Fig. 4.29, the monomer emission is normalized and fixed in order to clearly demonstrate the increase in the excimer emission relative to the monomer emission). At concentrations of pyrene of ~ 0.1 M and greater, only excimer emission is observed.

Let us apply Fig. 4.28 (D = A = pyrene, Py) to describe the formation of a Py excimer. The diagram indicates how the energy of two pyrene molecules varies as a function of their internuclear separation. For the ground-state pair at large distances of separation (~ 10 Å, Fig. 4.28, left) the energy of the pair is constant, since intermolecular interactions are weak at this separation distance. At a separation of ~ 4 Å, which is close to the equilibrium separation of the excimer, the energy of the ground-state collision complex Py/Py rises rapidly due to occupied π orbital repulsions (Fig. 4.28, top). From this figure it is easy to see why the Py excimer emission is structureless and why no absorption is observed that corresponds to Py/Py + $h\nu \rightarrow$ Py-*-Py transitions, since there is essentially a zero concentration of Py/Py pairs. The emission is structureless because the Py-*-Py $\rightarrow$ Py/Py + $h\nu$ emission is to an unstable, dissociative state (the molecule dissociates before it can complete a vibrational cycle). There is no measurable absorption corresponding to

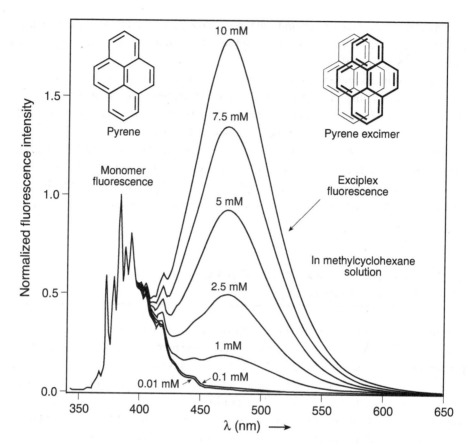

Figure 4.29   Experimental example of the excimer emission of pyrene in methylcyclohexane. The concentrations of pyrene in methylcyclohexane are shown.

the excimer ground state, because too few pairs of Py molecules are in a collision complex at any given instant (the concentration of ground-state complexes of two Py molecules is too small to be measured by absorption of light).

From a spectroscopic analysis of Py excimer emission, and a correlation of this with the emission of Py crystals, it was concluded that the structure of the "face-to-face" Py singlet excimer is favored. This structure is in agreement with expectations based on maximal overlap of π orbitals.

The electronic stabilization energy of the Py excimer is substantial ($\Delta H = -10$ kcal mol^{-1}), but the entropy of formation is quite negative ($\Delta S = -20$ eu; $T \Delta S = 6$ kcal mol^{-1} at room temperature in a nonpolar solvent).[41] Thus, at ambient temperatures, formation of the excimer is favorable, with $\Delta G \sim -4$ kcal mol^{-1}.

In contrast to the large solvent shifts observed for exciplex emission, the emission wavelength for excimers is usually not very solvent dependent. The reason is because CT interactions are not as pronounced for excimers as for exciplexes. This finding is to be expected from the inherently less polar nature of excimers.

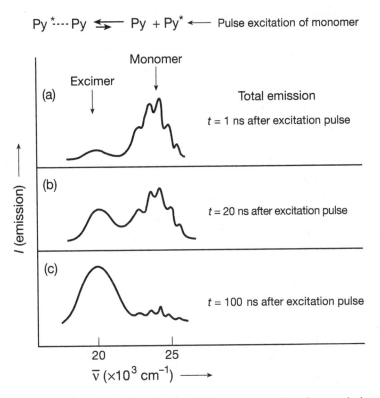

$$Py^* \cdots Py \; \rightleftharpoons \; Py + Py^* \; \longleftarrow \; \text{Pulse excitation of monomer}$$

Figure 4.30 The dynamic behavior of Py monomer and excimer emission. (a) After 1 ns, most of the excited Py exists as the monomer. (b) After 20 ms, comparable amounts of excimer and momomer are observed. (c) After 100 ns, most of the excited Py molecules exist in the excimer form.

The time dependence of emission from Py solutions (*time-resolved emission spectroscopy*) provides an excellent confirmation for the dynamic nature of excimer formation.[42] If a solution of pyrene in cyclohexane ($\sim 10^{-3}$ M) is excited with a brief pulse of light, only the excited monomer is produced (the Py/Py collision complex is too low in concentration to absorb). If the *total* emission spectrum is taken after $\sim 1 \times 10^{-9}$ s (~ 1 ns), the spectrum is mainly that of the *monomer* (Fig. 4.30a); that is, the diffusion of an excited monomer toward a ground-state Py and the formation of an excimer is required for excimer emission, and Py molecules can only move a few angstroms in 10^{-9} s. The small amount of excimer emission results from the formation of excimers by excited Py monomers that happen to encounter Py ground-state molecules during the time of the excitation pulse. After ~ 20 ns, substantial diffusional displacements have occurred, and the concentrations of excimer and monomer become comparable (Fig. 4.30b). After ~ 100 ns, the emission spectrum (Fig. 4.30c) is that which is normally seen under steady-state conditions.

Because of the weaker binding generally found for triplet excimers, excimer phosphorescence is not typically observed.

4.40   Exciplexes and Exciplex Emission[43]

As in the case of excimer fluorescence emission, exciplex fluorescence emission is usually observed as a broad structureless band at longer wavelengths relative to the monomer fluorescence emission (Fig. 4.31). The pyrene-diethylaniline system is an exemplar for exciplex formation and emission. For example, the monomer fluorescence of aromatic hydrocarbons, such as Py, is often quenched at the diffusional rate by electron donors, such as aniline and its derivatives. The quenching is accompanied by the appearance of a broad structureless band $\sim 5000 \text{ cm}^{-1}$ ($\sim 15 \text{ kcal mol}^{-1}$) to the red of the fluorescence of the hydrocarbon monomer. This new fluorescence (Fig. 4.31), assigned to the exciplex, increases in intensity as the electron-donor concentration is increased. There is no corresponding change in the absorption spectrum as the donor concentration is increased.

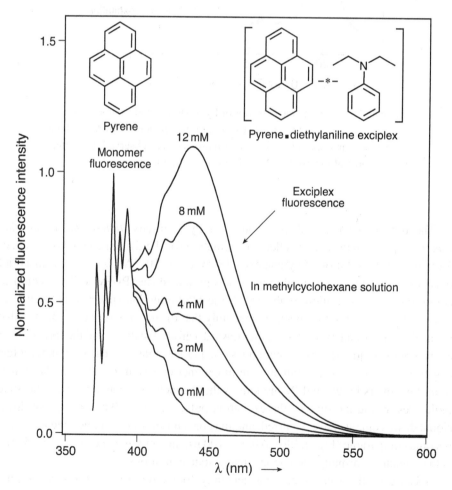

Figure 4.31   Pyrene fluorescence (dotted line) and pyrene–diethylaniline exciplex fluorescence (solid line).

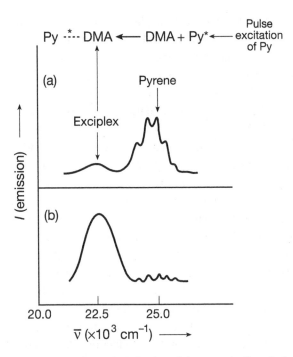

Figure 4.32 Dynamic behavior of the pyrene–dimethyl-aniline exciplex. (a) After 1 ns, most of the excited Py exists as the monomer. (c) After 100 ns, most of the excited Py molecules exist in the pyrene–dimethylaniline exciplex form.

The dynamic behavior of exciplex formation is nicely demonstrated by time-resolved emission spectroscopy.[43] Figure 4.32 shows the total emission of pyrene-plus-dimethylaniline in cyclohexane: (a) about 1 ns after an excitation of the Py chromophore, and (b) ~ 100 ns after the excitation. It is evident that immediately after the excitation, emission is mainly from the Py monomer, but as time goes on the exciplex begins to increase its contribution to the total.

4.41 Twisted Intramolecular Charge-Transfer States[44]

From the above discussion of radiative transitions, we conclude that, as a rule, for the majority of organic molecules only one fluorescence spectrum should be observed, that is, the $^*R(S_1) \rightarrow R(S_0) + h\nu$ radiative transition. Reactions in which an electronically excited state $^*R$ is converted into the excited state of a product ($^*P$) would correspond to exceptions to this rule, since both $^*R(S_1) \rightarrow R(S_0) + h\nu$ and a $^*P(S_1) \rightarrow P(S_0) + h\nu$ radiative transitions are possible, producing a *dual fluorescence*, from a single initial $^*R$. A $^*R \rightarrow {}^*P$ process occurs entirely on an electronically excited surface and is termed an "adiabatic" photochemical reaction. Excimer formation, $^*R + N \rightarrow R\text{-}^*\text{-}N$, is an example of a bimolecular adiabatic reaction. There are a number of cases of unimolecular adiabatic photoreactions that result from the twisting about single bonds. If spontaneous intramolecular rotation occurs in *R and the rotation leads to a product, *P that is in a spectroscopic minimum capable of detection by emission from *P, the photochemical process is detectable if the *P → P + hv process occurs, that is, if *P emits a measurable number of photons.

An exemplar[45] of the observation of dual fluorescence due to rotation about a single bond is found for 4-N,N-dimethylaminobenzonitrile, **11**. In the ground state R(**11**) possesses a more or less planar conformation R(**11p**). Upon rotation about the C—N single in bond for R(**11p**), a higher-energy twisted conformation R(**11t**) is produced. Franck–Condon photoexcitation of R(**11p**) produces a planar excited state *R(**11p**). It is not possible to directly photoexcite R(**11t**), since the later conformation is not significantly populated in the ground state. In a nonpolar solvent, a single fluorescence emission is observed from *R(**11p**) with a maximum at ∼ 350 nm. However, in polar solvents a second fluorescence emission is observed at ∼ 450 nm. The excited state *R(**11p**) may be stabilized by rotation about the C—N bond because of CT from the electron-donor amine group to the electron-acceptor cyano group (an example of a free rotor effect about a single bond). Thus, fluorescence at ∼ 350 nm is assigned to emission from the planar locally excited (LE) *R(**11p**) and the fluorescence at ∼ 450 nm is assigned to emission from the product of rotation about the C—N bond of *R(**11p**), the electronically excited twisted intramolecular CT structure, *P(**11t**). Thus, an adiabatic CT reaction *R(**11p**) → *P(**11t**) occurs upon rotation about the C—N bond on the excited surface. Because of its **t**wisted **i**ntramolecular **c**harge-**t**ransfer (TICT) characteristics, *R(**11p**) is termed a TICT state. The planar state has been termed a locally excited (LE) state. A schematic representation of the adiabatic reaction is shown in Fig. 4.33.

Evidence (Fig. 4.34) for the fluorescence assignments for *R (**11p**) and *P(**11t**) is found in the observance of a single fluorescence emission at ∼ 350 nm for **12** (which is structurally constrained to be planar) and of a single fluorescence emission at ∼ 450 nm for **13** (which is structurally constrained to be twisted). We conclude that a coplanar conformation for which the lone pair is nearly parallel to the carbon p-orbitals of the aromatic π system (**11p**) favors the emission at 350 nm. Twisted

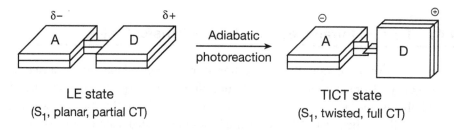

LE state
(S$_1$, planar, partial CT)

Adiabatic
photoreaction

TICT state
(S$_1$, twisted, full CT)

Figure 4.33   A model for formation of a TICT state through an adiabatic *R → *P charge-transfer process. A locally excited planar state (LE), possessing partial charge-transfer characteristics undergoes rotational relaxation toward a twisted conformation that is coupled with intramolecular electron transfer to produce the TICT state.

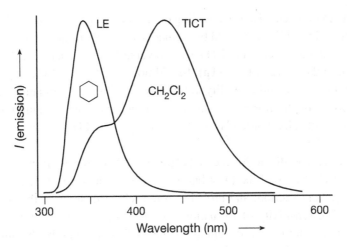

Figure 4.34   Example of the TICT fluorescence of **11t** and **11c**.

structures where the nitrogen lone pair is perpendicular to the π orbital system (**11t**) favor the band at 450 nm.

11p
Planar: 350 nm

11t
Twisted: 450 nm

12
350 nm only

13
450 nm only

The conversion from the planar to the twisted form may involve several barriers: (1) an intramolecular electronic energy reorganization barrier; (2) a thermodynamic barrier if the energy of *R(**11t**) is higher than that of *R(**11p**); and (3) a supramolecular barrier that depends on the friction and electronic reorganization involved in moving the solvent environment. Thus, depending on the structure of the molecule in nonviscous solvents, the TICT state could be formed irreversibly if the energy of *R(**11t**) is much lower than the energy of *R(**11p**), or reversibly if the energy of the *R(**11t**) is comparable to the energy of *R(**11p**). In addition, the degree of CT depends on the polarity of the solvent and its ability to promote or inhibit the formation of the TICT state. Finally, as the friction of the environment increases, the formation of the TICT state will be kinetically inhibited even if the energy of *R(**11t**) is much lower than the energy of *R(**11p**).

The reaction coordinate for the *R(**11p**) → *R(**11t**) transition involves not only the "gas-phase" energy of the twist but also a certain degree of CT, solvent dipolar relaxation, solvent friction, and probably some rehybridization at the N atom from sp^2 to sp^3. The sensitivity of formation for TICT states to the intramolecular and supramolecular features of a system allow the TICT state to probe the micropolarity and microviscosity of solvents. The large electronic, conformational changes makes them ultrasensitive to supramolecular (solvent cage) effects.

An important characteristic of the exemplar perpendicular TICT state is the structural orthogonality (zero overlap) of the electron donor n orbital for the dimethylamino group and the π orbital system for the acceptor benzonitrile group. This situation can be viewed as one that leads to a dipole moment that is near maximum for full electron transfer from the donor to the acceptor. This feature and the energy minimum for the perpendicular configuration are the essential features of a TICT state. Twisted single and double bonds can be understood within one single theoretical framework, that of diradicaloid states (Chapter 6). An estimation of the energy of *R(**11p**) relative to *R(**11t**) can be obtained from knowledge of electrochemical characteristics of the donor (amine) and acceptor (cyano) groups.

The essence of the TICT phenomenon is the adiabatic transfer of an electron (negative charge) during the *R → *P process. A related phenomenon is the transfer of a proton (positive charge) during the *R → *P process. The latter is termed an "excited state intramolecular proton transfer" (ESIPT). In appropriate cases, the TICT and ESIPT processes can be coupled. The intramolecular hydrogen abstraction of o-hydroxybenzophenone is an exemplar of ESIPT. In this case, the radiationless deactivation of *R(ESIPT) is much faster than emission.

4.42   Emission from "Upper" Excited Singlets and Triplets: The Azulene Anomaly

Because of the large number of organic molecules known to obey Kasha's rule[17] (in condensed phases, fluorescence is only observed from S_1 and phosphorescence only from T_1), claims of "anomalous" emission from the S_2, S_3, and so on, and the T_2, T_3, and so on, states of molecules should be viewed with suspicion. To date, examples of emission from $T_n (n > 1)$ are extremely rare. However, well-documented cases of $S_2 \rightarrow S_0 + h\nu$ fluorescence are found for azulene (**14**) and its derivatives.[46] The fluorescence spectrum of **14** reaches a maximum at ~ 374 nm, whereas its $S_0 \rightarrow S_1$

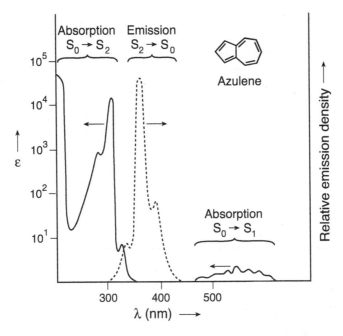

Figure 4.35   The anomalous $S_2 \rightarrow S_0$ fluorescence of azulene. The solid curve shows the $S_0 \rightarrow S_2$ (UV) and the $S_0 \rightarrow S_1$ (vis) absorption of azulene. The fluorescence of azulene (dashed curve) is an approximate mirror image of the $S_0 \rightarrow S_2$.

absorption maximizes at 585 nm (azulene is a blue organic compound). The 0,0 band of the fluorescence and the 0,0 band of $S_0 \rightarrow S_2$ absorption overlap and display an approximate mirror-image relationship to one another (Fig. 4.35). A reason for the observation of an $S_2 \rightarrow S_0$ emission is a large energy gap, which slows down the normally very rapid rate of $S_2 \rightarrow S_1$ internal conversion by decreasing the Franck–Condon factor for radiationless transitions coupled with a fast inherent rate of k_F^0 from S_2. Examples of S_2 emission are known that can be ascribed to thermal population of S_2 from S_1, followed by emission from S_2.

Interestingly, the "normal" $S_1 \rightarrow S_0$ fluorescence from azulene is extremely weak ($\phi_F < 10^{-4}$) and can be obtained only under special conditions.[47] The anomalous lack of normal fluorescence from azulene can be understood in terms of a relatively *small* energy gap between S_1 and S_0, which leads to a relatively fast internal conversion, or in terms of a conical intersection crossing near the S_1 minimum. Thus, this explains the anomalous observation of significant fluorescence from S_2 and the virtual lack of emission from S_1 (both are violations of Kasha's rule for emission): Because of its unusual molecular orbitals, the energy gap between S_2 and S_1 is unusually large, leading to a relatively slow internal conversion and the energy gap between S_1 and S_0 is relatively small, leading to a relatively fast internal conversion. As a result, internal conversion can occur in competition with a relatively slow inherent fluorescence.

Substitution of fluorine atoms on the azulene[48] can cause the $S_2 \rightarrow S_0$ fluorescence to increase to significant quantum yields. For example, the azulene **14F** possesses a value of $\Phi_F \sim 0.2$, a significant quantum yield for emission!

14
Azulene

14F

References

1. For excellent nonmathematical discussions of light and its interaction with molecules. (a) R. K. Clayton, *Light and Living Matter, The Physical Part,* McGraw-Hill, New York, 1970. See, for more rigorous treatments (b) H. H. Jaffe and M. Orchin, *Theory and Applications of Ultraviolet Spectroscopy,* John Wiley & Sons, Inc., New York, 1962.

2. For a more detailed discussion, the reader is referred to any elementary textbook of physics. D. Halliday and R. Resnick, *Physics,* John Wiley & Sons, Inc., New York, 1967.

3. For an excellent discussion of light as a wave: W. Kauzman, *Quantum Chemistry,* Academic Press, New York, 1957, p. 546ff.

4. (a) G. W. Robinson, in *Experimental Methods of Physics, Vol. 3.,* L. Marton and D. Williams, eds., Academic Press, New York, 1962, p. 154. (b) W. Heitler, *Quantum Theory of Radiation,* Clarendon Press, Oxford, UK, 1944. (c) G. N. Lewis and M. Calvin, *Chem. Rev.* **25**, 273 (1939).

5. (a) E. J. Bowen, *Quart. Rev.* **4**, 236 (1950). (b) *Chemical Aspects of Light,* Clarendon Press, Oxford, UK, 1946. (c) A. Maccoll, *Q. Rev.* **1**, 16 (1947). (d) D. R. McMillin, *J. Chem. Ed.* **55**, 7 (1978).

6. (a) G. Balavoine, A. Moradpour, and H. B. Kagan, *J. Am. Chem. Soc.* **96**, 5152 (1974). (b) W. Kuhn and E. Knoph, *Z. Phys. Chem.* **7B**, 292 (1930).

7. E. A. Braude, *J. Chem. Soc.* 379 (1950).

8. (a) F. Perrin, *J. Phys. Radium* **7**, 390 (1962). (b) I. B. Berlman, *Mole. Cryst.* **4**, 157 (1968).

9. (a) S. J. Strickler and R. A. Berg, *J. Chem. Phys.* **37**, 814 (1962). (b) W. R. Ware and B. A. Baldwin, *J. Chem. Phys.* **40**, 1703 (1964). (c) J. B. Birks and D. J. Dyson, *Proc. R. Soc.* **A275**, 135 (1963). (d) W. H. Melhuish, *J. Phys. Chem.* **65**, 229 (1961). (e) R. G. Bennett, *Rev. Sci. Instr.* **31**, 1275 (1960). (f) R. S. Lewis and K. C. Lee, *J. Chem. Phys.* **61**, 3434 (1974). (g) D. Phillips, *J. Phys. Chem.* **70**, 1235 (1966).

10. (a) G. N. Lewis and M. Kasha, *J. Am. Chem. Soc.* **67**, 994 (1945). (b) M. Kasha, *Chem. Rev.* **41**, 401 (1948). (c) G. N. Lewis and M. Kasha, *J. Am. Chem. Soc.* **66**, 2100 (1944). (d) S. P. McGlynn, T. Azumi, and M. Kasha, *J. Chem. Phys.* **40**, 507 (1964).

11. (a) B. S. Solomon, T. F. Thomas, and C. Sterel, *J. Am. Chem. Soc.* **90**, 2449 (1968). (b) D. A. Hansen and E. K. C. Lee, *J. Chem. Phys.* **62**, 183 (1975). (c) R. B. Condall and S. Ogilvie, in *Organic Molecular Photophysics,* Vol. 2, J. B. Birks, ed., John Wiley & Sons, Inc., New York, 1975, p. 33.

12. (a) B. S. Neporent, *Pure Appl. Chem.* **37**, 111 (1976). (b) B. S. Neporent, *Opt. Spectrosc.* **32**, 133 (1972). (c) H. Suzuki, *Electronic Absorption Spectra and the Geometry of Organic Molecules,* Academic Press, New York, 1967, p. 79.

13. See this reference for a more rigorous discussion. (a) G. Herzberg, *Spectra of Diatomic Molecules,* Van Nostrand, Princeton, NJ, 1950.

14. (a) S. P. McGlynn, T. Azumi., and M. Kinoshita, *Molecular Spectroscopy of the Triplet State,* Prentice Hall, Englewood Cliffs, NJ, 1969. (b) M. J. S. Deward and R. C. Dougherty, *The PMO Theory of Organic Chemistry,* Plenum, New York, 1974.

15. R. M. Hochstrasser and A. Marzzallo, *Molecular Luminescence*, E. Lim, ed., W. A. Benjamin, New York, 1969, p. 631.

16. The interested reader should see the following references: (a) J. B. Birks, *Photophysics of Aromatic Molecules*, John Wiley & Sons, Inc., New York, 1970. (b) C. A. Parker, *Adv. Photochem.* **2**, 305 (1964). (c) N. J. Turro, *Molecular Photochemistry*, Benjamin, New York, 1967. (d) R. Becker, *Theory and Interpretation of Fluorescence and Phosphorescence*, John Wiley & Sons, Inc., New York, 1969. (e) S. L. Murov, I. Carmichael, and G. L. Hug, *Handbook of Photochemistry,* 2ed., Marcel Dekker, New York, 1993.

17. M. Kasha, *Disc. Faraday Soc.* **9**, 14 (1950).

18. (a) H. H. Jaffe and M. Orchin, *Theory and Applications of Ultraviolet Spectroscopy*, John Wiley & Sons, Inc., New York, 1962. (b) C. A. Parker, *Photoluminescence in Solution*, Elsevier, Amsterdam, The Netherlands, 1968. (c) J. R. Lakowicz, *Principles of Fluorescence Spectroscopy*, Plenum, New York, 1999.

19. (a) F. Hirayama and S. Lipsky, *J. Chem. Phys.* **51**, 3616 (1969). (b) M. S. Henry and W. P. Helman, *J. Chem. Phys.* **56**, 5734 (1972). (c) F. Hirayama and S. Lipsky, *J. Chem. Phys.* **62**, 576 (1975).

20. (a) E. Havinga, *Chimia* **16**, 145 (1962). (b) J. Pusset, and R. Bengelmans, *Chem. Commun.* 448 (1974).

21. (a) M. S. Henry and W. P. Helman, *J. Chem. Phys.* **56**, 5734 (1972). (b) A. M. Halpern and R. M. Danziger, *Chem. Phys. Lett.*, **72** (1972).

22. W. W. Schloman and H. Morrison, *J. Am. Chem. Soc.* **99**, 3342 (1977).

23. (a) S. Sharafy and K. A. Muszkat, *J. Am. Chem. Soc.* **93**, 4119 (1971). (b) D. Gegion, K. A. Muszkat, and E. Fischer, *J. Am. Chem. Soc.* **90**, 12, 3097 (1968). (c) J. Saltiel, O. C. Aafirion, E. D. Megarity, and A. A. Lamola, *J. Am. Chem. Soc.* **90**, 4759 (1968). (d) C. D. De Boer and R. H. Schlessinger, *J. Am. Chem. Soc.* **90**, 803 (1968).

24. (a) R. Hochstrasser, *Electrons in Atoms*, W. A. Benjamin, San Francisco, 1966. (b) D. S. McClure, *J. Phys. Chem.* **17**, 905 (1949). (c) M. A. El-Sayed, *J. Chem. Phys.* **38**, 2834 (1963). (d) M. A. El-Sayed, *J. Chem. Phys.* **36**, 573 (1962); *J. Chem. Phys.* **41**, 2462 (1964).

25. (a) G. N. Lewis and M. Kasha, *J. Am. Chem. Soc.* **67**, 994 (1945). (b) G. M. Lewis and M. Kasha, *J. Am. Chem. Soc.* **66**, 2100 (1944). (c) A. Terenin, *Acta Physicochim. USSR* **18**, 210 (1943).

26. (a) S. R. LaPaglia, *Spectrochim. Acta* **18**, 1295 (1962). (b) N. J. Turro, K.-C. Liu, W. Cherry, M.-M. Liu, and B. Jacobson, *Tetrahedron Lett.*, 555 (1978). (c) R. E. Kellogg and N. C. Wyeth, *J. Chem. Phys.* **45**, 3156 (1966).

27. (a) D. Evans, *J. Chem. Soc.*, 1351 (1957); *J. Chem. Soc.*, 2753 (1959); (b) D. Evans, *J. Chem. Soc.* 1735 (1960); *J. Chem. Soc.*, 1987 (1961). (c) A. Grabowska, *Spectrochim. Acta* **20**, 96 (1966). (d) H. Tsubomura and R. S. Mulliken, *J. Am. Chem. Soc.* **82**, 5966 (1960).

28. G. Kavarnos, T. Cole, P. Scribe, J. C. Dalton, and N. J. Turro, *J. Am. Chem. Soc.* **93**, 1032 (1971).

29. (a) S. P. McGlynn, et al., *J. Phys. Chem.* **66**, 2499 (1962). (b) S. P. Mcglynn, et al. *J. Chem. Phys.* **39**, 675 (1963). (c) G. G. Giachino and D. R. Kearns, *J. Chem. Phys.* **52**, 2964 (1970). (d) G. G. Giachino and D. R. Kearns, *J. Chem. Phys.* **53**, 3886 (1963). (e) N. Christodonleas and S. P. McGlynn, *J. Chem. Phys.* **40**, 166 (1964). (f) D. S. McClure, *J. Chem. Phys.* **17**, 905 (1949).

30. M. Kasha, *J. Chem. Phys.* **20**, 71 (1952).

31. (a) M. R. Wright, R. P. Frosch, and G. W. Robinson, *J. Chem. Phys.* **33**, 934 (1960). (b) A. Grabowska, *Spectrochim. Acta.* **19**, 307 (1963).

32. (a) C. A. Parker and T. A. Joyce, *Trans. Faraday Soc.* **65**, 2823 (1969). (b) W. D. K. Clark, A. D. Litt, and C. Steel, *Chem. Commun.* 1087 (1969). (c) J. Saltiel, H. C. Curtis, L. Metts, J. W. Miley, J. Winterle, and M. Wrighton, *J. Am. Chem. Soc.* **92**, 410 (1970). (d) See this reference for a review of the factors allowing the observation of phosphorescence in fluid solution. N.J. Turro, K.C. Liu, M.F. Chow, and P. Lee, *Photochem. Photobio.* **27**, 500 (1978). (e) K. Kalyanasundaram, F. Grieser, and J. K. Thomas, *Chem. Phys. Lett.* **51**, 501 (1977). (f) H. Gatterman and M. Stockburger, *J. Chem. Phys.* **63**, 4341 (1975).

33. See this reference for a review of the method of flash spectroscopy. G. Porter, *Techniques of Organic Chemistry,* Vol. 8, John Wiley & Sons, Inc., New York, 1963, p. 1054.

34. See this reference for a review of T-T absorption. H. Labhart and W. Heinzelmann, in *Photophysics of Organic Molecules*, Vol. 1, J.B. Birks, ed., John Wiley & Sons, Inc., New York, 1973, p. 297.

35. See these references for reviews of excimers and excliplexes. (a) T. Forster, *Angew. Chem. Int. Ed.*

En. **8**, 333 (1969). (b) J.B. Birks, *Photophysics of Aromatic Molecules*, John Wiley & Sons, Inc., New York, 1970, p. 301. (c) H. Beens and A. Weller, in *Organic Molecular Photophysics*, Vol. 2, J. B. Birks, ed., John Wiley & Sons, Inc., New York, 1975, p. 159.

36. J.-M. Lehn, *Supramolecular Chemistry*, VCH, New York, 1995.

37. See this reference for a survey of CT phenomena, including absorption and emission: J. B. Birks, *Photophysics of Aromatic Molecules,* John Wiley & Sons, Inc., New York, 1970, p. 403.

38. M. P. Niemczyk, N. E. Schore, and N. J. Turro, *Mol. Photochem.* **5**, 69 (1973).

39. (a) P. C. Subudhi and E. C. Lim, *J. Chem. Phys.* **63**, 5491 (1975). (b) T. Takemura, M. Aikawa, H. Baba, and Y. Shindo, *J. Am. Chem. Soc.* 98, 2205 (1976). (c) S. O. Kajima, P. C. Subudhi, and E. C. Lim, *J. Chem. Phys.* **67**, 4611 (1977).

40. T. Forster and K. Kasper, *Z. Physik. Chem., N.F.* **1**, 275 (1954).

41. J. B. Birks, *Photophysics of Aromatic Molecules*, John Wiley & Sons, Inc., New York, 1970, p. 357.

42. K. Yoshihara, T. Kasuya, A. Inoue, and S. Nagakura, *Chem. Phys. Lett.* **9**, 469 (1971).

43. (a) A. Weller, *Pure Appl. Chem.* **16**, 115 (1968). (b) H. Knibbe, D. Rehm, and A. Weller, *Ber. Bunsen. Gesell.* **73**, 839 (1969); (c) *Ber. Bunsen. Gesell.* **72**, 257 (1968); *Ber. Bunsen. Gesell.* **73**, 834 (1969). (d) J. B. Birks, *Photophysics of Aromatic Molecules*, John Wiley & Sons, Inc., New York, 1970, p. 403.

44. See this refence for a review of the TICT phenomenon. Z. R. Grabowski, K. Rotkiewicz, and W. Rettig, *Chem. Rev.* **103**, 3899 (2003).

45. Z. R. Grabowski and J. Dobkowski, *Pure Appl. Chem.* **55**, 245 (1983).

46. (a) M. Beer and H. C. Longuet-Higgins, *J. Chem. Phys.* **23**, 1390 (1955). (b) G. Viswath and M. Kasha, *J. Chem Phys.* **24**, 757 (1956). (c) J. B. Birks, *Chem. Phys. Lett.* **17** 370 (1972). (d) S. Murata, C. Iwanga, T. Toda, and H. Kohubun, *Ber. Bunsen. Ges. Phys. Chem.* **76**, 1176 (1972).

47. (a) P. M. Rentzepis, *Chem. Phys. Lett.* **3**, 717 (1969). (b) G. D. Gillispie and E. C. Lim, *J. Chem. Phys.* **65**, 4314 (1976).

48. S. V. Shevyakov, H. Li, R. Muthyala, A. E. Asato, J. C. Croney, D. M. Jameson, and R. S. H. Liu, *J. Phys. Chem. A* **107**, 3295 (2003).

CHAPTER 5

Photophysical Radiationless Transitions

5.1 Photophysical Radiationless Transitions As a Form of Electronic Relaxation

Chapter 4 was concerned with the portion of the working paradigms of molecular organic photochemistry (Scheme 1.1) that involved the radiative processes of absorption and emission (Scheme 4.1). This chapter is concerned with the portion of the working paradigm that involves the *photophysical radiationless transitions* (Scheme 5.1). These transitions are of two types: (1) radiationless transitions between electronically excited states (**R $\rightarrow$ *R, where ** signifies an upper electronically excited state, $S_{n>1}$ or $T_{n>1}$, and * signifies the lowest electronically excited state of a given multiplicity, S_1 or T_1); and (2) radiationless transitions between the lowest excited states of a given multiplicity and the ground state (*R $\rightarrow$ R). Vibrational relaxation will be an important feature of all radiationless electronic transitions, although it is generally so fast it is rarely rate determining.

Radiationless transitions between electronic states are a form of electronic relaxation in which electronic energy is converted into the kinetic energy (KE) associated with nuclear vibrational motion.[1] This chapter is concerned with the answers to questions about photophysical radiationless transitions, such as

1. What factors determine the rates and efficiencies of internal conversions (vibrational and electronic radiationless transitions that involve no change of electron spin)?
2. What factors determine the rates and efficiencies of intersystem crossings (vibrational and electronic radiationless transitions that involve a change of electron spin)?
3. What are the relationships between the rates and probabilities of radiationless processes and the electronic configurations of the states undergoing these processes?

265

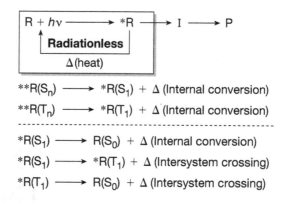

Scheme 5.1   Radiationless photophysical transitions of interest to molecular organic photochemistry. The symbol Δ indicates heat is released.

4. What are the relationships between the rates and probabilities of radiationless processes and quantum mechanical concepts?
5. How can radiationless processes be visualized in terms of molecular mechanisms and representative points on energy surfaces?
6. How are radiationless photophysical processes related to photochemical processes?

This chapter provides a structural and mechanistic basis, grounded in fundamental quantum mechanical principles, for answering the above questions. Photophysical radiationless transitions and photochemical primary processes may not always be sharply distinguished (Scheme 1.5). Photochemical primary processes can be treated as being analogous to photophysical processes, except that the conversion of electronic energy into nuclear energy causes such a distortion of the original ground-state structure that the molecule does not return to its original nuclear geometry of the ground-state spectroscopic minimum, S_0 (a photophysical process). Instead, a new species, I or P (Scheme 1.1), is formed (i.e., a photochemical process has occurred). The conclusions and generalities made for photophysical radiationless processes in this chapter are extended to photochemical radiationless transitions in Chapter 6.

5.2   Radiationless Electronic Transitions as the Motion of a Representative Point on Electronic Energy Surfaces

In quantum mechanical terms, a radiationless transition between electronic states corresponds to a radiationless transformation of two electronic wave functions, say, $\psi_1 \rightarrow \psi_2$, where ψ_1 is a starting state and ψ_2 is a final state. As we have done earlier, we seek quantum intuition to evaluate the plausibility of such radiationless electronic transitions by first appealing to a visualizable classical physics paradigm for radiationless transitions to obtain classical intuition, and then adjusting the classical interpretation with quantum features in order to obtain quantum mechanical intuition derived from classical physics intuition.

A classical physics intuition comes from the interpretation[2] of radiationless electronic transitions in terms of motion along (or jumps between) energy surfaces by a point on the energy surface representing the *instantaneous nuclear configuration of a molecule*. This point is termed the *representative point r*. Each *possible* nuclear configuration of a molecule corresponds to a representative point on a ground electronic energy surface or on an electronically excited energy surface (Section. 1.13). The representative point, r, will make a radiationless "jump" from one electronic surface to another at certain *critical nuclear configurations* (r_c). These critical geometries will generally correspond to nuclear configurations for which two (or more) energy surfaces come close together in energy, or for which there is a minimum on the excited electronic surfaces.[3] Classical intuition suggests radiationless transitions are *plausible* when (1) there is a common nuclear geometry for the representative point, r, for which two states are close in energy and for which some sort of resonance coupling can occur between the two states, ψ_1 and ψ_2, as the result of some appropriate interaction (electronic, vibronic, spin–orbit, etc.), or (2) there is a geometry on an excited surface that is an energy minimum, for which the representative point persists for a relatively long period of time and has an opportunity to search for the interaction that triggers a radiationless transition. On the other hand, when the representative point is located at any nonminimum position on an excited surface that is separated from all other energy surfaces by a large energy gap, a radiationless transition is not probable for any achievable geometry since the point moves rapidly from all nonminimum positions.

According to the classical theory of radiationless jumps between surfaces,[2] the probability (P) that the representative point will make a jump as it approaches r_c is given by Eq. 5.1:

$$P(\text{Probability of surface jump at } r_c) \sim \exp(-\Delta E^2/v\Delta s) \qquad (5.1)$$

In Eq. 5.1, ΔE is the energy separation between the surfaces involved in the transition at the nuclear geometry r_c, v is related to the velocity (related to the KE) of the nuclei as they approach r_c, and Δs is related to the difference in the values of the slopes of the two surfaces in the region near r_c. Note that dE/dr, the slope of an energy curve, is directly related to the magnitude of the nuclear *Coulombic forces*, the positive charges of the nuclei, that act on the negative electrons and that control the motions of the electrons. Note also from Eq. 5.1 that, since the energy term appears as a negative exponential, *the probability of a jump will decrease as ΔE increases and will increase as v decreases or as Δs decreases*. These mathematical features are the basis of a classical intuition for analyzing radiationless transitions. For example, we note from Eq. 5.1 that the limiting probability equals 1 for a surface jump will be approached when the energy difference, ΔE, between the two surfaces approaches 0, or when the velocity, v, on a surface is very small, or when the difference in slopes of the surfaces, Δs, is small (or some combination of these effects). Thus, Eq. 5.1 provides us with a powerful classical intuition with regard to the probability of radiationless transitions between electronic surfaces, particularly between an electronically excited-surface and a lower-energy surface, either a lower-energy excited state or ground-state surface.

The classical interpretation of radiationless transitions first ascribes a simplicity to the motion of a representative point in the regions where the surfaces are well separated: The point just keeps moving downhill in energy, searching for minima on the surface, transferring excess energy to the environment (solvent molecules) as it moves along the surface. When two energy surfaces are separated by a large energy gap (ΔE), the probability of a transition to the lower surface is very small according to Eq. 5.1 (because of the exponential dependence of the rate on $-\Delta E^2$). Thus, the motion of the representative point continues on the initial singlet surface until the nuclei attain a *critical geometry* (r_c) corresponding to a region where $\Delta E \sim 0$ and the value of $P \to 1$. From chemical intuition, we can deduce that when ΔE is close to 0 and two surfaces cross each other, the two states are not only close in electronic energy but are also close in structure, conditions that are ideal for interconversion of the two states, since very little reorganization of energy or structure will be required at the crossing point (r_c). In summary, from Eq. 5.1, we obtain the important classical intuition that in regions where $\Delta E \sim 0$, there is the greatest probability ($P \to 1$, as $\Delta E \to 0$) of the representative point jumping to a lower surface. We will seek to answer two questions: *(1) What are the pathways and geometries of the representative point that will bring two surfaces close together in energy, and (2) How can we deduce these pathways and geometries from consideration of the molecular electronic structure?*

The x-axis of an energy surface diagram (e.g., Fig. 5.1) corresponds to a specific change in nuclear configuration of the system. The lowest-energy path for the representative point is termed the "reaction coordinate." *We reserve the term reaction coordinate for the special lowest-energy pathway on the ground surface unless otherwise specified.* We emphasize (see hypothetical example in Scheme 1.5) that the lowest-energy pathway on the ground surface (reaction coordinate) for proceeding from R to I or to P may not be the lowest-energy pathway for proceeding from *R to I or to P.

Now, we start with the classical picture of radiationless transitions on an energy surface as a basis for developing a pictorial model of the quantum mechanical situation and the behavior of wave functions ψ_1 and ψ_2 as the representative point moves along a surface corresponding to the energy of one of the wave functions. Consider Fig. 5.1, which schematically shows the change of energy (y-axis) of two states ψ_2 and ψ_1 as functions of a changing nuclear geometry (x-axis). As the nuclear geometry changes from left to right, the energy of ψ_1 goes up continuously and the energy of ψ_2 goes down continuously. Suppose the representative point (r) starts in the state ψ_2. Figure 5.1 shows three possible exemplar surface situations for a representative point whose motion (represented by the arrows in Fig. 5.1) from left to right is indicated by the arrows on the surface.

Figure 5.1a corresponds to a "true" zero-order *surface crossing* situation for which the surfaces for the wave functions ψ_1 and ψ_2 cross at point r_c but do not mix (do not interact) at all. When there is such a surface crossing, the representative point that starts on the surface corresponding to the initial electronically excited state (corresponding

$r_c \equiv$ critical nuclear geometry corresponding to a zero-order surface crossing

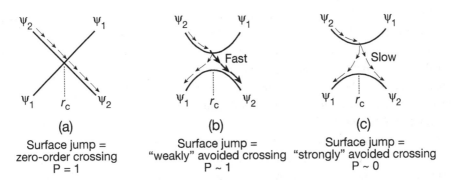

Figure 5.1   Representation of the dynamics of a representative point (motion represented by arrows) on an energy surface.

to the initial wave function ψ_2) maintains its electronic characteristics as it crosses the lower-energy electronic state ψ_1; that is, the states do not mix in the region near r_c, even in the case where the energies of the two states are identical (i.e., $\Delta E = 0$). In the case of a true surface crossing (no mixing at r_c), the probability (P) of a radiationless transition from the excited surface before r_c to the lower surface is 1.0. This means that the excited-state wave function ψ_2, which initially is higher in energy than ψ_1, becomes lower in energy than ψ_1 at r_c. As a result, even though the wave function ψ_2 has not changed its essential electronic character, *it has gone from being classified an excited-state wave function ψ_2 to a ground-state wave function, ψ_2, when passing through r_c!*

Figure 5.1b corresponds to a surface crossing at r_c that is a weakly avoided crossing; that is, a small degree of mixing of the states occurs at this nuclear geometry. According to the classical expression of Eq. 5.1, a surface crossing at r_c is highly plausible, since ΔE at r_c is small. As in the case of the surface crossing, the probability of a radiationless transition is close to 1.0. However, in contrast to the situation in Fig. 5.1a, depending on the speed at which the representative point approaches r_c, it may either continue on its trajectory "to the right of r_c" and "jump" to the state characterized by ψ_2, or "jump" to a certain extent "to the left of r_c" to the state characterized by ψ_1. The differences between the true surface crossing and a weak surface avoiding are subtle, but they both possess the same feature with respect to experimental interpretation: *radiationless processes corresponding to situations shown in Fig. 5.1a or b will be the fastest possible for electronically excited states and in the limit may be rate limited only by the rate of vibrational relaxation.* The important point is that for the weakly avoided surface crossing, in some trajectories along the surface, the wave function will continue to look like ψ_2 after passing r_c; in other trajectories, the wave function will change its character and look like ψ_1 after passing r_c. In Section 5.8, we provide experimental exemplars of radiationless processes at surface crossings or near-surface crossings. Chapter 6 introduces the idea of "conical intersections" that occur near surface crossings and discusses some interesting new features of radiationless transitions that occur when the representative point moves into a conical intersection.

In contrast to Fig. 5.1a and b, the surface situation in Fig. 5.1c corresponds to a strongly avoided crossing at r_c (i.e., ΔE is large at r_c). In quantum mechanical terms, this means that the original zero-order (Born–Oppenheimer) approximation for the wave functions ψ_1 and ψ_2 near geometries corresponding to r_c is not a good approximation. When first-order mixing is taken into account, the two wave functions mix strongly in the region near r_c, and this mixing generates a large splitting of the energies (ΔE) of the two states. A jump of the representative point to the lower energy surface from anywhere on the excited surface near r_c is relatively slow, because ΔE is large throughout the entire trajectory of the representative point on the excited surface, even in the region near r_c. Therefore we expect the representative point to spend a certain amount of time near the excited surface minimum produced by the avoided crossing near r_c. A jump from the upper to the lower surface may eventually occur from the region of this minimum of the excited surface, but the jump is expected to occur at a relatively slow rate (compared to the motion of the representative point in Fig. 5.1a and b) because of the large energy gap between the surfaces along the reaction coordinate. The reason for the slow rate may be viewed as resulting from the requirement for energy conservation during the transition, which will make the radiationless transition slow (a large amount of energy, ΔE, must be somehow dumped into the environment). The representative point is expected to reach a certain equilibrium at r_c and to undergo oscillatory motion back and forth, like a harmonic oscillator with r_c as its point of minimum potential energy (PE).

Because it is oscillating back and forth on an excited energy surface, as the result of some electronic interactions between ψ_1 an ψ_2, the jump of the representative point to lower energy may eventually occur in the direction of ψ_1 or ψ_2, as indicated by the arrows in Fig. 5.1c. The latter possibility indicates the intimate relationship between photophysical and photochemical radiationless transitions. For a moment, suppose that the jump to lower surface that occurs to the right in Fig. 5.1c corresponds to a radiationless *photochemical* reaction, $\psi_2(*R) \rightarrow \psi_1(I)$, and that the jump to the left in Figure 5.1c corresponds to a radiationless *photophysical* transition, $\psi_2(*R) \rightarrow \psi(R)$. In each case the system near r_c is making a radiationless transition to the ground state. In one case (jump to the right, *R $\rightarrow$ I), we classify the radiationless transition as a photochemical reaction; in the other case (jump to the left, *R $\rightarrow$ R), we classify the radiationless transition as a photophysical transition. Now, let us consider the wave mechanical interpretation of jumps between surfaces in order to clarify the basis for determining the magnitude of the "electronic coupling" that causes zero-order crossings to be avoided.

5.3   Wave Mechanical Interpretation of Radiationless Transitions between States[3]

As in classical mechanics, wave mechanics also treats the problem of radiationless transitions in terms of the motion of a representative point on energy surfaces along a reaction coordinate.[4] In order to calculate the electronic *potential energy* (PE) surfaces from wave functions, a number of simplifying assumptions are necessary. As discussed in Section 2.2, solution of the Schrödinger wave equation (Eq. 2.1) must be made under the assumption that nuclear and electronic motion are separable and that the electrons instantaneously respond to the changing nuclear motion (the Born–Oppenheimer approximation). This approximation allows the solution of the electronic wave equation to be formulated in terms of the motion of *nuclei* only, because for each nuclear geometry it is assumed that there is only one electronic distribution for the ground state (R) and another for each excited state (*R). This process of computing electron distributions using the Born–Oppenheimer approximation is also termed the *adiabatic approximation*, and energy surfaces generated under such assumptions are termed *adiabatic* surfaces.

As in the case of classical surfaces, in quantum mechanics the dynamics of nuclear motion on adiabatic surfaces are also treated in terms of the motion for a "representative point" indicating the instantaneous nuclear configuration that determines the electronic distribution. The adiabatic PE surfaces are determined by the solution of the electronic Schrödinger equation (Eq. 2.1) for a large number of nuclear geometries or representative points, r_c. The lowest-energy nuclear geometries create a ground-state energy surface (as mentioned above, this family of representative points for the lowest-energy surface is termed the *reaction coordinate*). The resulting energy surface represents the lowest-energy nuclear geometries for a given electronic state, whether it is the ground state (R) or an excited state (*R). Note that the reaction coordinate (lowest-energy geometries along a given surface) may be different for *R and R; that is, since the electronic distributions in R and *R are different, the lowest-energy geometries for each are not related and in general should be different.

If the changes in the electronic energy of a molecule brought about by nuclear motions are essentially adiabatic (i.e., if the Born–Oppenheimer approximation is valid), then the behavior of the motions of the electrons may be treated by solution of the wave equation for stationary nuclei. The problem of evaluating electron motion as a function of nuclear motion is the same as that for vibrations between nuclei of a molecule, or for the making and breaking of bonds between nuclei in chemical reactions. The differences between the vibrations and chemical reactions are only in the extent of nuclear motion and the occurrence of different equilibrium positions for nuclear motions of the reactants and products. If the electron motions are treated as *adiabatic* (i.e., completely, continuously, and instantaneously adjusting to changes in nuclear structure), and if the nuclear motions are treated as classical, one can in principle evaluate the electronic PE for all nuclear configurations. This is how one generates adiabatic electronic PE surfaces. *For the purposes of radiationless transitions, the actual motions of the nuclei follow the rules of classical mechanics, and the motions of the nuclei are completely subject to the control by the adiabatic surface.* Thus, the motions of nuclei are wholly determined by the PE surfaces "on which" the *representative point* (which represents the *nuclear structure of the molecule*) happens to be. Thus, the Born–Oppenheimer approximation justifies the use of classical ideas for radiationless transitions between energy surfaces as a plausible basis for developing an intuitive wave-mechanical interpretation of transitions between surfaces.

Figure 5.2 interprets the three exemplar situations of Fig. 5.1 in terms of quantum mechanics and adds two other important exemplars. The five exemplars of Fig. 5.2

$r_c \equiv$ "critical" nuclear geometry at which a radiationless transition occurs

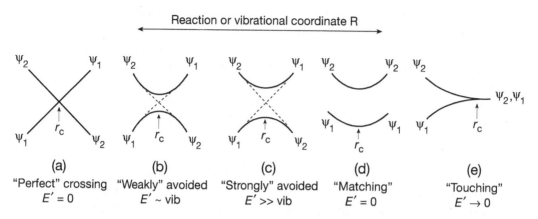

Figure 5.2   Five exemplar surface topologies for a two-surface system. (a) "Perfect" crossing for which $\psi_1 \to \psi_2$ transitions are strictly forbidden. (b) and (c) "Weakly" and "strongly" avoided crossings for which $\psi_1 \to \psi_2$ transitions are possible near r_c (see text). (d) "Matching" for which $\psi_1 \to \psi_2$ transitions are very improbable. (e) An extended "touching" for which the surfaces have similar energies after the representative point has reached r_c.

will provide a convenient framework for discussing common surface relationships and radiationless transitions between energy surfaces from a wave mechanical point of view: (a) a "perfect" zero-order crossing; (b) a weakly avoided crossing for which the magnitude of the avoiding is on the same order as vibronic coupling of electron motion; (c) a strongly avoided crossing for which the magnitude of the avoiding is much larger than vibronic coupling; (d) a "matching" of energy surfaces that are separated in energy by a large amount and also are not related electronically (cf. a perfect matching to perfect crossing); (e) an extended "touching" of the energy surfaces after the representative point reaches r_c. In Fig. 5.2, the energy E' refers to the interaction energies between the two surfaces at r_c. Note that for "perfect crossing," "perfect matching," and "touching," the value of $E' = 0$ or approaches 0 in the zero-order approximation.

In practice, an approximate electronic wave function (ψ) rather than the true electronic wave function (Ψ) must be employed in computing energy surfaces. Suppose that the representative point corresponds to a nuclear configuration of a molecule of interest from an initial state (either ψ_1 or ψ_2) into a region corresponding to a zero-order crossing at geometry r_c. Since the energy and structure of the two states are identical at r_c, the situation is optimal for mixing of zero-order states, that is, near r_c.

$$\text{Initial state } \psi_1 \quad \to \quad [\psi_1 \pm \psi_2] \quad \to \quad \psi_1 \text{ or } \psi_2 \quad \text{Final state} \qquad (5.2)$$

$$\text{(mixing near } r_c)$$

In words, Eq. 5.2 states that the zero-order states ψ_2 and ψ_1 are capable of engaging in electronic resonance near r_c. If resonance should occur, the electron motion is no longer adequately defined in terms of one of the zero-order functions alone; that is,

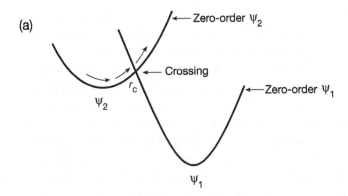

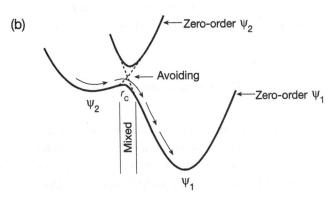

Figure 5.3   Schematic representations of the trajectories of a representative point. (a) The $\psi_2 \to \psi_1$ transition does not occur at r_c, because a surface crossing occurs between the two zero-order wave functions ψ_1 and ψ_2. (b) The $\psi_2 \to \psi_1$ transition does occur because a surface avoiding occurs between the two zero-order wave functions near r_c. The wave functions retain their zero-order character for geometries that are not near r_c.

the adiabatic (Born–Oppenheimer) approximation is no longer accurate, because the nuclear motion and motion of the representative point are no longer unambiguously controlled by a single surface defined by ψ_2 or ψ_1 but are now defined by an unstable, mixed state $[\psi_1 \pm \psi_2]$.

Figure 5.3 shows schematic exemplars for two limiting situations for surface crossing of ψ_1 and ψ_2 in terms of the electronic PE curves discussed in Chapter 3: (a) A surface crossing of ψ_1 and ψ_2 at r_c (the wave functions do not mix at r_c) and (b) a surface avoiding of ψ_1 and ψ_2 at r_c (the wave functions mix to some extent at r_c). In case (a), a trajectory of the representative point on ψ_1 from left to right is shown. As the representative point approaches r_c, since there is no interaction between ψ_1 and ψ_2, the point continues along the PE for ψ_1. We would imagine the representative point oscillating back and forth in the potential curve for ψ_1. In case (b), as the representative point approaches r_c, because the wave functions ψ_1 and ψ_2 mix at r_c, the representative point continues along the lower-energy curve that switches from being "ψ_1-like" to the left of r_c to "ψ_2-like" to the right of r_c. In other words, a radiationless transition from ψ_1 to ψ_2 has occurred as the representative point moves from left to right past r_c for case (b). The situations in Fig. 5.3 refer to any type of radiationless transition and can be used to describe the effect of vibrations in mixing ψ_1 and ψ_2 (e.g., Fig. 5.6) or the effect of spin–orbit coupling in mixing ψ_1 and ψ_2 (e.g., Fig. 5.7).

We can obtain some quantum insight into the time scale of electronic "mixing" of ψ_1 and ψ_2 by noting the relationship between frequency and energy (i.e., since $\Delta E = h\nu$, then $\nu = \Delta E/h$). Since the time τ (units s), of a vibration is the inverse of the frequency ν (units s^{-1}), Eq. 5.3 provides a relationship between the energy gap (ΔE) separating two electronic states, and the time it takes one state to "oscillate" its structure into that of the other. The frequency of these imaginary oscillations back and forth can be loosely termed an *electronic tautomerism*. In the limit of perfect resonance, the "lifetime" τ that the molecule spends in a tautomeric form is of the order of

$$\tau(\text{"lifetime" of an electronic tautomer in s}) = h/\Delta E \sim 10^{-13}/\Delta E \quad (5.3)$$

where ΔE is the resonance energy in kilocalories per mole. Clearly, if a switch from one electronic tautomer to another is to occur, it must happen within the time $\Delta\tau$ that the molecule is in the region of crossing of surfaces (i.e., $\tau < \Delta\tau$). The stronger the electronic interaction (ΔE), the faster the jump from one electronic state to another can occur.

This computation provides us with a numerical benchmark for the time scales of mixing of electronic states as the representative point moves along the reaction coordinate through the critical region near r_c. Strong resonance interactions (e.g., crossings of states of the same electronic and spin symmetries) are of the order of > 30 kcal mol^{-1}. In this hypothetical example, the tautometers may be imagined to be undergoing interconversions via nuclear perturbations occurring at a rate of $1/\tau(10^{14}\text{–}10^{15}$ s$^{-1})$.

If the relative velocity of the nuclei moving through the crossing region is $\sim 10^4\text{–}10^5$ cm s$^{-1} = 10^{12}\text{–}10^{13}$ Å s^{-1} (typical values for the vibrations of atoms), then the time ($\Delta\tau$) the nuclei spend in a given region (say, along a line of length of ~ 3 Å) is $\sim 10^{-12}\text{–}10^{-13}$ s. This rough order-of-magnitude calculation shows that the lifetime of the tautomer is *shorter* than the time it takes to cross the interaction region. Thus, the electronic tautomerization is complete before the nuclei can move out of the interaction region. In the case of weak resonance interaction (< 1 kcal mol^{-1}), the lifetime of the tautomer is $\sim 10^{-13}$ s or longer. Thus, electronic tautomerization may or may not occur in the interaction region, depending on the velocity of the representative point and the precise value of τ. Equation 5.4 relates the frequency, ν, of electronic tautomerization to the energy gap between two interacting states.

Frequency of electronic $\nu = 1/\tau = \Delta E/h = 10^{13}\Delta E$ (kcal mol^{-1})   (5.4)
tautomerization

5.4   Radiationless Transitions and the Breakdown of the Born–Oppenheimer Approximation

Radiationless transitions between surfaces are difficult at geometries for which the adiabatic electronic surfaces are far apart (more than several vibrational quanta) in energy.[5] On the other hand, *radiationless transitions are most plausible at geometries* (r_c) *corresponding to crossings of the zero-order adiabatic surfaces*. These geometries that take *R to a ground state correspond to "leaks" or funnels on the excited-state surfaces, and radiationless jumps are expected to occur with their highest probability when the representative point corresponds to a geometry in the region of a zero-order crossing. These are precisely the regions in which the electronic wave function is a rapidly changing (orbital and/or spin) function of nuclear geometry. As the representative point passes through such a region, the nuclear motion has a certain probability of being controlled by *either* electronic surface depending on available vibronic and spin–orbit interactions.

5.5   An Essential Difference between Strongly Avoiding and Matching Surfaces

An essential difference between the strongly avoiding (Fig. 5.2c) and matching (Fig. 5.2d) topologies is the occurrence of a zero-order linkage (dotted lines in Fig. 5.2) between the upper and lower surfaces in the strongly avoiding case; this does not occur in the matching case. The importance of this distinction is that for a strongly avoided crossing, the representative point will tend to become equilibrated and may jump from ψ_2 to ψ_1 in the region near r_c, because the dotted line (zero-order connectivity) provides a dynamic link between the upper and lower surfaces. When surfaces are matched near r_c, there is no dynamic link coupling them at the purely electronic level. In Chapter 6, we see that the strongly avoiding topology that occurs is important for geometries corresponding to transition states for allowed pericyclic reactions and that there tend to be conical intersections near these geometries.

5.6   Conical Intersections Near Zero-Order Surface Crossings

The crossing (or touching) of two energy surfaces, while a rare situation for diatomic molecules, turns out to be quite common for organic molecules. The energy surfaces shown in the figures in this chapter are simplified two-dimensional (2D) energy curves: More realistic energy surfaces are multidimensional and impossible to visualize. However, for a three-dimensional (3D) extension of the energy-curve concept, the immediate vicinity of the crossing point in a zero-order representation becomes a "double cone," one portion of which corresponds to the upper surface and the other portion to the lower surface. At the touching point of the two cones, the wave functions for the two surfaces are degenerate. This touching point is termed a "conical intersection" and has recently been found, through computational analysis,[6] to be an important concept in the examination of photochemical processes. Conical intersections serve as very efficient *funnels* for the representative point to move from an upper surface to a lower surface. The funnel (F) in the *R → F process of Scheme 1.1 is often a conical intersection. In favorable cases, motion through a conical intersection can occur at effectively the rate of vibrational relaxation. We shall postpone a detailed discussion of conical intersections until Section 6.12. However, here we point out that very fast *electronic* radiationless transitions, that is, those occurring at rates of vibrational transitions, are often assumed to occur through conical intersections in 3D (or, as described in this chapter, in 2D as the result of a true or near surface crossing).

5.7   Formulation of a Parameterized Model of Radiationless Transitions

Only a few parameters are necessary for a qualitative evaluation of the plausibilities (or relative probabilities) of radiationless transitions. As with radiative transitions (Section 4.26), we may consider the *experimental* probability of a radiationless transition (usually expressed in terms of an observed rate constant, k_{obs}) as the product of the rate constant for a hypothetical "fully allowed" (k_0) process and prohibition factors (f_i) due to electronic, vibrational, or spin features that contrive to reduce k_0 to the observed k_{obs}. The prohibition factor (f_i) is a parameter in the same spirit of the oscillator strength for radiative transitions and the various selection rules that lead to a decrease in observed oscillator strengths from an ideal, maximal value of 1.0. In the same spirit, we can suppose that k_{obs} may be parameterized, as shown in Eq. 5.5:

Observed rate: $k_{obs} = k_0 \times f_e \times f_v \times f_s$ maximal rate prohibition factors

$$(5.5)$$

In Eq. 5.5, f_e, f_v, and f_s represent the "prohibition factors" due to the electronic, nuclear, and spin configurational changes, respectively, that occur during the radiationless transitions.

The magnitude of term f_e is determined by the selection rules for electronic transitions, which is in turn related to the magnitude of the matrix element for the *pure electronic part* of the radiationless transition. The magnitude of f_e may be qualitatively evaluated by inspection of the positive overlap of the *orbitals that change* during the radiationless transition. The magnitude of f_v is related to the magnitude of the overlap of the nuclear wave functions for the initial and final states; that is, the magnitude of f_v is directly related to the *Franck–Condon factor* $< \chi_1 | \chi_2 >^2$. Finally, f_s is related to the spin multiplicity in the initial and final states. For organic molecules, the magnitude of f_s will depend on the spin–orbit coupling interactions that occur during an intersystem crossing.

Each one of the factors, f_i, represents a structural rearrangement, such as an electronic rearrangement (f_e), a vibrational change (f_v), or a spin rearrangement (f_s), that may be rate limiting for the radiationless transition. In addition, excess electronic energy released by the radiationless transition must be "accepted" somehow, either intramolecularly or intermolecularly, in order to obey the law of energy conservation. However, the energy release of excess vibrational energy is rarely rate determining. The acceptors of the excess electronic energy may be either the vibrations of the molecule undergoing the transition or the vibrational, rotational, and translation motions of solvent molecules. Thus, intramolecular vibrations and intermolecular collisions typically serve as a "heat bath" that rapidly soaks up the excess electronic energy originally localized in the molecule undergoing the radiationless transition. Now, let us show how the transfer of energy (electronic and vibrational) is involved in radiationless transitions of the type given in Scheme 5.1.

5.8 Visualization of Radiationless Transitions Promoted by Vibrational Motion; Vibronic Mixing

In the zero-order approximation, we assume that for a given, fixed nuclear geometry the electronic states may be classified in terms of a single electronic orbital configuration and a single spin type (multiplicity). We consider these zero-order states "electronically and spin pure" (e.g., pure n,π^* or π,π^* states and pure singlet or triplet states). In the first approximation, we consider various mechanisms for mixing of the states that do not involve a change in spin. In first order, n,π^* and π,π^* states are mixed to a certain extent by electron–electron interactions and vibrations of a molecule. Mixing of electronic states as the result of vibrations is termed "vibronic" mixing. This vibronic mixing "relaxes" selection rules for transitions that are strictly forbidden (matrix element equals zero) in zero order. In addition to vibronic interactions, spin–orbit interactions serve to mix the zero-order states that differ in spin.[7]

Let us use an $n,\pi^* \rightarrow \pi,\pi^*$ transition of a ketone as an exemplar of a vibrationally induced radiationless transition. In terms of orbitals, an $n,\pi^* \rightarrow \pi,\pi^*$ radiationless transition corresponds to a one-electron jump of a π electron into an n orbital (we assume that a π^* electron does not change its position in space during the transition). Thus, we consider the main electronic change in the transition to correspond to a $\pi \rightarrow n$ orbital jump (Fig. 5.4). For the electronic orbital jump to be isoenergetic, vibrational motion must cause a "switching" of the energetic ordering for the n and π levels. This switching causes the n,π^* and π,π^* states to oscillate back and forth in energy during the vibrations. It is not absolutely necessary that the π,π^* state be lower in energy than the n,π^* state, but only that a crossing point can be accessible during the lifetime of the excited state as it vibrates, and that the system then deactivates along the π,π^* surface. In some cases, thermal energy (provided by collisions with molecules in the environment) can be employed by the system in order for the crossing point to be reached; that is, the radiationless process will require an activation energy in order to go over an energy barrier. The reverse situation of a $\pi,\pi^* \rightarrow n,\pi^*$ transition can also be easily imagined based on the same simple orbital considerations. In this case, the system starts off in a π,π^* configuration and an $n \rightarrow \pi$ orbital transition occurs.

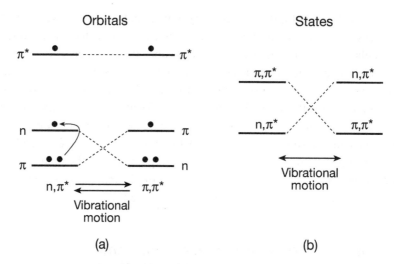

Figure 5.4   (a) Orbital and (b) state descriptions of $n,\pi^* \to \pi,\pi^*$ state switching as a result of vibrational motion.

From Fig. 5.4, we note that in certain cases, *the configuration of an electronically excited state can change during vibrations!*

From the standpoint of wave mechanics, the largest matrix elements for vibronic interactions are due to vibrations that generate oscillating electric dipoles in the same region of space as the transition dipoles for the states undergoing electronic transition. That this should be the case is plausible based on the quantum intuition available from discussions of the oscillating dipole interactions that modeled the interaction for light and electrons. The "best" vibrations for causing mixture of the n,π^* and π,π^* states are those nuclear motions that cause displacement of atoms possessing substantial n and π density.

Not all vibrations are effective in mixing n,π^* and π,π^* states (Fig. 3.1). For example, for a ketone in a planar geometry, any vibrations that do not disrupt planarity are not effective in generating electric dipoles near the oxygen atom, because the n and π orbitals are orthogonal (possess an electronic overlap integral $< n|\pi >= 0$) as long as the system is strictly planar. Nonplanar vibrations, on the other hand, cause the p orbitals on oxygen to rehybridize and to pick up s character (Fig. 5.5). Consequently, $< n|\pi >\neq$ zero for nonplanar geometries. However, the mixing of surfaces due to nonplanar vibrations may be weak.

Let us consider how "mixing" of states is viewed in terms of energy surfaces (Fig. 5.6). Suppose a zero-order crossing of an n,π^* and a π,π^* state of a ketone exists. Figure 5.6 shows the occurrences of a surface crossing (a) for a planar vibration and of a surface avoiding (b) for a nonplanar vibration. Note that the states are significantly mixed only for those geometries near the crossing point r_c, where the energies of the two zero-order states are nearly equal. Thus, for geometries near the energy minima on the lower surface in Fig. 5.6, the zero-order approximation of a pure n,π^* and a pure π,π^* state is still quite valid.

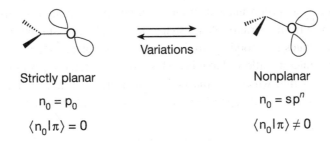

Strictly planar

$n_0 = p_0$

$\langle n_0 | \pi \rangle = 0$

Nonplanar

$n_0 = sp^n$

$\langle n_0 | \pi \rangle \neq 0$

Figure 5.5   Schematic for the "mixing" of s character into a p orbital as the result of nonplanar vibrations.

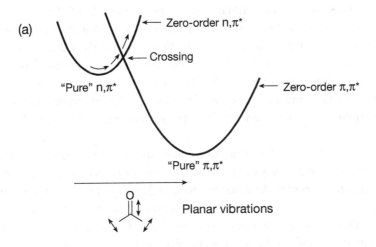

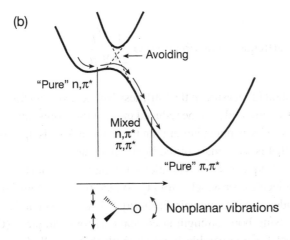

Figure 5.6   The effect of planar (a) and nonplanar (b) vibrations on a zero-order crossing of n,π^* and π,π^* surfaces.

As a result of a certain nuclear motion (vibrations), the nuclear structure may change (Fig. 5.5) from a strictly planar shape (for which the assumption of purity of n,π^* and π,π^* states is a good approximation) to a nonplanar shape (for which the n,π^* and π,π^* states are allowed to mix). This vibronic interaction removes the zero-order surface crossing, which is replaced by a first-order surface, avoided crossing. The magnitude of the avoiding is given by a matrix element of the type:

$$< n,\pi^*|P_{\text{vib}}|\pi,\pi^* > \tag{5.6}$$

where P_{vib} is an operator representing the perturbing vibrational motion (e.g., C—C—O bending vibration) that causes the mixing of the states. The magnitude of the matrix element of Eq. 5.6 may be estimated from the magnitude of the orbital overlap integral $< n,\pi^*|\pi,\pi^* >$. Since the overlap of the π^* orbital with itself is unity, then $< n,\pi^*|\pi,\pi^* > \sim < n|\pi >$; that is, *it is only the overlap of the n and π^* orbitals that determine the effective overlap in Eq. 5.6*. Although $< n|\pi^* >$ equals zero for planar vibrations (Fig. 5.5, left) $< n|\pi > \neq 0$ is for nonplanar vibrations, which introduces s character into the π and n orbitals. The selection rules for intersystem crossing based on the magnitude of matrix elements for types such as Eq. 5.6 are termed El-Sayed's rules.[11]

In discussing radiationless transitions, it is useful and important to recall the benchmarks for the frequency of vibrational motion for pairs of atoms that commonly occur in organic molecules. As noted in Chapters 2 and 3, vibrational motion between two atoms, X—Y, may be represented as the periodic motion of a harmonic oscillator. The oscillation *frequency* of a harmonic oscillator is related to the restoring force ($-k\Delta r$) and the masses of the particles involved in the oscillation by the expression given in Eq. 5.7:

$$\nu(\text{frequency of vibration}) = \frac{1}{2\pi}\sqrt{\frac{k}{\mu}} \tag{5.7}$$

where k is the force constant (a measure of the "stiffness" of the restoring force, or for a molecule, of the bond strength) and μ is the reduced mass of the atoms involved in the vibration. When the mass of X is much larger than the mass of Y, μ is approximately equal to the mass of Y, that is, $\mu \sim$ mass of the lighter nucleus.

Table 5.1 lists some frequencies of vibrations for pairs of atoms involved in vibrations in organic molecules for which strongly bonded pairs of atoms may often be treated as an independently vibrating pair. From Eq. 5.7, we expect and find that ν will *increase* with increasing bond strength at constant μ, for example, $\nu(C\equiv C) > \nu(C—C)$. We also expect for comparable bond strength that ν will *decrease* with increasing μ, for example, $\nu(C—H) > \nu(C—D)$ and $\nu(C—C) > \nu(C—Cl)$. In terms of numbers, we see that the highest frequencies are of the order of 10^{14} s^{-1} for C—H stretching vibrations and the lowest frequencies are of the order of (~ 1–2×10^{13} s^{-1}) for stretches involving at least one heavy atom or an all carbon bending.

Table 5.1 Some Common Bond Types and Associated Stretching Frequencies and Bond Strengths

Bond Type	Vibrational Type (cm^{-1})	Frequency $(\times 10^{13} s^{-1})$	Bond Strength $(kcal\ mol^{-1})$
C—Cl	Stretch 700	2.1	80
C≡C	Stretch 2200	5.6	100
C=O	Stretch 1700	5.1	180
C=C	Stretch 1600	4.2	165
N=N	Stretch 1500	4.0	110
C—C—H	Bend 1000	3.0	100
C—C	Stretch 1000	3.1	85
C—C—C	Bend 500	1.5	85
C—H	Stretch 3000	9	100
C—D	Stretch 2100	6	100

5.9 Intersystem Crossing: Visualization of Radiationless Transitions Promoted by Spin–Orbit Coupling

From the vector model for electronic spin (Sections 2.7 and 3.6) it was deduced that two possible mechanisms exist for intersystem crossing, a spin "flip" and a spin "rephasing." For organic molecules, spin–orbit interactions are by far the major mechanism for intersystem crossing,[7] whereas spin–spin interactions provide an alternate mechanism for certain radical pairs and biradicals.[9]

Consider Fig. 5.7 for a general schematic interpretation of intersystem crossing in terms of energy surfaces.[10] In the zero-order approximation (Fig. 5.7a), we assume there is no mechanism for intersystem crossing, resulting from our artificial separation of electronic and spin motions (cf. to Fig. 5.3a). In this approximation, if a molecule is in an initial singlet state, it will stay forever in the singlet state, or if it is in an initial triplet state, it will remain forever in the triplet state, even if a reaction coordinate exists such that a crossing of the single and triplet states occurs.

If spin–orbit coupling is present when the system is near or at r_c, there may be a mixing of the singlet and triplet states at the crossing point.[7] As a representative point r moves along an initially pure singlet (or pure triplet surface), intersystem crossing may occur near the nuclear geometry corresponding to r_c, *if certain conditions are met*: The interaction that mixes the spin states must be "turned on" and must be effective when the representative point is near r_c. This is, of course, the set of conditions that we have always encountered for resonance between two waves. Thus, there is a requirement not only that a magnetic force capable of changing the spin multiplicity must exist at r_c, but also that the interaction must operate effectively during the period when the representative point is in the region near r_c. These, of course, are precisely the conditions that are required for any type of radiationless transition,

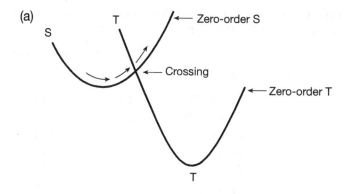

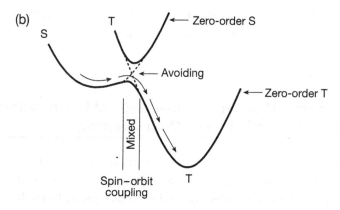

Figure 5.7   Intersystem crossing is strictly forbidden in the zero-order approximation (a), but becomes partially allowed when a spin-mixing mechanism is available near the crossing point of the energy curves for the S and T states (b).

whether it is electronic, vibrational, or spin and are the characteristic conditions of wave resonance.

5.10 Selection Rules for Intersystem Crossing in Molecules

We learned in Section 3.21 that spin–orbit interactions for organic molecules are effective in promoting intersystem crossing near r_c only if: (1) the orbital transition involved possesses the character of a $p_x \rightarrow p_y$ orbital jump to generate orbital angular momentum, and (2) the orbital transition is localized on a single atom. These conditions determine the selection rules for intersystem crossing in molecules (Eq. 5.6).[8]

Let us take a carbonyl group as an exemplar for our qualitative analysis of spin–orbit coupling. The singlet state of a carbonyl group may be derived from an n,π^* or a π,π^* configuration. Consider the four possible intersystem crossings from S_1 of a carbonyl group for the different starting and final electronic configurations shown in Eqs. 5.8–5.11. The corresponding transitions are shown schematically in Figs. 5.8 and 5.9.

$$^1n,\pi^* \to {}^3\pi,\pi^* \quad \text{(see Fig. 5.8a)} \tag{5.8}$$

$$^1n,\pi^* \to {}^3n,\pi^* \quad \text{(see Fig. 5.8b)} \tag{5.9}$$

$$^1\pi,\pi^* \to {}^3n,\pi^* \quad \text{(see Fig. 5.9a)} \tag{5.10}$$

$$^1\pi,\pi^* \to {}^3\pi,\pi^* \quad \text{(see Fig. 5.9b)} \tag{5.11}$$

If we assume that the rate of intersystem crossing is directly related to the magnitude of spin–orbit coupling (Eq. 3.22), then we can estimate the magnitude by inspecting several configurations of the initial state to determine whether there is finite first-order interaction that complies with the selection rules for spin–orbit coupling with any of the atomic orbital configurations of the final state (i.e., whether a $p_x \to p_y$ orbital jump localized on a single atom occurs for any of the limiting atomic orbital configurations).[10]

First, we consider a $^1n,\pi^* \to {}^3\pi,\pi^*$ intersystem crossing (ISC) (Eq. 5.8). Two major atomic orbital representations of the $^1n,\pi^*$ state appear in Fig. 5.8a. The $^1n,\pi^* \to {}^3\pi,\pi^*$ transition has a zero-order spin–orbit coupling in one of its major atomic orbital configurations (Fig. 5.8a, right): $\pi^*(\uparrow) \to n(\downarrow)$. As a result, *ISC can be triggered via a one-center* $p_x \to p_y$ *interaction on the oxygen atom.* In addition, the highly electrophilic, half-filled n orbital provides a substantial driving force for the electronic transition by attracting one of the electrons from the π orbital.[11] Thus, we have deduced a selection rule that the $^1n,\pi^* \to {}^3\pi,\pi^*$ intersystem crossing is allowed because of effective spin–orbit coupling.

Next, consider a $^1n,\pi^* \to {}^3n,\pi^*$ intersystem crossing (Eq. 5.9). In this case, ISC does not involve an orbital change. Figure 5.8b shows two major atomic orbital representations of the n,π^* state. A zero-order spin–orbit coupling is not possible in either representation, because neither the $n(\uparrow) \to n(\downarrow)$ nor the $\pi^*(\downarrow) \to \pi^*(\uparrow)$ electronic transitions generate orbital angular momentum along the bond axis; that is, no $p_x \to p_y$ single atomic orbital jump is involved. Thus, *there is no zero-order spin–orbit coupling for the* $^1n,\pi^* \to {}^3n,\pi^*$ *transition and ISC is forbidden in zero order.*

Now, let us consider the role of spin–orbit coupling for a $^1\pi,\pi^* \to {}^3n,\pi^*$ transition (Eq. 5.10) and for a $^1\pi,\pi^* \to {}^3\pi,\pi^*$ transition (Eq. 5.11). Starting from a $^1\pi,\pi^*$ state, we must consider three atomic orbital configurations that represent resonance forms of the state (Fig. 5.9). For a $^1\pi,\pi^* \to {}^3n,\pi^*$ transition, by inspection of Fig 5.9a, two of the atomic orbital configurations (middle and right configurations) have a zero-order one-center $p_y \to p_x$ spin–orbit coupling with the $^3n,\pi^*$ state. Thus, a $^1\pi,\pi^* \to {}^3n,\pi^*$ transition is allowed to occur by spin–orbit coupling and *ISC is allowed in zero-order*. On the other hand, by inspection of the contributing orbital representation of Fig. 5.9b, for the $^1\pi,\pi^* \to {}^3\pi,\pi^*$ transition, there is no zero-order spin–orbit coupling between any of the singlet configurations and the triplet configuration (Fig. 5.9b), so that *ISC is forbidden in zero order.*

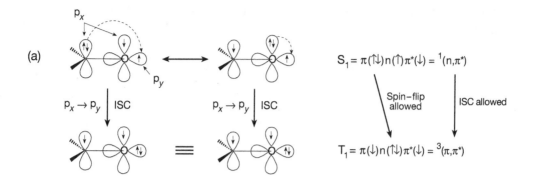

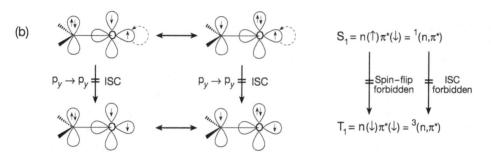

Figure 5.8   Qualitative orbital description of the basis for (the allowed $^1(n,\pi^*) \to ^3(\pi,\pi^*)$ and the forbidden $^1(n,\pi^*) \to ^3(n,\pi^*)$ intersystem crossings.

As a result of our analysis, we deduce the following selection rules (i.e., El-Sayed's rules)[8] for $S_1 \to T_1$ intersystem crossing of carbonyl groups in zero order:

$$S_1(n,\pi^*) \to T_1(n,\pi^*) \quad \text{Forbidden} \tag{5.12}$$

$$S_1(n,\pi^*) \to T_1(\pi,\pi^*) \quad \text{Allowed} \tag{5.13a}$$

$$S_1(\pi,\pi^*) \to T_1(n,\pi^*) \quad \text{Allowed} \tag{5.13b}$$

$$S_1(\pi,\pi^*) \to T_1(\pi,\pi^*) \quad \text{Forbidden} \tag{5.14}$$

The rules derived from these exemplars are general for n,π^* and π,π^* states and are not restricted to carbonyl compounds. They also hold for any system capable of n,π^* and π,π^* transitions.

As benchmarks for the rate of $S_1 \to T_1$ ISC for carbonyl compounds experimental values of k_{ST} for an alkyl ketone,[12] benzophenone,[13] and pyrenaldehyde[14] are $\sim 10^8 \text{ s}^{-1}$, $\sim 10^{11} \text{ s}^{-1}$, and $\sim 10^7 \text{ s}^{-1}$, respectively. Figure 5.10 summarizes the transitions involved for the cases given by Eqs. 5.12–5.14. The magnitude of the rate constants is related to the proximity of the n,π^* and π,π^* states that are available to mix the singlet and triplet states.

By extending the logic for spin–orbit coupling in $S_0 \to T_1$ transitions, we deduce that a $p_x \to p_y$ transition on a single atom is also optimal for a $T_1 \to S_0$ intersystem

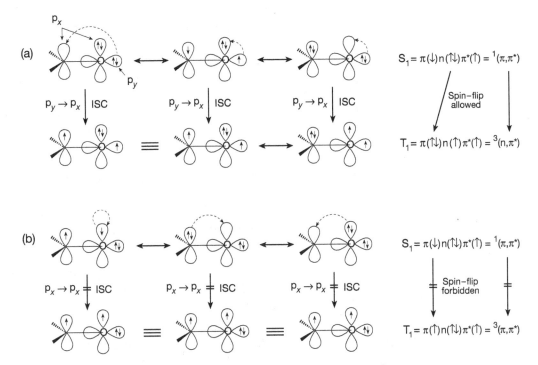

Figure 5.9   Qualitative orbital description of (a) the allowed $^1(\pi,\pi^*) \to {}^3(n,\pi^*)$ and (b) the forbidden $^1(\pi,\pi^*) \to {}^3(\pi,\pi^*)$ intersystem crossings.

crossing transition, so that we deduce the selection rules of Eqs. 5.15 and 5.16.

$$T_1 \to S_0 \qquad T_1(n,\pi^*) \to S_0(n^2) \quad \text{Allowed} \qquad (5.15)$$

Transitions

$$T_1(\pi,\pi^*) \to S_0(\pi^2) \quad \text{Forbidden} \qquad (5.16)$$

As exemplars, the $T_1(n,\pi^*) \to S_0(n^2)$ ISC transitions of benzophenone and acetone ($k_{TS} \sim 10–100 \text{ s}^{-1}$) are much faster than for $T_1(\pi,\pi^*) \to S_0(\pi^2)$ ISC transitions for pyrenaldehyde ($k_{TS} < 1 \text{ s}^{-1}$), which is in agreement with the above selection rules. Note that the actual magnitude of spin–orbit coupling for the excited states of organic molecules possessing light atoms is on the order of 0.3–0.001 kcal mol^{-1}; that is, spin–orbit coupling is a very weak electronic perturbation.[7]

We have seen that in zero order (Eq. 5.14) $S_1(\pi,\pi^*) \to T_1(\pi,\pi^*)$ ISC is forbidden because there is no spin–orbit coupling generated by this transition in zero order. However, some perturbation may introduce a finite amount of spin–orbit coupling to make the transition allowed to some extent. What are the effects of vibrations on spin–orbit coupling?[15] As an exemplar of the effect of vibrations on spin–orbit coupling, consider a situation for which there is a surface crossing between a π,π^* singlet and a π,π^* triplet state of ethylene or benzene (Fig. 5.11). The effect of vibrations on spin–orbit coupling is actually a *second-order* electronic coupling effect for which a weak interaction (spin–orbit coupling) is induced by a second weak coupling (vibronic

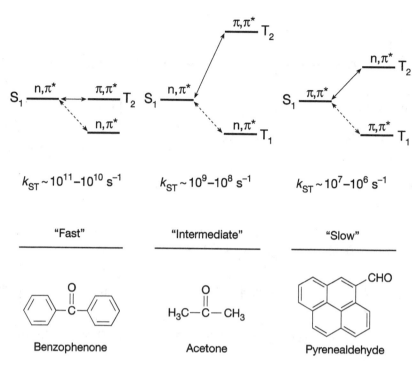

Figure 5.10   Examples of different rates of intersystem crossing for $S_1(n,\pi^*) \rightarrow$ $T(\pi,\pi^*)$, $S_1(n,\pi^*) \rightarrow T(n,\pi^*)$, and $S_1(\pi,\pi^*) \rightarrow T(\pi,\pi^*)$ of carbonyl compounds.

coupling). As usual, we look to orbital interactions to introduce the required $\sigma \rightarrow \pi^*$ single atomic orbital jump for spin–orbit coupling to be effective. In this case, since there are no n orbitals to take part in the process, we need to invoke σ and σ^* orbitals for interactions with the π,π^* states. This means that vibrations need to mix $\sigma \rightarrow \pi^*$ and $\pi \rightarrow \sigma^*$ transitions, which possess the character of $p_x \rightarrow p_y$ transitions. Since the $\sigma \rightarrow \pi^*$ and $\pi \rightarrow \sigma^*$ transitions are not one-center in nature (π and π^* electrons are strongly delocalized), and because the energy gap between the orbitals involved in the transitions are high in energy, the spin–orbit coupling mixed in by vibrations is typically very weak and transitions involving a change in spin are very slow in general.

Now, let us consider the effect of certain specific types of vibrations. In-plane vibrations generally do not mix π- and σ-type orbitals, and therefore do not cause mixing of singlet and triplet states of planar hydrocarbons (Fig. 3.1); on the other hand, out-of-plane vibrations do mix π- and σ-type orbitals and are therefore capable of (weakly) mixing singlet and triplet states.[16] Figure 5.11 schematically shows the surface situation for in-plane and out-of-plane vibrations with ethylene as an exemplar. In executing planar vibrations, the molecule may pass through a surface crossing geometry, but will not be able to "turn on" a spin–orbit interaction, because the planar vibration is an ineffective promoter for coupling spin and orbital motion. If the molecule is brought into a crossing geometry by an out-of-plane vibration, a finite but very weak spin–orbit interaction occurs, the surfaces avoid, and a $^1\pi,\pi^* \rightarrow {}^3\pi,\pi^*$

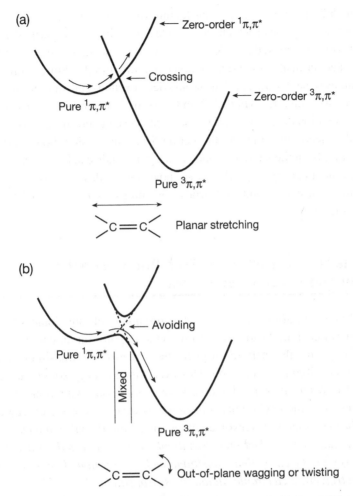

Figure 5.11   Vibronic interactions. (a) planar vibrations do not mix states, so that a zero-order crossing persists in first order. (b) Out-of-plane vibrations cause state mixing and a surface avoiding.

conversion is possible ($k_{ST} < 10^7$ s^{-1}). In effect, the "door is open" at a crossing geometry only if the molecule is brought to the crossing region by an appropriate vibration or nuclear motion.

5.11   The Relationship of Rates and Efficiencies of Radiationless Transitions to Molecular Structure: Stretching and Twisting as Mechanisms for Inducing Electronic Radiationless Transitions

Now, we describe a theory that relates the probability of radiationless transitions of electronically excited states to molecular electronic structure. From the discussions in Sections 5.2 and 5.3, an important feature of such a theory is the idea of "critical geometries" (r_c) from which radiationless transitions are particularly favored. Thus, a theory relating radiationless processes to molecular structure must also be able to connect critical geometries to molecular structure.[1-3] Surface crossings, surface touchings (Chapter 6), and shallow and deep excited-state minima all correspond to such critical geometries (Figs. 5.2 and 5.3). In Chapter 6, we consider how one may use correlation diagrams to relate the occurrence of zero-order surface crossings to molecular electronic structure. In Section 5.12, we consider how "surface touchings" resulting from the stretching and twisting of bonds can be related to molecular electronic structure. In Section 5.13, we see how radiationless transitions from excited-state minima that correspond to "surface matchings" may be related to molecular electronic structure.

5.12   The "Loose Bolt" and "Free-Rotor" Effects: Promoter and Acceptor Vibrations

It is possible for certain vibrations to act as *promoters* of radiationless transitions if specific vibrations both "carry" the representative point on an energy surface to a region near r_c, and also help to "trigger" the radiationless transition by providing an appropriate vibronic interaction. However, since energy conservation must be maintained in detail after a radiationless transition, some vibrations (or collisions with molecules in the environment) must act as acceptors of the energy difference between the electronic states involved in the transition. *If a vibration can act as both a promoter inducing an electronic radiationless transition and an acceptor of the excess energy produced by that transition, such a vibration should be particularly effective in triggering electronic radiationless transitions.*[17]

Two exemplar promoter and acceptor vibrations of organic molecules are the stretching of a C—C σ bond and the twisting of a C=C π bond. Both types of vibrations in the extreme (breaking of a stretching C—C bond and severe twisting of a C=C bond) cause a surface touching. A schematic representation of a surface touching that is induced (a) by stretching a σ bond and (b) by twisting a π bond is shown in Fig. 5.12a and b, respectively. A stretching or twisting vibration may escort the representative point on an excited surface (wave function, $^*\psi$) to a region on the surface near r_c. If this same vibration also causes vibronic mixing of $^*\psi$ and the ground-state wave function (ψ), a radiationless transition from $^*\psi$ to ψ near r_c then becomes plausible. In some cases, the same vibration serves as a means of accepting excess energy that is involved in the transition between $^*\psi$ and ψ. The stretching vibration is analogous to a "loose bolt" in some moving part of a machine. The loose bolt tends to be set in motion by other moving parts of the machine[17a] and thereby "takes up" KE produced by other moving parts. In the case of the molecule, the excess energy may lead to complete dissociation of the bond; that is, the promoting stretching vibration turns into an irreversible separation of the atoms involved in the vibration.

In the case of the twisting motion of a double bond (Fig. 5.12b), instead of a loose bolt, the situation is more analogous to a "free rotor" that accepts the excess energy and

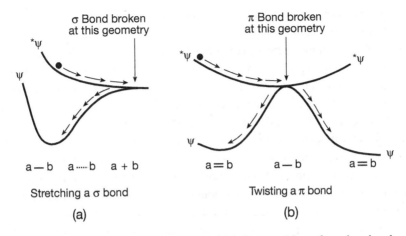

Figure 5.12   Schematic representation of (a) the stretching of a σ bond and (b) the twisting of a π bond. At a 90° twist, the double bond is broken and becomes a single bond.

begins to rotate. In this situation, the twisting motion leads to a cis–trans isomerization in the case of twisting about C=C π bonds.

Experimental exemplars of the free-rotor and loose-bolt effects on radiationless transitions, such as *internal conversions*, are available from fluorescence analysis of conjugated aromatic compounds. For example, at 25 °C, *trans*-stilbene (**1**) is weakly fluorescent ($\Phi_F = 0.05$), and *cis*-stilbene (**2**) is nonfluorescent ($\Phi_F \sim 0.00$).[18] On the other hand, under the same experimental exemplars the structurally constrained stilbene derivatives **3** and **4** are strongly fluorescent ($\Phi_F \sim 1.0$).[19]

1	**2**	**3**	**4**
$\Phi_F = 0.05$	$\Phi_F = 0.00$	$\Phi_F = 1.0$	$\Phi_F = 1.0$

Steric interactions between the phenyl groups of *cis*-stilbene provide the molecule with an inherent torque in *R (for both S_1 and T_1) that provides a tendency to twist about the central C=C π bond, introducing a free rotor along the π,π^* surface. In S_1, the state energy is rapidly lowered by twisting (see Fig. 5.12b; for a more detailed discussion, see Section 6.17). The twisting motion brings S_1 to a geometry r_c that is favorable for radiationless conversion to S_0. Although **1** can also twist about the C=C bond in S_1, it lacks steric interactions, which produce a torque that could lead to an efficient free rotor effect. As a result, the twisting motion about the C=C bond in *R is slower for **1** than it is for **2**. Fluorescence of **1** can now compete with both internal conversion and intersystem crossing.

For cyclic compounds **3** and **4**, the twisting motion about the C=C bond is severely hindered by structural constraints. Consequently, these molecules are relatively rigid and are unable to adopt nuclear geometries that differ substantially from the initial geometry of S_0. The representative point on the S_1 surface cannot move to regions near r_c that correspond to twisted nuclear geometries favorable for radiationless conversions. As a result, fluorescence completely dominates internal conversion and intersystem crossing ($\Phi_F \sim 1.0$). In addition to structural constraints, low temperature and a rigid intermolecular environment can inhibit twisting motions that promote radiationless conversions. If small energy barriers ($\sim$ 3–5 kcal mol^{-1}) separate the representative point from the lower-energy twisted geometries, at low temperatures ($\sim$ 77 K) these barriers may not be surmounted during the excited-state lifetime, and efficient emission results. A rigid environment may be viewed as perturbing the PE surface corresponding to rotation by introducing an energy barrier to twisting.[20] These barriers imposed by the environment (e.g., solvent) are due to the requirement that molecules in the neighborhood of the twisting C=C bond must be displaced if the twisting is to be substantial, a difficult process in a rigid environment.

The free-rotor effect may also operate to facilitate radiationless transitions of singlets (internal conversion to S_0 and intersystem crossing to T_1) or triplets (intersystem crossing $T_1 \rightarrow S_0$). As an illustration,[21] the triplet state of 1-phenylcycloheptene (**5**) undergoes a very rapid intersystem crossing at room temperature relative to the more rigid and structurally constrained 1-phenylcyclobutene (**6**). This result may be interpreted to be due to a surface touching (Fig. 5.12b; see Section 5.7 for a discussion) that results from twisting about the C=C bond. Since **5** is much more flexible than **6** with respect to this motion, in T_1 it may move toward the twisted geometry at a faster rate, and therefore undergo more rapid intersystem crossing.

Easy $\longrightarrow$ [structure]
C$_6$H$_5$

1-Phenylcycloheptene
5
$k_{TS} = 4 \times 10^9$ s^{-1}

Difficult $\longrightarrow$ [structure]
C$_6$H$_5$

1-Phenylcyclobutene
6
$k_{TS} = 6 \times 10^7$ s^{-1}

An example of a loose-bolt effect[17a] on internal conversion is available from data on the radiationless decay of alkyl benzenes.[22] For example, the fluorescence yield of toluene (**7**) is $\Phi_F \sim 0.14$, and its rate of internal conversion is $\sim 10^7$ s^{-1}, whereas the fluorescence yield of *tert*-butylbenzene (**8**) is $\Phi_F \sim 0.032$, and its rate of internal conversion is 10^8 s^{-1}, an order of magnitude faster.

[structure] —CH$_3$

Toluene **7**

$\Phi_F = 0.14$, $k_{IC} = 10^7$ s^{-1}

[structure]
CH$_3$
|
—C—CH$_3$
|
CH$_3$

tert-Butylbenzene **8**

$\Phi_F = 0.032$, $k_{IC} = 10^8$ s^{-1}

It was shown that the mechanism for decrease in Φ_F in going from **7** to **8** was not due to intersystem crossing or to photoreaction. Thus, it appears that the σ-bonds of the *tert*-butyl group serves as a "loose bolt" to accelerate internal conversion by a factor of ~ 10 (possibly via a mechanism related to Fig. 5.12a). Deuteration of the *tert*-butyl group does not affect the rate of internal conversion, supporting the conclusion that it is the C—C bond between the benzene ring and the *tert*-butyl group that is the loose bolt triggering deactivation, not the C—H bonds of the *tert*-butyl group.

In addition, the observation that **7** is much more strongly phosphorescent (relative quantum yield, $\Phi_p^{rel} \sim 1.0$) than **8** ($\Phi_p^{rel} \sim 0.00$) at 77 K provides further support of the loose-bolt effect on $T_1 \to S_0$ intersystem crossing.[23]

5.13   Radiationless Transitions between "Matching" Surfaces Separated by Large Energies

Figure 5.2d shows a hypothetical exemplar for a "matching" of two energy surfaces. By matching, we mean that the two surfaces are not electronically related. Such adiabatic surfaces correspond to essentially pure wave functions ψ_1 and ψ_2. In the hypothetical example, there are no geometries along the reaction coordinate for which the two surfaces approach the same energy. Is a radiationless transition between the two surfaces still possible? The answer is yes, but the transition will be relatively slow. The transition will occur to a measurable extent only if a longer lifetime of the excited state ψ_2 makes up for the weaker coupling between ψ_1 and ψ_2, because the representative point finds itself in an energy minimum at r_c. However, keep in mind the qualification that radiationless transitions between matching surfaces are generally expected to be much slower than those at critical geometries corresponding to situations (a)–(c) and (e) in Fig. 5.2. In the case of matching surfaces, the representative point on the excited surface is viewed as eventually moving to, and residing for a period in, the minimum (r_c) of the upper surface as it awaits a perturbation that will trigger the transition to the lower surface. Of course, if it takes too long for the perturbation to occur, radiative transition from the upper to the lower surface may occur if the Franck–Condon (FC) factors (Chapter 4) are favorable or the representative point will pick up energy and move over barriers to other regions of the excited-state energy surface. What determines the magnitude of the Franck–Condon factor,[24] $f_v = <\chi_i | \chi_f>^2$, for radiationless transitions from an excited surface that is matched with a lower-energy surface? The value of f_v (Eq. 5.5) in general follows the energy-gap law given by Eq. 5.17.

$$f_v \sim \exp(-\Delta E) \qquad (5.17)$$

Thus, the rate of a radiationless transition from r_c (for a matching) will fall exponentially as ΔE increases, if f_e and f_s are not rate determining.

As an exemplar of how the FC factor operates to control the rate of spin-allowed ($f_s = 1$) processes corresponding to Fig. 5.2d, consider the radiationless process of internal conversion between two electronically excited states. Suppose that two excited states, S_2 and S_1, possess potential curves that undergo a zero-order intersection

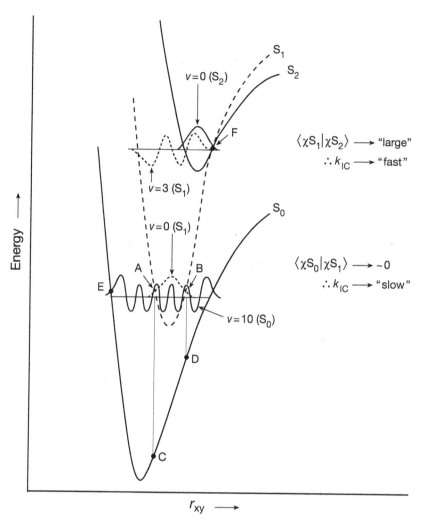

Figure 5.13   Representation of internal conversions between S_2 and S_1 and between S_1 and S_0.

at point F (Fig. 5.13), but the ground-state potential curve is matching with respect to both S_1 and S_2. A transition from $S_2 \rightarrow S_1$ can occur without an appreciable alteration of position or momentum of the nuclei if the transition occurs at geometries near point F, because the Franck–Condon factor is favorable near F (Fig. 5.13). Such an internal conversion is expected to occur relatively rapidly. However, the internal conversion corresponding to the direct radiationless transition $S_2 \rightarrow S_0$ is highly improbable. The surfaces of the two states do not cross; that is, the S_0 and S_2 states are matched. In addition, the energy gap between S_2 and S_0 is very large. As a result, the interactions between the two states will be very weak. Note that the slow internal conversion of $S_2 \rightarrow S_0$ is almost always competing with the extremely rapid (10^{12}–10^{13} s^{-1}) internal conversion between excited states ($S_2 \rightarrow S_1$). This finding is, of course, why we expect photochemical and photophysical processes to start at

S_1, the basis of Kasha's rule, which states that all photophysical and photochemical processes usually start in S_1 or T_1, irrespective of which excited state or vibrational level is initially produced.[1f]

Now, consider what happens when a molecule reaches the lowest vibrational levels of S_1 and becomes vibrationally equilibrated in the $v = 0$ level. From Figure 5.13 at the energy of the $v = 0$ level of S_1, the vibrational eigenfunction for χ_{S_0} (a high-energy vibrational level of S_0) oscillates rapidly from negative to positive values in the region where χ_{S_1} (the $v = 0$ state of S_1) is always positive. This situation leads to mathematical cancellation of the overlap integral $< \chi_{S_1} | \chi_{S_0} >$. Thus, the overlap integral $< \chi_{S_1} | \chi_{S_0} >$, which determines the magnitude of the Franck–Condon factor, will be very small, making the rate of the $S_1 \rightarrow S_0$ transition very slow. This finding contrasts with the situation in the region of interaction about point F. In this case, χ_{S_1} and χ_{S_2} (in the v = 0 state) are situated so that the overlap integral $< \chi_{S_2} | \chi_{S_1} >$ has a significant positive value and the rate of internal conversion is relatively rapid (see Sections 4.8, 4.12, and 4.13 for a review.) This situation is common for organic molecules, so the $S_2 \rightarrow S_1$ transition is generally much faster and more probable than the $S_1 \rightarrow S_0$ transition, if the electronic parts of the molecular wave functions for the two transitions are comparable.

As discussed above, organic molecules that possess rigid cyclic structures, for which twisting (free rotor) and stretching (loose bolt) of bonds is inhibited, tend to fluoresce strongly (e.g., structures **3** and **4** discussed above). Now, we may use quantum intuition to rationalize this result from the standpoint of the FC principle for radiationless transitions. This principle tells us that for rigid structures, the radiationless conversions $S_1 \rightarrow S_0$ and $T_1 \rightarrow S_0$ will be difficult, because the restraints placed on the molecule tend to hold the nuclei close together in a small region of space. In effect, the S_1 states of such systems are constrained in their exploration of the excited surface as they seek to find "loose bolts," "free rotors," or spin–orbit interactions. Thus, they fluoresce with high efficiency and $\Phi_F \sim 1$ (e.g., **3** and **4**).

5.14   Factors That Influence the Rate of Vibrational Relaxation

The advent of pulsed picosecond (ps) and femtosecond (fs) laser spectroscopy has allowed the direct measurement of electronic and vibrational relaxation processes, which occur in the picosecond and femtosecond time range in fluid solutions at room temperature. For organic molecules,[25,26] values of k_{vib} (rate constant for vibrational period) are typically $\sim 10^{12}$–10^{14} s^{-1} (Table 5.1). Why is the transfer of excess vibrational energy to the environment so rapid? The vibrations within a molecule and the vibrations of the molecules in the environment have the ability to rapidly accept the energy of nuclear motion of a molecule (in both its ground and an excited state) and convert it into many different degrees of vibrational motion. Because the number of vibrational, rotational, and translational energy levels of the solvent environment of *R is for all intents and purposes continuous, these energy levels may serve as a classical heat bath, capable of taking up any amount of vibrational energy that the excited molecule needs to dispose.[26]

Direct measurements of vibrational relaxation have been made for both the ground state of organic molecules and of electronically excited states[25]. The general qualitative observations can be summarized as follows.

1. For a molecule in a given electronic state, vibrational relaxation occurs in two time domains. The first domain (typically 10–0.1 ps, 10^{11}–10^{13} s^{-1}) corresponds to *intramolecular vibrational relaxation* (IVR). The second, the longer time domain, is a slower process (typically 1000–10 ps, 10^9-10^{11} s^{-1}) and corresponds to *intermolecular vibrational energy transfer* (VET) from the vibrationally excited molecule to the solvent.

2. Electronic relaxation is generally not rate limiting for IVR from **R to *R (where **R is an upper-level excited state such as S_2 or S_3).

3. Intramolecular vibrational relaxation is slightly faster in excited states (*R) than in ground states (R).

4. Most organic solvents have comparable efficiency in accepting excess vibrational energy.

A pictorial description of vibrational relaxation of an electronically and vibrationally excited molecule is shown in Fig. 5.14, with formaldehyde initially in its *equilibrated* S_1 state as a concrete exemplar. In the equilibrated n,π^* excited state, the electron motion and position are responsible for the excess electronic energy of the molecule. However, the vibrations of S_1 are not excited and the local solvent molecules are "cool"; that is, their translational and vibrational motion is equilibrated with the temperature of the solvent. Imagine that internal conversion occurs from S_1 (or intersystem crossing occurs from T_1) and that radiationless transition is promoted by the vibronic interaction of the stretching C=O vibrational motion that couples electronic and vibrational motions. This means that the electronic motion and position change (e→e) and the C=O vibration (e→v) are coupled, and for the radiationless process from S_1 to S_0 a "vibrationally hot" ground state ($^\ddagger S_0$, where the ‡ represent *excess* vibrational excitation) exists. Each of the transitions, $S_1 \rightarrow {}^\ddagger S_0$ or $T_1 \rightarrow {}^\ddagger S_0$, after they occur isoenergetically, requires excess vibrational energy to be dissipated before they can "relax" to S_0. Where does the energy go after the isoenergetic electronic transition has occurred? We imagine the energy to be dissipated as follows (Fig. 5.14): The electronic transition from a π^* orbital to an n orbital produces a "hot" ‡C=O vibration in S_0, since the transition is induced by a C=O vibration. In other words, the C=O stretch serves as a loose bolt that promotes radiationless transition to the ground state. The vibrationally excited ‡C=O group first transfers some of its energy intramolecularly to the C—H bending vibrations (IVR, 10–0.1 ps, 10^{11}–10^{13} s^{-1}) and later the excess vibrational energy is then transferred on a longer time scale to the solvent (intermolecular VET, 1000–10 ps, 10^9–10^{11} s^{-1}).[25] The excited vibrations of the formaldehyde "cool down" and the local microscopic temperature around the molecule "heats up"; that is, the solvent molecules in the immediate vicinity of the formerly excited molecule have a higher translational and vibrational energy than the average for the macroscopic temperature.

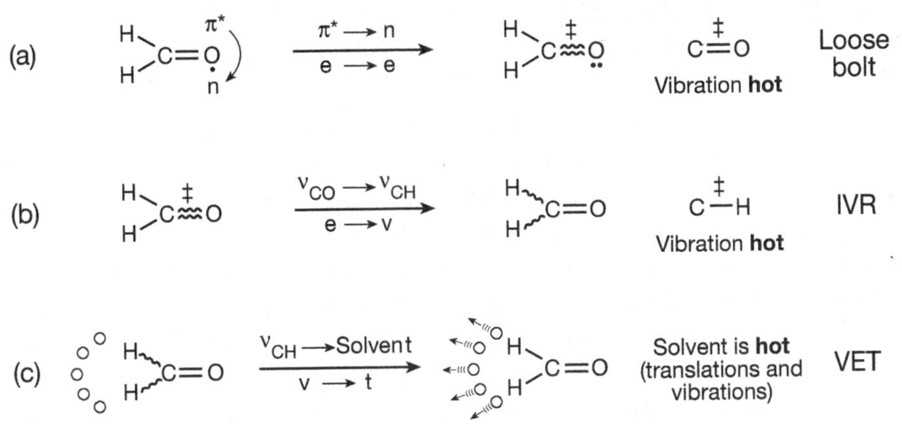

(a) $\text{H}_2\text{C}=\text{O}$ (with π^*, n levels) $\xrightarrow[\text{e} \rightarrow \text{e}]{\pi^* \rightarrow n}$ $\text{H}_2\text{C}\overset{\ddagger}{\approx}\text{O}$ $\text{C}\overset{\ddagger}{=}\text{O}$ Vibration **hot** Loose bolt

(b) $\text{H}_2\text{C}\overset{\ddagger}{\approx}\text{O}$ $\xrightarrow[\text{e} \rightarrow \text{v}]{\nu_{\text{CO}} \rightarrow \nu_{\text{CH}}}$ $\text{H}_2\text{C}=\text{O}$ $\text{C}\overset{\ddagger}{-}\text{H}$ Vibration **hot** IVR

(c) $\text{H}_2\text{C}=\text{O}$ $\xrightarrow[\text{v} \rightarrow \text{t}]{\nu_{\text{CH}} \rightarrow \text{Solvent}}$ $\text{H}_2\text{C}=\text{O}$ Solvent is **hot** (translations and vibrations) VET

Figure 5.14   Schematic description of energy dissipation for formaldehyde. (a) The conversion of electronic excitation into vibration for formaldehyde, initially localized as excitation (wiggly lines) between the C and O atoms, is followed by (b) vibrational energy transfer to the C—H bonds, and (c) transfer to translational motion of the solvent.

In fluid solution or rigid matrices, the take-up of energy by the solvent is rarely if ever rate determining for a radiationless transition.[25,26] At this point, one might ask whether emission of infrared (IR) radiation is a significant path for transitions between the rotational and vibrational levels. Experimentally, there is no significant competition between IR emission (loss of a few quanta at a time) and radiationless deactivation, but there is between UV and vis emission. This result follows theoretically from the relationship:[27]

$$k = 3h/64\pi^4\bar{\nu}^3|H_{21}|^2 \sim < H_{21} >^2 \tag{5.18}$$

where $\bar{\nu}$ is the wavenumber of the photon that is emitted upon passing from state 2 to state 1, H_{21} is the electric dipole matrix element for the transition (Section 5.3), and k is the rate constant for spontaneous emission.

Since $\bar{\nu}$ is 1000–3000 cm^{-1} for IR transitions, while $\bar{\nu} \sim 30{,}000$ cm^{-1} for UV transitions, the $\bar{\nu}^3$ term places a prohibition factor of $\sim 10^{-3}$ or greater on the relative rate of IR emission as compared to UV–vis emission. Furthermore, the dipole moment changes involved in pure vibrational transitions are usually small compared to those that occur upon passing from one electronic state to another. This factor may apply another order-of-magnitude prohibition on the rate of IR emission.

For example,[28] an IR emission of CO_2 occurs at ~ 1000 cm^{-1} and possesses a transition dipole of ~ 0.1 D. This finding results in a vibrational emission rate constant $k \sim 10^2$ s^{-1}. Since the rates of vibrational deactivation of molecules in solution[25] are on the order of 10^{12}–10^{13} s^{-1}, we see that vibrational fluorescence will generally be a minor pathway for vibrational deactivation in condensed phases. For comparison, the transition dipole of allowed electronic transitions are on the order of several debye and the transitions occur at much higher values of $\bar{\nu}$.

5.15   The Evaluation of Rate Constants for Radiationless Processes from Quantitative Emission Parameters

In general, a combination of experimentally determined emission lifetimes (τ_e) and emission quantum yields (Φ_e) provides a convenient means of computing the unimolecular rate constants for the radiationless processes of internal conversion (k_{IC}) and intersystem crossing (k_{ST} and k_{TS}). Knowledge of the rates of interconversions and lifetimes of excited states is of great importance in analyzing photochemical problems. Equations 5.19–5.24, based on the working state energy diagram of Fig. 1.4, incorporates all of the important unimolecular photophysical transitions that are expected to occur after absorption of light (Eq. 5.19) and provide a means of estimating rate constants from spectral data alone. For simplicity, the absence of irreversible photochemical reactions and specific bimolecular quenching is assumed (a good assumption for data taken at 77 K). Such processes are readily included if desired.

Reaction	Step	Rate	Eq. Number
$h\nu + S_0 \rightarrow S_1$	Excitation	I	(5.19)
$S_1 \rightarrow S_0 + \Delta$	Internal conversion	$k_{IC}[S_1]$	(5.20)
$S_1 \rightarrow T_1 + \Delta$	Intersystem crossing	$k_{ST}[S_1]$	(5.21)
$T_1 \rightarrow S_0 + \Delta$	Intersystem crossing	$k_{TS}[T_1]$	(5.22)
$T_1 \rightarrow S_0 + h\nu$	Phosphorescence	$k_P^0[T_1]$	(5.23)
$S_1 \rightarrow S_0 + h\nu$	Fluorescence	$k_F^0[S_1]$	(5.24)

The steady-state approximation for the concentration of excited singlet states, $[S_1]$, assumes that the rate of absorption of photons (I_{abs}) is equal to the total rate of deactivation of S_1, and leads to Eq. 5.25:

$$I_{abs} = (k_{ST} + k_F^0 + k_{IC})[S_1] \qquad (5.25)$$

where I_{abs} is the rate of absorption of light in einsteins $L^{-1} s^{-1}$ (photons $L^{-1} s^{-1}$), and $[S_1]$ is the instantaneous concentration of excited singlets. Similarly, the steady-state approximation assumes that the rate of formation of triplets is equal to the total rate of deactivation of T_1, leading to Eq. 5.26, where $[T_1]$ is the instantaneous concentrations of triplets.

$$k_{ST}[S_1] = (k_{TS} + k_P^0)[T_1] \qquad (5.26)$$

Rearranging Eq. 5.26 to solve for $[T_1]$ leads to Eq. 5.27:

$$[T_1] = k_{ST}[S_1]/(k_P^0 + k_{TS}) \qquad (5.27)$$

Under the conditions of steady-state excitation, the efficiency or quantum yield (Φ) of a process from S_1 or T_1 is simply the ratio of the rates of the process of interest to the total deactivation rate of the state (for T_1, the efficiency of formation of T_1 from S_1 needs to be taken into account). From the conventional state-energy diagram

(Scheme 1.4), we can express the quantum yields (Φ) for transitions from each state as shown in Eqs. 5.28–5.32:

$$\Phi_F = k_F^0/(k_F^0 + k_{ST} + k_{IC}) \tag{5.28}$$

$$\Phi_{IC} = k_{IC}/(k_F^0 + k_{ST} + k_{IC}) \tag{5.29}$$

$$\Phi_{ST} = k_{ST}/(k_F^0 + k_{ST} + k_{IC}) \tag{5.30}$$

$$\Phi_P = \Phi_{ST} \times k_P^0/(k_P^0 + k_{TS}) \tag{5.31}$$

$$\Phi_{TS} = \Phi_{ST} \times k_{TS}/(k_P^0 + k_{TS}) \quad . \tag{5.32}$$

For S_1, Φ_F is equal to the ratio of the rate of fluorescence (k_F^0) to the total rate of deactivation of the S_1 state. Similarly, Φ_{ST} is equal to the ratio of the rate for intersystem crossing (k_{ST}) to the total rate of deactivation for S_1. However, the value of Φ_P not only depends on the ratio of the rate of phosphorescence to the total deactivation rate for T_1 but also depends directly on Φ_{ST}, the probability that T_1 is formed from S_1. For example, if Φ_{ST} is zero, there can be no measurable phosphorescence upon direct excitation of S_1, because the probability of production of T_1 from S_1 is zero. (However, as we see in Chapter 7, even when Φ_{ST} is zero, there is an indirect means of producing T_1 by energy transfer from "triplet sensitizers.")

The singlet lifetime τ_S is equal to the inverse of the sum of all rates that deactivate S_1 (Eq. 5.33) and the triplet lifetime τ_T is equal to the inverse of the sum of all rates that deactivate T_1 (Eq. 5.34).

$$\tau_S = 1/(k_F^0 + k_{ST} + k_{IC}) \quad \text{Experimental } S_1 \text{ lifetime} \tag{5.33}$$

$$\tau_T = 1/(k_P^0 + k_{TS}) \quad \text{Experimental } T_1 \text{ lifetime} \tag{5.34}$$

Thus, with the definition of the inherent radiative lifetimes $\tau_F^0 = (k_F^0)^{-1}$ and $\tau_P^0 = (k_P^0)^{-1}$ the expressions for the key quantum yields of a state energy diagram may be written as shown in Eqs. 5.35–5.39:

$$\Phi_F = k_F^0 \tau_S \tag{5.35}$$

$$\Phi_{IC} = k_{IC} \tau_S \tag{5.36}$$

$$\Phi_{ST} = k_{ST} \tau_S \tag{5.37}$$

$$\Phi_P = \Phi_{ST} k_P^0 \tau_T \tag{5.38}$$

$$\Phi_{TS} = \Phi_{ST} k_{TS} \tau_T \tag{5.39}$$

Experimentally, values of τ_S and τ_T may be evaluated directly by flash photolytic measurement of the decay of S_1 and T_1 as a function of time. In the case of S_1, the most convenient method to monitor $[S_1]$ is usually by directly measuring the fluorescence intensity emitted from S_1 as a function of time after a flash excitation pulse. Thus, the measured value of τ_S is not the pure radiative lifetime τ_F^0, which applies only to situations for which $\Phi_F = 1.0$. Similarly, measurement of the phosphorescence lifetime, τ_P, provides a direct measure of τ_T. In the case of triplet states, $[T_1]$ also may be measured by flash absorption spectroscopy.

Measurement of Φ_F, Φ_P, Φ_{ST}, τ_S, and τ_T allow evaluation of the rate constants k_F^0, k_P^0, k_{IC}, k_{ST}, and k_{TS}. The measurement of Φ_{ST} requires special methods.[29] For certain systems, such as rigid aromatic hydrocarbons at 77 K, internal conversion from S_1 can be neglected, in which case $\Phi_{ST} = 1 - \Phi_F$; that is, every singlet that does not fluoresce is assumed to undergo intersystem crossing. The validity of this assumption depends on the absence of photoreactions, surface crossings, or bimolecular quenching processes of S_1.

5.16 Examples of the Estimation of Rates of Photophysical Processes from Spectroscopic Emission Data

As an example[30] of the computation of radiationless rate constants from spectroscopic emission data, consider the state diagram for 1-chloronaphthalene (Fig. 5.15). At 77 K (see Fig. 4.18c for the emission spectrum of 1-chloronaphthalene), this molecule shows a weak fluorescence ($\Phi_F = 0.06$) and a strong phosphorescence ($\Phi_P = 0.54$). The measured fluorescence and phosphorescence lifetimes are $\sim 10 \times 10^{-9}$ s and ~ 0.3 s, respectively.[30] About 40% of the absorbed photons are not accounted for by emitted photons ($\Phi_F + \Phi_P = 0.60$). Internal conversion ($S_1 \rightarrow S_0$) is not expected to compete with fast intersystem crossing ($S_1 \rightarrow T_1$) for a rigid aromatic molecule, such as naphthalene. With the usual assumption that the quantum yield for internal conversion, $\Phi_{IC} = 0$, from the expression $\Phi_{ST} = 1 - \Phi_F$, we compute $\Phi_{ST} = 1 - 0.06 = 0.94$. Since $\Phi_{ST} = 0.94$, but Φ_P is only 0.54, we deduce that $\Phi_{TS} = \Phi_{ST} - \Phi_P = 0.40$.

From these data and use of Eqs. 5.37 and 5.39, we can compute the rate constants, k_{ST} and k_{TS}, as shown in Eq. 5.40 and 5.41, respectively:

$$k_{ST} = \Phi_{ST}/\tau_S = 0.94/10^{-8} = 9.4 \times 10^7 \text{ s}^{-1} \tag{5.40}$$

$$k_{TS} = \Phi_{TS}/\Phi_{ST}\tau_T = 0.40/0.94)(0.3) = 1.4 \text{ s}^{-1} \tag{5.41}$$

If k_{IC} is at least 10 times smaller than the major rate-determining deactivation pathway of S_1 (i.e., k_{ST}), we may estimate an upper limit to k_{IC} as

$$k_{IC} < 0.1\,k_{ST} \quad \text{or} \quad k_{IC} < 0.1\,k_F^0 \tag{5.42}$$

The inherent or radiative rate constants for emission, k_F^0 and k_P^0, may now be calculated from Eqs. 5.35 and 5.38.

$$k_F^0 = \Phi_F/\tau_S = 0.06/(10^{-8}\text{s}) = 6 \times 10^6 \text{ s}^{-1} \tag{5.43}$$

$$k_P^0 = \Phi_P/\Phi_{ST}\tau_P = 0.54/(0.94)(0.3\text{s}) = 1.9 \text{ s}^{-1} \tag{5.44}$$

As a second example, consider benzophenone (state energy diagram shown in Fig. 5.16). The spectrum of benzophenone is shown in Fig. 4.19a. This molecule is

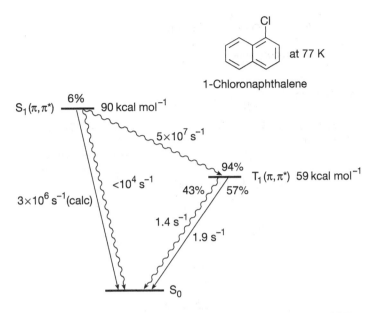

Figure 5.15   State energy diagram for 1-chloronaphthalene at 77 K.

nearly nonfluorescent ($\Phi_F < 10^{-4}$) and possesses a very short singlet lifetime ($\tau_S \sim 10^{-11}$ s). At 77 K, benzophenone[31] shows a strong phosphorescence ($\Phi_P = 0.90$) with a lifetime of 6×10^{-3} s. Since 90% of the photons absorbed by benzophenone are accounted for by phosphorescence, a maximum of 10% of the S_1 molecules can be undergoing $S_1 \rightarrow S_0$ internal conversion. In fact, it appears that nearly every molecule in S_1 undergoes intersystem crossing to T_1. In other words,

$$k_{ST} = \Phi_{ST}/\tau_S = 10^{11} \text{ s}^{-1} \tag{5.45}$$

This remarkably fast intersystem crossing rate is typical of certain carbonyl compounds possessing $S_1(n,\pi^*)$ states with a close-lying $T(\pi,\pi^*)$ state.

The values of k_{TS} and k_P for benzophenone are given by

$$k_{TS} = \Phi_{TS}/\Phi_{ST}\tau_T = 1.7 \times 10 \text{ s}^{-1} \tag{5.46}$$

$$k_P^0 = \Phi_P/\Phi_{ST}\tau_T = 1.5 \times 10^2 \text{ s}^{-1} \tag{5.47}$$

If Φ_F is too weak to measure accurately, the value of k_F^0 may be determined indirectly, that is, by application of the equation relating ε_{max} to τ_F^0 (Eq. 4.18). The value of $\tau_F^0 = 10^{-6}$ s for the pure fluorescence rate, k_F^0, of benzophenone is obtained in this manner.

Experimentally, the observed unimolecular rate constant for a radiationless transition k_{obs} is often found to be composed of a temperature-independent rate constant (k_{obs}^0) and a temperature-dependent rate constant (k_{obs}^T), as shown in Eq. 5.48.

$$k_{obs} = k_{obs}^0 + k_{obs}^T \exp(-E_a/RT) \tag{5.48}$$

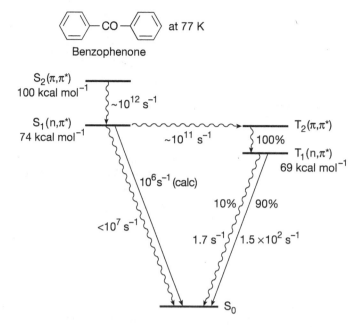

Figure 5.16   State diagram for benzophenone at 77 K.

The temperature-independent part of k_{obs} may be viewed as due to radiationless transitions that occur during the zero-point motion of the molecule. Because quantum zero-point motions cannot be stopped, such transitions occur even at temperatures approaching 0 K! The temperature-dependent part of k_{obs} may be viewed as due to radiationless transitions that require an activation energy (E_a). Such transitions usually obey an Arrhenius relationship, $\exp(-E_a/RT)$. The activation energy may, for example, be associated with an energy barrier required to achieve a loose bolt or free rotor motion that is required to trigger a radiationless transition.

5.17   Internal Conversion ($S_n \rightarrow S_1$, $S_1 \rightarrow S_0$, $T_n \rightarrow T_1$)

The three most important classes of internal conversion commonly encountered for organic molecules are

1. Radiationless transition from an upper excited singlet state (S_n, n > 1) to the lowest excited singlet state, $S_n \rightarrow S_1$(rate constant = k_{IC}^{SS});
2. Radiationless transition from an upper excited triplet state to the lowest-lying triplet state, $T_n \rightarrow T_1$(rate constant = k_{IC}^{TT});
3. Radiationless transitions from the lowest-energy singlet state to the ground singlet state, $S_1 \rightarrow S_0$(rate constant = k_{IC}).

Now, we present some examples of internal conversions and discuss the relationship of the rate of internal conversion to the structures of electronically excited states.

5.18   The Relationship of Internal Conversion to the Excited-State Structure of *R

Let us consider some experimental information concerning internal conversion in aromatic hydrocarbons. We begin with data taken at 77 K in a rigid glass matrix. Under these conditions, photoreactions and bimolecular quenching generally may be completely avoided, and both fluorescence and phosphorescence are readily observed, often with significant quantum yields. In addition, the rigid matrix inhibits radiationless transitions that require loose bolts and free rotors, and the low temperature inhibits radiationless transitions that require passing over energy barriers. Table 5.2 summarizes some pertinent data for the quantum yields of fluorescence (Φ_F), intersystem crossing (Φ_{ST}), and the energy of the lowest singlet (E_{S_1}). The following three sweeping generalizations have been noted for rigid aromatic hydrocarbons and serve as a basis for discussion:[32a]

1. Fluorescence occurs only from S_1 to S_0; phosphorescence occurs only from T_1 to S_0; S_n and T_n emissions are extremely rare (known as Kasha's rule).
2. The quantum yield of fluorescence and the quantum yield of phosphorescence are independent of initial excitation energy (known as Vavilov's rule).
3. The sum of the quantum yields for fluorescence and phosphorescence is approximately equal to unity (known as Terenin's rule).

Table 5.2   Quantum Yields for Fluorescence and Intersystem Crossing of Organic Molecules[a]

Molecule (Configuration of S_1)	Φ_F	Φ_{ST}	$1 - (\Phi_F + \Phi_{ST})$[b]	$E_{S_1}^c$
Benzene (π,π^*)	0.05	0.25	0.70	110
1,3-Dimethylbenzene (π,π^*)	0.3	0.65	<0.05	100
Naphthalene (π,π^*)	0.20	0.80	<0.05	92
Anthracene (π,π^*)	0.70	0.30	<0.05	76
Tetracene (π,π^*)	0.1	0.65	0.20	60
Pentacene (π,π^*)	0.10	0.15	0.75	50
Azulene (π,π^*)	0.000	Low	Low	50
Pyrene (π,π^*)	0.6	Low	<0.05	77
Acetone (n,π^*)	0.001	~1.0	0.05	85
Biacetyl (n,π^*)	0.002	~1.0	0.05	65
Benzophenone (n,π^*)	0.000	~1.0	0.05	75
5-Methyl-2-heptanone (n,π^*)	0.000	0.10	0.90	85
Cyclobutanone (n,π^*)	0.000	0.00	1.0	80
1,3-Pentadiene (π,π^*)	0.000	0.00	1.0	100

a. Values depend on solvent and are temperature dependent. They are therefore only approximate and intended to be representative.

b. A lower limit of 5% is placed on the experimental uncertainty of measurements of Φ. This quantity sets an upper limit to Φ_{IC}.

c. Singlet energy (0,0 energy) in kilocalories per mole.

These rules result from the generally very fast internal conversion from upper electronically excited states to S_1 and or T_1 and very fast vibrational relaxation (10^{12}–10^{13} s^{-1}) within an electronically excited state ($S_n \rightarrow S_1$ and $T_n \rightarrow T_1$ processes). On the other hand, internal conversion from $S_1 \rightarrow S_0$ is much slower (large energy gap and poor Franck–Condon factors), and molecules that are relatively rigid due to structures of environment cannot compete with fluorescence and intersystem crossing. Now, let us see how such a conclusion may be deduced from experimental data. The lack of measurable fluorescence from $S_n (n > 1)$ means that the fluorescence emission yields, $\Phi_F(S_n)$, from these states are $< 10^{-4}$. From the value (Eq. 4.23) of the extinction coefficient $\varepsilon(S_0 \rightarrow S_n)$, we may estimate the radiative rate constant, $k_F^0(S_n)$, for the $S_n \rightarrow S_0 + h\nu$ transition. Knowledge of $k_F^0(S_n)$ and an experimentally detectable limit of Φ_F allow us to compute a limit for the radiationless rate $S_n \rightarrow S_1$.

Consider anthracene (Fig. 4.9) as an example for the estimation of the rate of internal conversion from upper excited singlet states. We can estimate the rate of fluorescence from upper singlet states (S_n) even though such fluorescence is not observed experimentally. For example, the $S_0 \rightarrow S_3$ absorption of anthracene maximizes at 252 nm with $\varepsilon_{max} \sim 2 \times 10^5$. From Eq. 4.27, we have Eq. 5.49.

$$k_F^0 \sim 10^4 \varepsilon_{max} \sim 2 \times 10^9 \text{ s}^{-1} \tag{5.49}$$

Since

$$\Phi_F^{S_2} = k_F^0 / k_D < 10^{-4} \tag{5.50}$$

$$k_D > 10^4 k_F^0 \sim 2 \times 10^{13} \text{ s}^{-1} \tag{5.51}$$

In other words, S_3 deactivates to S_1 with a rate constant of $\sim 10^{13}$ s^{-1}, which is on the order of the rate for IVR. Since *only emission* from S_1 is observed when the S_3 of anthracene is excited, we must conclude that $k_{IC}(S_3 \rightarrow S_1) \sim 10^{13}$ s^{-1}. For most organic molecules, $k_{IC}(S_n \rightarrow S_1)$ falls in the range 10^{11}–10^{13} s^{-1}. Evidently, electronic relaxation by internal conversion from the upper levels is rate limited only by nuclear vibrational motion and not by electronic relaxation. This result in turn suggests that zero-order crossings are common for $S_n (n > 1)$ states and that critical geometries may be readily achieved during vibrational motion of S_n in its $v = 0$ level.

Because the fluorescence quantum yield for anthracene is significant ($\Phi_F \sim 0.7$) from S_1, the $S_1 \rightarrow S_0$ internal conversion must, at best, be weakly competitive only with other modes of decay from S_1. In Table 5.2, we noted that at 77 K, $\Phi_F + \Phi_{ST} \sim 1$ for many aromatic hydrocarbons possessing rigid structures. Consequently, the internal conversion $S_1 \rightarrow S_0$ cannot occur to more than a few percent (the experimental error of measuring Φ) for these compounds, as expected from earlier discussions.

As another exemplar,[32b] since the singlet decay of pyrene (Fig. 4.15) is $\sim 10^6$ s^{-1}, and since $\Phi_F + \Phi_{ST} \sim 1.00$, we must conclude that $k_{IC}(S_1 \rightarrow S_0) < 10^6$ s^{-1}. Thus, a factor of $\sim 10^6$ separates the typical rate of an $S_n \rightarrow S_1$ internal conversion from the $S_1 \rightarrow S_0$ internal conversion for pyrene.

Relatively few data exist on the direct measurement of the rates of $T_n \rightarrow T_1$ internal conversions. The reason for this lack of data may be technical rather than theoretical in nature. One might expect that $T_2 \rightarrow T_1$ *fluorescence* should occur with a *range* of efficiencies, as does $S_1 \rightarrow S_0$ fluorescence. However, $T_2 \rightarrow T_1$ energy gaps generally are on the order of 30 kcal mol^{-1} or *smaller*. This means that $T_2 \rightarrow T_1$ fluorescence must compete with efficient internal conversion, which can now be quite fast because of the relatively small energy gap between T_2 and T_1 (Section 5.19). In addition, even if such fluorescence occurred efficiently, it would appear at wavelengths greater than ~ 800 nm, a wavelength that is technically difficult to detect with high sensitivity.

Despite the difficulty in observing $T_2 \rightarrow T_1$ fluorescence experimentally, a very weak $T_2 \rightarrow T_1$ fluorescence has been observed in a few cases. For example,[33] 9,10-dibromoanthracene (DBA) displays a weak ($\Phi_F \sim 10^{-6}$) $T_2 \rightarrow T_1$ fluorescence. From the extinction coefficient for $T_1 \rightarrow T_2$ absorption, the value of $k_F^0(T_2 \rightarrow T_1)$ can be derived, and it is found to be $\sim 10^5$ s^{-1}. From the value of Φ_F and Eq. 5.27, we deduce that $k_{IC}(T_2 \rightarrow T_1) \sim 10^{11}$ s^{-1}, a very reasonable value considering the rather large energy gap between the interconverting states. Indirect evidence also supports a k_{IC} value of $\sim 10^{11}$ s^{-1} for DBA.[34]

5.19   The Energy Gap Law for Internal Conversion ($S_1 \rightarrow S_0$)

In the absence of a zero-order surface crossing between S_1 and S_0, an $S_1 \rightarrow S_0$ internal conversion must occur via a "Franck–Condon *forbidden*" mechanism; that is, the nuclei in one state must undergo a rather drastic change in position and momentum as a result of the transition, since the net overlap of vibrational wave functions in both states is small (Figs. 3.5 and 5.13).[24,35] For such situations, the $S_1 \rightarrow S_0$ internal conversion is generally rate limited by the FC factor, $< \chi | \chi >^2 = f_\nu$. If we take 10^{13} s^{-1} as an order-of-magnitude estimate of the maximum rate of internal conversion, then from Eq. 5.5, we have Eq. 5.52.

$$k_{IC} \sim 10^{13} f_\nu \qquad (5.52)$$

It is possible to estimate f_ν from spectral data.[36] Both theoretical and experimental evidence demonstrate that f_ν is a very sensitive function of the *energy gap*, ΔE, between the zero-point vibrational levels of the states undergoing internal conversion.[35] An expression for k_{IC} is obtained from Eq. 5.17 and Eq. 5.52:

$$k_{IC} \sim 10^{13} \exp(-\alpha \Delta E) \qquad (5.53)$$

(where α is a proportionality constant). The energy gap law can be attributed to the changes in the Franck–Condon factors, which become increasingly unfavorable (at an exponential rate) with increasing energy separation (ΔE).

Experimentally, as we have seen from the data in Table 5.2, $S_1 \rightarrow S_0$ internal conversion is usually negligible relative to fluorescence or intersystem crossing for nonphotoreactive, relatively rigid molecules if $\Delta E(S_1 \rightarrow S_0)$ is larger than

~ 50 kcal mol^{-1}. For example, for $\Delta E \sim 100$ kcal mol^{-1}, $f_v \sim 10^{-8}$, so that $k_{IC} \sim 10^5$ s^{-1}. For $\Delta E \sim 50$–60 kcal mol^{-1}, $f_v \sim 10^{-5}$, so that $k_{IC} \sim 10^8$ s^{-1}. Since the *slowest* rates of $S_1 \rightarrow S_0$ fluorescence are generally $> 10^6$ s^{-1}, and since the slowest rates of $S_1 \rightarrow T_1$ intersystem crossing are generally $> 10^6$ s^{-1}, we see that internal conversion is unlikely to compete favorably with fluorescence or intersystem crossing if $\Delta E(S_1 \rightarrow S_0)$ is > 50 kcal mol^{-1}, unless there is some vibrational motion that is favorable for inducing the radiationless transition (i.e., a free rotor or loose bolt in the molecule).

From the above discussion, we have a basis for Ermolev's rule (Eq. 5.54)[37].

$$\Phi_F + \Phi_{ST} = 1 \text{ because } 1 - (\Phi_F + \Phi_{ST}) \sim \Phi_{IC} \qquad (5.54)$$

For example, tetracene and pentacene (Table 5.2) possess relatively low singlet energies (~ 60 and 50 kcal mol^{-1}, respectively) and undergo significant internal conversion from S_1($\Phi_{IC} \sim 0.20$ and 0.75, respectively). The large value of Φ_{IC} for benzene (0.80) is probably due to a reversible photoreaction or a structurally specific surface crossing of the S_1 and S_0 surfaces. In the case of the last three entries in Table 5.2 (5-methyl-2-heptanone, cyclobutanone, and 1,3-pentadiene), photochemical reaction from S_1 probably accounts for a significant fraction of Φ_{IC}. In an extreme case, azulene ($E_S \sim 40$ kcal mol^{-1}) undergoes internal conversion from S_1 with a rate on the order of $\sim 10^{12}$ s^{-1}. In this case, a relatively small $S_0 - S_1$ energy gap and presence of a nearby conical intersection contribute to the unusually rapid rate of internal conversion.

5.20 The Deuterium Isotope Test for Internal Conversion

There is an experimental deuterium "isotope test" available for the origin of the energy gap law of Eq. 5.51. The Franck–Condon factors are generally *greatest* for high-frequency vibrations,[24,35] which is the case because the higher the energy of a vibration, the fewer the number of quanta is required to match an electronic gap with vibrational energy. The highest-frequency vibrations in organic molecules (Table 5.1) often correspond to C—H stretching motions (~ 3000 cm^{-1}). Thus, we expect that electronic–vibrational energy transfer will be fastest for "leakage" through C—H vibrations that serve as loose bolts. If C—H vibrations are replaced by lower-energy C—D vibrations (~ 2200 cm^{-1}), the rate of electronic to vibrational energy transfer should be slowed down substantially. Thus, we predict that if $S_1 \rightarrow S_0$ occurs via an electronic–vibronic mechanism, the lifetime of $S_1(\tau_S)$ will be *increased* by the substitution of C—D for C—H, because k_{IC} will decrease (Eq. 5.33). Thus, we have an *isotope test* for internal conversion: If k_{IC} is rate determining in deactivating *R, the substitution of D for H in the molecule will lead to a decrease in the rate of deactivation of *R.

Experimentally, the replacement of C—H bonds by C—D bonds in aromatic hydrocarbons[38] generally does not change the singlet-state lifetime or the fluorescence yield significantly, confirming that k_{IC} is not a significant contributor to the deactivation of S_1. For example, at 77 K both pyrene-h_{10} and pyrene-d_{10} possess a fluorescence yield of 0.90 and singlet lifetimes (τ_S) of 450 ns.[38c] Since $\tau_S = (k_F^0 + k_{ST} + k_{IC})^{-1}$ and τ_S (pyrene-h_{10}) = τ_S (pyrene-d_{10}), we may conclude that $k_F^0 + k_{ST} \gg k_{IC}$, because a large decrease in k_{IC} is expected upon going from the h_{10} to d_{10} compound. The important point here is that although there is no significant deuterium isotope effect on τ_S or Φ_F for pyrene, internal conversion ($S_1 \rightarrow S_0$) cannot contribute significantly to the decay of S_1, as expected from the energy gap law (Section 5.19). In Section 5.27, we see that a large deuterium effect does operate on $T_1 \rightarrow S_0$ intersystem crossing.

In contrast to the small influence of the substitution of D for H on τ_S and Φ_F for aromatic hydrocarbons, this substitution may cause a significant enhancement for ketones and aldehydes. The effect of deuteration on Φ_F is most dramatic for aldehydes, especially in the vapor phase.[39]

For example, the fluorescence quantum yield of formaldehyde increases by a factor of ~ 20 upon going from $H_2C{=}O$ to $D_2C{=}O$.[39] Evidentially, the substitution of D for H greatly decreases the magnitude of k_{ST} or k_{IC}. It is not yet clear whether a spin–orbit or FC inhibition is involved.

Although the effect of deuteration on acetone[40] is less striking (τ_S is 1.17 ns for acetone-h_6 and 2.3 ns for acetone-d_6), it is nonetheless significant. In this case, it appears that intersystem crossing may be specifically slowed down by deuteration, possibly because of a decrease in the value of the FC term f_v in Eq. 5.5.

5.21   Examples of Unusually Slow $S_n \rightarrow S_1$ Internal Conversion

Azulene (**9**) and its derivatives provide a striking exception to the general rule that $S_2 \rightarrow S_1$ internal conversion completely dominates fluorescence (in condensed phases).[41] For **9**, Φ_F for $S_2 \rightarrow S_0$ fluorescence is ~ 0.03 (see Section 5.14), in violation of Kasha's rule. For some fluorinated derivatives,[42] the value of Φ_F for $S_2 \rightarrow S_0$ fluorescence is ~ 0.2! The observation of $S_2 \rightarrow S_0$ fluorescence for azulene results from the fact that its rate of internal conversion ($S_2 \rightarrow S_1$) is exceptionally slow ($k_{IC} \sim 7 \times 10^8$ s^{-1}). This finding is consistent with the exceptionally large energy gap of ~ 40 kcal mol^{-1} between the 0,0 levels of S_2 and S_1 for azulene (Eq. 5.53); that is, the $S_2 \rightarrow S_1$ internal conversion possesses a relatively poor FC factor (Eq. 5.52).

Azulene **9**

Azulene and its derivatives are also remarkable for their exceptionally rapid rate of $S_1 \rightarrow S_0$ internal conversion (Table 5.2). Direct measurements[43] indicate that $k_{IC}(S_1 \rightarrow S_0)$ is $\sim 10^{12}$ s^{-1}. Such a large rate constant is consistent with a small value of E_{S1} and/or a surface crossing of the S_1 and S_0 surfaces near the $v = 0$ level of S_1.

5.22   Intersystem Crossing from $S_1 \rightarrow T_1$

Table 5.3 shows the "spread" of measured values for k_{ST}, the rate constant for intersystem crossing from S_1 to T_1. First, note that the smallest values of k_{ST} ($\sim 10^6$ s^{-1}) occur for aromatic hydrocarbons (pyrene, naphthalene) as expected from El-Sayed's rule,[11] because of weak spin–orbit coupling for systems undergoing $S_1(\pi,\pi^*)$ to $T_1(\pi,\pi^*)$ ISC (Section 5.10). The largest values of k_{ST} ($\sim 10^{10}$–10^{11} s^{-1}) occur for molecules containing heavy atoms (e.g., bromonaphthalene) or for molecules possessing $S_1(n,\pi^*)$ states that can mix effectively with $T_n(\pi,\pi^*)$ states (e.g., benzophenone). However, other factors in addition to the "heavy atom" or "n,π^*" effects also influence the value of k_{ST}. In particular, as an extension of the idea behind Eq. 5.53, k_{ST} is expected to depend on the energy gap, ΔE_{ST}, between S_1 and the state to which intersystem crossing actually occurs (i.e., T_1 or some *upper* triplet, T_n).

The $S_1 \rightarrow T_1$ transition may occur via (1) direct spin–orbit coupling of S_1 to the upper vibrational levels of T_1, or (2) spin–orbit coupling of S_1 to an upper T_n state followed by rapid $T_n \rightarrow T_1$ internal conversion.

For case 1, we expect the measured value of k_{ST} to depend on the energy gap between S_1 and T_1, whereas for case 2 the energy gap between S_1 and T_1 should not be significant in determining the value of k_{ST}. In addition to the singlet–triplet energy gap, we expect that vibrational motion in the S_1 state is required to allow the molecule to explore different conformations as it searches for an effective spin–orbit coupling mechanism.

Table 5.3   Representative Values of Intersystem Crossing Rates ($S_1 \rightarrow T_1$) and Singlet–Triplet Energy Gaps

Molecule	k_{ST}(s^{-1})[a]	ΔE_{ST} (kcal mol^{-1})	Transition
Naphthalene	10^6	30	$S_1(\pi,\pi^*) \rightarrow T_1(\pi,\pi^*)$
Pyrene	10^6	30	$S_1(\pi,\pi^*) \rightarrow T_1(\pi,\pi^*)$
Triphenylene	5×10^7	30	$S_1(\pi,\pi^*) \rightarrow T_1(\pi,\pi^*)$
1-Bromonaphthalene	10^9	30	$S_1(\pi,\pi^*) \rightarrow T_1(\pi,\pi^*)$
9-Acetoanthracene	$\sim 10^{10}$	~ 5	$S_1(\pi,\pi^*) \rightarrow T_2(n,\pi^*)$
Perylene	$< 10^8$	~ 30	$S_1(\pi,\pi^*) \rightarrow T_1(\pi,\pi^*)$
3-Bromoperylene	$< 10^8$	~ 30	$S_1(\pi,\pi^*) \rightarrow T_1(\pi,\pi^*)$
Acetone	5×10^8	5	$S_1(n,\pi^*) \rightarrow T_1(n,\pi^*)$
Benzophenone	10^{11}	5	$S_1(n,\pi^*) \rightarrow T_2(\pi,\pi^*)$
Benzil	5×10^8	5	$S_1(n,\pi^*) \rightarrow T_1(n,\pi^*)$
Biacetyl	7×10^7	5	$S_1(n,\pi^*) \rightarrow T_1(n,\pi^*)$
Anthracene	1×10^8	small	$S_1(\pi,\pi^*) \rightarrow T_2(\pi,\pi^*)$
9,10-Dibromoanthracene	1×10^8	30	$S_1(\pi,\pi^*) \rightarrow T_1(\pi,\pi^*)$
			$S_1(\pi,\pi^*) \rightarrow T_2(\pi,\pi^*)$
9-Chloroanthracene	5×10^8	small	$S_1(\pi,\pi^*) \rightarrow T_2(\pi,\pi^*)$
9,10-Dichloroanthracene	1×10^8	30	$S_1(\pi,\pi^*) \rightarrow T_1(\pi,\pi^*)$
[2.2.2]-Diazabicyclooctane	$\sim 10^6$	25	$S_1(n,\pi^*) \rightarrow T_1(n,\pi^*)$

5.23   The Relationship Between $S_1 \to T_1$ Intersystem Crossing and Molecular Structure

How can k_{ST} be related to molecular structure, electronic configurations, and energy levels? Let us start with aromatic hydrocarbons as exemplars, since in this case we are basically dealing only with π,π^* excited states for *R. An important empirical observation (Table 5.3) is that, at 77 K, nearly *all* measured values of k_{ST} for aromatic *hydrocarbons* fall in the range $\sim 10^6$–10^8 s^{-1}. This range of rates is comparable to that of fluorescence (k_F^0) from aromatic hydrocarbons, $\sim 10^6$–10^9 s. As a result, most aromatic hydrocarbons exhibit a measurable amount of fluorescence. Significant intersystem crossing occurs if k_F^0 is not maximal and if the measurements are made at low temperatures in a rigid medium.

To exemplify how factors influencing the efficiency of $S_1 \to T_1$ transitions operate, let us compare the values of k_{ST} for the aromatic hydrocarbons anthracene[44] and pyrene.[45] In each case, an $S_1(\pi,\pi^*) \to T_n(\pi,\pi^*)$ transition occurs. However, a variation factor of ~ 100 is noted in the value of k_{ST} (k_{ST} for anthracene $\sim 10^8$ s^{-1}; k_{ST} for pyrene $\sim 10^6$ s^{-1}). The possibilities for this variation in k_{ST} are (a) differing degrees of electronic coupling between S_1 and the triplet state to which intersystem crossing occurs, (b) a larger size of the energy gap between S_1 and the triplet state, T_n, to which intersystem crossing occurs for pyrene (which would, of course, reduce any spin–orbit coupling compared to anthracene), and (c) differing degrees of spin–orbit coupling between S_1 and the triplet state to which crossing occurs. The variation may be qualitatively explained on the basis of (b). In the case of pyrene, it appears that S_1 crosses directly to an excited vibrational level of T_1.[45] The energy gap ΔE_{ST} is ~ 30 kcal mol^{-1}. In the case of anthracene and substituted anthracenes, S_1 *may cross to T_2, which is nearly isoenergetic with S_1.*[46] Thus, in anthracene a small energy gap and consequently a favorable Franck–Condon factor exist for intersystem crossing. Because the energy gap between S_1 and T_2 of anthracene is very small, substituents on the anthracene and solvent effects on a given anthracene can cause significant variations in the quantum yield of fluorescence even thought the radiative rate of fluorescence does not change significantly with structural variations or solvent. The reason for the wide variation is that k_{ST} varies significantly, depending on whether T_2 is higher or low in energy than S_1, whereas the value of k_F, which is independent of spin, remains constant. For example, anthracene and 9,10-dibromoanthracene have similar values of k_{ST}. This result is surprising, since there is no heavy-atom effect on k_{ST}. As surprising is the observation that although k_{ST} is larger for 9-chloroanthracene than for anthracene (as expected for a heavy atom effect), the value of k_{ST} *decreases* for 9,10-dichloroanthracene. These results are explained as the result of the variation of the energy of T_2 as a function of substituent and the effect of the energy of S_1 versus the energy of T_2. Solvent effects can also influence the value of T_2 relative to S_1, which makes the fluorescence of certain anthracenes solvent dependent through the variation of k_{ST} relative to k_F.

According to theory,[47] mixing of π,σ^* and σ,π^* triplet states with $S_1(\pi,\pi^*)$ can produce spin–orbit coupling and thereby imbue S_1 of aromatic hydrocarbons with "triplet character." This mixing must be vibronically induced so that mixing is relatively ineffective for molecules possessing a rigid structure in S_1. However, the energies of the π,σ^* and σ,π^* triplet states are generally too high for effective mixing with $S_1(\pi,\pi^*)$. This poor mixing causes small rate constants for $S_1 \rightarrow T_1$ (and $T_1 \rightarrow S_0$) intersystem crossing for most aromatic hydrocarbons.

With respect to $S_1(n,\pi^*)$ states, a similar situation appears to hold with respect to the value of k_{ST} and the energy gap when $S_1(n,\pi^*) \rightarrow T_1(n,\pi^*)$ transitions (forbidden by El-Sayed's rule,[11] Fig. 5.8) are involved. Recall that $S_1(n,\pi^*) \rightarrow T_1(n,\pi^*)$ transitions do not generate spin–orbit coupling in zero order. Thus, $S_1(n,\pi^*) \rightarrow T_1(n,\pi^*)$ intersystem crossings for both acetone ($k_{ST} = 5 \times 10^8$ s^{-1})[48] and biacetyl ($k_{ST} = 1 \times 10^8$ s^{-1})[49] involve "forbidden" spin–orbit mechanisms in spite of a small (~ 5 kcal mol^{-1}) ΔE_{ST} and (presumably) a favorable FC factor. Recall that a small value of k_{ST} leads to the expectation of a strong fluorescence for aromatic hydrocarbons. Aliphatic ketones (e.g., acetone) possess a relatively small fluorescence rate constant ($k_F \sim 10^5$ s^{-1}), due to the orbital forbiddenness of the $n \rightarrow \pi^*$ transition. The net result for aliphatic ketones is that typically $k_{ST} \gg k_F^0$; as a result, $\Phi_{ST} \sim 1$ (Table 5.3). However, cyclic azoalkanes that possess a large energy gap between S_1 and T_1 ($\Delta E_{ST} \sim 25$ kcal mol^{-1}) encounter a poorer Franck–Condon factor, and k_{ST} is thereby considerably smaller.[50] Consequently, some cyclic azoalkanes exhibit a very high Φ_F because of a relatively slow rate of intersystem crossing.

The largest values of k_{ST} for organic molecules not possessing heavy atoms are found for systems undergoing $S_1(n,\pi^*) \rightarrow T_1(\pi,\pi^*)$ transitions (allowed by El-Sayed's rule,[11] Fig. 5.8) with small energy gaps, for example, benzophenone[13,51] [$k_{ST} = 10^{11}$ s^{-1}, $S_1(n,\pi^*) \rightarrow T_1(\pi,\pi^*)$] and 9-benzoyl anthracene[52] [($k_{ST} = 10^{10}$ s^{-1}, $S_1(\pi,\pi^*) \rightarrow T_1(n,\pi^*)$].

The magnitudes of k_{ST} for alkyl ketones are sensitive to subtle features of molecular structure. For example, for acetone[48] $k_{ST} \sim 5 \times 10^8$ s^{-1}, whereas for di-*tert*-butyl ketone,[53] $k_{ST} \sim 10^8$ s^{-1}, and for perfluoroacetone,[48] $k_{ST} \sim 10^7$ s^{-1}. This variation suggests a decreasing amount of spin–orbit coupling or decreasing Franck–Condon factor as one proceeds from acetone to *tert*-butyl-ketone to perfluoroacetone. The observation of a deuterium isotope effect[40] on k_{ST} and the rather large influence of substitution of fluorine are consistent with reduced FC factors.

5.24   Temperature Dependence of $S_1 \rightarrow T_n$ Intersystem Crossing

The fluorescence yield (Φ_F) and singlet lifetimes (τ_S) of organic molecules are sometimes found to vary with temperature. Since k_F^0 is generally temperature independent,[32] some radiationless process from S_1 must be temperature dependent. Indeed, photoreactions from S_1 commonly have small energy barriers, and therefore will possess temperature-dependent rate constants. Intersystem crossing ($S_1 \rightarrow T_n$) or internal conversion ($S_1 \rightarrow S_0$) also may be temperature dependent if upper vibrational levels of S_1 possess a different mechanism for radiationless transition than the $v = 0$ level.

For example, the rate constant for intersystem crossing can be expressed (cf. this equation to Eq. 5.48) as

$$k_{ST}^{OBS} = k_{ST} + A\exp(-E_a/RT) \qquad (5.55)$$

where A is the frequency factor. Experimentally, k_{ST}^{OBS}, the observed rate constant, is sometimes found to be essentially temperature independent below a certain temperature and to follow Eq. 5.55 above that temperature. A common mechanism for this temperature dependence is thermally activated $S_1 \rightarrow T_n (n \neq 1)$ intersystem crossing.

The rate of intersystem crossing in certain anthracene derivatives is temperature dependent.[46] This observation has been explained in terms of a temperature-dependent rate of intersystem crossing from S_1 to T_2 described in Section 5.23. For example, the value of k_{ST} for 9,10-dibromoanthracene may be expressed as $k_{ST} \sim 10^{12} \exp(-E_a/RT)$, where $E_a \sim 4.5$ kcal mol^{-1}. A large energy gap between S_1 and T_1 serves to slow down the direct $S_1 \rightarrow T_1$ intersystem crossing because of an unfavorable Franck–Condon factor. That T_2 is populated, rather than an upper vibration level of T_1, is supported by triplet–triplet absorption measurements, which demonstrate that T_2 lies ~ 4–5 kcal mol^{-1} above S_1. Since the value of E_a is ~ 4.5 kcal mol^{-1}, it is logical to suppose that an activated $S_1 \rightarrow T_2$ process is involved in the temperature-dependent intersystem crossing of DBA (see Section 5.29 for further discussion of the $S_1 \rightarrow T_2$ intersystem crossing in anthracenes).

It is interesting to note that Φ_F and τ_S are usually not very temperature dependent for temperatures below ~ 100 K. For example,[54] Φ_F of naphthalene is ~ 0.3 at both 77 and 4 K. This implies that $k_{ST}(S_1 \rightarrow T_1)$ is temperature independent in the range from 4 to 77 K. This finding, in turn, suggests that the energy term in Eq. 5.55 becomes small relative to k_{ST} at temperatures < 100 K.

5.25   Intersystem Crossing ($T_1 \rightarrow S_0$)

Of the three key important radiationless processes, $S_1 \rightarrow S_0$, $S_1 \rightarrow T_1$, and $T_1 \rightarrow S_0$, only in the last case is there no electronic state that lies between the initial and final states, T_1 and S_0. Thus, phosphorescence and intersystem crossing are both derived from a common electronic $T_1 \rightarrow S_0$ transition.

5.26   The Relationship between $T_1 \rightarrow S_0$ Intersystem Crossing and Molecular Structure

As in the case of $S_1 \rightarrow S_0$ internal conversion, we expect an energy gap law (Eqs. 5.17 and 5.53) and a deuterium isotope effect if $T_1 \rightarrow S_0$ occurs via an electronic–vibrational (vibronic) mechanism. In addition to vibronic effects that are responsible for the energy gap law, spin–orbit coupling may be important. It is not *a priori* obvious whether Franck–Condon (f_v) or spin–orbit factors (f_s) will determine the ultimate value of k_{TS}.

Table 5.4   Some Representative Values of Triplet Energies, Phosphorescence Radiative Rates, Intersystem Crossing Rates, and Phosphorescence Yields[a]

Molecule	E_T	k_P^0	k_{TS}	Φ_P
Benzene-h_6	85	~0.03	0.03	0.20
Benzene-d_6	85	~0.03	<0.001	~0.80
Naphthalene-h_8	60	~0.03	0.4	0.05
Naphthalene-d_8	60	~0.03	<0.01	~0.80
$(CH_3)_2C{=}O$	78	~50	1.8×10^3	0.043
$(CD_3)_2C{=}O$	78	~50	0.6×10^3	0.10

a. In organic solvents at 77 K. E_T in kcal mol^{-1}, k, in s^{-1}.

However, for aromatic hydrocarbons, a clear-cut relationship between k_{ST} and the energy of the triplet state, $E(T_1)$, is found.[35] As for the $S_1 \rightarrow S_0$ process, we expect the high-frequency C—H vibrations to serve as the major acceptor vibrations for leakage of electronic into vibrational energy.[24] Indeed, a plot of the log k_{ST} versus $E(T_1)$ (corrected for the number of C—H vibrations) is linear.[35] This result provides strong evidence that the electronic energy of T_1 "leaks out" via C—H vibrations for rigid aromatic hydrocarbon.

In addition to FC factors, the value of k_{ST} (Table 5.4) can be influenced by increased spin–orbit coupling due to the heavy-atom effect, or by the occurrence of a $T_1(n, \pi^*) \rightarrow S^0$ transition. Thus, for naphthalene, $k_{TS} \sim 0.4$ s^{-1}, while for 1-bromonaphthalene $k_{TS} \sim 100$ s^{-1}. Ketones possessing $T_1(n,\pi^*)$ undergo the fastest $T \rightarrow S_0$ crossings yet measured ($k_{TS} \sim 10^3$ s^{-1}) for organic molecules that do not possess heavy atoms.

5.27 The Energy Gap Law for $T_1 \rightarrow S_0$ Intersystem Crossing: Deuterium Isotope Effects on Interstate Crossings

Importantly, the *same mathematical form of the energy gap law* for internal conversion (an exponential dependence on the energy gap, ΔE, Eq. 5.53) also applies to T_1 to S_0 intersystem crossing.[35] Accordingly, deuterium isotope effects are expected on the rate constant k_{TS}. Indeed, dramatic deuterium isotope effects are found for radiationless $T_1 \rightarrow S_0$ transitions at 77 K.[24,35] These results contrast sharply with those for $S_1 \rightarrow T_1$ transitions. Presumably, the difference lies in the much smaller energy gap required for $S_1 \rightarrow T_1$ transitions compared to $T_1 \rightarrow S_0$ transition, and the greater probability of surface crossings between S_1 and T_1. For example (Table 5.4), the lifetimes of naphthalene triplets at 77 K increase from ~2 s to ~20 s upon substitution of C—H for C—D, and the lifetimes of acetone triplets increase from 0.56 to 1.7 ms upon substitution of C—H for C—D.

The radiative lifetime of the triplet state of deuterated hydrocarbons nearly equals the maximum radiative lifetime (i.e., in the deuterated materials nearly every triplet emits, whereas only a small fraction of the aromatic triplets that possess C—H bonds emit.)[55,56] This striking result derives from FC factors; in other words, f_ν is much smaller for C—D vibrations than for C—H vibrations. The value of f_ν for C—D vibrations is ~ 20–30 times smaller than f_ν for C—H vibrations.[8b] For example, the triplet states of both C_6H_6 and C_6D_6 lie at 85 kcal mol^{-1} above S_0. This finding corresponds to ~ 10 vibrational quanta for C—H vibrations. The lower amplitude of the C—D vibrations requires a larger number of vibrational quanta to equal 85 kcal mol^{-1}. Therefore, the vibrational level of S_0 that is reached when the deuterated material converts from T_1 to S_0 possesses a large vibrational amplitude, and intersystem crossing from T_1 to S_0 is inhibited. In terms of the language of the loose bolt, a C—H stretching vibration is a "looser bolt" than a C—D stretching vibration. This finding means that the C—H vibration is better able to trigger the $T_1 \rightarrow S_0$ intersystem crossing than a C—D vibration.

5.28   Perturbation of Spin-Forbidden Radiationless Transitions

The term *heavy-atom effect* has been coined to describe the influence of heavy-atom substitution on spin-forbidden radiative and radiationless transitions.[19] For practical purposes, the atoms in the first full row of the periodic table (e.g., C, N, O, F) are considered to be "light" atoms. It is usually assumed that the dominant influence of the heavy-atom effect influences, to a certain extent, all photophysical spin-forbidden transitions (radiationless, $S_1 \rightarrow T_1$, $T_1 \rightarrow S_0$, and radiative, $S_0 \rightarrow T_1$, $T_1 \rightarrow S_0$) but does not significantly influence spin-allowed transitions (internal conversion, $S_1 \rightarrow S_0$, and fluorescence, $S_1 \rightarrow S_0$). In this approximation, the rate constants k_{ST}, k_{TS}, and k_P^0 should all be increased to a certain extent by the heavy-atom effect, but k_F^0 and k_{IC} should not. In addition, the value of the extinction coefficient for singlet–triplet absorption, $\varepsilon(S_0 \rightarrow T_1)$, should be significantly enhanced by the heavy-atom effect, but the extinction coefficient for singlet–singlet absorption $\varepsilon(S_0 \rightarrow S_1)$ should not be significantly influenced by the heavy-atom effect.

From Eqs. 5.35–5.39, we expect that the heavy-atom effect will generally:

1. Decrease Φ_F (when k_{ST} is competitive with k_F^0 for the light-atom analogue).
2. Increase Φ_{ST} (when k_{ST} is competitive with k_F^0 compared to the light-atom analogue).

On the other hand, Φ_P and Φ_{TS} may either increase or decrease with heavy-atom substitution, depending on whether k_P^0 or k_{TS} (both of which are influenced by the heavy-atom effect) is influenced to the greater extent.

From this analysis, we deduce that heavy-atom effects will not be universal. They will occur only when the heavy atom increases k_{ST} (k_P^0 or k_{TS}) to a value that alters τ_S (or τ_T). Whether the heavy-atom effect is manifested in an experimental system depends on whether the additional spin–orbit coupling introduced by the heavy atom is comparable to the inherent spin–orbit coupling in a molecule, and on the rates of deactivation of the light-atom analogue.

Empirically, the maximum heavy-atom effect that is typically induced on k_{ST} by bromine atom substitution corresponds to a rate on the order of 10^8–10^9 s^{-1}. As a result, if k_F^0 or k_{ST} are on the order of 10^9 s^{-1} or greater, a bromine atom may not produce a significant effect on the fluorescence lifetime or quantum yield. In practice, this means that heavy-atom effects on $S_1 \to T$ intersystem crossing tend to be minimal for states that already possess substantial spin–orbit coupling (e.g., n,π^* states coupled to π,π^* states), possess very fast fluorescence rates, or have no nearby triplet states with which to mix (e.g., perylene). A striking example of the latter is the fact that tetrabromoperylene has a fluorescence yield of ~ 1.0.

5.29   Internal Perturbation of Intersystem Crossing by the Heavy-Atom Effect

As an example of the heavy-atom effect on k_{ST}, k_{TS}, and k_P^0, consider the data in Table 5.5. The heavy-atom effect is most pronounced on S_1 when the comparison light-atom structure possesses both inherently slow intersystem crossing and a slow rate of fluorescence deactivation. A classical exemplar is given by naphthalene and its halogen derivatives. Both k_F^0 and k_{ST} are $\sim 10^6$ s^{-1} for the light-atom parent naphthalene. These are relatively small values for each of these processes. Substitution of F (considered a light atom) for H has, as expected, relatively little effect on the emission efficiencies or rate constants of fluoronaphthalene relative to naphthalene. However, substitution of Cl, then Br, then I leads to an ever-increasing decrease in Φ_F, and an accompanying increase in k_{ST} and k_P^0. The effect on Φ_P is not as readily predictable because both k_P^0 and k_{TS} are influenced by the heavy atom.

As a second exemplar of the *lack of a heavy-atom effect*, perylene possesses a very fast and efficient fluorescence ($\Phi_F \sim 0.98$, $k_F^0 \sim 10^8$ s^{-1}).[45] If the heavy-atom effect were the same for naphthalene and for perylene, then substitution of Br for H would increase k_{ST} to $\sim 10^9$ s^{-1}. The lack of a heavy-atom effect on k_F^0 when going from perylene to bromoperylene can be understood in terms of a very fast inherent fluorescence, which dominates the usual heavy enhancement of k_{ST}.

In going from anthracene to 9-bromoanthracene, a dramatic decrease in Φ_F and increase in k_{ST} is noted, as expected, from the heavy-atom effect. Remarkably, substitution of a second bromine atom (9-bromoanthracene $\to$ 9,10-dibromoanthracene, DBA) results in an *increase* in Φ_F and a *decrease* in k_{ST}. This "inverse" heavy-atom effect is explained in terms of the influence of halogen substitution on the position of T_2 relative to S_1 for anthracenes. In anthracene, k_{ST} occurs via an $S_1 \to T_1$ mechanism because T_2 is higher in energy than S_1. However, for 9-bromoanthracene, T_2 is lowered in energy so that it falls below or closer to the energy of S_1. Thus, an $S_1 \to T_2$ mechanism for intersystem crossing becomes available, and the heavy-atom effect operates.

For DBA, however, the second bromine atom causes the T_2 state to become higher in energy than the S_1 state.[46] Thus, in spite of the presence of two bromine atoms, DBA still possesses a modest fluorescence yield ($\Phi_F \sim 0.05$). This implies that the heavy-atom effect does not dominate and bring about very rapid intersystem crossing. Furthermore, the fluorescence yield of DBA is found to be extremely solvent dependent. This peculiar behavior is understandable when it is realized that T_2 lies ~ 5 kcal mol^{-1} above S_1 for DBA and that solvent interactions can cause the energy

Table 5.5   The Internal Heavy-Atom Effect on Transitions between States[a]

Molecule	k_F^0	k_{ST}	k_P^0	k_{TS}	Φ_F	Φ_P
Naphthalene	10^6	10^6	10^{-1}	10^{-1}	0.55	0.05
1-Fluoronaphthalene	10^6	10^6	10^{-1}	10^{-1}	0.84	0.06
1-Chloronaphthalene	10^6	10^8	10	10	0.06	0.54
1-Bromonaphthalene	10^6	10^9	50	50	0.002	0.55
1-Iodonaphthalene	10^6	10^{10}	500	100	0.000	0.70
Perylene	2×10^8	10^7			0.98	
3-Bromoperylene	2×10^8	10^7			0.98	

a. Data for rigid solution at 77 K. At room temperature, k_{TS} is often dominated by bimolecular deactivation of T_1 or by reactions of T_1. Rate constants are approximate in s^{-1}.

difference between S_1 and T_2 to vary so that in some cases T_2 is lower in energy than S_1 and intersystem crossing is fast; in other cases, T_2 is higher in energy than S_1 and intersystem crossing is slow.

The energetic spacing of S_1 and T_2 require that DBA either (1) undergoes activated intersystem crossing from S_1 to T_2, or (2) undergoes direct intersystem crossing from S_1 to an upper vibrational level of T_1. Either mechanism will cause k_{ST} to be slowed down. The solvent effect is due to the shifting of the position of S_1 and T_2 as a function of solvent. In solvents such that $E(T_2) > E(S_1)$, the value of Φ_F is at a maximum.

5.30   External Perturbation of Intersystem Crossing

Examples of enhancement of the overall $S_1 \rightarrow T_1$ process by molecular oxygen (a paramagnetic species, containing no heavy atoms),[57] xenon,[58] organic halides,[59] and organometallics[60] are known. In the case of oxygen, several mechanisms are possible, including enhancement of paramagnetic induced spin–orbit coupling and energy transfer (to produce a simultaneous $S_1 \rightarrow T_1$ transition in the perturbed molecule and a $T_0 \rightarrow S_1$ transition in the oxygen molecule). The efficiency of the oxygen effect depends on the oxygen concentration and may be expressed as $k_{ST}^{O_2}[O_2]$, where $k_{ST}^{O_2}$ is a bimolecular rate constant for oxygen perturbation and $[O_2]$ is the concentration of oxygen in the sample. The overall or observed rate of intersystem crossing (k_{ST}^{obs}) is given in Eq. 5.56:

$$k_{ST}^{obs} = k_{ST} + k_{ST}^{O_2}[O_2] \tag{5.56}$$

Typical values of $k_{ST}^{O_2}$ are $\sim 10^{10} - 10^9$ M^{-1} s^{-1}.[57] The solubility of O_2 in many organic solvents (< 1 atm of O_2) is $\sim 10^{-2}$ M. Thus, $k_{ST}^{O_2}[O_2] \sim 10^8 - 10^7$ s^{-1} and the effect will be noticeable only if $k_{ST} \sim 10^8$ s^{-1} or less. For example, $\Phi_{ST} \sim 0.3$ for pyrene ($k_{ST} \sim 10^7$ s^{-1}) in the absence of O_2 and increases to ~ 1.0 under 1 atm of O_2. Similar effects are noted for xenon as an $S_1 \rightarrow T$ perturber.[58]

Although O_2 and Xe are known to be efficient quenchers of T_1 states, it is clear that not only enhancement of $T_1 \rightarrow S_0$, but also other quenching pathways (including reaction) may occur.

External heavy-atom effects for heavy atoms that are not directly connected to aromatic cores are well established. As an illustration, consider the naphthalene derivatives **10** and **11**.[61]

10	**11**	**12**
$k_{ST} = 2 \times 10^6 \, s^{-1}$	$k_{ST} = 300 \times 10^6 \, s^{-1}$	$k_{ST} = 500 \times 10^6 \, s^{-1}$
$k_{TS} = 2 \times 10^{-1} \, s^{-1}$	$k_{TS} = 40 \times 10^{-1} \, s^{-1}$	$k_{TS} = 600 \times 10^{-1} \, s^{-1}$

The effect of the "external" bromine in **11** is to enhance both k_{ST} and k_{TS}. The enhancement of k_{ST} is comparable to the "internal" effect of bromine on **12**. Note, however, that k_{TS} is much larger for **12** than **11**. This finding may arise from the somewhat different surface situation in **12**, which may result in better FC factors for better intersystem crossing.

5.31 The Relationship between Photophysical Radiationless Transitions and Photochemical Processes

In this chapter we have considered the photophysical radiationless pathways by which an electronically excited molecule can "find its way" back to its original ground state. If we view radiationless transitions in the general sense as a conversion of electronic energy into nuclear motion, then the distinction between photophysical and photochemical processes becomes blurred (as implied in Scheme 1.7).[62] Indeed, we can imagine (Fig. 5.17) that they differ only in the degree of nuclear geometric change. If the distortion from an original ground-state geometry is not too severe, return to the original geometry via radiationless transition(s) is possible. Suppose this transition takes place via the funnel on the excited surface shown in Fig. 5.17. Relatively small changes in nuclear shape ("to the right" of the funnel minimum) will tend to deliver the molecule back to the ground state in a nuclear configuration that will favor formation of products rather than reactants. Thus, a photophysical transition that takes place through the funnel via transitions "to the left" of the minimum and returns the excited molecule to its original ground state may not differ qualitatively from photochemical transitions that produce products. In Chapter 6, we will see that these funnels sometimes correspond to avoided crossings and conical intersections connecting the excited surface with the ground-state surface.

The notions of photophysical radiationless transitions and photochemical reactions are intimately related via the common theory of energy surfaces.

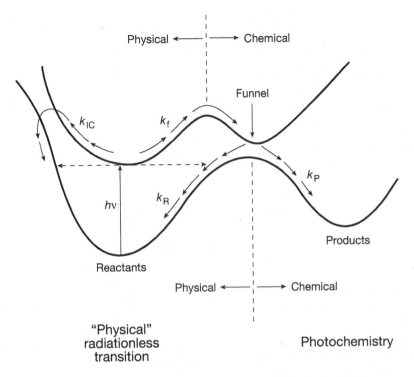

Figure 5.17   Schematic surface description of a possible relationship between "photophysical" and "photochemical" radiationless processes.

References

1. See the following references for a discussion of the theory of radiationless processes with a historical perspective. (a) G. W. Robinson, *Excited States*, Vol. 1, E. Lim, ed., Academic Press, New York, 1974, p. 1. (b) M. Kasha, *Light and Life*, W. D. McElroy and B. Gloss, eds., Johns Hopkins Press, Baltimore, MD, 1961, p. 31. (c) M. Kasha, *Comparative Effects of Radiation*, M. Burton, J. S. Kirby-Smith, and J. J. Magee, John Wiley & Sons, Inc., New York, 1960, p. 72. (d) M. Kasha, *Rad. Res., Supl. 2*. 243 (1960). (e) J. B. Birks, *Photophysics of Aromatic Molecules*, John Wiley & Sons, Inc., New York, 1970, p. 142. (f) M. Kasha, *Disc. Faraday Soc.* **9**, 14 (1950). (g) B. R. Henry and M. Kasha., *Annu. Rev. Phys. Chem.* **19**, 163 (1968).

2. (a) C. Zener, *Proc. R. Soc. London* **A137**, 696 (1939); *Proc. R. Soc.* **A140**, 660 (1968). (b) R. B. Bernstein, *Molecular Reaction Dynamics*, Clarendon, Oxford, UK, 1974. (c) E. E. Nikitin, *Russ. Chem. Rev.* **43**, 905 (1974).

3. For excellent texts on photochemistry that integrate spectroscopy and quantum mechanics. (a) M. Klessinger and J. Michl, *Excited States and Photochemistry of Organic Molecules*, VCH Publishers, New York, 1995. (b) J. Michl and V. Bonacic-Koutecky, *Electronic Aspects of Organic Photochemistry*, John Wiley & Sons, Inc., New York, 1990.

4. (a) M. J. S. Dewar and R. C. Dougherty, *The PMO Theory of Organic Chemistry*, Prentice Hall, Englewood Cliffs, NJ, 1975. (b) J. Michl, *Mol. Photochem.* **4**, 253 (1972). (c) L. Salem, C. Leforestier, G. Segal, and R. Wetmore, *J. Am. Chem. Soc.* **97**, 479 (1975). (d) L. Salem, *J. Am. Chem. Soc.* **96**, 3486 (1974). (e) A. Devaquet, *Pure Appl. Chem.* **41**, 535 (1975); (f) A. Devaquet, *Topics Chem.* **54**, 1 (1975). (g) W. G. Dauben, L. Salem, and N. J. Turro, *Acc. Chem. Res.* **8**, 41 (1975).

5. (a) W. Kauzman, *Quantum Chemistry*, Academic Press, New York, 1957, p. 534. (b) E. Teller, *Israel J. Chem.* **7**, 227 (1969).

6. See the folloing for reviews and discussions of the concept of conical intersections in photochemistry. (a) K. B. Lipkowitz and D. B. Boyd, eds., *Reviews in Computational Chemistry*, John Wiley & Sons, Inc., New York, 2000, Chap. 2. (b) M. A. Robb, *Pure Appl. Chem.* **67**, 783 (1995). (c) F. Bernardi and M. Olivcci, *Chem. Soc. Rev.* 321 (1996).

7. C. H. Ting, *Photochem. Photobio.* **9**, 17 (1969).

8. (a) T. Azumi and K. Matsuzaki, *Photochem. Photobio.* **25**, 315 (1977). (b) E. W. Schlag, S. Schneider, and S. F. Fischer, *Annu. Rev. Phys. Chem.* **22**, 465 (1971). (c) B. R. Henry and W. Sisbrand, *Organic Molecular Photophysics*, J. B. Birks, ed., John Wiley & Sons, Inc., New York, 1969. (d) R. S. Becker, *Theory and Interpretation of Fluorescence and Phosphorescence*, John Wiley & Sons, Inc., New York, 1969.

9. (a) E. C. Lim, Y. H. Li, and R. Li, *J. Chem. Phys.* **53**, 2443 (1970). (b) N. Kanamaru and E. C. Lim, *J. Chem. Phys.* **62**, 3252 (1975).

10. See the following references for excellent reviews of spin–orbit coupling. (a) S. P. McGlynn, T. Azumi, and M. Kinoshita, *Molecular Spectroscopy of the Triplet State*, Prentice-Hall, Englewood Cliffs, NJ, 1969, p. 183. (b) L. Salem and C. Rowland, *Angew. Chem. Int. Ed. Eng.* **11**, 92 (1972). (c) L. Salem, *Pure Appl. Chem.* **33**, 317 (1973).

11. (a) M.A. El-Sayed, *J. Chem. Phys.* **38**, 2834 (1963). (b) M. A. El-Sayed, *J. Chem. Phys.* **36**, 573 (1962). (c) M. A. El-Sayed, *J. Chem. Phys.* **41**, 2462 (1964).

12. A. Halpern and W. R. Ware, *J. Chem. Phys.* **53**, 1969 (1970).

13. R. W. Anderson, R. M. Hochstrasser, H. Lutz, and G. W. Scott, *J. Chem. Phys.* **61**, 2500 (1974).

14. K. Bredereck, T. Forster, and H. G. Oesterlin, *Luminescence or Organic and Inorganic Materials*, John Wiley & Sons, Inc., New York, 1962, p. 161.

15. B. R. Henry and W. Siebrand, *J. Chem. Phys.*, **54**, 1072 (1971).

16. S. L. Madej, S. Okajima, and E. C. Lim, *J. Chem. Phys.* **65**, 1219 (1976).

17. (a) G. N. Lewis and M. Calvin, *Chem. Rev.* **25**, 272 (1939). (b) S. H. Lin and R. Bersohn, *J. Chem. Phys.* **48**, 2732 (1968). (c) G. Calzaferri, H. Gugger, and S. Leutwyler, *Helv. Chim. Acta* **59**, 1969 (1976). (d) S. H. Lin, *J. Chem. Phys.* **44**, 3759 (1969).

18. (a) S. Sharafy and K. A. Muskat, *J. Am. Chem. Soc.* **93**, 4119 (1971). (b) D. Gegion, K. A. Muskat, and R. Fischer, *J. Am. Chem. Soc.* **90**, 12, 3097 (1968).

19. (a) C. D. DeBoer and R. H. Schlessinger, *J. Am. Chem. Soc.* **90**, 803 (1968). (b) J. Saltiel, O. C. Zafirious, E. D. Megarity, and A. A. Lamola, *J. Am. Chem. Soc.* **90**, 4759 (1968).

20. D. Dellinger and M. Kasha, *Chem. Phys. Lett.* **38**, 9 (1976).

21. H. E. Zimmerman, K. S. Kamm, and D. P. Werthemann, *J. Am. Chem. Soc.* **97**, 3718 (1975).

22. W. W. Schloman and H. Morrison, *J. Am. Chem. Soc.* **99**, 3342 (1977).

23. P. M. Froehlich and H. Morrison, *J. Phys. Chem.* **76**, 3566 (1972).

24. (a) G. W. Robinson and R. P. Frosch, *J. Chem. Phys.* **38**, 1187 (1963). (b) G. W. Robinson and R. P. Frosch, *J. Chem. Phys.* **37**, 1962 (1962). (c) D. L. Dexter and W. B. Fowler, *J. Chem Phys.* **47**, 1379 (1967). (d) R. F. Borkman, *Molec. Photochem.* **4**, 453 (1972).

25. See the following references for reviews of vibrational relaxation in solution. (a) L. K. Iwaki and D. D. Dlott, in *Encyclopedia of Chemical Physics and Physical Chemistry*, Vol. III, J. H. Moore and N. D. Spenser, (eds.) Institute of Physics Publishing, Philadephia, 1999. (b) L. K. Iwaki, J. C. Deak, S. T. Rhea, and D. D. Klott, in *Ultrafast Infrared and Raman Spectroscopy*, M. D. Fayer, ed., Dekker, New York, 2001.

26. See the following references for experimental examples of vibrational relaxation. (a) Coumarin derivative. J. P. Maier, A. Seilmeier, A. Loubereau, and W. Kaiser, *Chem. Phys. Lett.* **46**, 527 (1977). (b) Rhoamine G. G. D. Ricard and J. Ducuing, *J. Chem. Phys.* **62**, 3616 (1975). (c) Coronene. C. F. Shank, E. P. Ippen, and O. Teschke, *Chem. Phys. Lett.* **45**, 291 (1977). (d) Carotene. A. N. Macpherson and T. Gillbro, *J. Phys. Chem.* **102**, 5049 (1998). (e) Perylene. Y. Jiang and G. J. Blanchard, *J. Phys. Chem.* **98**, 9417 (1994).

27. G. Herzberg, *Spectra of Diatomic Molecules*, 2nd ed. Van Nostrand, Princeton, NJ, 1950.

28. H. Statz, C. L. Tang, and G. F. Foster, *J. Appl. Phys.* **37**, 4278 (1966).

29. F. Wilkinson, *Organic Molecular Photophysics*, J. B. Birks, ed., John Wiley & Sons, Inc., New York, 1975, p. 95.

30. V. L. Ermolaev and K. J. Svitaskev, *Opt. Spectr.* **7**, 399 (1959).

31. (a) E. H. Gilmore, G. E. Gibson, and D. S. Mc-Clure, *J. Chem. Phys.* **20**, 829 (1952). Correction. (b) E. H. Gilmore, G. E. Gibson, and D. S. Mc-Clure, *J. Chem. Phys.* **23**, 399 (1955).

32. (a) J. B. Birks, *Photophysics of Aromatic Molecules*, John Wiley & Sons, Inc., New York, p. 143ff, 1970. (b) J. B. Birks,*Photophysics of Aromatic Molecules*, John Wiley & Sons, Inc., New York, p. 128.

33. G. D. Gillispie and E. C. Lim, *J. Chem. Phys.* **65**, 2022 (1976).

34. R. O. Campbell and R. S. H. Liu, *J. Am. Chem. Soc.* **95**, 6560 (1973).

35. (a) W. Siebrand, *J. Chem. Phys.* **46**, 440 (1967). (b) W. Siebrand, *J. Chem. Phys.* **47**, 2411 (1967). (c) W. Siebrand, *Symposium on The Triplet State*, Cambridge University Press, Cambridge, UK, 1967, p. 31.

36. J. P. Byrne, E. F. McCoy, and I. G. Ross, *Aust. J. Chem.* **18**, 1589 (1965).

37. V. L. Ermolaev, *Sov.et Phys. Uspekhi* **80**, 333 (1963).

38. (a) J. D. Laposa, E. C. Lim, and R. E. Kellogg, *J. Chem. Phys.* **42**, 3025 (1965). (b) J. B. Birks, *Photophysics of Aromatic Molecules*, John Wiley & Sons, Inc., New York, 1970, p. 122. (c) N. Kanamaru, H. R. Bhattacharjie, and E. C. Lim, *Chem. Phys. Lett.* **26**, 174 (1974).

39. R. C. Miller and E. K. C. Lee, *Chem. Phys. Lett.*, **41**, 52 (1976).

40. (a) A. M. Halpern and W. R. Ware, *J. Chem. Phys.* **54**, 1271 (1971). (b) A. C. Luntz and V. T. Maxson, *J. Chem. Phys.* **26**, 553 (1974).

41. S. Murata, C. Iwanga, T. Toda, and H. Kokubun, *Chem. Phys. Lett.* **15**, 152 (1972).

42. N. Tetreault, R. S. Muthyala, R. S. Liu, R. P. Steer, *J. Phys. Chem. A* **103**, 2524 (1999).

43. (a) J. P. Heritage and A. Penzkofer, *Chem. Phys. Lett.* **44**, 76 (1976). (b) E. P. Ippen, C. V. Shank, and R. L. Woerner, *Chem. Phys. Lett.* **46**, 20 (1977).

44. R. G. Bennett and P. J. McCartin, *J. Chem. Phys.* **31**, 251 (1975).

45. H. Dreekamp, E. Koch, and M. Zander, *Chem. Phys. Lett.* **31**, 251 (1975).

46. (a) A. Kearvill and F. Wilkinson, *J. Chem. Phys.* **125** (1969). (b) A. Kearvill and F. Wilkinson, *Molec. Crystals* **4**, 69 (1968).

47. R. B. Henry and W. Siebrand, *Chem. Phys. Lett.* **54**, 1072 (1971).

48. A. Halpern and W. R. Ware, *J. Chem. Phys.* **53**, 1969 (1970).

49. M. Almgren, *Mol. Photochem.* **4**, 327 (1972).

50. B. S. Solomon, T. F. Thomas, and C. Steel, *J. Am. Chem. Soc.* **90**, 2449 (1968).

51. (a) J. M. Morris and D. F. Williams, *Chem. Phys. Lett.* **25**, 312 (1974). (b) M. A. El-Sayed and R. Leyerle, *J. Chem. Phys.* **62**, 1579 (1975). (c) M. Batley and D. R. Kearns, *Chem. Phys. Lett.* **2**, 423 (1968).

52. T. Matsumoto, M. Sato, and S. Hiroyama, *Chem. Phys. Lett.*, **13**, 13 (1972).

53. D. A. Hansen and K. C. Lee, *J. Chem. Phys.* **62**, 183 (1975).

54. T. F. Hunter, *Photochem. Photobio.* **10**, 147 (1969).

55. J. B. Birks, T. D. S. Hamilton, and J. Najbar, *Chem. Phys. Lett.* **39**, 445 (1976).

56. R. H. Clark and H. A. Frank, *J. Chem. Phys.* **65**, 39 (1976).

57. (a) B. Stevens and B. E. Algar, *J. Phys. Chem.* **72**, 3468 (1968). (b) B. Stevens and B. E. Algar, *Chem. Phys. Lett.* **1**, 58, (1967). (c) B. Stevens and B. E. Algar, *Chem. Phys. Lett.* **1**, 219, (1967).

58. A. R. Horrocks, A. Kearvill, K. Tickle, and F. Wilkinson, *Trans. Faraday Soc.* **62**, 3393 (1966).

59. F. H. Quina and F. A. Carroll, *J. Am. Chem. Soc.* **98**, 6 (1976).

60. E. Vander Donckt, and C. Vogels, *Spectrochim. Acta* **27A**, 2157 (1971).

61. G. Kavarnos, T. Cole, P. Scribe, J. C. Dalton, and N. J. Turro, *J. Am. Chem. Soc.* **93**, 1032 (1970).

62. (a) D. Phillips, J. Lemaire, C. S. Burton, and W. A. Noyes, *Adv. Photochem.* **5**, 329 (1968). (b) G. S. Hammond, *Adv. Photochem.* **7**, 373 (1969).

CHAPTER 6

A Theory of Molecular Organic Photochemistry

6.1 Introduction to a Theory of Organic Photoreactions

The theory of organic photochemistry uses the molecule and its electrons, vibrating nuclei, and electron spins as the key intellectual units for providing an understanding of the nature of photophysical and photochemical processes. This theory employs the laws and methods of quantum mechanics, with its wave functions, operators, and matrix elements, to generate energy surfaces and to visualize and to track representative points as a molecule moves along surfaces and makes transitions between excited- and ground-state surfaces. Chapters 3–5 outlined ways of describing photophysical transitions. This chapter is concerned with the elaboration of the reaction mechanisms shown in the paradigm of Scheme 1.1. We describe the overall *R → P transitions via the primary photochemical processes listed in Scheme 6.1, where *R = a thermally equilibrated electronically excited state, I = an equilibrated *ground-state* reactive intermediate, *I = an equilibrated electronically excited state of a reactive intermediate, F = a "funnel" (a structure that has no ground-state equivalent, to be described in detail in Section 6.11), and *P = an equilibrated electronically excited state of the product, P. The paradigm of Scheme 6.1 provides a conceptual basis for understanding the mechanisms of organic photochemical reactions. This chapter reviews the connection between theory and reaction mechanisms in terms of energy surfaces.

Consider an overall ground-state reaction, R → P. Figure 6.1 represents an energy surface for a hypothetical *elementary chemical step* in a reaction pathway. An elementary chemical step is a transformation along an energy surface from one energy minimum (R) to a second energy minimum (P) through a *single energy maximum*, a transition state (TS). A single-elemental reaction step corresponds to a *concerted reaction,* which does not involve a reactive intermediate (I). An organic reaction mechanism for an elementary chemical step describes (1) the temporal sequence of the changes for the critical molecular structures along the reaction pathway, and (2) the energetics of these structures. The *representative point (r)* labels the nuclear structure

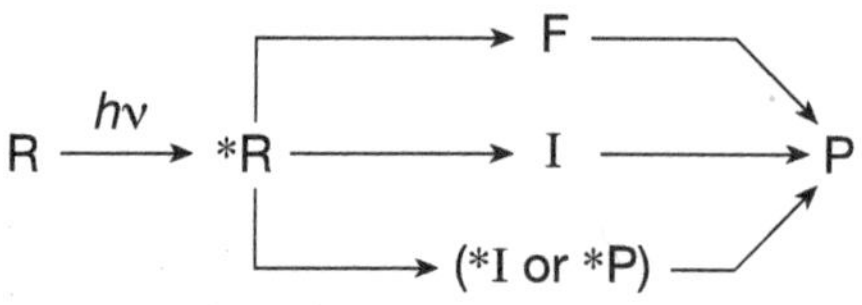

F = funnel from excited to ground-state surface
I = ground-state reactive intermediate
*I = excited state of a reactive intermediate
*P = excited state of product

Scheme 6.1   Global paradigm for the plausible overall pathways from ground-state reactant(s), R, to isolated product(s), P. The reactive intermediate, I, may be a diradical, I(D), or a zwitterion, I(Z). The funnel F may be a Franck–Condon (FC) minimum, an avoided crossing (AC), or a conical intersection (CI).

and energy of every point (the y-axis) along a reaction pathway (the x-axis). A plot of the potential energy (PE) change of a reacting system along the reaction coordinate is termed an *electronic energy surface* for a reaction. A two-dimensional (2D) "slice" of this surface that describes only the reaction coordinate is termed a *PE curve*. In general, there is a unique, lowest-energy reaction pathway corresponding to a sequence of representative points for the overall R $\rightarrow$ P process, called the *reaction coordinate (RC)*, which connects the structures along the reaction path in time. Thus, Fig. 6.1 is a PE curve that represents the reaction coordinate for a hypothetical elementary step, R $\rightarrow$ P. The overall general net flow of R to P is indicated by the arrows on the energy surface in Fig. 6.1.

If a reactive intermediate (I) exists on the pathway between R and P, the reaction pathway consists of *two consecutive elementary steps*, R $\rightarrow$ I and I $\rightarrow$ P (Fig. 6.2). In this case two energy maxima corresponding to two transition states (TS_1 and TS_2) and a minimum corresponding to the reaction intermediate (I) occur along the reaction coordinate. This chapter develops mechanisms that employ electronic energy curves to describe the plausible pathways for the *overall* process *R $\rightarrow$ P, which may proceed by one of the pathways shown in Scheme 6.1.

It is informative to compare the description of an elementary step for a thermally induced ground-state reaction R $\rightarrow$ P with that for a photochemical reaction *R $\rightarrow$ P. The ground-state reaction R $\rightarrow$ P is modeled (Fig. 6.1) accurately and reliably by the assumptions that (1) the reactants (R) start in an energy minimum on the electronic ground state; (2) thermal energy from the surroundings causes the representative point (r), characterizing instantaneous nuclear geometries, to move along the reaction coordinate (the lowest-energy nuclear configurations that are on the pathway R $\rightarrow$ P); (3) the representative point passes through a critical geometry (r_c), which corresponds to a maximum of energy (the TS) along the reaction coordinate; and (4) thermal

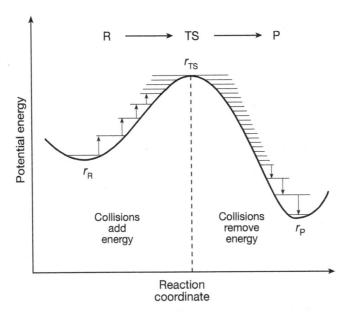

Figure 6.1   A PE curve describing an elementary chemical step for the reaction R → P. The curve has the following critical features: (1) a minimum corresponding to R; (2) a maximum corresponding to TS; and (3) a minimum corresponding to P. If each of the points on the curve is the lowest energy for the pathway from R → P, the PE curve corresponds to the reaction coordinate (lowest-energy path) for the reaction.

energy propels the representative point over the transition state to an energy minimum corresponding to the product (P). Thus, an elementary step of a ground-state thermal reaction can be described well theoretically in terms of a *single electronic energy surface (or curve)*.

Figure 6.2 schematically shows an energy surface corresponding to a thermal reaction, R → I → P, which involves a reactive intermediate (I). In this particular exemplar, the barrier for I → TS_1 is shown as higher than the barrier for I → TS_2, a situation that is common for many reactive organic intermediates. As a result of the lower barrier from I to P, I proceeds to P faster than it reverts back to R. Thermal energy from collisions with solvent molecules is responsible for the movement of the representative point along the reaction coordinate. The overall general net flow of R to P is indicated by the arrows on the energy surface in Fig. 6.2.

In contrast to the single energy surface that is sufficient to model an elementary step of a thermal ground-state reaction, the modeling of the overall photochemical reaction *R → P requires the use of at least two energy surfaces: a ground-state surface with minima for R and P and an excited-state surface on which a primary photochemical process from *R is launched.[1] Minima and maxima on the excited-state surface do not have any necessary correspondence with the minima and maxima on the ground surface, since the nuclear motion and structural stability of *R(HO)^1(LU)^1, where

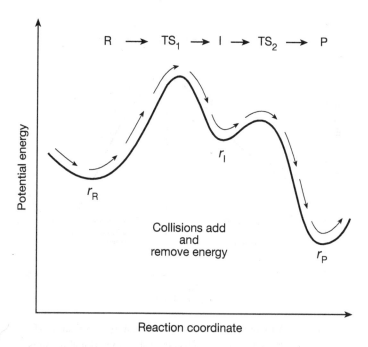

Figure 6.2   An energy curve describing a reaction coordinate involving a reactive intermediate (I) and two transition states (TS_1 and $\mathbf{TS_2}$).

HO = the highest occupied and LU = the lowest unoccupied molecular orbital, is controlled by a different electronic orbital configuration from the one controlling the nuclear motion and structural stability of $R(HO)^2$. This finding also means that the reaction coordinate (lowest-energy path) for the ground-state path from R → P may not have any necessary relationship to the reaction coordinate (lowest-energy path) for the excited-state path *R → P. However, since our exemplars will be qualitative and emphasize the general features of the energy surface, we assume that both the excited- and ground-state surfaces correspond to approximate minimum energy paths and are therefore approximate reaction coordinates for a similar overall reaction R → P.

6.2   Potential Energy Curves and Surfaces

An electronically excited molecule (*R) in a particular electronic state (S_1 or T_1) may exist with various configurations of its nuclei, each configuration in space corresponding to a particular value for the PE of the system. A representative point corresponds to a particular energy and specific nuclear geometric configuration on the energy surface of the state. For a diatomic molecule the situation is particularly simple, since the internuclear separation (r) is the only variable describing the nuclear geometry of the system; thus, r serves as a very well-defined representative point of the energy and structure for any state of a diatomic molecule. A plot of the lowest PE for the nuclear geometry of a diatomic molecule as a function of the separation between the two nuclei is a simple 2D *PE curve*, approximated by the parabolic curve of a harmonic oscillator (e.g., Figs. 2.3 and 2.4). Since essentially all organic molecules are polyatomic, their instantaneous nuclear geometry is a much more complicated function of the nuclei position in space than the PE curve of a diatomic molecule. Indeed, the PE of an organic molecule as a function of the nuclear geometry is not a simple 2D curve but a complex *multidimensional surface*.

A multidimensional plot of the electronic PE for an organic molecule versus the nuclear configuration for a given electronic state is called a *PE surface* for that state.[1] Although it is not as accurate, or as realistic a representation as an energy surface for polyatomic molecules, a 2D energy *curve* (e.g., Figs. 6.1 and 6.2) is still an excellent starting point for a discussion of complex energy surfaces, since energy curves are much more readily visualized and interpreted than a three-dimensional (3D) or multidimensional energy surface. This simplification is reasonable since the PE curve of a diatomic molecule contains most of the important qualitative features relating energy and structure and their changes encountered, even those for complex energy surfaces. Thus, we start with PE curves and modify them as necessary so that they may serve as a useful approximation of the more realistic multidimensional PE surfaces.

In going from a diatomic to a multiatomic molecule, we can replace the concept of a single internuclear separation (r) of two nuclei with the notion of a single *center of mass that represents the entire nuclear geometry of an organic molecule* (r_c). This center of mass replaces the single geometric coordinate, the internuclear separation, as the familiar *representative point* that moves along an energy surface with the same characteristics as a single point or (classically) particle. The center of mass for a system is known, from elementary physics, to depend only on the masses of the particles for the system and their positions relative to one another. Since the relative positions of the atoms in a molecular structure correspond to the nuclear geometry, we see that the center of mass is an appropriate variable to represent the nuclear geometry.

When a complicated array of bound particles (the aggregate of nuclei that make up the positive electric field of a molecule) moves under the influence of external forces, *the center of mass moves in the same way that a single particle subject to the same external forces would move.* The concept of the center of mass as a representative point allows us to readily visualize an energy trajectory for a complex system of particles executing very complicated motions along an energy surface. The important topological (i.e., qualitative geometric) features of a PE curve may be generalized to deduce the topological features of PE surfaces. A PE curve or 3D energy surface serves as an excellent approximation to the more complex multidimensional energy surfaces and allows an easy visualization, providing an insight into many problems of importance in molecular organic photochemistry.

6.3   Movement of a Classical Representative Point on a Surface[2]

The behavior of a representative point on a PE curve or surface possesses certain features that are analogous to that of a marble rolling along a curved surface (imagine the arrows in Figs. 6.1 and 6.2 representing the marble's motion). In addition to the

PE of position, the marble possesses kinetic energy (KE) of motion as it moves along the energy surface. The marble is held onto a real surface by the force of Earth's gravitational field; in analogy, the representative point of an energy curve is evidently "held" onto the PE surface by some sort of force. If the marble leaves the surface as the result of an impulse (a force delivered in a short time period) from some external source, it makes a momentary departure from the surface, but gravity provides a restoring force that quickly attracts the particle back onto the surface. What is the nature of the analogous restoring force that attracts the representative point on the energy curve? That force is the simple result of the Coulombic (electrostatic) attractive force of the negative electrons to the positive electrical field provided by the nuclear framework. *The restoring electrostatic forces that keep the representative point "sticking" to the surface are analogous to the force of gravity that "holds" the particle on a surface that represents the lowest energy of the system (reaction coordinate).*

From elementary physics,[2] the magnitude of both the gravitational and the Coulombic forces acting on the representative point of an energy surface have an analogous mathematical form relating the force on the particle to the change in PE with change in structure (Eq. 6.1).

$$\text{Force acting on the particle at } r \qquad F = -d\text{PE}/dr \qquad \text{Slope of the curve at } r \tag{6.1}$$

Equation 6.1 is of great importance in understanding the behavior of the representative point, since it relates the magnitude of the force (F) acting on the representative point to the *instantaneous slope* ($-d\text{PE}/dr$) or the *"steepness"* of a PE curve at a given nuclear geometry (r). *The steeper the slope, the stronger the "pull," or force, of the positive nuclei on the negative electrons. The shallower the slope, the weaker the pull and the smaller the attraction.*

Recall that the classical theory (Eq. 5.1) for "jumps" between electronic surfaces assumes that two distinct forces operate on the representative point (i.e., on the nuclear configurations) as the point approaches a surface crossing, since near the crossing two different surfaces compete to control the motion of the representative point. The nature of these two forces can be deduced from the simple classical theory of jumps between energy surfaces in two dimensions. In this case the probability of a surface jump occurring when the representative point approaches a surface crossing whose geometry is r_c is given by Eq. 6.2.

$$\text{P (probability of surface jump at } r_c) = \exp(-\Delta E^2/v\Delta s) \tag{6.2}$$

In Eq. 6.2, ΔE is the energy gap between the energy surfaces at the critical geometry (r_c), v is the velocity of the nuclear motion along the deactivation or reaction coordinate, and Δs is the difference in the gradient or slope of the two surfaces as the representative point approaches r_c. From the classical picture, we are provided with the intuition that two factors, the relative steepness for the slope of the two surfaces and the velocity of the representative point, are important in determining the probability of a jump between surfaces, in addition to the energy gap between the surfaces at r_c.

From quantum mechanical theory for jumps between surfaces, two independent factors also operate on the representative point as it moves along an energy surface. The change in the shape of a wave function, as a critical geometry is approached, is equivalent to the difference in slopes (Δs) in Eq. 6.2. This result is reasonable because the gradient ($d\text{PE}/dr$) is the force acting on the nuclei, so that the steeper the gradient, the greater the force acting on the nuclei and the greater the influence of the force on the trajectory of the representative point. Suppose the difference in the slopes is small and the shape of the wave functions is similar as the representative point approaches r_c; in this case the representative point will not experience an abrupt change in its motion as it approaches r. In quantum theory, the term v in Eq. 6.2 is replaced by a term that considers the strength of vibrations that mix the wave functions as a critical geometry is approached.

6.4   The Influence of Collisions and Vibrations on the Motion of the Representative Point on an Energy Surface

What forces cause the representative point (r) to move on a PE surface and change the molecular geometry? Consider the effect of collisions on a molecule in solution. Such collisions may be considered as impacts (forces delivered in short periods of contact) that a molecule experiences as a result of interactions with other molecules in its immediate neighborhood (e.g., the molecules in the immediate environment that form the wall of a solvent cage, to be discussed in detail in Section 7.34). The magnitude of these impacts depends on temperature, varies over a wide distribution of energies, and follows a Boltzmann distribution of energies. Near room temperature, the average thermal energy per impact is ~ 0.6 kcal mol^{-1}. The energies associated with collisions are nearly continuous. Thus, near room temperature, collisions can be considered to provide a reservoir of continuous energy that will match vibrational energy gaps without difficulty and allow rapid, efficient energy exchange (typical rates of intermolecular vibrational energy transfer are of the order of 10^9–10^{11} s^{-1}, Section 5.14). These collisions and the energy exchange with the environment are what causes the representative point to move along the energy surface as shown in Figs. 6.1 and 6.2.

6.5   Radiationless Transitions on PE Surfaces: Surface Maxima, Surface Minima, and Funnels on the Way from *R to P

The concept that a PE surface can control the motion of the nuclei for a molecular system can be considered from several points of view. From the classical point of view, the PE surface reflects the work (force times distance) that must be done to bring the constituent atoms from infinite separation to the specified nuclear geometry. From the quantum mechanical point of view, a PE surface, when represented by the appropriate quantum mechanical description (wave functions, operators, matrix elements) of the nuclear motion, reproduces experimental observations and applies to a wide variety of experimental situations. Many phenomena can be interpreted as being controlled by a single PE surface. For example, reactions in the ground state generally involve a single electronic PE surface that possesses one or more maxima (transition states) that separate the reactant R from reactive intermediates I and products P (Figs. 6.1 and 6.2). The energy surfaces for photochemical reactions possess a feature that is completely absent for ground-state surfaces: *Somewhere on the way from *R → P, surfaces may touch, cross, or avoid each other.* In addition, a minimum on the way from *R → P serves as a funnel (F) to the ground-state surface. When we use the superscript *, we mean that the representative point is on an excited electronic energy surface described by a wave function (*Ψ). When the electronic excitation label (*) "disappears" and we write R, I, or P, we mean that the representative point is on a surface for an electronic ground state. Clearly, *the overall process from *R → P must be described in terms of more than one energy surface, that is, at least one electronically excited surface, determined by the pathway that the representative point carries *R, and one ground-state surface, eventually leading to P.*

In ground-state chemical reactions, the reactant R starts in a minimum on the PE surface and proceeds to various reactive intermediates, such as I, or products P via passage through a pathway-specific transition structure (Figs. 6.1 and 6.2). On the other hand, most photochemical reactions start in a Franck–Condon (spectroscopic) minimum (*R) on an electronically excited surface, and radiationless transitions eventually bring the system to a reactive intermediate, say I, or passes through a funnel (F) to a ground-state product (P) without forming a reactive intermediate (see the hypothetical PE curve, Scheme 1.5). Evidently, for all photochemical reactions there is a region of energy and structure for which the system undergoes an electronic transition and switches or jumps from an excited to a ground-state surface. These regions are the "funnels" (F) that determine the direction and motion of the representative point as it zooms toward the ground-state surface, toward the products that will result and be isolated.

6.6   A Global Paradigm for Organic Photochemical Reactions

This chapter uses electronic energy surfaces to develop a global paradigm that will provide conceptual tools and methods required to qualitatively describe and visualize the important types of *primary photochemical* processes initiating from *R, as shown in Scheme 6.1 (*R → I, *R → F, *R → *I, *R → *P). Spin is not included at this point for simplicity, since each of these processes is assumed to be an elementary chemical step that must preserve spin in the step; however, it is understood that spin is implicit and will need to be considered explicitly for any experimental system for which a spin change occurs, for example, intersystem crossing or phosphorescence. Also note that the symbol "R" does not necessarily represent a single molecule but represents *all* of the reactants involved in the primary photochemical process of interest. In addition, the symbol "*R" represents *all* of the excited states expected from a state energy diagram (e. g., S_1 and T_1 in particular), and "P" represents all of the isolated products that are relatively stable at room temperature and that can be "put in a bottle." The latter is an arbitrary distinction, and one could consider the formation of I, *I, or *P as a product that just happens to be too short-lived to be bottled.

Thermal reactions take place totally on a single (ground-state) surface, whereas *every* photochemical reaction requires *R to make a transition from an excited-state surface to a ground state (I or P) *somewhere along the reaction coordinate*. This means the representative point describing the reaction pathway must undergo a transition from an excited- (initially that of *R) to a ground-state surface (that of I or P) at some point r_c. A jump from one electronic state to another can be visualized in terms of jumps between electronic energy surfaces. As seen in Chapter 5, such jumps between electronic states correspond to a breakdown of the Born–Oppenheimer approxima-tion (Section 5.3), which assumes that the electrons can always "instantaneously" adjust their position in space to follow nuclear motion along the reaction coordinate. However, in the region of the jump, the electrons are passing into a region where an excited- and a ground-state surface both compete to control nuclear motion. Indeed, immediately after the jump, the forces acting on the nuclei are suddenly different and are governed by the new electronic surface and not by the old one. Therefore, for a cer-tain instant in the region of structures near r_c, the nuclei are "confused" about which energy surface is in charge of controlling their motion. This feature of jumping from one surface to another is not found in ground-state chemistry and is responsible for some of the unusual features of photochemical reactions compared to ground-state re-actions, especially the features related to "funnels" (F) that occur on the excited-state surface and have no ground-state equivalent.

What exactly are these "funnels" (F) that are shown in Schemes 1.1 and 6.1? By a funnel, we mean any critical nuclear geometry (r_c) on an electronically excited energy surface from which *R can make a jump to a ground-state surface (or more generally, to any different lower-energy surface, excited or ground state). Thus, the term "funnel" describes the critical region of an *excited*-state surface at which the representative point makes a radiationless transition to a lower-energy surface (either the ground or another lower-energy excited-state surface). The radiationless transition involving funnels may be photophysical (*R $\rightarrow$ F $\rightarrow$ R) or photochemical (*R $\rightarrow$ F $\rightarrow$ I or *R $\rightarrow$ F $\rightarrow$ P).

The transition *R $\rightarrow$ I (Scheme 6.1) is one of the most commonly encountered pri-mary photochemical reactions in all of organic photochemistry. This primary process involves the conversion of the electronically excited-state *R into a conventional ther-mally equilibrated ground-state reactive organic intermediate (I) (typically a radical pair, RP, a biradical, BR, or a zwitterion, Z). In general, the reactive intermediate I can be chemically trapped or can be detected directly by spectroscopic methods. Two rel-atively rare primary photochemical processes are also shown in Scheme 6.1: *R $\rightarrow$ *I and *R $\rightarrow$ *P, the formation of an excited state of a reactive intermediate (*I) and the excited state of a product (*P), respectively. Such processes are termed *adiabatic* pho-tochemical reactions.[3] These reactions indicate that the chemistry that converts *R to

*P occurs entirely on a single electronically excited surface. These processes occur when the excited and ground surface do not come close to one another anywhere along the entire reaction coordinate for the *R → P transformation (the surface-matching exemplar of Fig. 5.2), in other words, when there are no funnels between *R and *P or *R and *I. Among the most common examples of the primary process *R → *I are the formation of excimers (R-*-R) and exciplexes (R-*-N), which are supramolecular excited-state complexes generated by the complexing of a ground-state R with *R or by complexing a ground state of a different molecule M with *R, respectively (Section 4.38). The electronically excited state of a reactive intermediate *I is an electronically excited state that is distinct from *R, and will have a similar number of options for photophysical and photochemical transitions to *R. The primary photochemical process *R → F (Scheme 6.1) does not involve an analogous conventional equilibrated reactive organic intermediate I on the way to P. Section 6.9 will discuss funnels in detail.

As in the case for the description of radiationless photophysical processes in Chapter 5, our initial strategy for a description of radiationless photochemical processes will be to produce an understanding of several *exemplar* energy surfaces that allow the development of general and fundamental qualitative portraits of energy surfaces. These results can be used to describe the various transitions in Scheme 6.1, without regard to the details of the structures or energies involved. Within this strategy we use some specific theoretical tactics to connect the qualitative picture of the reaction coordinate to actual common chemical processes (e.g., the stretching of σ bonds and the twisting of π bonds) of the key fundamental functional-group chromophores of organic photochemistry (carbonyls, alkenes, enones, and aromatics). Our tactics will appeal to the principles of frontier molecular orbital theory[4] and the initial interactions of the HO and LU for guidance on plausible primary photochemical processes and plausible reaction coordinates that lead from an initial excited state, *R to I, F, *I, or *P. We will use the stereochemical implications of orbital interactions and the requirement for conservation of orbital symmetry for assumed geometries to generate correlation diagrams that will provide a basis for describing the essential and critical characteristics of the PE surface implied in Scheme 6.1.

6.7   Toward a General Theory of Organic Photochemical Reactions Based on Potential Energy Surfaces

In our theory of the overall *R → P process, we are particularly interested in certain features of the excited energy surface relative to the ground energy surface. These critical features along the reaction coordinate correspond to

1. The molecular structures (geometries) at r_c, on the excited surface that serve as funnels (F), because these geometries are those from which the representative point jumps from the excited surface to the ground surface.
2. The energy barriers that separate *R from the funnels (F) that lead to the ground state, because these barriers determine the competition between deactivation

of *R by other photochemical and photophysical pathways that compete with passage through the funnel.

3. The molecular geometries along the reaction coordinate correspond to the *fastest* overall pathway from *R → P, because these geometries are the most probable pathways, of the all plausible pathways, followed by the representative point.

4. The "deactivation" coordinates characterizing the funnels (F) leading from the excited to the ground state, because these structures, according to the Franck–Condon principle, will determine the ground-state structure produced immediately after passing through the funnel.

5. The barriers that separate *R from I, because such barriers will determine the competition between the formation of I and other pathways for deactivation of *R.

Ideally, our theory will allow us to visualize the structural and energetic details of the reaction pathways and structural connections all along the entire photochemical reaction pathway, "from cradle to grave" (from R + $h\nu$ to P). In addition, the theory should be able to account in a natural way for the *photophysical* radiationless processes, *R → R + Δ, which compete with the primary photochemical processes along the reaction coordinate. There should be no fundamental difference between the way the theory treats radiationless photophysical and photochemical processes, since both involve pathways involving the movement of a representative point initially along an excited-state surface. The essential distinction is that the photophysical transition *R → R is "physical" only in the sense that the geometries of *R and R are similar (with no significant chemical change in the geometric sense), and the photochemical transitions *R → I, F, *I, *and* *P are "chemical" in the sense that the geometries of *R and I (or F or *I or *P) are significantly different from R, and therefore would naturally be considered different chemical species (Fig. 5.17).

Recent quantum mechanical computational methods have allowed the calculation of excited energy surfaces with considerable confidence.[5] These methods have provided valuable insight into the nature of photochemical reactions for regions of PE surfaces that are difficult or impossible to investigate by conventional experimental methods. We will see that these advances have allowed the theoretical confirmation for the existence of conical intersections (CIs) described in Section 5.6. Although they are not directly observable through spectroscopy, the existence of CIs may be investigated indirectly through spectroscopic experiments, guided by computations of where they may exist on excited energy surfaces.

The above discussion is mainly applicable to radiationless transitions that do not involve a spin change. A large number of the known organic photochemical reactions proceed via a *R(T$_1$) → ^{3}I(D) primary photochemical process. Since the isolated product (P) is generally a singlet state, *there must be an ISC step somewhere in the overall process* ^{3}I(D) → ^{1}I → ^{1}P. Thus, in addition to a model for orbital interactions that will allow us to determine the plausible photochemical processes corresponding to *R → I, we need a theoretical model for spin interactions and for ISC, particularly the ^{3}I → ^{1}I process, which must take place before P can be formed. The vector model for spin used to describe transitions between spin states discussed in Chapter 3 can be applied to visualizing intersystem crossing between spin states of diradicals, I(D).

6.8   Determining Plausible Molecular Structures and Plausible Reaction Pathways of Photochemical Reactions

In providing answers to questions for the plausibility of primary photochemical processes, we seek to visualize intermediates and pathways along the energy surfaces that connect the starting molecular structures (*R) to the possible final equilibrated molecular structures (reactive intermediates, I, and stable products, P). Visualization of these energy surfaces provides maps or networks of geometric structures and pathways that would allow us to immediately recognize the important nuclear geometries involved in organic photoreactions.

Now, we shall review some important features of *PE curves and surfaces* and then develop concepts and methods for a theory of photochemical reactions that will assist in a qualitative understanding of the photochemical reactions pathway. Then, we review how *frontier orbital HO–LU interactions* can provide an understanding of initial motions of *R along the reaction coordinate.[4] These initial trajectories provide insight into the lowest energy barriers for the initial pathways from *R toward I or F. Then, we develop general exemplars for two common *surface touchings* that result from the *stretching of σ or the twisting of π bonds.*[1] Next, we describe specific *surface crossings* or *avoidings* that result from orbital and state correlations of exemplar primary photochemical reactions, based on certain reaction symmetries.[6]

6.9   The Fundamental Surface Topologies for "Funnels" from Excited Surfaces to Ground-State Surfaces: Spectroscopic Minima, Extended Surface Touchings, Surface Matchings, Surface Crossings, and Surface Avoidings[1]

A mechanistic question central to all of photochemistry is how to determine the location and describe the electronic characteristics of any "funnels" (F) on the excited surface that separate the part of the reaction occurring on the excited surface (*R) from the part of the reaction path that lies on the ground surface (I or P). For most purposes, five common limiting surface topologies are encountered in the most common primary photochemical processes of organic molecules and will serve as exemplars for all of the commonly encountered organic photochemical reactions. These topologies are shown as 2D curves in Fig. 6.3a–e:

(a) An equilibrated excited-state surface minimum of *R occurs close to a geometry (r_c) that also corresponds to a ground-state minimum (Fig. 6.3a) of the reactant (R). Such minima on the excited surface have geometries that are similar to the ground state and are referred to as *spectroscopic minima* or *FC minima*, from which both radiative and radiationless photophysical processes of *R can originate. When such surfaces are well separated in energy, they are both described by the Born–Oppenheimer approximation. For each of the four other topologies shown in Fig. 6.3, it is assumed that the *R states are *initially* produced in a spectroscopic FC minimum, but then the representative point overcomes some small energy barrier and, propelled by thermal collisions (Fig. 6.1), begins moving along a reaction coordinate from left to right in the figures. One can expect FC minima in the excited-state surface for *R at geometries close to the minima in R for excited states for which the $R + hv \rightarrow$ *R process does not significantly change the overall bonding picture of *R. This would be true, for example, for the π,π^* states of rigid aromatic compounds (which have many bonding π electrons even in the excited state, as well as no readily available distortional motions). On the other hand, upon electronic excitation, simple C=C bonds undergo substantial bonding changes (a π electron removed, a π^* electron added) so that twisting about the carbon–carbon bond is facilitated and the geometry corresponding to the minimum of *R is likely to be quite different from that of R, as in case (e) discussed below. The photophysical processes of internal conversion and emission from *R occur from FC minima of excited aromatic molecules (Chapters 4 and 5).

(b) An *extended surface "touching"* (ST), for which an excited-state surface and a ground-state energy surface touch (become degenerate or nearly so in energy) at a critical point (r_c, Fig. 6.3b). After reaching r_c, the excited- and ground-state surfaces are isoenergetic (or very nearly so) at a structure I for an extended geometry change; in other words, the two surfaces "touch" (have the same energy) for an extended change in geometry. This surface topology is a signature of a σ bond stretching that eventuates in bond rupture. The primary photochemical product of this surface topology is often a *diradical intermediate*, I(D) (to be discussed in Section 6.17), and corresponds to a *R $\rightarrow$ I(D) primary photochemical process. Exemplars of such a situation are *all* of the primary photochemical reactions of the n,π^* states of carbonyl compounds and *any* primary process that produces a diradicaloid species by stretching a σ bond (Fig. 6.8). In the case of surface touchings, the transition from the excited energy surface is smooth and continuous (adiabatic) and does not involve an abrupt jump between energy surfaces; in a certain sense, the excited- and ground-state surfaces have "merged" and become indistinguishable. Note that at r_c the system may either continue toward I or, in special cases, revert to R. Hydrogen-atom abstraction and α-cleavage of ketones (Fig. 1.1) proceed through extended surface touchings.

(c) An *extended-surface "matching"* (SM), for which an excited-state surface and a ground state are separated by a large energy gap ($> 40\ \text{kcal mol}^{-1}$) and do not interact significantly after the representative point has moved along the excited surface and reached a certain critical geometry (r_c, Fig. 6.3c). In this topology, the representative point continues on an excited surface for the duration of the primary photochemical reaction. The primary photochemical product may be the excited state of a reactive intermediate (*I) or of a product (*P), that is, (R $\rightarrow$ *I or *R $\rightarrow$ *P). The important feature of such topologies

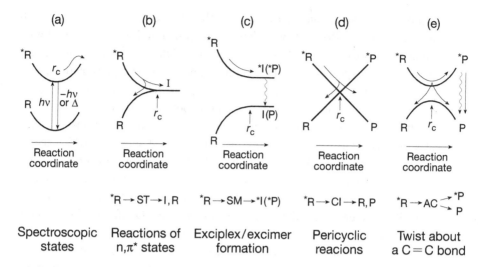

Figure 6.3   Two-dimensional PE curve descriptions of five limiting exemplars of pathways leading from an electronically excited to a ground-state surface. (a) A ground-state and Franck–Condon, or spectroscopic, excited surface with minimum at r_c. (b) An extended surface touching at r_c. (c) An extended surface matching at r_c. (d) A surface crossing at r_c. (e) An excited-state minimum over a ground-state maximum corresponding to an avoided crossing at r_c.

is that *there are no funnels to the ground state between *R and the product *I (or *P) so that the reactions are adiabatic photoreactions.*[3] Exciplexes, excimers (Section 4.38), formation of twisted internal charge transfer state (Section 4.42), and excited-state proton transfers are common examples of *R → *I pathways.[3]

(d) A *surface "crossing"* (SC), for which an excited- and ground-state surface cross one another at a critical geometry (r_c, Fig. 6.3d). Such surface crossings serve as funnels to take excited states from an *excited surface* to the ground-state surface so rapidly that they occur at a rate that is often limited by vibrational relaxation (10^{12}–10^{13} s^{-1}). For 3D surfaces, such surface crossings take the form of a CI at r_c. (CIs will be discussed in detail in Section 6.12.) The determination of the nuclear geometries for which CIs are found on an excited surface usually requires computational studies.[5] Surface crossings that are predicted at some zero-order approximation may persist to a certain extent at higher levels of approximation. When the representative point passes through a CI, *the system is not equilibrated,* and therefore the Born–Oppenheimer approximation is not valid. Intramolecular vibrational relaxation that occurs on the 1–10 ps time scale (Section 5.14), rather than electronic relaxation, often limits the rate of passage of the representative point through a CI.

(e) An equilibrated *excited-state surface minimum* occurring close to a geometry (r_c) above a ground-state maximum (Fig. 6.3e) that serves as an equilibrated funnel to the ground-state surface. The excited- and ground-state surfaces in the vicinity of r_c are adiabatic for this topology. The topology of Fig. 6.3e is typical of a $*R \rightarrow F$ process for which F is not a conical intersection resulting from a surface crossing but is a *strongly avoided surface crossing* (ASC) at r_c. Strongly avoided crossings (AC) occur when the surface crossings predicted in zero order undergo large electronic mixing at r_c in first order (in other words, the selected zero-order approximation is a poor one). The large electronic mixing leads to a large energy splitting of the zero-order states. Importantly, however, there is an electronic "heritage" between the wave functions of the ground and excited states, which is absent in the case of matching energy surfaces. This heritage means that, in spite of a large energy separation, the excited and ground surfaces can mix weakly. The ASC topology is typical of pericyclic reactions, that are forbidden in the ground state and is also typical of twisting about double bonds in the excited *singlet* state. The concepts of surface crossing and avoiding will be discussed further in the next sections. Representative points in a ASC minima undergo a transition to some lower state at a rate dictated by Fermi's golden rule (Eq. 3.8).

In summary, the rates of surface jumps from thermally equilibrated (adiabatic) funnels (Figs. 6.3a, c, and e) depend on the energy gap between the surfaces. The larger the energy gap, the slower the radiationless jump from the excited- to the ground-state surface; such a situation implies a very weak interaction between excited and ground surfaces, so that Fermi's golden rule (Eq. 3.8) is a valid approximation for determining the rates of radiationless transitions. For the surface touching (Fig. 6.3b) and crossing (Fig. 6.3d), the energy gap at r_c is zero. In such cases, the golden rule does not apply and the rate of transition between the excited- and ground-state surfaces may be limited by the rate of vibrational relaxation on the surface if there is no spin change involved [e.g., $*R(S_1) \rightarrow {}^1I$ or Z or $*R(T_1) \rightarrow {}^3I$ primary photochemical processes]. For the cases where there is a significant energy separation between the surfaces (Figs. 6.3a, c, and e), the decay to the ground state may be sufficiently slow to allow complete equilibration on the excited surface. This means that the species corresponding to the minima of the excited surface should be directly detectable by spectroscopic methods.

6.10   From 2D PE Curves to 3D PE Surfaces: The "Jump" from Two Dimensions to Three Dimensions

Consider Fig. 6.4, which shows 3D PE surface representations of the five exemplar 2D energy curves of Fig. 6.3. These 3D representations more accurately capture the ability of an energy surface to control and direct the motion of a representative point that enters a certain region of the energy surface as the result of initial frontier orbital interactions within *R and also indicate the greater range of structural and energetic possibilities that exist on the ground and excited surfaces. Note that in reality the motion of the representative point is really in a multidimensional space that cannot be visualized readily. Fortunately, the 3D representation is still a satisfactory approximation for analyzing most commonly encountered photochemical systems.

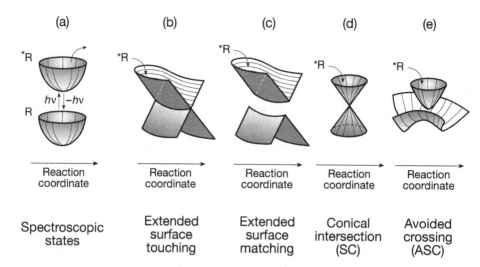

| (a) | (b) | (c) | (d) | (e) |

Reaction coordinate (×5)

| Spectroscopic states | Extended surface touching | Extended surface matching | Conical intersection (SC) | Avoided crossing (ASC) |

Figure 6.4   Three-dimensional PE surface descriptions of five limiting cases of funnels leading from an electronically excited surface to a ground-state surface. (a) A ground-state and Franck–Condon, or spectroscopic, excited surface with minima. (b) An extended surface touching. (c) An extended surface matching. (d) A surface crossing. (e) An excited-state minimum over a ground-state maximum due to a strongly avoided crossing.

6.11   The Nature of Funnels Corresponding to Surface Avoidings and Surface Touchings Involved in Primary Photochemical Processes

First, let us consider the *Franck–Condon (spectroscopic) minima* (Fig. 6.4a) that correspond to geometries from which *vertical* photophysical processes (e.g., R + $hv \rightarrow$ *R and *R $\rightarrow$ R + Δ) take place. Both *R and R correspond to *minima* of similar nuclear geometry (similar value of r_c) on both the excited- and ground-state surfaces. Radiative transitions from such excited state minima are considered "vertical" in the sense that the transition does not involve a significant change in the nuclear geometry; that is, these radiative transitions obey the FC principle (Sections 3.8 and 4.21). Equilibrated minima on the excited surface may be considered as starting points for all photophysical processes (fluorescence, phosphorescence, internal conversion, and ISC) for which the excited state *R(S_1 or T_1) ends up in the original minimum, R, from which light absorption took place. Photochemical reactions involve the motion of the representative point on the excited surface from *R from the FC minimum to other *geometries serving as "funnels" that take an excited- to a ground-state reactive intermediate, I, or conical intersection, F (*R $\rightarrow$ I or *R $\rightarrow$ F, respectively).* In contrast to FC minimum, I and F will possess geometries very different from those on the ground-state surface for R, since both I and F are very different chemical species than *R. (We are ignoring the *R $\rightarrow$ *I and *R $\rightarrow$ *P processes for simplicity at this point, since there are no new principles involved in these transformations.) Note that in the topology of an avoided crossing (Fig. 6.4e), the geometry, r_c, of the ASC corresponds to the geometry of the ground-state maximum! Thus, the *ASC funnel is *not* a FC minimum since the geometry at r_c on the ground surface is a maximum, not a minimum (we use the * label since this minimum clearly appears on an electronically excited surface). At geometries near r_c on the excited surface, transitions to the ground state are slow because of very poor FC factors at this geometry (Section 3.5). Pictorially, the excited-state vibrations are near the $v = 0$ level, whereas the ground-state vibrational levels are at values of $v >> 0$, resulting in very poor overlap of the vibrational wave functions (Chapter 3). Vibronic interactions are required to take *ASC to the ground-state surface. The specific vibrations (e.g., bending, twisting) that couple *ASC to the ground state will determine the geometries that will be produced when the transition occurs, and these geometries will in turn determine the structures of the ground-state products, P.

Surface touchings (Fig. 6.4b) are the most common surface topologies for primary processes involving n,π^* states. The primary process corresponding to this topology leads to diradical intermediates (radical pairs and biradicals) as the primary products in a *R $\rightarrow$ I(D) step. Surface touchings, crossings, and avoided crossings are common topologies for primary photochemical reactions involving n,π^* and π,π^* states. In the following sections, we will describe exemplars for each of these topologies.

6.12   "The Noncrossing Rule" and Its Violations: Conical Intersections and Their Visualization

The "noncrossing rule" applies to two energy curves if there is, in zero order, a geometry (r_c) for which two electronic energy curves possess *exactly the same symmetry, energy, and nuclear geometry*. The rule states that energy surfaces corresponding to two states of exactly the same symmetry do not cross but "avoid" each other at r_c. The idea behind the rule is that for an adiabatic surface (one that follows the Born–Oppenheimer approximation), when two states have the same symmetry, energy, and geometry, one might expect that if the nuclei are not moving too rapidly, there will always be some quantum mechanical mixing of the wave functions of *Ψ (the excited state) and Ψ (the ground state) to produce two adiabatic surfaces. One state will be of higher energy and the other will be of lower energy; that is, the surfaces avoid each other as a result of the quantum mechanical mixing of the wave functions *Ψ and Ψ at r_c. For example, in a zero-order approximation, two curves for a diatomic molecule may cross (Fig. 6.3d), but in a higher approximation, where actual distortions of a molecule and loss of idealized symmetry occur, the two curves will show an avoiding (Fig. 6.3e).

The noncrossing rule applies strictly *only* to molecules possessing very high symmetry all along a reaction coordinate at r_c. In practice, such a high-symmetry situation holds only for diatomic molecules, which are axially symmetric. However, polyatomic molecules, such as organic molecules, with many possible vibrations of different shapes and vibrational frequencies, generally possess little or no symmetry with respect to local geometries along the reaction coordinate, so that the noncrossing rule is no longer a strict selection rule; as a result, two electronic states of the same formal (spatial, vibrational, spin) symmetry may truly cross. This feature of surfaces was pointed out a number of years ago[7], but before the 1990s, there was little or no computational or experimental evidence as to whether surface crossings were real in general, or what the rules might be for occurrence of surface crossings. For example, the plausibility of true surface intersections between singlet-state surfaces in organic photochemistry was suggested by computations of excited-state surfaces in the 1970s.[1] In the 1990s, more advanced computations were possible because of advances in computers and software,[1,5] and the development of algorithms for computing energy surfaces showed convincingly that intersections of two PE surfaces are very common. These computations emphasized that in 3D the energy surfaces in the immediate vicinity of the touching point have the form of a double cone, which possesses a common vertex at r_c (as shown in Fig. 6.4d). *The double cone is termed a conical intersection (CI),* resulting from the touching of two electronic PE surfaces when plotted along two axes (energy and reaction coordinate).[7] The generation of the CI can be visualized as resulting from the sweeping of the intersection created by a surface crossing in 3D about the symmetry axis. It is now widely accepted by theorists that crossings between singlet surfaces and associated CIs are common and accessible from *R.

An important feature of a CI is that *if it is accessible from *R, it serves as a very efficient funnel, F, that takes the representative point rapidly from an excited to the ground-state surface toward products, P (photochemistry), or back to the ground state, R (photophysics).* Movement of the representative point along a "wall" of a conical intersection is essentially a vibrational relaxation of the system, and therefore provides a very efficient pathway from an excited to a ground surface; as such, it can be an efficient funnel leading from an excited- to the ground-state surface. The trajectories of the representative point through the tip of the cone may follow a steep slope of the cone wall (a strong driving force) and effectively convert the electronic energy of the excited surface into nuclear motion (vibrations) on the ground surface. *A conical intersection will not generally be a rate-determining electronic bottleneck in a radiationless pathway between surfaces of the same multiplicity.* The return from a conical intersection to the ground state can occur with unit probability when the representative point enters the region of the intersection. Thus, the rate of reaction through a CI is limited by the rate at which the representative point passes from *R to the CI. If barriers exist between *R and the CI, the rate of passing over these barriers will determine the rate of the surface crossing.

In the classical expression for the probability (P), a jump from an excited state to a ground state is proportional to the negative exponential of the square for the energy gap between the two states undergoing transition (Eq. 6.2). Since the energy gap is 0 at the common point of a conical intersection that connects one excited surface with another excited-state surface or with a ground-state surface, the classical probability for a surface transition is unity, that is, $P = 1$ and a transition between excited surfaces with a probability of 100%. In such a case, if there is no spin prohibition on the motion of the representative point, the rate of transition from the excited to the ground surface will be limited only by the time scale of intramolecular vibrational relaxation (IVR), which is $\sim$ 100 fs–10 ps, or 10^{-13}–10^{-11} s, for typical organic molecules. Such processes are termed as *ultrafast* photochemical and photophysical transitions. Thus, as a rule of thumb, the experimental observation of an ultrafast radiationless process can be considered as experimental evidence for a *R $\rightarrow$ F(CI) process. For example, the internal conversion after S_1 excitation of ethylene has been computed[8] to take place on the order of tens of femtoseconds ($\sim 10^{-14}$ s) from the broadening of the bands of ethylene's ultraviolet (UV) spectrum (internal conversion is so rapid that fluorescence is not detectable). This time scale is on the order of a torsional twisting vibration about the C=C bond in the ground state. Since the C=C bond is much weaker in the $S_1(\pi,\pi^*)$ state because of the loss of a π electron and the gain of a π^* electron, we can conclude that internal conversion to $S_0(\pi)^2$ takes place in the first twisting attempt on the singlet surface. Indeed, numerous observations of processes from *R that take place on the order of 10–100 fs have served as evidence for the occurrence of CIs, because such processes indicate rates on the order of vibrational relaxation (10^{13}–10^{14} s^{-1}), indicating a "barrierless free fall" from *R to some other lower energy state in the paradigm of Scheme 6.1.

In closing, note that "hot" ground-state reactions might be considered as plausible for the case in which conversion from a funnel to the ground state surface occurs to form a primary product that retains excess kinetic energy ($^\dagger P_1$). This excess KE can allow the representative point to overcome substantial energy barriers for reaction on the ground surface to form a new product P_2: *R $\rightarrow$ F $\rightarrow$ $\neq P_1$ $\rightarrow$ P_2. Such a process is not plausible, because the rate of IVR is so fast that the excess energy from the excited to the ground surface cannot be "stored" by $\neq P_1$ and be used to overcome significant energy barriers on the ground-state surface.

6.13 Some Important and Unique Properties of Conical Intersections

What are the essential differences between a CI (Fig. 6.4d) and an ASC (Fig. 6.4e)? The excited-surface minimum of an ASC corresponds to a geometry (r_c) that is thermally equilibrated; thus, the representative point oscillates about r_c with a characteristic zero-point motion. However, the representative point passes through a CI at r_c to the ground state so rapidly that there is not enough time for thermal equilibrium of the system. In addition, when the representative point approaches a CI, its molecular electronic structure becomes controlled by forces that are absent when we consider equilibrated species that obey the Born–Oppenheimer approximation. In the latter case, specifying the positions of the nuclei (r) completely characterizes the electronic structure and the reaction coordinate. In contrast to the reaction coordinate that is a trajectory along an equilibrated energy surface, there exists *deactivation coordinates* that consist of "seams" that connect CIs along an excited surface.[5] The position of a deactivation coordinate has no necessary natural electronic connection with the reaction coordinate (lowest-energy pathway) along the excited surface. The representative point may encounter CIs as it makes its trajectories along the reaction coordinate of an excited state. When the representative point happens to intersect a deactivation coordinate, a CI has been reached and a rapid jump from an excited-state to a lower-state surface occurs.

Clearly, it is very important to be able to predict the molecular structures that correspond to the deactivation coordinate (i.e., the molecular geometries that correspond to a family of connected CIs). However, such predictions are complex because, in order to predict the behavior of the representative point when it encounters a deactivation coordinate, one must specify not only the coordinates of the representative point (r), but also the *direction* and the *velocity* of the representative point as it approaches the deactivation coordinate. Currently, the information required to specify the behavior of the representative point as it approaches a CI requires an explicit quantum mechanical computation,[1,5] the details of which are beyond the scope of this book. However, we will see in later sections that zero-order surface crossings can be predicted without computation from qualitative state correlation diagrams and that these surface crossings serve as a guide to the molecular geometries of CIs.

The reaction coordinate for *R on the excited surface and the deactivation coordinate for the *R → CI process may correspond to completely different molecular geometries. The path for deactivation of *R will depend on the accessibility of a path from *R → CI, that is, on whether there are any barriers along the reaction coordinate between the initial spectroscopic minimum in which *R is formed and the deactivation coordinate. Thus, when a CI is involved, the rate of reaction may take only femtoseconds when the path is direct from the minimum of *R, or it may take much longer if there is a barrier between the minimum of *R and the CI. In the latter case, the rate of deactivation is determined by the time it takes the representative point to find the CI.

Figure 6.5 elaborates on Fig. 6.4 by showing the movement of the representative point along the reaction coordinate for primary reactions after the point has moved from its initial spectroscopic minimum (Fig. 6.5a). In Fig. 6.5b and c, the situation is straightforward, and the *R → I and *R → *I (or *P), respectively, are readily visualized in terms of energy surfaces.

In Fig. 6.5e, the *R → ASC is more interesting since the representative point may either proceed to a product P or return to R. The interpretation is straightforward, because the representative point becomes equilibrated on the excited surface, and in its oscillations near the minimum, *R may undergo deactivation (1) when an oscillation is moving the representative point toward the structure of R on the ground surface (resulting in a photophysical internal conversion from *R to R), or (2) when an oscillation is moving the representative point toward the structure of P on the ground surface (resulting in a photochemical primary process to produce P). Now, we have refined our view of the manner in which *R can undergo radiationless processes that are related, one leading to photochemistry and the other to photophysical deactivation (Fig. 5.17).

Perhaps the most interesting and nonintuitive situation is given in Fig. 6.5d, where *R is shown as entering a CI and then producing several products (P_1 and P_2) in addition to undergoing an internal conversion to R. This sort of behavior may be general for the representative point when it passes through the CI because of the fundamental nature of the forces that exist in a CI but are absent for adiabatic (thermally equilibrated) surfaces. We can obtain a qualitative insight into the origin of these forces[1,5], which lead to a multiplicity of products from a CI, by considering the vibrational mixing that occurs between a ground and an excited state in the region near a CI (a more rigorous basis refers to two independent forces or vectors, as described

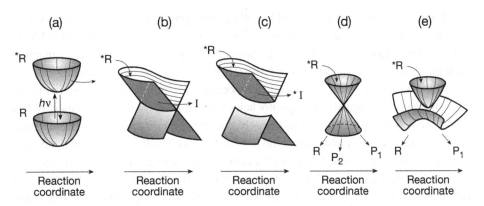

Figure 6.5   Passage of the representative point regions of an energy surface. (a) A ground-state and FC, or spectroscopic, excited surface with minima. (b) An extended-surface touching. (c) An extended-surface matching. (d) A surface crossing (a conical intersection). (e) An excited-state minimum over a ground-state maximum, due to a strongly avoided crossing.

below). In general, more than one vibrational mode will be able to mix excited- and ground-state surfaces. The vibrational mode that mixes the excited and ground states will direct the representative point in a specific direction (favored by the mixing vibration) that is along a particular reaction coordinate toward a particular product (say, R, P_1, or P_2 in Fig. 6.5d), as the point exits the bottom of the CI and leaves the excited-state surface.

This analysis[5] leads to the expectation that if several vibrational modes mix states in the vicinity of the CI, then *when the representative point passes through the CI, because of competing vibronic mixing, it will have certain probabilities of following more than one reaction coordinate, even though it may have followed only one reaction coordinate entering the CI.* This result is the basis for the possibility of forming more than one product, even if the initial trajectory of the representative point is along a specific reaction coordinate and if the deactivation pathway is through a single CI. A visualization of this situation appears in Fig. 6.5d, where three products (R, P_1, and P_2) are produced from a single *R. The nature of the reaction on the ground-state surface after passing through the CI will be controlled by the direction of the velocity of the representative point and by the nature of the ground-state minima that are accessible for "capture" of the representative point.

It is informative to compare a CI, which corresponds to a critical transition geometry that serves as a funnel on an excited surface, to the well-known transition state of thermal reactions (Fig. 6.1), which corresponds to the geometry of a maximum on the ground-state energy surface. On the ground state, r must pass through the transition state on the way to P. On the excited surface, r must pass through a funnel somewhere along the *R → P reaction pathway. On the ground state, the transition state can be characterized by a single vector that corresponds to the motion of the representative point through the maximum energy point along the reaction coordinate. In contrast, a CI is characterized by two mutually *independent* vectors (one corresponding to the gradient dPE$/dr$, which is a force that pushes or pulls the representative point toward certain structures, and the other corresponding to a state-mixing vector that makes structures that mix better, and are more likely to form). The interactions between these two vectors provide the possibility for multiple *independent* reaction path directions after passing through the CI. Knowledge of the geometries of a CI and of the directions of the gradient difference and the state-mixing vector are of central importance in understanding photochemistry that involves the involvement of CIs along the reaction pathway. Unfortunately, the usual qualitative methods employed for structural-reactivity correlations do not work for a CI; full computational methods must be used to determine whether a CI is plausible along a reaction coordinate.

The CIs may correspond to nuclear geometries that are far removed from initial FC minima on an excited surface (as is the case for surface touchings and avoided crossings). If the representative point, in exploring an excited surface, crosses a deactivation coordinate and falls into a CI, its rate of passage from the excited surface will be on the order of vibrational relaxation. Because of the competing forces the representative point experiences, when it approaches the apex of the CI, its trajectory is easily directed toward more than one product on the ground-state surface. When a CI is involved in a photochemical process, the rate-determining step is exploration of the excited surface by the representative point before it finds and crosses a deactivation coordinate and falls into the CI. The process $^*R \rightarrow CI \rightarrow P$ is the excited-state analogue of a concerted reaction or elementary step, that is, a process for which there is no true reactive intermediate involved on the path from $^*R$ to P. Thus, from a mechanistic point of view, a CI that serves as a funnel on the excited surface plays a similar role to a transition state on the ground-state surface. Both a CI and a TS describe the nuclear geometry of a transition structures, a fleeting species whose lifetime is on the order of vibrations. In a thermal reaction, the ground TS corresponds to the point of PE for which the probability of passage from the reactant to the product is maximal. In a photochemical reaction, the CI corresponds to the point where the probability of transition to the ground state is maximal.

Figure 6.6 schematically compares two of the most common situations for radiationless deactivation for the overall $^*R \rightarrow P$ process and indicates important differences between hypothetical reaction paths that passes through a CI (Fig. 6.6, left) and those that proceed via a reactive intermediate (I) through a surface touching (Fig. 6.6, right). The middle of Fig 6.6 shows the radiative transition between the FC minimum of R and $^*R$.

Consider the two separate hypothetical reaction coordinates from the FC minimum of $^*R$; Path 1 (Fig. 6.6, left) is a $^*R \rightarrow CI \rightarrow P_2 + P_3$ pathway. If the representative point experiences a lower energy barrier in moving to the left from the initial FC geometry of $^*R$, it moves toward the region of the CI. A deactivation coordinate (the lowest-energy points corresponding to all connected CIs) is shown crossing with the reaction coordinate r_c on the way from $^*R$ to P. When the representative point enters the CI region from the excited surface, it is possible that a bifurcation will occur (i.e., two paths will be followed), due to the separate forces of the gradient differences of the excited- and ground-state surfaces and of the mixing of the wave functions of the excited and ground states. As a result, there is a certain probability that upon leaving the CI, the representative point will be directed toward the product P_2, which is on the original reaction coordinate. There is also a certain probability that upon leaving the

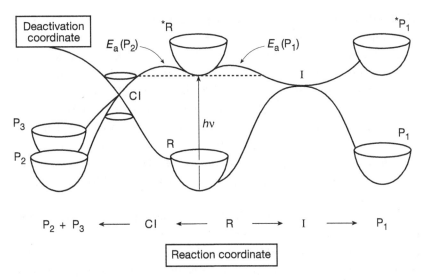

Figure 6.6   Schematic representation of two primary photochemical processes. Left: *R → CI→P$_2$ (and/or P$_3$). Although the original reaction coordinate may move the representative point toward P$_2$, after passing through the CI, P$_3$ may be formed as a result of a bifurcation from the motion of the representative point in passing through the canonical intersection CI. Right: *R → I → P$_1$. In this case, only P$_1$ is formed.

CI, the representative point will be directed toward a new product, P$_3$. In fact, there is even a probability that some of the trajectories of the representative point might take it toward R, resulting in a net photophysical radiationless transition *R → CI → R.

Figure 6.6 (right) shows a second hypothetical reaction coordinate for a *R → I → P$_1$ process. The representative point is shown moving to the right from the geometry of *R toward the geometry of a reactive intermediate, I, which is a thermally equilibrated species. The process is direct, and once the surface touching is reached, I is formed. There is no abrupt change in the electronic structure of *R on its way to P in this case and only one product (P$_1$) is formed. Whether the representative point moves to the left or right is determined by the presence and/or size of the energy barriers, E_a, between *R and the CI or I.

In some cases, the two topologies shown in Fig. 6.6 may "merge." The essential feature of a process involving a CI is the intersection of the reaction and deactivation coordinate (the "energy seam" of CIs). When such a deactivation coordinate crosses near the geometry of I, the representative point may fall into a CI, making the two cases in Fig. 6.6 merge into very similar descriptions of the reaction.

6.14   Diradicaloid Structures and Diradicaloid Geometries[1,9]

The terms "diradical" and "biradical" have sometimes been used interchangeably in the literature to describe the reactive intermediate (I). In this section, we review the definitions that will be used throughout the remainder of this text. We use the term *diradical* (D) to represent any reactive intermediate that can be viewed as possessing two independent radical centers, that is, two half-filled MOs. The most common examples of such species in organic photochemistry are RP, for which the radical centers occur on two molecular fragments, and BR, for which the radical centers occur on a single molecular fragment. The homolytic cleavage of a C—C bond in propane produces a *methyl–ethyl radical pair* ($CH_3 \bullet + \bullet CH_2CH_3$), whereas the cleavage of the C—C bond of cyclopropane produces a 1,3-trimethylene *biradical* ($\bullet CH_2CH_2CH_2 \bullet$). For the purposes of this text, *both the methyl–ethyl RP and the 1,3-trimethylene BR are considered diradicals (D)*. Thus, for the remainder of the text, it is understood that a reactive intermediate that is labeled I(D) may be either an I(RP) or an I(BR).

The terms "diradicaloid" (and "biradicaloid") has been used in the photochemical literature to describe the electronic characteristics of certain structures possessing two electrons that occupy two orbitals of similar or equal energy. The text will employ just one of these terms, *diradicaloid,* to describe the electronic structure of molecules that possess two nearly equal energy, nonbonding orbitals containing two electrons in some configuration. The importance of the diradicaloid concept is that *many minima on excited surfaces correspond to diradicaloid structures.* We will see that diradicaloid structures, in addition to exhibiting the expected diradical characteristics, sometimes exhibit *zwitterionic* characteristics depending on the structure of the system. Furthermore, diradicaloid structures can be correlated with and derived from equilibrium geometries by twisting of double bonds or stretching of single bonds. A *perfect diradical* is defined as a diradical that satisfies two conditions: *its two nonbonding orbitals have exactly the same energy and do not interact electronically at all.* A *diradicaloid* may be considered as any diradical that does not meet both of these conditions: *either the two orbitals have (slightly) different energies or interact weakly (or both).* Let us describe some of the important features of diradicaloid structures.

Real molecules never possess the expected properties of perfect diradicals, but if their electronic structure contains two nonbonding orbitals that can be occupied by two electrons, we can use the characteristics of a perfect diradical as a zero-order approximation of a diradicaloid. The structure of a molecule that corresponds to a diradicaloid is termed a *diradicaloid geometry.* Radical pairs and biradicals will generally possess diradicaloid geometries. Importantly, from the standpoint of the theory of photochemical reactions, diradicaloid geometries often can be correlated with and related to the geometries of surface touchings (Fig. 6.4b), CIs (Fig. 6.4d), or avoided crossing minima (Fig. 6.4e). As such, the species possessing these geometries are important both because they are inherently chemically reactive and because they serve as funnels from the electronically excited to the lower excited states and to the ground state as the representative point proceeds along the reaction pathway.

An important electronic feature of diradicaloid geometries is the general possibility of *four* low-lying *electronic configurations and states* that could result from occupancy of both electrons in a single nonbonding orbital. Figure 6.7 shows the relationship between the HO and LU of *R (left), a perfect diradical (middle), and a diradicaloid (right). The relative energies are not to scale.

In the case of a perfect diradical (Fig. 6.7, middle), the energies of the two NB orbitals are exactly equal ($NB_1 = NB_2$) and there is no orbital overlap between

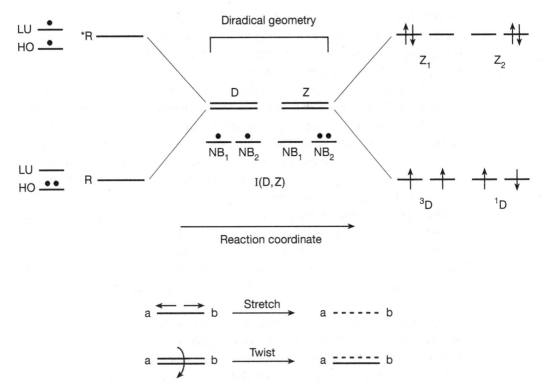

Figure 6.7   Schematic description of the surface relationships of excited states, diradicals, and zwitterions.

NB_1 and NB_2. As a result, there is no energy difference between the configurations $(NB_1)^1(NB_2)^1$, $(NB_1)^2$, and $(NB_2)^2$. Note that the first configuration is that of a classical diradical and the latter two configurations resemble that of zwitterions. When the energy of NB_1 and NB_2 differ and when they overlap slightly, the electronic system switches from a perfect diradical to a diradicaloid (Fig. 6.7, right). For the diradicaloid, the energies of the possible configurations differ, with the classical diradical configuration $D = (NB_1)^1(NB_2)^1$ generally of lower energy than either of the zwitterion configurations $Z_1 = (NB_1)^2$ and $Z_2 = (NB_2)^2$. However, the energy difference between the D and Z states may be small, so that mixing between states may occur. This potential for mixing is a characteristic feature of diradicaloid structures.

The Z configurations must be spin-paired singlets because for the Pauli principle: $Z_1 = NB_1(\uparrow\downarrow)NB_2(\ )$ and $Z_2 = NB_1(\ )NB_2(\uparrow\downarrow)$, where $NB(\ )$ means there are no electrons in NB. Since both Z configurations have their electrons paired, they must be singlets, so a superscript spin label is unnecessary. However, there are two distinct states since the electrons may be paired in either NB_1 or NB_2 to yield Z_1 or Z_2, respectively.

There are two possible *spin* configurations for a D configuration: (1) the singlet, $^1D = NB_1(\uparrow)NB_2(\downarrow)$, and (2) the triplet, $^3D = NB_1(\uparrow)NB_2(\uparrow)$. In MO terms, such configurations are termed *covalent*.

A very important conclusion from the above analysis is that a *diradicaloid* structure (when NB_1 and NB_2 have slightly different energies or orbitals that slightly overlap) will always have the possibility of possessing a set of *four* states (^1D, ^3D, Z_1, and Z_2;

Fig. 6.7, right).[6] In the text, the symbol Z will refer to the set of two zwitterionic states, Z_1 and Z_2, and the symbol D will refer to the set of two diradical states, ^1D and ^3D. The ranking of the energies and the energy separations of these states will depend on the nature of NB_1 and NB_2. For carbon-centered, nonbonding orbitals, the energies of NB_1 and NB_2 will typically be very similar, and the ^1D and ^3D covalent configurations will generally be lower in energy than the charge-separated Z configurations (carbon atoms do not like to pile up or give up charge unless they are attached to some electron-releasing or electron-withdrawing group, respectively). Thus, ignoring spin considerations for the moment, carbon-centered diradicaloids will possess two low-energy D states and higher-energy Z states. In most cases, the classical diradical structure will suffice to represent the D state.

Many photoreactions produce diradicals or diradicaloid structures as primary photochemical products, that is, conform to a $*R \rightarrow I(D, Z)$ primary photochemical process, where I is a diradicaloid structure. For example, the $*R \rightarrow I$ primary process for organic molecules involving n,π^* states typically produces a diradicaloid geometry that may be viewed in more detail as a $*R(n,\pi^*) \rightarrow I(D)$ process ($D = {}^1D$ or ^3D). *All triplet states (n,π^* or π,π^*) generally follow the pathway $*R(T_1) \rightarrow I(^3D)$, since a $*R(T_1) \rightarrow I(^1D)$ process would violate spin selection rules.* On the other hand, the primary photochemical processes of many singlet π,π^* states may be viewed as $*R(\pi,\pi^*) \rightarrow I(Z)$. We will see that these ideas provide a tremendous simplification of the number of plausible mechanisms that need to be considered in describing the pathways of photochemical reactions.

6.15 Diradicaloid Structures Produced from Stretching σ Bonds and Twisting π Bonds[1]

Two of the simplest and most fundamental exemplars of diradicaloid geometries may be reached from normal equilibrium geometries of R or *R (Fig. 6.8) by (1) the stretching and breaking of a σ bond of R and *R, and (2) the twisting and breaking of a π bond of R and *R. In spite of their simplicity, these two exemplars provide a nearly universal conceptual basis for the interpretation, classification, and understanding of the primary photochemical processes of the basic chromophores of organic chemistry (carbonyl compounds, olefins, enones, and aromatic compounds). Specific exemplars of primary photochemical processes corresponding to σ-bond stretching and π-bond twisting will be described in Sections 6.40 and 6.41.

6.16 An Exemplar for Diradicaloid Geometries Produced by σ-Bond Stretching and Bond Breaking: Stretching of the σ Bond of the Hydrogen Molecule

The stretching and breaking of the single σ bond of a hydrogen molecule (H—H) into two hydrogen atoms (H• + H•) is an exemplar for the diradicaloid geometry produced by stretching and breaking *any* σ bond (including the C—C bonds of organic

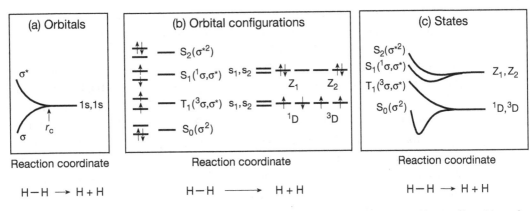

Figure 6.8   Orbitals, orbital configurations, and state correlation diagrams for stretching and breaking of a σ bond for a hydrogen molecule.

molecules). As the H—H bond stretches and the distance between hydrogen nuclei increases, the representative point (r) eventually reaches a geometry (r_c) for which the bond can be considered as completely cleaved and a diradicaloid structure is produced (Eq. 6.3).[1,6,9] On the ground-state surface, this corresponds to a thermal R → I(D) process, and on the excited-state surfaces this corresponds to a photochemical *R → I(D, Z) process.

$$H—H \rightarrow \quad H\text{——}H \qquad \rightarrow H\bullet + H\bullet \qquad (6.3)$$

Stretched σ-bond   Cleaved σ-bond
diradicaloid   diradical

When the σ bond is nearly completely broken, two degenerate hydrogen atomic 1s orbitals are formed. This geometry corresponds to our definition of diradicaloid geometry (two nonbonding orbitals weakly overlapping and two electrons available to occupy them), so that the four electronic states of the diradicaloid are possible (^{1}D, ^{3}D, Z_1, and Z_2; Fig. 6.8). Figure 6.8 shows the qualitative behavior of the stretching of the σ bond of H_2 in terms of (a) the behavior of energy for the bonding σ (HO) and antibonding σ^* (LU) orbitals, (b) the possible orbital configurations and states that can be constructed from these orbital configurations, and (c) the behavior of the four electronic surfaces corresponding to the four electronic states as the bond is stretched and broken. The qualitative description of the stretching and breaking of the carbon–carbon bond of CH_3—CH_3 can also be represented in Fig. 6.8 by replacing the HO and LU of H_2 with the HO and LU of CH_3—CH_3.

Let us examine in detail the behavior of the σ and σ^* MOs as the H—H bond stretches and breaks. The electronic *orbital* correlation diagram is shown in Fig. 6.8a. When the H nuclei are close to their equilibrium separation of ~ 0.5 Å, the σ HO orbital is very low in energy relative to the σ^* LU orbital, but as the bond begins to stretch, the energy gap between the ground- and excited-state orbitals decreases; finally, as the bond is broken, two nonbonding (NB) 1s orbitals of equal energy are produced. This corresponds to the geometry of the diradicaloid for an extensively stretched (i.e., essentially σ bond broken) H_2.

The electronic *configurations* and *states* that can develop from the diradicaloid geometry resulting from the breaking of a σ bond are shown in Fig. 6.8b. In terms of orbital configurations, the four lowest-energy electronic states that can be derived for the equilibrium nuclear geometry of H_2 are $S_0(\sigma)^2$, $T_1(\sigma, \sigma^*)$, $S_1(\sigma, \sigma^*)$, and $S_2(\sigma^*)^2$ (Fig. 6.8b). As the H—H bond stretches, the energy of the σ HO orbital increases and the energy of the σ^* LU orbital decreases (Fig. 6.8a). When the bond has stretched considerably and the H nuclei are far apart ($\sim$ 2–3 Å), both the σ and σ^* orbitals will approach the same energy and correlate with a pair of NB atomic 1s orbitals, one on each H atom. The latter geometry (r_c) corresponds to a diradicaloid geometry, and as we have seen, four electronic states (^1D, ^3D, Z_1, and Z_2) will exist at the diradicaloid geometry (Fig. 6.8b). The four states derived from completely separated atoms H + H, in the order of increasing energy (ignoring electron-exchange effects), are $^3D(1s_1, 1s_2) \sim {}^1D(1s_1, 1s_2) \ll Z_1(1s_1)^2 = Z_2(1s_2)^2$. The ^3D and ^1D states will have similar energies, but the ^3D will generally be slightly lower in energy than the ^1D state because of exchange interactions (Chapter 2).

Moving the representative point to larger separations transforms the diradicaloid structure into a thermally equilibrated intermediate, I(D), a diradical, which consists of two hydrogen atoms (Fig. 6.8b). For an organic molecule, I(D) will typically represent a carbon-centered RP (e.g., the α-cleavage of an alkanone) or BR (e.g., the α-cleavage of a cyclanone) reactive intermediate. For the one-electron level of approximation that we are using, the energies of the two D and two Z states are equal. For a higher level approximation, application of Hund's rule (Section 2.12) places ^3D at a lower energy than ^1D. The two Z states will split apart in energy for any asymmetrical nonbonding orbital situation that drops the energy of one of the nonbonded orbitals relative to the other. The variation of the energy for a diradical with structure has been discussed in Section 2.12.

The *state* correlation diagram[1] of the reaction coordinate for breaking a H—H bond is shown in Fig. 6.8c. In a correlation diagram, states of reactants are correlated to the product states produced through the correlation of the orbitals as a function of a specified reaction coordinate; the latter is the stretching of the H—H bond. The σ orbital will correlate with a 1s orbital on each of the atoms and lead to a $D(1s_1, 1s_2)$ state in the product. A qualitative correlation can be made as follows (correlation diagrams are considered in more detail in Section 6.23 and following): since the ground-state S_0 is a singlet, it correlates with the lowest-energy $^1D(1s_1, 1s_2)$ state of the diradicaloid. The $T_1(\sigma, \sigma^*)$ state must correlate with the $^3D(1s_1, 1s_2)$ state of the diradicaloid, since the latter is the only triplet state of the product. By exclusion, *both* $S_1(\sigma, \sigma^*)$ and $S_2(\sigma^*)^2$ must correlate with Z_1 or Z_2.

The state correlation diagram shown in Fig. 6.8c is an exemplar for the behavior of the energy for an energy surface that tracks the cleavage of any σ bond. The situation shown in Fig. 6.8c, in fact, corresponds closely to the actual surfaces for the hydrogen molecule, where it is known from experiment and computation that the surface along the S_1 and S_2 possesses shallow minima just before the diradicaloid geometry is reached. These minima represent an energetic compromise arising from the competing tendency to minimize the energy between the σ and σ^* orbitals as the bond stretches and to minimize the energy required for charge separation in the zwitterionic states, which is favored for small nuclear separations. Thus, the stabilizing attraction of positive charge for negative charge is slightly greater than the destabilizing decrease in the bonding due to separation of the atoms. Such shallow minima can be assumed to be general for all simple σ-bond cleavages. In our classification of surface topologies in Figs. 6.3 and 6.4, we indicated that the adiabatic *R $\to$ *I processes were possible. In terms of diradicaloid geometries, we can see that $S_1 \to Z$ reactions would correspond to an adiabatic *R(S_1) $\to$ *I(Z) process, since the Z state is an excited state of the reactive intermediate I.

In contrast to the singlet state, which possesses a shallow minimum just before r_c, *the triplet state does not possess a minimum for any geometry* but eventually "flattens out" energetically in large separations for which the triplet surface "touches" the ground-state surface (Fig. 6.3b). We can classify the latter as a surface touching corresponding to a *R(T_1) $\to$ I(D) process. The parameter S_0, of course, possesses a deep minimum corresponding to the stable ground-state geometry of the molecule, which corresponds to a much shorter internuclear distance than the diradicaloid geometry as the result of the two electrons *both* going into the strongly bonding HO.

In summary, the exemplar for the surface behavior corresponding to a simple σ-bond cleavage (Fig. 6.8) possesses the following surface characteristics, which are useful in the interpretation of many situations where the major photochemical process involves stretching and breaking for a σ bond of *R:

1. The stretching of the σ bond produces a diradicaloid geometry when the bond is essentially broken, and at that geometry the corresponding cluster of associated four diradicaloid states (^{3}D, ^{1}D, Z_1, Z_2) results.

2. Along the ground surface (S_0) the bond is stable at all geometries except for large separations. The ground state correlates with ^{1}D. A large activation energy is required for thermal stretching and cleavage of the σ bond.

3. Along the triplet surface (T_1) the bond is *unstable* at all geometries, and little or no activation energy is required for photochemical cleavage of T_1 to produce I(^{3}D).

4. Along the S_1 and S_2 surfaces, the bond is unstable and stretches spontaneously but possesses a shallow minimum for geometries of a very stretched, but not completely broken, bond (resulting from weak zwitterionic attractions between positive and negative charges); from this diradicaloid geometry, a small energy barrier must be overcome for complete cleavage to produce the I(Z).

The stretching and breaking for the σ bond of the hydrogen molecule serve as an exemplar for the details of *all* photoreactions involving extensive surface touching (Figs. 6.3b and 6.4b). This model will be an excellent starting point for reactions of the *R $\to$ I(D) type, which includes all of the reactions of n,π^* states of carbonyl compounds and many reactions of *triplet π,π^** states.

6.17   An Exemplar for Diradicaloid Geometries Produced by π-Bond Twisting and Breaking: Twisting of the π Bond of Ethylene

Now, let us consider an exemplar for tracking the behavior of the energy surfaces for the pathway of twisting and breaking of a C=C π bond, a very common process in the photochemistry of olefins. For this process, we use the exemplar for the twisting of an ethylene molecule (CH_2=CH_2) π bond, which provides an exemplar for twisting and breaking of a π bond. As the ethylene molecule is twisted, it eventually arrives at a nuclear geometry for which the two methylene groups are mutually perpendicular (90° geometry), as shown in Eq. 6.4.

At close to the 90° geometry, assuming sp^2 hybridization of the carbon atoms is preserved during rotation, the two CH_2 groups are mutually perpendicular, and the π bond is broken. Two degenerate nonbonding p orbitals are produced so that this structure corresponds to a diradicaloid geometry and a 1,2-diradical is produced, as shown in Eq. 6.4.

$$(6.4)$$

Planar geometry	Twisted geometry
Both CH_2 groups	CH_2 groups are
in the same plane	mutually perpendicular

As the π bond twists, the energy of the π orbital sharply increases, and the energy of the π^* orbital sharply decreases (Fig. 6.9a) because the bonding overlap between the two π orbitals decreases. Since two orthogonal (non-overlapping) nonbonding orbitals of equal energy are produced at the 90° geometry, a "perfect" diradical would be produced if the sp^2 hybridization of the carbon atoms is maintained; however, any molecular distortion (e.g., a pyrimization of one or both carbon atoms) that removes the degeneracy of the orbital energy or causes an overlap of the orbitals will cause the perfect diradical to become a diradicaloid. At the diradicaloid geometry, as in the case of a highly stretched σ bond, two different electronic configurations (D, Z) and four different electronic states (^1D, ^3D, Z_1, and Z_2) are possible (Fig. 6.9b). At 90° of twist (the diradicaloid configuration), the π and π^* orbitals have transformed into two orthogonal nonbonding p orbitals, one on each carbon atom.

The four electronic states derived from the possible orbital–spin configurations for the planar and twisted geometries are given in Eqs. 6.5 and 6.6.[1,9]

$$\text{Planar ethylene:} \quad S_0(\pi)^2,\ T_1(\pi,\pi^*),\ S_1(\pi,\pi^*),\ \text{and } S_2(\pi^*)^2 \quad (6.5)$$

$$\text{90° Twisted ethylene:} \quad {}^3D(p_1, p_2),\ {}^1D(p_1, p_2),\ Z_1(p_1)^2,\ \text{and } Z_2(p_2)^2 \quad (6.6)$$

Twisting about the C=C bond of a *R(π,π^*) state of ethylene sharply relieves electron–electron repulsion resulting from the π^* electron. Furthermore, twisting

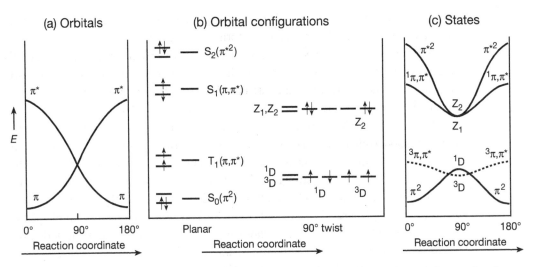

Figure 6.9   Orbitals, orbital configurations, and state correlation diagrams for twisting a π bond.

is easy in *$R(\pi,\pi^*)$ because the removal of an electron from the π orbital has considerably weakened the π bond, and the promotion of an electron to the π^* orbital has further weakened and essentially broken the π bond that existed in R. Notice from Fig. 6.9 that twisting about the carbon–carbon bond *will tend to lower the energy for all of the excited states of ethylene.* The reason for the energy lowering is due to the stabilizing character of the zwitterionic structures as the twisting occurs in S_1 and S_2. Thus, the electronic energies of S_2, S_1, and T_1 all drop rapidly as a function of ethylene twisting, because electronic excitation has effectively broken the π bond, and the bonding between carbon atoms in a *$R(\pi,\pi^*)$ state is similar to that of a carbon–carbon single bond for which free rotation occurs. On the other hand, the electronic energy of S_0 *increases* as the molecule is twisted, because the π bond is being broken as the p orbitals of the π bond are being decoupled.

The state correlations $S_0 \rightarrow {}^1D$, $T_1 \rightarrow {}^3D$, $S_1 \rightarrow Z_1$, and $S_2 \rightarrow Z_2$ may be made by inspection (Fig. 6.9c) on the basis of simple orbital-symmetry considerations. The symmetry element that brings the starting planar geometry into the twisted (diradicaloid) geometry is the one associated with the rotation of one CH_2 group.[9] The overall state symmetries must be defined in terms of this symmetry element. Although the rigorous state correlation is best done by use of group theory and point-group analysis, the following qualitative description will indicate the basis of the correlation.

The wave function for the $S_0(\pi)^2$ configuration (Fig. 6.9b) at the planar geometry is essentially covalent in character (two electrons are shared, but there is essentially one electron on each carbon atom)[1]; that is, there is very little ionic character to planar ethylene, and the wave function for π^2 has (in terms of atomic orbitals) the form $p_1(\uparrow)p_2(\downarrow)$. This means that for $S_0(\pi)^2$, at all times there is always only one p electron near carbon 1 and one near carbon 2 (i.e., the bond is covalent), and the two electrons have paired spins. Thus, there is very little Z character to S_0. For the $T_1(\pi,\pi^*)$ configuration at the planar geometry, there can never be two electrons placed on one carbon in the same p orbital (a violation of the Pauli principle), since the electrons have parallel spins. The T_1 state is thus *purely covalent* and has absolutely no ionic character, and its wave function has the form $^3(\pi,\pi^*) = p_1(\uparrow)p_2(\uparrow)$.

The wave functions for $S_1(\pi,\pi^*)$ and $S_2(\pi^*)^2$ differ from that of $S_0(\pi)^2$ and this difference must reflect the basis for the high-energy content of the states. Computation indicates that the former two states are best described by *zwitterionic* wave functions and that the zwitterionic character increases as the degree of twist increases.[1] As twisting about the C=C bond occurs, the Z states become degenerate (in the one-electron approximation) at the 90° geometry. A more sophisticated approach to the state correlation diagram indicates that $S_0(\pi)^2$ and $S_2(\pi^*)^2$ correlate with each other in zero order, but there is a strongly ASC of the energy surfaces at the 90° geometry. This finding leads to an ASC minimum in both $S_1(\pi,\pi^*)$ and $S_2(\pi^*)^2$ and an ASC maximum in S_0. There is considerable experimental evidence that this simple picture is consistent with the known π,π^* photochemistry of ethylenes and their derivatives and aromatic hydrocarbons and their derivatives.

In summary, the important qualitative features of the state correlation diagram for a twist about the ethylene double bond (Fig. 6.9c) are

1. The occurrence of *minima* that serve as *funnels* on the $S_2(\pi^*)^2$, $S_1(\pi,\pi^*)$, and $T_1(\pi,\pi^*)$ surfaces at the 90° diradicaloid geometry, which correspond to Z_2, Z_1, and ^3D, respectively.

2. The occurrence of a *strongly avoided crossing nature* of the minimum on the excited surface at Z_2. That is, a zero-order crossing of $S_0(\pi)^2$ and $S_2(\pi^*)^2$ surfaces is strongly avoided, and the adiabatic surfaces show a minimum in $S_2(\pi^*)^2$ and a maximum in $S_0(\pi)^2$ as the result of the strong avoiding.

3. The geometry corresponding to the maximum in the $S_0(\pi)^2$ surface corresponds to the ^1D geometry.

4. The $S_0(\pi)^2$ and $T_1(\pi,\pi^*)$ states, and the S_2 and S_1 states, respectively, "touch" at the diradicaloid 90° geometry. Importantly, at the diradicaloid geometry the $S_2(\pi^*)^2$ state is degenerate with the $S_1(\pi,\pi^*)$ state. Notice that this touching only occurs near the 90° geometry and is not an extended touching, as in the case of the σ-bond cleavage.

5. The zwitterionic (closed-shell) electronic character of $S_2(\pi^*)^2$ and $S_1(\pi,\pi^*)$ for all geometries corresponding to significant twisting.

6. The diradical electronic structure of $T_1(\pi,\pi^*)$ holds for all geometries, independent of the extent of twisting.

Geometries with large average distances between the nonbonding orbitals are termed "loose" diradicaloid geometries; e.g., $^*R(T_1)$ of H—H. Those with small average distances between the nonbonding radical centers are termed "tight" diradicaloid geometries (e.g., $^*R(S_1)$ of H—H). As an exemplar for the diradicaloid geometries of S_1 and T_1 states in general, funnels from diradicaloid $^*R(S_1)$ and $^*R(T_1)$ states may occur at different geometries. This difference in initial geometries can result in different photochemistry from $^*R$ states that possesses the same orbital configurations but differ in spin. Furthermore, because the diradicaloid structures are by definition nonbonding orbitals, the value of the exchange integral, J, will typically be very small for such structures when the orbitals are oriented at nearly 90° to one another.

In Section 6.24, we see that the state correlation diagrams for *thermally forbidden* ground-state pericyclic reactions have the *same* general topology as those for a twisting about the double bond of ethylene (Fig. 6.9). The exemplar for twisting about a simple π bond has far-reaching utility in the interpretation of photochemical reactions of many other systems containing C=C bonds. By creating funnels, twisting is also implicated in creating a CI.

6.18   Frontier Orbital Interactions
As a Guide to the Lowest-Energy Pathways
and Energy Barriers on Energy Surfaces

The simple energy surfaces based on the stretching of σ bonds (Fig. 6.8) and the twisting of π bonds (Fig. 6.9) provide qualitative exemplars of how energy surfaces behave for two important geometry changes that are common for photoexcited organic molecules. These energy surfaces describe the behavior expected for the ground state of R in addition to the excited states *R(S_1, S_2, and T_1). Now, we develop a more general approach to examine how energy surfaces behave as a function of reaction-path geometries for the two important primary processes *R $\rightarrow$ I (formation of a diradical, D) and *R $\rightarrow$ F $\rightarrow$ P. For the latter, *R does not form a conventional reactive intermediate (I), but rather proceeds from *R to a funnel (F) and then directly to P. We start with a description of how the use of orbital interactions can provide selection rules for the lowest-energy pathways for photochemical reactions of electronically excited states (*R) of organic molecules. Inspection of orbital interactions will allow us to create a set of selection rules or, loosely speaking, a set of plausible primary photochemical reaction pathways for *R. Then, we analyze some exemplars for *R $\rightarrow$ I and *R $\rightarrow$ F $\rightarrow$ P processes in terms of orbital and state surface-energy correlation diagrams, which will serve as a basis for describing the reaction coordinate and for the occurrence of energy maxima and minima along the primary photochemical reaction pathway.

One may obtain a qualitative intuition for the occurrence of energy barriers and minima on electronically excited surfaces by starting with the concepts of so-called "frontier orbital interactions" in order to deduce the lowest-energy reaction coordinate of a selected reaction, and then using an idealized symmetry to represent this reaction coordinate, which will serve as the basis for the generation of an orbital or state correlation diagram.[4] The theory of frontier orbital interactions assumes that the reactivity of organic molecules is determined by the *very initial interactions* that result from the transfer of electrons in an occupied orbital to an unoccupied (or half-occupied) orbital. *The most important orbitals in the frontier orbital analysis are the HO and LU of the ground state of an organic molecule.* As a zero-order approximation, the same LU of the ground state is used for excited states. Thus, R has a frontier orbital configuration of $(HO)^2(LU)^0$, and *R has a frontier orbital electron configuration of $(HO)^1(LU)^1$.

Two important features of the interacting frontier orbitals determine the extent of favorable charge-transfer (CT) interaction from the electrons in the HO to the vacant LU orbital: *(1) the energy gap between the two orbitals,* ΔE_{HO-LU}, *and (2) the degree of positive orbital overlap (bonding) between the two orbitals, that is, value of the overlap integral* $< HO|LU >$. These are the same two features that we have seen many times as a variation of Eq. 3.4 from perturbation theory: Small energy gaps and positive overlap integrals are the best conditions for constructive resonance between two states, and the degree of constructive resonance determines the degree of mixing that stabilizes the states. For a comparable energy gap between the orbitals, significant positive (in-phase, constructive interference of wave functions) overlap of the interacting HO and LU orbitals of reactants signals a small energy barrier to CT and a low activation barrier to reaction. On the other hand, negative (out-of-phase, destructive interference of wave functions), or zero, net overlap of the HO and LU orbitals signals a large energy barrier to CT and a large barrier to reaction.

Suppose that the representative point starts on a given electronically excited surface from *R and that there are two reaction-pathway choices (say *R → I$_1$ and *R → I$_2$) that possess different energy barriers for reaction. We may consider the reaction-coordinate pathway with the smaller energy barrier to be plausible (or allowed) and the pathway with the larger energy barrier to be implausible (or forbidden). In effect, we postulate that reactions prefer to proceed via pathways that involve transition structures that have obtained the most favorable positive orbital overlap and the smallest energy gap between the interacting orbitals. Thus, consideration of initial or frontier orbital interactions provides a basis for *selection rules of plausibility* (termed *allowed* if plausible or *forbidden* if implausible) for *possible* reaction pathways.

After hypothesizing a reaction pathway through the use of frontier orbitals, one can obtain qualitative information concerning the maxima and minima on an excited surface from orbital correlation diagrams and state correlation diagrams. Constructing a qualitative correlation diagram requires a model with a certain level of known (or assumed) symmetry. Within the framework of correlation diagrams, if an initial orbital (or state) makes an endothermic (uphill) correlation with a high-energy product orbital (or state) along one reaction coordinate, a higher energy barrier is expected along the reaction coordinate compared to the situation for which an initial orbital (or state) makes an exothermic (downhill) correlation with a low-energy product orbital (state) along another coordinate. Thus, from the correlation diagram we can obtain qualitative information on correlation-imposed energy barriers. Therefore, we have basic selection rules for plausible reaction paths based on frontier orbital considerations: An uphill process expected from a correlation diagram is termed *forbidden,* and the downhill process expected from a correlation diagram is termed *allowed.* In effect, we have created a selection rule: Namely, that the movement of the representative point along a surface corresponding to a reaction coordinate that possesses *an orbital* (or *state) correlation-imposed energy barrier* will be less probable (a forbidden reaction) than movement along a surface that does not possess correlation-imposed barriers (an allowed reaction). Note that the terms *forbidden* and *allowed* refer to relative rates and relative plausibility and are not intended to be absolute terms.

6.19   The Principle of Maximum Positive Orbital Overlap for Frontier Orbitals[1,4]

Now, let us consider how the principle of *maximum positive orbital overlap* of frontier orbitals allows the prediction of an *allowed or plausible* set of (low-energy) reaction pathways from an initial state (R or *R). According to quantum theory, MOs are wave functions that have spatial or stereochemical directiveness, relative to an associated fixed nuclear framework. As a result, if a reaction is to be initiated by positive orbital overlap, certain stereochemical positions of nuclei (and their associated electron clouds) in space will be favored over others provided that they correspond to a greater degree of positive orbital overlap (for comparable energy gaps between the orbitals). The principle of stereoelectronic control of reaction pathways, an application of frontier molecular orbital (FMO) theory, postulates that (for comparable energy gaps between orbitals) reaction rates are controlled by the degree of positive overlap of orbitals in space (i.e., certain nuclear geometries are easier to achieve along a reaction coordinate then others because of the greater positive orbital overlap accompanying one nuclear orientation relative to another). *In summary, the principle of maximum positive overlap postulates that reaction rates are proportional to the degree of positive (bonding) overlap of frontier molecular orbitals, and that this overlap depends on the details (e.g., stereochemistry) of the orbital interactions.* Although basically qualitative in nature, this principle is a powerful basis for analyzing photochemical reactions and for quickly sorting out plausible and implausible reaction pathways. The theoretical importance of frontier orbitals originates from one of the quantum mechanical rules that strongest bonding results from the overlap of orbitals when the orbitals have similar energies (a requirement for resonance).

6.20   Stabilization by Orbital Interactions: Selection Rules Based on Maximum Positive Overlap and Minimum Energy Gap[4]

Now, we follow the consequences of the postulate that the chemical reactivity of *R is determined by the FMOs corresponding to the electronic configuration of *R. In this case the key FMOs usually correspond to the HO and the LU. Since the HO has the highest energy of all the occupied orbitals in the ground state, it is most readily deformed and most readily gives up electron density to electrophilic (electron-seeking) sites (i.e., LUs) in the environment. The HO possesses the highest polarizability and the lowest ionization potential of any orbital that is occupied in the ground state (R). In other words, the HO is the most nucleophilic orbital and can be considered an electron donor.

The LU, which is unoccupied in the ground state, is capable of accepting an electron (i.e., is electrophilic) and is most capable of accepting electron density with minimum increase in the total molecular energy. We postulate that the initial perturbation in a chemical (or photochemical) reaction is assisted by CT of the HO electrons toward a LU; with this postulate we can readily visualize how the transfer of charge from one orbital to another actually occurs and how stereochemistry influences the plausibility of a reaction pathway.

Since we are dealing with qualitative considerations only, we assume that the HO and LU of the ground-state molecule (R) is a good zero-order approximation for the HO and LU of the excited state *R.

The fundamental underlying principle of frontier HO–LU interactions as a means of understanding chemical reactivity of ground-state reactions is the assumption that a majority of chemical reactions should take place most easily (i.e., have the lowest activation enthalpy) at the position of, and in the direction of, maximum positive overlap of the HO and the LU of the interacting species. Although the same basic principle applies for reactions of excited states, for *R(S_1 or T_1), it is important to note that for *R the HO and LU are half-filled (i.e., singly occupied). A singly occupied MO (termed SO) produced by electronic excitation may play the role of either a HO or a LU, or both.

Perturbation theory (Section 3.5) predicts that the stabilization energy E_{stab} due to overlap of FMOs is given qualitatively by Eq. 6.7, which is equivalent to the golden rule expression, Eq. 3.4.

$$E_{stab} \sim <HO|LU>^2/\Delta E \qquad (6.7)$$

The magnitude of stabilization (bonding) will depend directly on the square of the magnitude for the net positive overlap of the FMOs and inversely on the term ΔE, which measures the energy difference between the pertinent interacting FMOs. Because *R possesses two singly occupied orbitals (a singly occupied HO and LU), a number of possible CT interactions are possible from or to the HO or LU of another molecule (or some other groups in the same molecule).

6.21   Commonly Encountered Orbital Interactions in Organic Photoreactions

Since we are concerned with CT interactions, we label (Fig. 6.10) the HO of *R as an electron acceptor (eA) orbital, because it is a half-filled *bonding* orbital and will be stabilized by accepting a second electron into the half-filled HO; we label the LU of *R as an electron donor (eD) orbital, because it is a half-filled *antibonding* orbital and will be stabilized by donating its electron to a lower energy orbital that is seeking an electron. We now consider the initial CT interaction of *R with another molecule, M, as being determined by the interactions of the HO and LU of *R with the HO and LU of M. Figure 6.10 schematically shows the two possible orbital interactions that can occur when *R interacts with M in the hypothetical situation for which the energies of the HOs of *R and M and the LUs or *R and M are similar: (a) M is an electron donor, so that the HO of *R removes charge from the HO of M, and (b) M is an electron acceptor, so that the LU of *R transfers charge to the LU of M. *These two exemplars demonstrate the ability of *R to serve simultaneously as a strong oxidant (removing an electron from the* HO *of* M) *and a strong reductant (adding an electron to the* LU

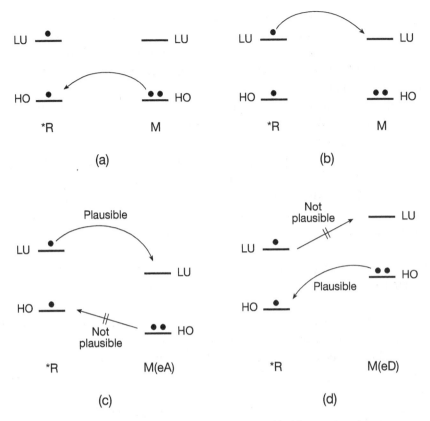

Figure 6.10   Schematic representation of frontier orbital interactions between an electronically excited molecule (*R) and a ground-state molecule (M). (a) Electron transfer from the HO of M to the half-filled HO of *R. (b) Electron transfer from the LU of *R to the LU of M. (c) Plausible electron transfer from the LU of *R to the LU of M(eA). (d) Plausible electron transfer from the HO of M(eD) to the half-filled HO of *R. In (c), the electron transfer from the HO of M(eA) to the half-filled HO of *R is not plausible. Likewise in (d), the electron transfer from the LU of *R to the LU of M(eA) is not plausible.

of M). These "redox" attributes of *R will be of great importance when we consider photoinduced electron transfer in Section 7.13.

We might ask, which of the two situations shown in Fig. 6.10a and b is more plausible in any actual case? Focusing on the excited-state *R, let us assume that its electron-donating and electron-accepting characteristics will remain constant as we vary the electron-donating and electron-accepting characteristics of M. Qualitatively, we can use simple energy considerations to decide on plausibility of frontier orbital interactions: Whether *R serves as an electron donor (half-filled LU donates an electron) or an electron acceptor (half-filled HO accepts an electron) will depend on the electronic characteristics and energies of the HO and LU of M relative to the HO and LU of *R. If the transfer of charge is downhill energetically, the process can be considered plausible (allowed). If the process is uphill, the process is considered implausible (forbidden).

Similar to the notation for *R, we use the notation M(eD) to describe M when it is an *electron donor* and M(eA) when M is an *electron acceptor*. To characterize M(eA) as electron accepting means that such a molecule has a low-lying empty LU (relative to the LU of *R) and a low-energy, doubly occupied HO (relative to the HO of *R) as shown in Fig. 6.10c. To characterize M(eD) as electron donating means that such a molecule has a high-energy, doubly filled HO and a high-energy, unfilled LU (Fig. 6.10d). We therefore expect that the major frontier interactions for *R + M(eA) will be transfer of charge from the half-filled LU of *R to the empty LU of M(eA), that is, the plausible charge-transfer LU(*R) → LU(M), as shown in Fig. 6.10c. We also expect that the major frontier interactions for *R + M(eD) will be transfer of charge from the filled HO of M(eD) to the half-filled HO of R, that is, the plausible charge-transfer HO(M) → HO(*R), as shown in Fig. 6.10d. For these orbital interactions to be most effective, Eq. 6.5 requires that the interacting orbitals be as close as possible in energy and that the overlap of the orbitals be as large and positive as possible. In addition, there will always be a thermodynamic aspect to orbital interactions such that the reaction that is downhill thermodynamically is always favored over a reaction that is uphill.

From Fig. 6.10 and the criteria of maximum positive orbital overlap and minimum energy gap, we can postulate the following recipe for deciding how orbital interactions will determine the favored nuclear motions for a given set of photochemical reaction pathways:

1. After setting up the MOs of the reactants according to their relative energies, identify the half-filled HO and LU orbitals of the electronically excited moiety *R and the filled HO and unfilled LU of the molecular reactant (M). This requires some knowledge or assumption as to the ability of M to serve as an electron acceptor, M(eA), or as an electron donor, M(eD). This knowledge is available from experimental electrochemical oxidation and reduction potential data (Chapter 7).

2. For a given, assumed reaction pathway, consider the possible orbital interactions between an HO and LU of the electronically excited moiety and the HO or LU of the unexcited moiety, taking into account the electron-accepting or electron-donating characteristics of M.

3. Determine which stereochemical orbital interactions, as indicated in Fig. 6.10, lead to the best positive overlap and whether the interaction arrow for these orbital interactions points up (thermodynamically unfavorable) or down (thermodynamically favorable).

4. Evaluate the positive orbital overlap and thermodynamic factors to determine qualitatively the more favorable reaction pathways.

5. Assign pathways having significant positive orbital overlap (bonding) as plausible (allowed) and pathways having significant negative orbital overlap (antibonding) as not plausible (forbidden).

This information provides a qualitative guide to the most favored reaction coordinate(s) of photochemical reactions, such as *R → I and *R → F → P.

6.22   Selection of Reaction Coordinates from Orbital Interactions for *R → I or *R → F → P Reactions: Exemplars of Concerted Photochemical Reactions and Photochemical Reactions That Involve Diradicaloid Intermediates

Now, we consider two of the most important classes of photochemical reactions of organic molecules: the processes *R → F → P and *R → I → P. Exemplars of the *R → F → P reaction include *cis–trans isomerization about a* C=C *bond* and the *photochemical "formally concerted" pericyclic reactions* initiated from $S_1(\pi,\pi^*)$ states of ethylenes, conjugated polyenes, and aromatic hydrocarbons. The most common examples of photochemical pericyclic reactions are electrocyclic ring openings and ring closures, cycloaddition reactions, and sigmatropic rearrangements. These reactions follow the Woodward–Hoffmann rules for photochemical pericyclic reactions.[10] Concerted photochemical pericyclic reactions must be initiated from $S_1(\pi,\pi^*)$, since a spin change is required if the reaction is initiated in $T_1(\pi,\pi^*)$, and the reaction cannot therefore occur in a concerted manner to produce P, since the latter is a singlet state.

Exemplars of the *R → I processes are the *set of plausible photochemical reactions* of the $S_1(n,\pi^*)$ *or* $T_1(n,\pi^*)$ states of ketones. These reactions include the following: hydrogen atom abstraction, electron abstraction, addition to ethylenes, α-cleavage reactions, and β-cleavage reactions. The *R → I process is also common for the $T_1(\pi,\pi^*)$ states of ethylenes and aromatic hydrocarbons. We will see that consideration of orbital interactions creates selection rules such that the set of plausible photochemical reactions of n,π^* states are the same and depend only on orbital interactions, not on spin. However, the initial energy of a state determines the thermodynamic feasibility of a primary photochemical reaction; since the $S_1(n,\pi^*)$ state possesses a higher energy than the $T_1(n,\pi^*)$ state, there is generally greater thermodynamic driving force for a reaction from $S_1(n,\pi^*)$ than from $T_1(n,\pi^*)$.

In analyzing a photochemical reaction, such as *R → I or *R → F → P, theoretically, one must select the appropriate reaction coordinate (or coordinates) that describes the nuclear geometry changes accompanying the transformation of reactants to products. In principle, all possible reaction coordinates should be analyzed. In practice, we seek to select only the lowest-energy reaction pathways from a given initial excited state. One may qualitatively identify these pathways by the use of orbital interactions and then analyze them by assuming that a certain symmetry is maintained along the reaction coordinate.

6.23   Electronic Orbital and State Correlation Diagrams[6,11]

Valuable information concerning the funnels or barriers on excited PE surfaces can be extracted from a study of state-energy correlation diagrams of exemplars for primary photochemical processes. Although they are qualitative, orbital and state-energy correlation diagrams for primary photochemical processes that preserve a specific symmetry element along the reaction coordinate are relatively easy to construct and can provide important insights into the locations of funnels and barriers in electronically excited state surfaces. An orbital correlation diagram is constructed by considering the orbital energies and symmetries for the beginning and end of an elementary step along a reaction pathway, for example, the very common *R → I primary process. We have already described exemplars for orbital and state correlation diagrams for the case of the stretching and breaking of the σ bond of H—H and the twisting of a C=C bond (Figs. 6.8 and 6.9).

The first step in a photochemical primary process can be analyzed in terms of orbital and state correlation diagrams. The correlation diagrams are very useful, in combination with FMO theory, in providing a qualitative set of selection rules for the plausibility of certain pathways of primary photochemical reactions. The analysis of correlation diagrams requires the recognition of certain symmetries of the initial and final states of a primary photochemical process. The basic assumption of a correlation diagram is that electronic symmetry is related to nuclear geometry and that electronic systems will more easily follow pathways for which electronic symmetry is preserved, or at least modified to the least extent, along the reaction pathway.

6.24   An Exemplar for Photochemical Concerted Pericyclic Reactions: The Electrocyclic Ring Opening of Cyclobutene and Ring Closure of 1,3-Butadiene

As an exemplar of concerted pericylic reactions, which generally occur from $S_1(\pi,\pi^*)$ states, we consider the photochemical electrocyclic ring opening of cyclobutene and the related photochemical electrocyclic ring closure of 1,3-butadiene (Fig. 6.11). The concepts developed from this exemplar can be readily extended to other examples of photochemical electrocyclic reactions and to other "concerted" photochemical peri-cyclic reactions, such as cycloadditions and sigmatropic reactions. Since the subject of concerted thermal and photochemical reactions is covered in most elementary organic chemistry courses, only a brief review is given here, and the reader is referred to more extensive discussions elsewhere.[4,10]

Consideration of orbital interactions leads to the stereochemical selection rules for pericyclic reactions. For example, the key aspect of concertedness for an electrocyclic reaction is the stereochemistry associated with the ring opening and closure, which follow the Woodward–Hoffmann rules (Fig. 6.11). According to the positive overlap and energy-gap selection rules of orbital interactions, we are led to the prediction that, for reactions initiated from the $S_1(\pi,\pi^*)$ states of cyclobutene, both $\sigma(HO) \to \pi(HO)$ and $\pi^*(LU) \to \sigma^*(LU)$ are the best CT possibilities between MOs and therefore should contribute most significantly.

Inspection of the orbital symmetry for *disrotatory and conrotatory processes*[10] (Fig. 6.11, right) shows that for *R, the disrotatory process is favored by the rule of maximum positive overlap. When the orbitals rotate in the disrotatory manner, the $\sigma(HO) \to \pi(HO)$ and $\pi^*(LU) \to \sigma^*(LU)$ charge transfers correspond to positive overlap, whereas the conrotatory motion corresponds to negative overlap. The predic-

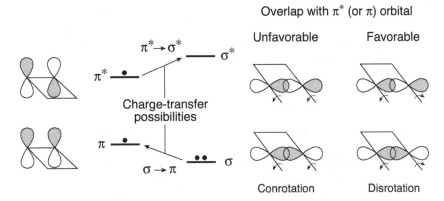

Figure 6.11   Orbital interactions for the conrotatory and disrotatory ring opening of the π,π^* state of cyclobutene to form 1,3-butadiene. The σ and π HOs are closest in energy, and therefore should have the best CT interaction. Similarly, the π^* and σ^* orbitals are closest in energy, and therefore should have the best CT interactions.

tions for the stereochemistry of the electrocyclic reactions of 1,3-butadiene are shown in Eqs. 6.8 and 6.9.

$$\xrightarrow[\text{Conrotatory}]{h\nu} \qquad \text{``Forbidden'' by} \qquad (6.8)$$
$$\text{orbital interactions}$$

$$\xrightarrow[\text{Disrotatory}]{h\nu} \qquad \text{``Allowed'' by} \qquad (6.9)$$
$$\text{orbital interactions}$$

Thus, simply from a consideration of frontier orbital interactions, we expect that for this four-electron system (for thermochemically feasible reactions) the *disrotatory interconversions are photochemically allowed* (favorable frontier orbital interactions), whereas the *conrotatory interconversions are photochemically forbidden* (unfavorable frontier orbital interactions). These conclusions are important, since they allow a theoretical process for selecting a small set of plausible *initial* interactions of π,π^* states that determine the stereochemistry of concerted pericyclic reactions. To provide a more fundamental theoretical basis for the selection rules and to confirm that these initial interactions are carried over in the $^*R \rightarrow F \rightarrow P$ reaction, we will need to generate orbital and state diagrams for the overall process, which will be done in Section 6.28.

6.25   Frontier Orbital Interactions Involving Radicals as Models for Half-Filled Molecular Orbitals

Since a radical is defined as a species possessing a half-filled orbital, the orbital interactions of a radical with a molecule will be a good model for the interactions

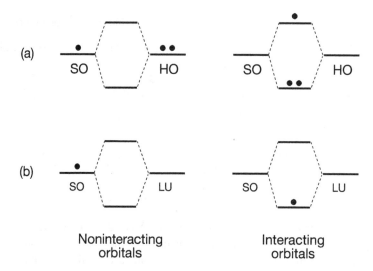

Figure 6.12   (a) Interaction of a SO of a radical with the filled HO of a molecule. (b) Interaction of a SO radical with the unfilled LU of a molecule. Note that the interaction of a SO with either a doubly occupied HO or an empty LU stabilizes the overall system electronically.

of the half-filled HO and LU of *R interacting with molecules (Fig. 6.10).[12] The half-occupied HO of *R can be modeled with an electrophilic (electron-accepting) radical and the half-filled LU of *R can be modeled with a nucleophilic (electron-donating) radical. In Fig. 6.12, the half-filled orbital is labeled "SO" for "singly occupied" to differentiate it from the HO and LU of a molecule. Consider the half-filled SO of a radical interacting with the filled HO (Fig. 6.12a) or the unfilled LU (Fig. 6.12b) of an organic molecule.

According to quantum mechanics, any time two orbitals interact, the result will be two new orbitals, one of lower energy (constructive interference) and one with higher energy (destructive interference). In Fig. 6.12a as the radical approaches the molecule, the SO brings in one electron and the HO of the molecule brings in two electrons as the orbitals mix to produce the two new orbitals. For qualitative purposes, the energy of the SO and HO are shown as equal in energy. We see that the interacting system is lower in energy than the noninteracting system because two of the three electrons are stabilized (two bonding interactions and one antibonding interaction yield net bonding). In Fig. 6.12b, the radical approaches the molecule, the SO brings in one electron, and the LU of the molecule is empty. The interacting system is again lower in energy because the single electron is placed in the lower-energy orbital.

From this very qualitative description of orbital interactions, we conclude that *the interaction of a radical with either the filled HO or the empty LU of a molecule leads to a more stabilized system.* In effect, we conclude that radicals will tend to form a "complex" with all molecules; the only issue is the degree of stabilization. Since *R has a $(HO)^1(LU)^1$ electron configuration (two half-filled orbitals), we can conclude that *electronically excited states will tend to form complexes with ground-state molecules.*

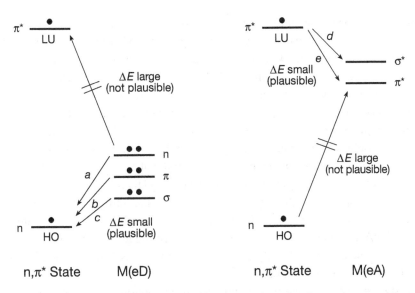

Figure 6.13   A hypothetical set of orbital interactions for the n,π^* state with substrates. Note that the s* orbital of M(eA) is shown as being lower in energy than the p* orbital of the n,p* state. This hypothetical situation is for a very powerful electron-attracting M(eA) molecule and is not the general case.

In Section 4.38, we saw that in some cases the stabilization is sufficient to lead to the formation of excited complexes (i.e., exciplexes and excimers).

With the above discussion of the interactions of radicals with molecules in mind, let us consider the photoreactions of *R for systems for which both HO and LU behave as radical centers, that is, for *R → I(D) reactions. As an exemplar of *R → I(D) photochemical reactions, we examine the photochemistry of the *R(n,π^*) states (both S_1 and T_1) of ketones. The objective of our examination is to develop a set of plausible primary photochemical processes of an n,π^* state of a ketone. Thus, one needs to survey all of the orbital interactions that are possible from an n,π^* state with the HO and LU of another molecule, M (intermolecular reactions), or with groups within the electronically excited molecule possessing (intramolecular reactions). Figure 6.10c and d show two limiting cases of the orbital interactions of *R with the orbitals of M. Figure 6.13 lists these possibilities, where *R is an n,π^* state: (1) CT interactions in which one of the electrons from the HO of M(eD) is transferred to the electrophilic half-filled n_0 HO of the n,π^* state, and (2) CT interactions from the nucleophilic half-filled π^* LU of the n,π^* state to a vacant LU of M(eA).

What are the most common orbitals of M that correspond to the nucleophilic (electron-donating) HO and to the electrophilic (electron-attracting) LU? For organic molecules, M, the three most common HOs correspond to the valence σ,π, and n orbitals, while the two most common LUs correspond to π^* and σ^* orbitals. In Figure 6.13 (left), the donation of HO electrons of M(eD) to the half-filled n orbital of the n,π^* state is shown schematically, with the typical energetic ordering being $n > \pi > \sigma$. Of course, this ordering will depend on the actual structure of M, and occasionally an n orbital may be lower in energy than a π orbital. In general, the energetic ordering of the antibonding orbitals is $\sigma^* > \pi^*$. Fig. 6.13 (right) shows the donation of the π^* electron of the n,π^* state to a σ^* or π^* LU of M(eA). In Section 7.14, experimental and theoretical guides to the quantitative aspects of electron-transfer processes will be discussed in more detail. However, for the qualitative purposes of this chapter, the issue of energetics of the initial orbital interactions is clear: energetically downhill transfer of charge is plausible; uphill transfer of charge is implausible.

An example of a σ HO orbital that can serve as an eD is the σ_{CH} orbital associated with a nucleophilic CH bond; an example of a π HO orbital that can serve as an eD (electron-donor group) is a nucleophilic $\pi_{C=C}$ orbital of a C=C $\leftarrow$ eD system (where eA is an electron-donating group); an example of an n HO orbital that can serve as an eD is the nucleophilic n_N HO orbital associated with the nitrogen atom of an amine. An example of an electrophilic π^* LU orbital is an electrophilic $\pi^*_{C=C}$ associated with a C=C $\rightarrow$ eA system (where eA is an electron-attracting group). An example of a σ^* LU orbital is the electrophilic σ^*_{C-X} associated with a C–X (carbon halogen or other electron-withdrawing group) bond.

According to the selection rules for orbital interactions, we expect (Fig. 6.13), the low-energy or allowed reaction pathways for the half-filled HO (n_0) orbital of the n,π^* state, to involve the following possible CT interactions for which charge is donated to the half-filled n_0 orbital: $\sigma CH \rightarrow n_0$, $\pi_{C=C} \rightarrow n_0$, and $n_N \rightarrow n_0$. In the case of the half-filled LU (π^*) orbital for the n,π^* state, we expect the allowed pathways to involve the following CT interactions for which charge is donated from the half-filled π^* orbital: $\pi^*_{C=O} \rightarrow \pi^*_{C=C}$ and $\pi^*_{C=O} \rightarrow \sigma^*_{C-X}$. When the thermodynamics are favorable, these orbital interactions define the orbital requirements that must be met for a reaction to be considered a member of the plausible set of photochemical reactions of an n,π^* state. This conclusion is rather important, since it illustrates the theoretical process for selecting a small set of plausible *initial* interactions of n,π^* states with potential reagents, both intermolecular and intramolecular. To confirm that these initial interactions are carried over in the *R $\rightarrow$ I reaction, we need to consider the more fundamental orbital and state diagrams for the overall process.

6.26   Orbital and State Correlation Diagrams[6,10,11]

Both orbital and state correlation diagrams are useful theoretical tools for obtaining insights into the details of energy surfaces and for analyzing transformations of one type of electronic orbital configuration into another. It is convenient and informative first to consider a correlation diagram in terms of the key orbitals involved in a reaction path, that is, to construct a working *orbital correlation diagram*. This orbital correlation diagram is then employed as the basis for generation of a working *state correlation diagram*. One has the choice of switching on mixing interactions when generating the orbital correlation diagram first, and then constructing the state correlation diagram, or constructing the state correlation diagram from a noninteracting orbital correlation diagram and then switching on mixing interactions at the state correlation level. In either case, we can then employ orbital and state correlation diagrams to deduce the nature of the energy surfaces that connect reactants to primary photochemical prod-

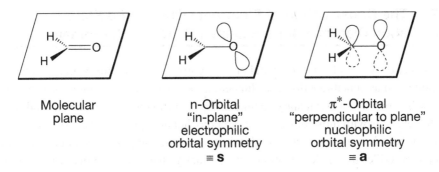

Molecular plane	n-Orbital "in-plane" electrophilic orbital symmetry $\equiv$ **s**	π^*-Orbital "perpendicular to plane" nucleophilic orbital symmetry $\equiv$ **a**

Figure 6.14   The symmetry plane of formaldehyde. The n_0 orbital lies in the symmetry plane and is termed symmetric (s) with respect to reflection through the symmetry plane. The π^* (and the π) orbital lies above and below the symmetry plane and is termed antisymmetric (a) with respect to reflection through the symmetry plane.

ucts, search for surface crossings, and determine whether the reactions are allowed or forbidden, that is, plausible or implausible.

The protocol for the generation of orbital and state correlation diagrams depends heavily on the concept of molecular and electronic symmetry (either for the complete molecule or a local portion of it that is relevant to the reaction of interest). It is assumed that the reader is familiar with concepts of chemical symmetry, so only a brief review of the terms used to describe orbital and state symmetry will be given here. The interested reader who wishes to probe more deeply into the topic is referred to some excellent discussions in the literature.[4,10,12]

The idealized symmetry properties of MOs with respect to a reference plane are of critical importance when considering the plausible photochemical reactions of n,π^* states.[6,11] If a molecule possesses a reference plane of symmetry, all of its MOs must be either symmetric (**s**) or antisymmetric (**a**) with respect to reflection through the reference symmetry plane. By convention, the symbol for symmetric orbital symmetry is **s**; the conventional symbol for antisymmetric orbital symmetry is **a**; the conventional symbol for a symmetric state symmetry is **S**; and the conventional symbol for antisymmetric state symmetry is **A**. Symmetric (**s**) means that the (mathematical) sign of the wave function does not change upon reflection from one side of the reference plane to the other. Antisymmetric (**a**) means that the wave function changes sign upon reflection from one side to the other. For example, for the planar formaldehyde molecule (Fig. 6.14), the n orbital has **s** symmetry (the wave function does not change sign above or below the molecular plane), and the π orbital, as well as the π^*, has **a** symmetry (the wave function changes sign above and below the molecular plane). In other words, reflection of the n orbital through the symmetry plane does not change the sign of the wave function (**s** symmetry), but reflection of the π (or π^*) orbital through the plane does change the sign (**a** symmetry). The product of the orbital symmetries for the half-occupied orbitals determines the state symmetries according to the rules: **a** $\times$ **a** (orbitals) = **S** (state); **s** $\times$ **s** (orbitals) = **S** (state); **s** $\times$ **a** (orbitals) = **A** (state). The reader is referred to texts and reviews for more detailed information on the assignment of symmetry elements to orbitals and states.[13]

6.27   The Construction of Electronic Orbital and State Correlation Diagrams for a Selected Reaction Coordinate

The construction of electronic orbital or state correlation diagrams starts with the selection of an assumed reaction coordinate that possesses some element of symmetry that will allow, for an elementary chemical step, a correlation of the orbitals and states of the reactants with those of the primary photochemical product. This selected coordinate is based on consideration of orbital interactions, computations, or experimental data; it describes the nuclear motions and geometry changes that transform the initial reactant into a primary photochemical product (*R → I). First, a zero-order correlation of the orbitals of the reactant with the "natural" orbitals of the primary product is made.[14]

The natural orbital correlations are based on preservation of the phase (orbital symmetry) relationships for the local electronic distributions in order to correlate orbitals of the reactant that correlates with (looks like) the orbitals for the product by simple inspection of their phase relationships. In the correlation diagrams employing natural orbitals, orbitals of the same symmetry are allowed to cross in order to identify the *intended correlations*. The state correlations are then determined from the orbital correlations. The state correlation diagram is then generated by using state symmetry to connect the states of reactants to the states of the primary product. The intended correlation may provide important insight to the source of small energy barriers or energy wells because they indicate "where the MOs try to go" before symmetry avoided crossings are considered and before the final-state correlation diagram is constructed.

6.28   Typical State Correlation Diagrams for Concerted Photochemical Pericyclic Reactions

The simple orbital interaction arguments for the conrotatory and disrotatory ring opening of cyclobutene gives the same predictions as a detailed orbital and state correlation diagram. The reader is referred to excellent reviews for the details of the orbital and state correlations.[10,12] We will summarize the important results that allow a generalization for all pericyclic reactions.

6.29   Classification of Orbitals and States for the Electrocyclic Reactions of Cyclobutene and 1,3-Butadiene: An Exemplar Concerted Reaction

A working state correlation diagram for the electrocyclic ring opening of cyclobutene and ring closure of 1,3-butadiene is constructed as shown in Fig. 6.15. Of the surfaces shown, only singlet surfaces need be considered for concerted pericyclic reactions, since triplet concerted reactions, requiring a spin change along the reaction coordinate, are implausible. For a complete and realistic correlation diagram to be constructed,

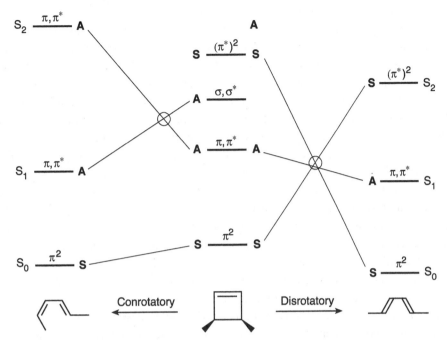

Figure 6.15   Simplified state correlation diagram for the concerted ring opening of cyclobutene to 1,3-butadiene. The state correlations are based on orbital symmetry considerations. In the case of disrotary motion, the orbital symmetries cause the ground state of **S** symmetry to correlate with an excited state of **S** symmetry.

high-energy states, such as σ,σ^*, and $(\pi^*)^2$, must be included. Consider the photochemical ring opening of 3,4-dimethylcyclobutene by either a conrotatory motion or by a disrotatory motion. The **S** and **A** labels are the state symmetry designations for the two motions.[6] The important features of Fig. 6.15 are that, for the conrotatory pathway, the correlation diagram from the $S_1(\pi,\pi^*)$ state correlates with a high-energy $S_n(\pi,\pi^*)$ state of the product 1,3-diene. Since this is an uphill energy correlation, the reaction is not plausible and is termed forbidden. On the other hand, for the disrotatory reaction pathway, the correlation diagram from the $S_1(\pi,\pi^*)$ of 3,4-dimethylcyclobutene correlates with a lower energy $S_1(\pi,\pi^*)$ of the product 1,3-diene. Since this is a downhill energy correlation the reaction is plausible and is termed allowed.

The topology of these surfaces, for this specific exemplar of an electrocyclic reaction, has been shown to be general for concerted pericyclic reactions.[6] Thus, *all ground-state forbidden pericyclic reactions can be expected to have a surface topology qualitatively equivalent to the disrotatory ring opening of cyclobutene to butadiene; all ground-state allowed pericyclic reactions may be expected to have a surface topology qualitatively equivalent to the conrotatory ring opening of cyclobutene to butadiene.*

The state correlation diagram of Fig. 6.15 shows several surface crossings (a zero-order surface crossing is indicative of the geometry of a CI). Crossing states of the same symmetry may become an ASC, if appropriate electronic interactions are available. From the correlation diagram, note that there are surface crossings between the singlet surfaces. Thus, there are two possible reaction mechanisms for a photochemical pericyclic reaction: either a *R(S$_1$) → CI → P or a *R(S$_1$) → ASC → P pathway would explain the results. Several lines of evidence favor the CI description:[15] (1) the rate of reaction of *R in certain pericyclic reactions[13] is in the range of 10^{13}–10^{14} s^{-1}, far too fast for reaction from an equilibrated ASC (which should follow Fermi's golden rule and possess rates on the order of 10^9 s^{-1} or less); (2) in certain pericyclic photochemical reactions, the strict stereospecificity expected from the Woodward–Hoffmann rules is not observed, which is a possibility for the representative point that falls into a CI; and (3) in many cases, several types of pericyclic products are formed, which is consistent with the bifurcation expected with reactions that proceed through a CI (Fig. 6.6, left). Since the distinction between a CI and ASC has not been accessible to direct experimental verification, computation is currently the only method available for investigating the mechanism of photochemical concerted reactions. For our purposes, we accept that either the CI or ASC possibilities are plausible in the absence of specific experimental or computational evidence to the contrary. However, the observation of pericyclic processes that occur on the order of femtoseconds is taken as evidence for the occurrence of a barrierless *R → IC process.

Figure 6.16 presents a further simplification of the state correlation diagram for Fig. 6.15. The essential features of any "concerted" pericyclic reaction are preserved. For the ground-state allowed ring-opening reaction (Fig. 6.16, left), there is a low barrier to reaction, but a significant barrier exists in the excited-state, ring-opening pathway.[14] For the ground-state, forbidden reaction (Fig. 6.16, right), there is a high-energy barrier to the ring-opening reaction in the ground state and a low barrier to a CI in the excited-state pathway.

The geometry for the transition state for a *forbidden* ground-state reaction (Fig. 6.16, right), like the 90° geometry in twisting about a π bond, corresponds to a diradicaloid structure.[1,9] From the general rules for radiationless transitions, such a structure corresponds to a critical geometry for a funnel from S$_1$, whether it was a CI or ASC at the diradicaloid geometry.

To a first approximation, the topology of the state correlation diagram shown in Fig. 6.16 may be extended to all concerted pericyclic reactions as follows. For four-electron (or more generally, 4N) concerted pericyclic reactions, the disrotatory (or stereochemically equivalent) pathway corresponds to motion from the center of the diagram to the right, and the conrotatory (or stereochemically equivalent) pathway corresponds to motion from the center to the left. We see that photochemical disrotatory four-electron (or 4N) concerted pericyclic reactions are generally allowed, in the sense that they are plausible because the stereochemistry required for favorable orbital overlap is readily achieved and leads to a funnel to the ground state through an avoided crossing or a CI.

The situation is quite different for six (or more generally 4N + 2) electron concerted pericyclic photochemical reactions, where the disrotatory (or stereochemically equivalent) pathway corresponds to motion from the center of the diagram to the

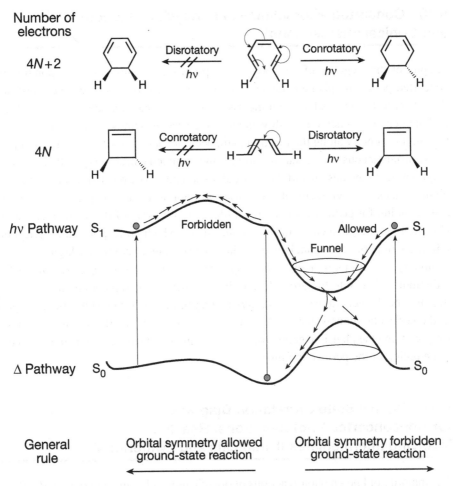

Number of electrons

$4N+2$ Disrotatory ← $h\nu$ Conrotatory → $h\nu$

$4N$ Conrotatory ← $h\nu$ Disrotatory → $h\nu$

$h\nu$ Pathway S_1 Forbidden   Allowed S_1 Funnel

Δ Pathway S_0 S_0

| General rule | Orbital symmetry allowed ground-state reaction | Orbital symmetry forbidden ground-state reaction |

Figure 6.16 A simplified, general schematic description of the two lowest singlet surfaces for a concerted pericyclic reaction. The selection rules are shown for $4N$-electron and for $4N + 2$-electron reactions ($N = 0$ or an integer, and $4N$ or $4N + 2$ is the number of electrons involved in bond making or breaking). See text for discussion. The energies of the surfaces are not to scale and are intended to show the general features of concerted pericyclic reactions.

left, and the conrotatory pathway corresponds to motion from the center to the right. Thus, concerted disrotatory photoreactions are forbidden, or more precisely: Concerted photochemical $4N + 2$ electron reactions from S_1 are implausible because they require an unfavorable stereochemical geometry in order to achieve the positive overlap required for concerted bonding along the reaction coordinate. This unfavorable geometry leads to an energy barrier for a concerted reaction. Other more rapid pathways for deactivation of S_1 compete favorably with the concerted process. Thus, the reaction for the $4N + 2$ electron systems are predicted to occur in a conrotatory manner, passing through a CI or ASC form, as shown from the center to the right of Fig. 6.16.

6.30   Concerted Photochemical Pericyclic Reactions and Conical Intersections[1]

In spite of the elegance of the above description for the surfaces corresponding to concerted pericyclic photochemical reactions, we need to remember that construction of these surfaces is based on the assumption of a reaction path that preserves a high degree of symmetry. In reality, the reacting system is unlikely to strictly possess this high symmetry along the reaction path, because factors, such as substitution patterns, instantaneous molecular vibrations, and symmetry destroying collisions with neighboring molecules, distort the molecular geometry. As a result, the representative point for these nonsymmetrical geometries will not necessarily experience a significant avoiding for geometries that are similar to those for r_c of the avoided crossing. Nonetheless, the geometries in the vicinity of r_c, which correspond to a pericyclic minimum or pericyclic funnel on an excited surface, but do not have a high degree of symmetry, are still more likely to correspond to CIs that are *very effective funnels* for radiationless decay to the ground state. Indeed, detailed computations indicate that this may be the general case. An understanding of the electronic states in the region of the pericyclic funnels that are deduced from correlation diagrams provides a basis for predicting the plausible geometries for CI funnels that can be assumed to be in the same region of the pericyclic funnel.

6.31   Typical State Correlation Diagrams for Nonconcerted Photoreactions: Reactions Involving Intermediates (Diradicals and Zwitterions)[1,9]

The majority of known photoreactions of organic molecules are not *concerted* (*R $\rightarrow$ F $\rightarrow$ P) in nature. Rather, photochemical reactions tend to involve reactive intermediates along the reaction pathway (*R $\rightarrow$ I). The most common photochemical intermediates (I) are species that are not fully bonded and possess two electrons in two nonbonding orbitals of nearly comparable energy. These reactive intermediates correspond to diradicals (D); radical pairs (RP), and biradicals (BR) or zwitterions (Z). Figures 6.8 and 6.9 capture the essence of the electronic features of the *R $\rightarrow$ I process and the nature of I(D,Z), showing the surface topologies associated with the stretching and breaking of a σ bond and the twisting and bending of a π bond, respectively. Figure 6.7 summarizes the relationships between these two fundamental surface topologies and the D and Z structures.

6.32   Natural Orbital Correlation Diagrams[14]

Recall that the process of constructing state diagrams corresponding to adiabatic surfaces may be performed in either one of two ways: (1) a zero-order intended or "natural orbital" correlation is used first with interactions causing surface avoiding at the orbital crossing points, and then the resulting adiabatic orbital correlations are used to construct the configuration and adiabatic state correlation diagrams; or (2) a zero-order intended or natural orbital correlation may be made *without interactions between orbitals*, leaving the orbital surface crossings; and then the resulting diabatic orbital correlations are used to construct the orbital correlation diagram. Finally, the electronic interactions are turned on to generate the adiabatic state correlation diagram. The second procedure that employs natural orbitals is useful in producing insight into the presence of small barriers, which are typical of many photochemical reactions, even those that are allowed by thermodynamic considerations and allowed according to the state correlation diagram. The basic idea is that the orbitals in initial interactions will tend to maintain their general symmetry characteristics before they encounter a nuclear geometry that induces mixing and causes avoiding at zero-order orbital surface crossings. The important point is that if the natural correlations from *R are to higher-energy orbitals, the orbital at the avoided crossing will have a higher energy than the starting orbital and will cause a barrier along the reaction pathway.

6.33   The Role of Small Barriers in Determining the Efficiencies of Photochemical Processes

Because photochemical primary processes (*R → I and *R→F→P) must compete with relatively fast photophysical processes (*R → R + $h\nu$ or Δ), photochemical primary processes can be efficient only if small or no energy barriers (energy maxima) exist along the *R → I and *R → F → P pathways. The qualitative adiabatic correlations at the *state* correlation level often fail to reveal the (often small) maxima that arise on PE surfaces as the result of avoided crossings at the *orbital* level. Orbitals tend to follow a natural change of shape along a reaction coordinate; that is, the wave functions that the orbitals represent have a natural tendency to conserve their *local* phase relationships and *local* electronic distributions in addition to the conservation of overall state symmetry properties. Furthermore, in a natural correlation diagram, the lines associated with the correlation of orbitals in the initial (*R) and final state (I or P) are always allowed to cross in order to indicate and emphasize the *intended* natural correlations of the orbitals before any interactions are allowed to mix the orbitals and produce the adiabatic correlation diagram. These crossings may provide insight into the source of small energy barriers (or energy minima that may result in either avoided crossings or CIs) because they indicate where the orbitals want to go naturally in the absence of mixing. Thus, if mixing is weak for any reason, the natural orbital correlation is basis for predicting barriers or minima. In a sense, the natural correlation maxima or minima are "memories of avoided crossings" in the state correlation diagram. Also note that these memories of avoided crossings may be used to predict the geometries at which surface crossings, that is, CIs, will occur along a photochemical pathway.

6.34   An Exemplar for the Photochemical Reactions of n,π^* States

We use the hydrogen abstraction reaction of the n,π^* state of ketones (Eq. 6.10, where X—H is a generalized structure for a hydrogen-atom donor) as an exemplar to derive prototypical orbital and state symmetry correlation diagrams for $*R \rightarrow I$ reactions that produce radical-pair and biradical intermediates, I(D).

$$
\begin{array}{ccc}
\overset{\displaystyle *O}{\underset{\displaystyle \underset{H}{\overset{\displaystyle \parallel}{\underset{\displaystyle C}{\diagup}}\diagdown}_{H}}{}}\;\; H\!-\!X & \longrightarrow & \overset{\displaystyle OH}{\underset{\displaystyle \underset{H}{\overset{\displaystyle \mid}{\underset{\displaystyle \overset{\displaystyle \bullet}{C}}{\diagup}}\diagdown}_{H}}{}}\;\; \bullet X \\[2em]
*R & & I(D)
\end{array}
\tag{6.10}
$$

Since saturated C—H bonds have very little electrophilic character, it is reasonable to assume that the hydrogen abstraction is initiated by the frontier interaction of the half-filled HO orbital on the n atom of the n,π^* state and the filled HO orbital of the σ_{CH} bond. This reaction is expected to be initiated by a $\sigma(HO) \rightarrow n_0(HO)$ charge-transfer orbital interaction (Fig. 6.13). The results of the orbital and state correlation diagrams may be immediately applied to reactions involving $n_N(HO) \rightarrow n_0(HO)$ and $\pi_{C=C}(HO) \rightarrow n_0(HO)$ interactions (electron abstraction from amines and addition to $C=C\pi$ bonds).

The photoreaction of $*R(n,\pi^*)$ excited ketones with alcohols (Fig. 6.17 and Eq. 6.11) involves primary photochemical hydrogen abstraction, which produces a radical pair intermediate as the primary photochemical product, I(D).[17] The radical pair has several reaction options for the I(D) $\rightarrow$ P path: radical–radical combination and disproportionation reactions or reaction with molecules to form new radicals that form P in secondary thermal reactions (Chapters 8 and 9).

The state correlation diagram (Fig. 6.17) for the reaction of the $*R(n,\pi^*)$ state of a ketone with a hydrogen donor is a prototype for reactions of a $*R(n,\pi^*)$ state with a molecule M(eD) to produce radical pairs and biradical intermediates. Thus, Fig. 6.17 can serve as an exemplar of *all* photochemical reactions that are initiated by the n_0 orbital (the HO) of the n,π^* state. For most alkanones, such as acetone, $S_1 = n,\pi^*$ and $T_1 = n,\pi^*$; that is, $*R$ is n,π^* in both cases. Thus, our correlation diagram will seek to connect $S_1 = n,\pi^*$ and $T_1 = n,\pi^*$, and $S_0 = \pi^2 n^2$ with the appropriate states of the product I(D, Z). We need to select a reaction coordinate in order to construct an orbital or state correlation diagram. What is a proper reaction coordinate for hydrogen abstraction by an n,π^* state? To determine the most likely reaction coordinate, we must select the most favorable orbital interactions, search for elements of symmetry in the geometry of the reactants that lead to the most favorable orbital interactions, and then relate that symmetry element to establish a correlation along the reaction coordinate of the reactant to the primary product transformation.

For concreteness, we analyze the exemplar reaction of formaldehyde with a hydrogen donor XH (Eq. 6.10), where the bond strength of X—H is such that the overall

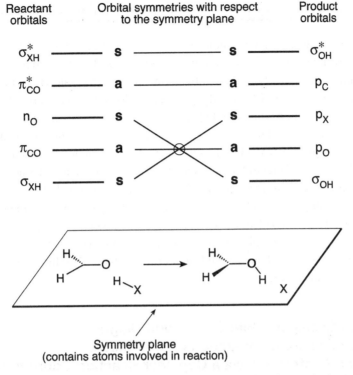

Figure 6.17   Natural orbital correlation diagram for coplanar hydrogen abstraction by formaldehyde.

process $^*R(n,\pi^*) \to I(D)$ is exothermic, and therefore thermodynamically favorable. In addition, if X—H possesses several different X—H bonds, the one that reacts most rapidly is the X—H bond that is relatively electron rich, that is, the bond that corresponds to a high energy σ MO. Since an X—H bond of a hydrogen-atom donor is being broken in the reaction and a HO bond is being made in the ketone, we select the $\sigma_{XH}(HO) \to n_0(HO)$ orbital CT as the reaction initiating orbital interaction. Since the degree of positive overlap of these two orbitals changes only slightly as a function of the orientation of the two molecules involved in the reaction, in constructing a state correlation diagram, we select a geometry for which all of the atoms involved in the reaction are in the same plane (Fig. 6.17, bottom). Therefore they will possess a symmetry element above and below the plane.

We generate an orbital and state correlation diagram *assuming* that the strictly planar approach shown in Fig. 6.17 represents the reaction coordinate. This assumption of an idealized coplanar reaction should provide a reasonable, qualitative zero-order description of the surfaces. In general, the *plane* containing the pertinent reaction centers will be a *discriminating symmetry element* for selecting the reaction coordinates of n,π* states that lead to diradicaloid geometries. Now, let us consider the consequences of this assumption in greater detail.

6.35   The Symmetry Plane Assumption: Salem Diagrams[6]

An extension of the idealized symmetry plane, shown at the bottom of Fig. 6.17, need only refer explicitly to a plane containing the nuclei *directly* involved in the electronic configuration approximation, that is, the atoms in the C=O and H—X bonds. In the case of $H_2C=O$ and H—X bonds, the pertinent atoms possess the orbitals associated with the CT interactions that initiate the photochemical hydrogen abstraction. Under these assumptions, the correlation between states of the same symmetry may be made by a simple electron-classification and electron-counting procedure. In turn, this classification and electron count may be simplified by using classical resonance structures for describing the electronic states involved (Fig. 6.18). These resonance structures can only possess electrons in idealized orbitals that are either symmetric (**s**) (do not change sign upon idealized reflection in the symmetry plane), or antisymmetric (**a**) (change sign upon idealized reflection in the symmetry plane). The postulate of an idealized symmetry plane demands that orbitals possesses either **a** or **s** symmetry with respect to that plane.

6.36   An Exemplar State Correlation Diagram for n-Orbital Initiated Reaction of n,π^* States: Hydrogen Abstraction via a Coplanar Reaction Coordinate

In our exemplar reaction, we have assumed that the primary photochemical reactions of n,π^* states are initiated by CT from a molecule M that can donate electrons from its HO to the half-filled HO of the n,π^* state. Experimentally, CT interactions from such electron-donating molecules to the electrophilic n orbital dominate the initial orbital interactions of n,π^* states. For example, the four most important reactions of the n,π^* states of ketones (hydrogen-atom abstraction, electron abstraction, addition to double bonds, and α-cleavage) are typically initiated (Fig. 6.12) by interaction of the half-filled n orbital with a σ bond (hydrogen-atom abstraction and α-cleavage), with a π bond (addition to double bonds), or with an unshared pair of electrons (electron abstraction). As a result, each of these primary photochemical reactions can be described in terms of a similar orbital and state correlation diagram, such as Fig. 6.18.

The pertinent orbitals for analysis of *coplanar* hydrogen abstraction are shown in Fig. 6.17 (top). *The coplanar geometry is selected as the reaction coordinate because it represents the best frontier orbital interaction (positive overlap) between the n_o and σ_{XH} orbitals, and because the assumption of a planar geometry allows a convenient classification of the pertinent orbitals symmetry.* The orbitals of the reactants (Fig. 6.17, left) are classified relative to the symmetry plane in a conventional notation (Fig. 6.14). Any orbital that lies in the plane must be of **s** symmetry with respect to reflection in the symmetry plane (i.e., the σ_{XH}, n_O, and σ_{XH}^* orbitals). This means that the wave function that describes the orbital does not change its sign when reflected from below to above the plane. Any orbital that exists above and below the plane with a node in the plane must be of **a** symmetry (i.e., the π_{co} and π_{co}^* orbitals).

The pertinent product orbitals are σ_{OH} and σ_{OH}^* (both of **s** symmetry) and the p orbitals on carbon (p_c above and below, **a**), the p orbital on X (p_x in the plane, **s**), and the p orbital on O (p_o above and below, **a**). From Fig. 6.17, the initial (zero-order) orbital correlations are maintained during reaction, which is expected at the very beginning of the interaction, as expected from the natural-orbital correlation of the orbitals. Note that the $\sigma_{XH} \to p_X$ and $n_o \to \sigma_{OH}$ correlations cross so that there is the possibility of a surface avoiding.

Completely filled orbitals are always totally symmetric (**S**) with respect to a symmetry element (i.e., both **a** $\times$ **a** = **S** and **s** $\times$ **s** = **S**), but two half-filled orbitals may be **A** (i.e., **s** $\times$ **a** or **a** $\times$ **s** = **A**) or **S** (**a** $\times$ **a** = **S** or **s** $\times$ **s** = **S**). Since the state symmetry is the composite (overlap) of all filled and unfilled orbitals, we can deduce the state symmetry of reactants by evaluating the product of the symmetry of appropriate reactant and primary product orbitals.

The state symmetries of the reactants and products are readily deduced from symmetries of the orbitals of Fig. 6.17. For example, S_0 must be of **S** symmetry because it possesses only doubly occupied orbitals (**a** $\times$ **a** = **S** or **s** $\times$ **s** = **S**). The symmetry of the n,π^* state is **s** (n orbital) $\times$ **a** (π^* orbital) = **A**; the symmetry of the π,π^* state is **a** (π orbital) $\times$ **a** (π^* orbital) = **S**; the symmetry of the D (p_x,p_c) state is **s**(p_x) $\times$ **a** (p_c) = **A**; and the symmetry of a Z state must be **S** because it possesses only doubly occupied orbitals (**a** $\times$ **a** = **S** or **s** $\times$ **s** = **S**).

Now, we may proceed to the zero-order state correlation diagram, or *Salem correlation diagram* (Fig. 6.18), by connecting states of the same symmetry, following the simple rule that we connect each state of the reactants to the *lowest* state of the same symmetry of the products until all the reactant states have been connected. For practical purposes, we need correlate only the pertinent reactant states S_1 and T_1 with product states; however, in Fig 6.18, the correlations of the $S_2(\pi,\pi^*)$ and $T_2(\pi,\pi^*)$ are shown for the sake of completeness. From Fig. 6.18, we see that following state symmetries, both $S_1(n,\pi^*)$ and $T_1(n,\pi^*)$ states, correlate directly with the lowest energy states of the product, I(D); that is, $S_1 = {}^1n,\pi^*$ correlates with ${}^1I(D)$ and $T_1 = {}^3n,\pi^*$ correlates directly with ${}^3I(D)$. We say that coplanar hydrogen abstraction to form ketyl radicals from $S_1(n,\pi^*)$ or $T_1(n, \pi^*)$ is *state symmetry allowed in the sense that the reactants correlate with the lowest possible product state*. By this we mean that in zero order, there is no electronic symmetry-imposed energy barrier on the surface connecting the initial excited state n,π^* of a given spin and the lowest-energy primary (diradical) product of the same spin.

On the other hand, the S_2 and T_2 states (both π,π^*) correlate with excited states of the product, I. These excited states are expected to have very high energies relative to S_2 and T_2. As a result, if S_2 or T_2 were to attempt to participate in coplanar hydrogen abstraction, a symmetry-imposed energy barrier would have to be overcome. We say that coplanar hydrogen abstraction to form ketyl radicals is symmetry forbidden from $S_2(\pi,\pi^*)$ or $T_2(\pi,\pi^*)$ because the symmetry of the orbital correlation electronically imposes an energy barrier.

In zero order (Fig. 6.18), there is a surface crossing of the S_0 and S_1 state along the reaction coordinate. In first order, we must consider the fact that the approach of the reactants will not be perfectly planar. Since the symmetries of the two states'

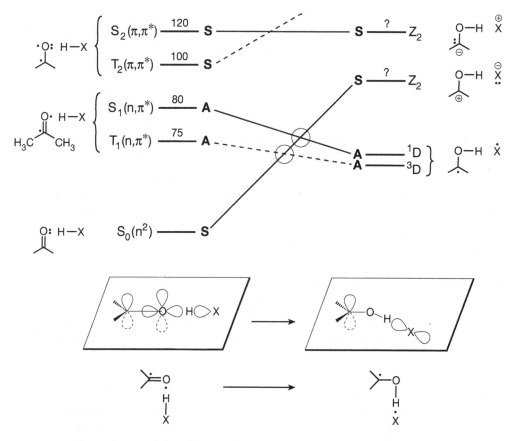

Figure 6.18   First-order correlation diagram for coplanar hydrogen abstraction. State energies in kilocalories per mole. Note that the geometry of symmetry-imposed crossings may be similar to the geometry for conical intersections. Therefore, the symmetry-imposed crossings of a correlation diagram provide a guide for the location and geometry of conical intersections.

crossings are different (since $S_0 = S$ and $S_1 = A$), the crossings either remain or are weakly avoided. This means that the situation is close to a real crossing and may correspond to a CI. If this is the case, when the representative point approaches the region of the crossing along the reaction coordinate, the system may continue the trajectory toward the product or be reflected back toward the reactants (Fig. 6.6).

In first order, the $T_1 \rightarrow {}^3D$ crossing will remain, since the multiplicity (spin symmetry) of the crossing surfaces is still different and spin–orbit coupling in addition to electronic interactions would be required for the surfaces to avoid each other.

From Fig. 6.18, we conclude that there are two low-energy pathways from the n,π^* states for the representative point in coplanar hydrogen-abstraction reactions. These two pathways (assuming an exothermic primary process) are

1. From the $S_1(n,\pi^*)$ state, the representative point will move along the reaction coordinate, decreasing its energy until it reaches the geometry corresponding to the surface crossing. This crossing is either a weakly ASC or a CI. What are the consequences of this ASC or CI? Note that the occurrence of ASC or CI has

no effect on *reactivity*; that is, the rate of reaction is determined by the energy barriers near the S_1 minimum. However, the *efficiency* of reaction from S_1 may be decreased, since entry into the ASC or CI allows partitioning from the excited surface to either S_0 of ^1D (by providing a funnel for radiationless transition from S_1 to S_0 or D_1), whereas in zero order only passage from S_1 to ^1D was allowed.

2. From the $T_1(n,\pi^*)$ state the representative point may decrease its energy by moving directly to ^3D, that is, proceeding through the crossing of T_1 and S_0 surfaces. Of importance is the fact that the reactivity and efficiency of T_1 are the same in both first and second order, since the surface crossing involves a spin change and can be only weakly avoided at best.

3. Finally, note that small barriers may arise from the neglect of intended correlations that are revealed in the natural orbital correlation diagrams.

6.37   Extension of an Exemplar State Correlation Diagram to New Situations

A combination of the methods for orbital interactions and state correlation diagrams allows us to generalize the exemplar correlation diagram for hydrogen abstraction to other reactions of an n,π^* state. At the orbital level, the key electronic features of hydrogen abstraction are the CT from a σ H—X σ orbital to the half-filled n_0 orbital (Fig. 6.17). We may postulate that any reaction whose electronic mechanism is dominated by electrophilic attack by the n_0 orbital will have the same surface topology as that deduced for hydrogen abstraction (Fig. 6.18).

For example, n,π^* states of ketones are known to (1) abstract electrons from amines (and other electron donors), (2) add to electron-rich ethylenes, and (3) transfer energy to electron-rich unsaturated compounds. On the basis of orbital interactions, we can expect electrophilic attack by the n_0 orbital of the n,π^* state to dominate each of these reactions. In each case, a RP or BR or radical ion pair intermediate is possible, and the same generalizations and expectations made for hydrogen abstraction are also possible. In other words, the state correlation topology of Fig. 6.18 may be employed for the three reactions of n,π^* states of ketones discussed above.

6.38   State Correlation Diagrams for α-Cleavage of Ketones[18]

The cleavage of a σ bond that is α to the excited carbonyl group (Eq. 6.11), as hydrogen-atom abstraction, is a prototype of the major primary photoreaction types for n,π^* states.[11] For example, the n,π^* states of acetone undergo α-cleavage to acyl and methyl radicals, as shown in Eq. 6.11.

$$\underset{\displaystyle CH_3CCH_3}{\overset{\displaystyle \overset{O}{\underset{\displaystyle \|}{}}}{}} \quad \overset{h\nu}{\longrightarrow} \quad \underset{\underset{\displaystyle n,\pi^*}{\displaystyle CH_3CCH_3}}{\overset{\displaystyle \overset{O}{\underset{\displaystyle \|}{}}}{}} \quad \longrightarrow \quad \underset{\underset{\displaystyle I(RP)}{\displaystyle CH_3C\cdot + \cdot CH_3}}{\overset{\displaystyle \overset{O}{\underset{\displaystyle \|}{}}}{}} \qquad (6.11)$$

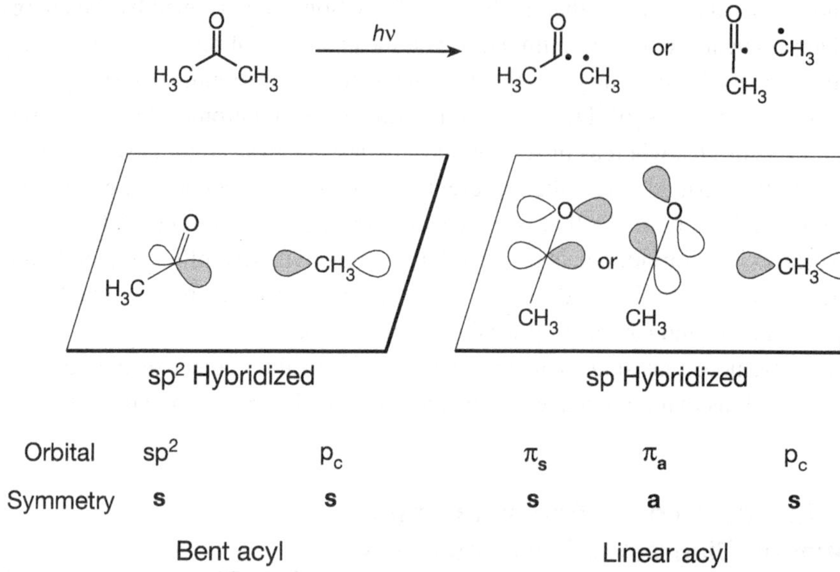

Figure 6.19   Orbital symmetries for α-cleavage relative to a plane of symmetry.

The reaction coordinate in this case is essentially the distance separating the carbonyl and methyl carbon atoms. The state symmetry of S_1 and of T_1 are n,π^* and are therefore both **A** with respect to the characteristic symmetry plane associated with the reaction (Fig. 6.18). The primary product states at the diradicaloid geometry can be described in terms of two limiting types: (1) those corresponding to a bent $CH_3CO \bullet$ group, and (2) those corresponding to a linear $CH_3CO \bullet$ group. The two lowest-energy electronic states of these two structures have different electronic symmetries relative to the characteristic symmetry plane. The bent form of $CH_3CO \bullet$ will be sp^2 hybridized at the carbonyl carbon, whereas the linear $CH_3CO \bullet$ will be sp hybridized at the carbonyl carbon (Fig. 6.19). The radical site (sp^2 orbital) generated in the bent acyl group remains in the symmetry plane and is therefore **s** with respect to coplanar α-cleavage. The linear acyl group possesses a π_s orbital that is in the symmetry plane and a π_a orbital that is perpendicular to the symmetry plane. Thus, the π_s orbital is **s** and the π_a orbital is **a** with respect to coplanar cleavage. Thus, two degenerate electronic states correspond to the linear acyl–methyl radical pair. One state (π_s,p) is **S** and the other state (π_a,p) is **A**. The number of distinct radical sites developed during a reaction is termed the reaction topicity.[6] Since α-cleavage to produce a bent acyl and alkyl radical produces two distinct radical sites (sp^2 and as here p_c orbitals), it is said to have a topicity equal to 2. On the other hand, α-cleavage to produce a linear acyl and alkyl radical produces three distinct radical sites (p_c, π_s, and π_a orbitals) and is said to have a topicity of 3. In the latter case, an odd electron may be in the π_s or π_a orbital, leading to two different diradical pairs, one with overall **S** symmetry (p_c and π_s) and one with overall **A** symmetry (p_c and π_a).

Topicity can have a major influence on the topology of a correlation diagram. Let us construct the correlation diagram for α-cleavage by proceeding in the usual manner.

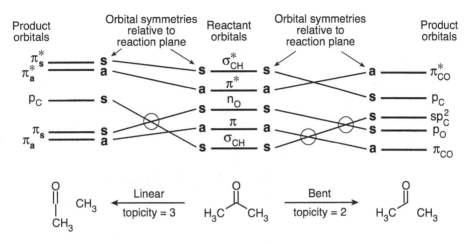

Figure 6.20   Orbital correlation diagram for α-cleavage of the orbital symmetries shown in Fig. 6.19.

First, we identify the symmetry of the key orbitals involved in the transformation. The orbital correlation is given in Fig. 6.20. Notice that the half-filled orbital on CH_3 is assumed to be a p orbital contained by the discriminating reaction plane; that is, this orbital is of **s** symmetry.

In the case of cleavage to form a linear acyl fragment, the lowest excited states $S_1(n,\pi^*)$ and $T_1(n,\pi^*)$ correlate directly to low-lying $^1D(\pi_a,p_c)$ and $^3D(\pi_a,p_c)$ states. However, for the cleavage to form a bent acyl radical, the $S_1(n,\pi^*)$ and $T_1(n,\pi^*)$ states correlate with an excited state of D, namely, $^*D(sp^2,\pi_{co}^*)$. Consequently, the initial slope of the surface for cleavage from an n,π^* state to a bent acyl rises steeply as this state tries to correlate with some diradical excited state ($^*D$).

From these considerations, we deduce that α-cleavage of n,π^* states to yield a linear acyl fragment is symmetry allowed, but that α-cleavage of n,π^* states to yield a bent acyl fragment is symmetry forbidden.[18] Thus, the pathway of higher topicity is allowed, whereas the pathway of lower topicity is forbidden.

Now, let us consider a state correlation diagram (Fig. 6.21) that depicts α-cleavage for acetone. The state and diradical energies are also shown for acetone as an exemplar. The situation for cleavage to the linear fragment is essentially the same. However, weakly avoided crossings occur along the surfaces for cleavage to the bent acyl fragment. Notice, however, that the $T_1(n,\pi^*)$ surface is subject to a surface crossing at an earlier point than $S_1(n,\pi^*)$. This earlier surface crossing may lead to a lower-energy, symmetry-imposed energy barrier for cleavage of $T_1(n,\pi^*)$ relative to $S_1(n,\pi^*)$.

From Fig. 6.21, we expect that α-cleavage of acetone will be an activated process no matter which pathway is followed. However, if the energy of the linear radical pair is lowered by structural effects (e.g., release of ring strain in a cyclic ketone or stabilization of the acyl or alkyl radical sites), the linear pathway may be specifically favored, since a symmetry-imposed barrier does not occur along this reaction coordinate.

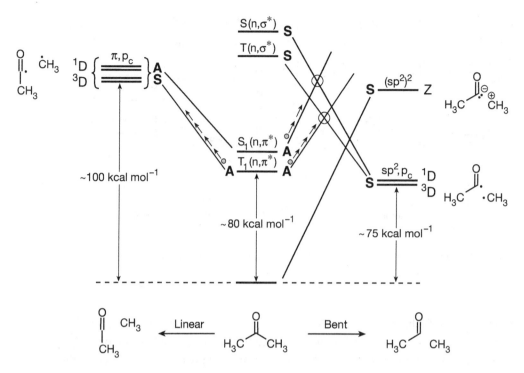

Figure 6.21   State correlation diagram for α-cleavage for the orbital correlation diagram of Fig. 6.20: topicity of 3 (linear path) shown at left and topicity of 2 (bent path) shown at right.

6.39   A Standard Set of Plausible Primary Photoreactions for π,π^* and n,π^* States

Note that the most commonly encountered lowest-energy excited states of organic molecules may be classified as $S_1(\pi,\pi^*)$, $T_1(\pi,\pi^*)$, $S_1(n,\pi^*)$, or $T_1(n,\pi^*)$. In this chapter, we have seen how theory can lead to (1) a prediction of the plausible (i.e., low-energy) primary photochemical processes via the consideration of frontier orbital interactions, and (2) the generation of the network for surface (reaction coordinate) pathways via the maps that result from orbital and state correlation diagrams. Now, we proceed to create a list of *all the expected plausible primary photoreactions of* S_1 *and* T_1 of common functional groups of organic molecules (ketones, olefins, enones, aromatic compounds), based on the above theoretical considerations.

6.40   The Characteristic Plausible Primary Photochemistry Processes of π,π^* States

If the C=C bond can be twisted and if mixing with $S_2(\pi^*)^2$ occurs during the twisting process, $S_1(\pi,\pi^*)$ states will often possess a substantial zwitterionic character that is further enhanced as twisting occurs (Fig. 6.9). As a result, the formation of zwitterions, not radical pairs and/or biradicals, is expected when *R is $S_1(\pi,\pi^*)$. Thus, an $S_1(\pi,\pi^*)$

excited state is expected to behave as a zwitterion, that is, exhibit the reactions of a carbonium ion or carbanion. We can characterize the primary process of $S_1(\pi,\pi^*)$ as a $S_1(\pi,\pi^*) \to I(Z)$ elementary step. Thus, the plausible set of primary photochemical reactions of $S_1(\pi,\pi^*)$ includes proton or electron-transfer reactions, nucleophilic or electrophilic additions, and carbonium or carbanion rearrangements that proceed to isolated products, P; that is, $S_1(\pi,\pi^*) \to I(Z) \to P$ (via carbonium ion or carbanion chemistry of Z).

In addition to zwitterionic reactions reflecting carbonium ion–carbanion characteristics, from the above discussion of orbital interactions and PE surfaces, we expect $S_1(\pi,\pi^*)$ states to undergo "concerted" pericyclic photoreactions whose stereochemistry follows the Woodward–Hoffmann rules. The favored stereochemical pathways of these reactions can be predicted by considering orbital interactions, and the prototype surface topology for such reactions will be analogous to Fig. 6.16. We understand that the concept of "concerted" as intended for ground-state reactions, is replaced with the concept of a surface crossing or conical intersection when describing excited singlet-state reactions.

From the energy surface of Fig. 6.9, $S_1(\pi,\pi^*)$ of ethylenes and polyenes serves as an exemplar for the formation of twisted zwitterionic intermediates and/or cis–trans isomerization.

From these simple theoretical considerations, we may now tabulate a list of plausible primary photochemical reactions that are initiated in $S_1(\pi,\pi^*)$:

1. Concerted pericyclic reactions (electrocyclic rearrangements, cycloadditions, cycloeliminations, sigmatropic rearrangements, etc.), which follow the Woodward–Hoffmann rules.
2. Reactions characteristic of carbonium ions (carbonium ion rearrangements, addition of nucleophiles, etc.) and of carbanions (addition to electrophilic sites, protonation, etc.).
3. Cis–trans isomerization.
4. Electron transfer.

Granted that $S_1(\pi,\pi^*)$ has the possibility of the above set of characteristic reactions, the actual rate of any one of these reactions will depend on the details of the reactant structure and the reaction conditions. The probability of any reaction from $S_1(\pi,\pi^*)$ will depend on a competition between the rate of reaction and the rate of other photophysical or photochemical pathways from $S_1(\pi,\pi^*)$. For example, the plausibility of electron transfer (from a carbanion center of Z or to the carbonium ion center of Z) will depend on the redox properties and the energy of $S_1(\pi,\pi^*)$. If the energetics of an electron transfer are exothermic, the electron transfer is plausible; if the energetics of an electron transfer are endothermic, the electron transfer is implausible. Electron transfer is discussed in detail in Chapter 7.

Because of spin selection rules that require conservation of spin in an elementary step, $T_1(\pi,\pi^*)$ *is not expected to undergo concerted pericyclic reactions* (unless the product can be produced in a triplet state or if a very good spin–orbit coupling mechanism is available along the reaction coordinate, both of which are improbable). Thus, a concerted pericyclic reaction is not considered a member of the plausible set of reactions from $T_1(\pi,\pi^*)$. Indeed, the reactions for $T_1(\pi,\pi^*)$ should be typical of those of a carbon radical, and the major pathway of reaction of $T_1(\pi,\pi^*)$ is expected to be formation of a triplet radical pair or triplet diradical; that is, $T_1(\pi,\pi^*) \rightarrow {}^3 I(D)$. The *typical reactions of a carbon-centered radical are qualitatively equivalent to the reactions of an oxygen-centered radical such as the n-orbital of a n,π^* state.* Of course, the rates of the reactions for the two types of radicals may differ dramatically, but the types of reactions are identical. Using the n_0 as a model for an oxygen radical leads to the following plausible set of primary photochemical reactions initiated from $T_1(\pi,\pi^*)$:

1. Hydrogen-atom abstractions.
2. Electron abstraction or donation.
3. Addition to unsaturated bonds.
4. Homolytic fragmentations.
5. Rearrangement to a more stable carbon-centered radical.

In addition to the above five primary photochemical processes, $T_1(\pi,\pi^*)$, like $S_1(\pi,\pi^*)$, will generally possess an inherent driving force to twist about double bonds. This process could lead to cis–trans isomerization or twisted diradical intermediates, which in turn can undergo transition to a strained twisted ground state. Note that ISC from $S_1(\pi,\pi^*)$ to $T_1(\pi,\pi^*)$ is very inefficient when the gap between $S_1(\pi,\pi^*)$ and $T_1(\pi,\pi^*)$ states is large (the general energy-gap rule for radiationless transitions, Eq. 5.53). This situation is typical of olefins. However, since the formation of $T_1(\pi,\pi^*)$ can be photosensitized by electronic energy transfer (Chapter 7), $T_1(\pi,\pi^*)$ states are readily accessible through this indirect method of sensitized excitation.

Many examples for the validity of the above set of primary photochemical reactions for $S_1(\pi,\pi^*)$ states are well established in the photochemistry of organic molecules.

6.41   The Characteristic Plausible Primary Photochemical Processes of n,π^* States

The photochemistry of n,π^* states can be expected to contrast with that of π,π^* states in two major respects:

1. The photochemistry of $S_1(n,\pi^*)$ and $T_1(n,\pi^*)$ for a given molecule should be qualitatively similar and differ quantitatively only in terms of rates (the singlet state possesses more energy than the triplet state, which may drive reactions where thermodynamic considerations for the exothermicity of a primary photochemical process are important). This situation contrasts with reactions expected from $S_1(\pi,\pi^*)$ and $T_1(\pi,\pi^*)$ that differ qualitatively, that is, zwitterionic and/or concerted versus diradicaloid and nonconcerted, respectively.
2. The photochemistry of n,π^* states only produces diradicaloids to a good approximation; that is, only n,$\pi^* \rightarrow I(D)$ processes are plausible.

Based on an orbital interaction analysis (Fig. 6.13), and the postulate that all n,π^* reactions proceed preferentially via D states, we conclude that the primary photochemical processes of n,π^* will produce radicals and that the overall photoreactions will mimic radical chemistry that is either initiated by the n orbital or the π^* orbital. From these considerations, we deduce that the plausible primary processes expected from a theoretical standpoint are

n-Orbital Initiated	π^*-Orbital Initiated
Atom abstraction	Atom abstraction
Radical addition	Radical addition
Electron abstraction	Electron donation
α-Cleavage	β-Cleavage

Although both atom abstraction and radical addition are expected in theory to be initiated by either the n or the π^* orbitals, the former will exhibit *electrophilic* and the latter *nucleophilic* characteristics. The reactions initiated by the n orbital will be analogous to those of an alkoxy radical ($RO^\bullet$), and the reactions initiated by the π^* orbital will be analogous to those of a ketyl radical ($R_2\overset{\bullet}{C}OR$). Furthermore, the stereoelectronic dispositions of the reactions will differ, depending on which frontier orbital dominates the electronic interactions with the substrate. For example, if the half-filled n orbital initiates the interaction, the reaction will be sensitive to steric factors influencing the approach of the substrate in the plane of the molecule and near the "edges" of the carbonyl oxygen. On the other hand, if the π^* orbital initiates the reaction, the reaction will be sensitive to steric factors that influence the approach of the substrate above and below the "faces" of the carbonyl function.

Since the π^*_{CO} orbital is delocalized, attack may be initiated at the carbon or at the oxygen atom. Ignoring the specifics of the substrate, if the reaction is initiated by the π^* electron, only the addition to carbon produces a low-energy diradical state. Attack of the π^* orbital to produce a bond to oxygen produces an electronically excited diradical and will therefore encounter a symmetry-imposed energy barrier.

6.42   Summary: Energy Surfaces as Reaction Graphs or Maps

In this chapter the global paradigm for overall pathways from electronic states (*R) to products (P) is described qualitatively in terms of electronic energy surfaces and electronic orbital and state correlation diagrams. These energy surfaces provide a satisfying pictorial description of the trajectory of a representative point along reaction coordinates and the possible and plausible fates of the representative point as it approaches surface crossings, conical intersections, and excited state minima. Salem diagrams provide intuition concerning the surface possibilities for the representative point on n,π^* states and Woodward–Hoffman-derived surface diagrams provide intuition concerning the surface possibilities for concerted pericyclic reactions. Two fundamental molecular motions, single-bond stretching and double-bond twisting, are shown to be excellent starting points for describing reaction coordinates for a range of primary photochemical processes. These two motions were discussed in Chapter 5 as the "loose bolt" and "free rotor" motions that could trigger radiationless electronic deactivations. These qualitative considerations have been confirmed to be valid in many cases by more detailed quantitative quantum mechanical computations.

References

1. M. Klessinger and J. Michl, *Excited States and Photochemistry of Organic Molecules*, VCH Publishers, New York, 1995. (b) J. Michl, *Mol. Photochem.* **4**, 243 (1972). (c) Ref. 1b, p. 257. (d) Ref. 1b, p. 287. (e)*Top. Curr. Chem.* **46**, 1 (1974).

2. For a discussion, see any introductory physics text. The following reference is an example. D. Halliday and R. Resnick, *Physics*, John Wiley & Sons, Inc., New York, 1966.

3. N. J. Turro, J. McVey, V. Ramamurthy, and P. Lechtken, *Angew. Chem., Int. Ed. Engl.* **18**, 572 (1979).

4. See the following references for reviews and discussions of the frontier orbital interactions and orbital symmetry control of organic reactions. (a) K. Fukui, *Top. Curr. Chem.* **15**, 1 (1970). (b) K. Fukui, *Acc. Chem. Res.* **4**, 57 (1971). (c) R. F. Hudson, *Angew. Chem. Int. Ed. Engl.* **12**, 36 (1973). (d) N. D. Epiotis, *Angew. Chem. Int. Ed. Engl.* **13**, 751 (1974). (e) N. J. S. Dewar, *Angew. Chem. Int. Ed. Engl.* **10**, 761 (1971). (f) H. E. Zimmerman, *Acc. Chem. Res.* **4** 272 (1971).

5. M. A. Robb, M. Garavelli, M. Olivucci, and F. Bernardi, **15**, 87 (2000). (b) F. Bernardi, M. Olivucci and M. A. Robb, *Chem. Soc. Rev.* **20**, 321 (1996).

6. (a) L. Salem, *J. Am. Chem. Soc.* **90**, 3251 (1968). (b) W. G. Dauben, L. Salem, and N. J. Turro, *Acc. Chem. Res.* **8**, 41 (1975) and references cited therein.

7. (a) E. J. Teller, *Phys. Chem.* **41**, 109 (1937). E. Teller, *Israel J. Chem.* **7**, 227 (1969).

8. M. Ben-Nun and T. J. Marttinez, *Chem. Phys.* **259**, 237 (2000).

9. See the following references for reviews of the role of diradicals and zwitterions in photoreactions. (a) L. Salem and C. Rowland, *Angew. Chem. Int. Ed. Engl.* **11**, 92 (1971). (b) L. Salem, *Pure Appl. Chem.* **33**, 317 (1973).

10. (a) R. B. Woodward and R. Hoffmann, *Angew. Chem. Int. Ed. Engl.* **8**, 781 (1969). (b) R. B. Woodward and R. Hoffmann, *Acc. Chem. Res.* **1**, 78 (1968). (c) H. C. Longuet-Higgins and E. W. Abrahamson, *J. Am. Chem. Soc.* **87**, 2046 (1965).

11. (a) A. Devaquet, *Top. Curr. Chem.* **54**, 1 (1975). (b) A. Devaquet, *Pure Appl. Chem.* **41**, 535 (1975).

12. I. Fleming, *Frontier Orbitals and Organic Chemistry*, John Wiley & Sons, Inc., New York, 1977.

13. F. A. Cotton, *Chemical Applications of Group Theory*, 3rd ed., John Wiley & Sons, Inc., New York, 1990.

14. B. Bigot, A. Devaquet, and N. J. Turro, *J. Am. Chem. Soc.* **103**, 6 (1981).

15. (a) M. O. Trulson and R. A. Mathies, *J. Phys. Chem.* **94**, 5741. (b) P. J. Reid, S. J. Doig, S. D. Wickhma, and R. A. Mathies, *J. Am. Chem. Soc.* **115**, 4754 (1993). (c) L. A. Walker et al., *Chem. Phys. Lett.*, **242**, 415, 1995.

16. W. T. A. M. Van der Lugt and L. J. Oosterhoff, *J. Am. Chem. Soc.* **91**, 6042 (1969).

17. See the following reference for a discussion. N. J. Turro, C. Dalton, K. Dawes, G. Farrington, R. Hautala, D. Morton, M. Niemczyk, and N. Schore, *Acc. Chem. Res.* **5**, 92 (1972) and references cited therein.

18. N. J. Turro, W. E. Farneth, and A. Devaquet, *J. Am. Chem. Soc.* **98**, 7425 (1976).

CHAPTER 7

Energy Transfer
and Electron Transfer

7.1 Introduction to Energy and Electron Transfer[1]

In Chapter 6, we learned that the overlap of frontier molecular orbitals (FMOs) can result in *weak electronic exchange interactions* between an electronically excited state (*R) and a second molecular species (M) (or between *R and reactive groups that are accessible intramolecularly). The interactions between the half-filled HO (highest-occupied molecular orbital) of *R with the HO of M (and the half-filled LU (lowest-unoccupied molecular orbital) of *R and the LU of M) is *generally stabilizing*, but the degree of stabilization depends on the degree of positive overlap and the energy differences between the interacting orbitals (Section 6.25). To a good approximation, the FMO interactions determine the reaction coordinate (the lowest potential energy, PE, pathway on an energy surface) for both thermal and photochemical reactions. *Electron* and *electronic energy transfer* are two of the most important interactions between *R and M. This transfer can be described in terms of similar initial frontier orbital overlap and electron exchange interactions. This chapter[1] will consider electronic energy- and electron-transfer processes together since *both processes possess a common frontier orbital interaction, namely, orbital overlap and electron exchange description.* Whether electron charge transfer (CT) or energy transfer will occur as the result of FMO interactions will depend on a number of factors that are described in this chapter. Also discussed is a very important electronic energy-transfer process that does not involve overlap of orbitals in space, but occurs *through space* resulting from the interaction of *an oscillating dipolar electric field* of *R that induces an electric oscillation of M and results in the formation of *M.

Consider the four cases of Scheme 7.1, which provide paradigms for the energy- and electron-transfer processes. Energy transfer may occur either by an electron exchange interaction that requires orbital overlap or by a dipole–dipole interaction that occurs through an oscillating electric field in space. Electron transfer occurs only by electron exchange interactions and requires orbital overlap. Cases 1 and 2

383

are *electronic energy-transfer* processes. For case 1, the energy-transfer results from orbital overlap between *R and M; electron exchange occurs during orbital overlap and provides the interaction responsible for the energy-transfer event. For case 2, energy-transfer results from the interaction of an *oscillating electric dipole field* that surrounds the space about *R and the electrons of M.

1. Exchange electronic energy transfer, where *R is an energy donor, M is an energy acceptor. $^*R + M \rightarrow R + {}^*M$
2. Dipole–dipole electronic energy transfer, where *R is an energy donor, M is an energy acceptor, where $^*R + M \rightarrow R + {}^*M$
3. Electron transfer, where *R(eD) behaves as an electron donor, M(eA) behaves as an electron acceptor; $I(R^{\bullet+}, M^{\bullet-})$ is a radical ion pair (RIP). $^*R(eD) + M(eA) \rightarrow I(R^{\bullet+}, M^{\bullet-})$
4. Electron transfer, where *R(eA) is an electron acceptor, M(eD) is an electron donor; $I(R^{\bullet-}, M^{\bullet+})$ is a RIP. $^*R(eA) + M(eD) \rightarrow I(R^{\bullet-}, M^{\bullet+})$

Scheme 7.1   The mechanistic paradigm for electronic energy transfer and electron transfer.

A critical difference between electron exchange and dipole–dipole interactions is that dipole–dipole interactions of two electric fields does not involve orbital overlap: Since it involves the interactions of fields, dipole–dipole interactions can occur through empty or molecularly occupied space, just as the dipole field of a magnet can occur through space filled with molecules or empty space. Thus, cases 1 and 2 represent two completely different electronic interactions. Each interaction will possess very different rate constant dependencies on factors, such as the distance separation and optical properties of *R and M (in particular their oscillator strengths and transition dipoles). For energy-transfer processes, *R is always the energy donor and M is always an energy acceptor.

Cases 3 and 4 are *electron-transfer* processes. In an electron-transfer process, *R may be an electron donor (case 3) or acceptor (case 4), as we saw in Chapter 6. To distinguish the two situations in cases 3 and 4, we use the symbol "eD" to indicate when *R (or M) is an *electron donor* and the symbol "eA" to indicate when *R (or M) is *an electron acceptor*. For case 3, electron transfer is triggered by frontier orbital overlap interactions (Fig. 6.10) that result in electron transfer from the LU of *R(eD) to the LU of M(eA). For case 4, electron transfer is triggered by frontier orbital interactions that result in an electron transfer from the HO of M(eD) to the HO of *R(eA). We use the terms "electron exchange" and "orbital overlap" interchangeably when discussing energy- and electron-transfer processes.

Some of the most important parameters in photoinduced electron and photoinduced energy transfer are the rate constant (k) of the primary photochemical process and the dependence of k on features, such as the distance of separation of *R and M, solvent polarity, the structure of *R and M, the electronic excitation energy possessed by *R and the redox potentials of *R and M. Let E_{*R} be the electronic excitation energy of *R and E_{*M} be the electronic excitation energy of *M. The bimolecular rate con-

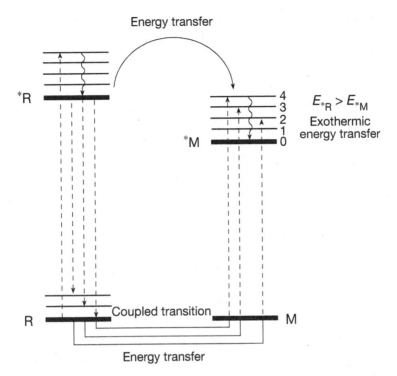

Figure 7.1 Energetically favorable conditions for the energy-transfer process; $^*R + M \rightarrow R + {}^*M$. The darker lines indicate the lowest vibrational level for each electronic state, and the lighter lines indicate the excited vibrational levels for each electronic state. Since $E_{*R} > E_{*M}$, some vibrations of M are excited in order to conserve energy during the energy-transfer step.

stant (k) of the primary photochemical process of *energy transfer* $^*R + M \rightarrow R + {}^*M$ (by either the exchange or the dipole–dipole interaction) will depend strongly on whether the overall energy-transfer process is energetically downhill (termed exothermic, $E_{*R} > E_{*M}$) or uphill (termed endothermic, $E_{*R} < E_{*M}$) with respect to the thermodynamics of the overall energy-transfer process (Fig. 7.1). Strictly speaking, the terms exergonic and endergonic should be used for *free energy changes* (ΔG) and the terms exothermic and endothermic, respectively, should be used for changes in *enthalpy* (ΔH). However, since they are probably more familiar to the student, we will use the terms exothermic and endothermic to describe reactions for which the reaction proceeds with a negative free energy (or enthalpy) and with a positive free energy (or enthalpy), respectively.

In fluid solution, an energy-transfer process usually proceeds at a rapid rate, often close to the rate of diffusion, if the overall energy-transfer process $^*R + M \rightarrow R + {}^*M$ is exothermic, that is, $E_{*R} > E_{*M}$ (Fig. 7.1). In the latter case the electronic excitation energy available to $^*R$ is sufficient to electronically excite $^*M$ without requiring significant thermal activation, and therefore can proceed at the rate at which $^*R$ and M diffuse together and interact. In order to conserve energy during the energy-transfer elementary step, since the electronic energy of *R is greater than the electronic energy of *M, the energy transfer will require excitation of vibrations of *M ($v = 4$ in the case shown in Fig. 7.1).

Photoinduced electron transfer corresponds to a primary photochemical process, a *R $\rightarrow$ I(RIP)$_{gem}$ step for which I is a *geminate (gem) radical ion pair*: *R + M $\rightarrow$ I(R$^{\bullet+}$, M$^{\bullet-}$)$_{gem}$ or (R$^{\bullet-}$, M$^{\bullet+}$)$_{gem}$. The radical ion pair that is produced by electron transfer is termed a "geminate pair" since the two partners of the radical ion pair were *born together* (are gem) as the result of an electron-transfer event. (We develop a convenient terminology for the description of various situations involving radical pairs in Section 7.26.) The rate constant (k) of the *electron-transfer* process is determined not only by the electronic excitation energy of *R, which can be employed to drive the electron-transfer process, but also by the thermodynamics of the electrochemical reduction–oxidation (redox) characteristics of the overall electron-transfer process. Although energy- and electron-transfer processes of cases 1, 3, and 4 have qualitative similarities at the initial stages of FMO overlap and of electron exchange interactions, the two processes become quantitatively different in some important respects as the representative point proceeds farther along the reaction coordinate.

An important difference between electron and energy transfer is that energy transfer from *R to M does not usually create a large redistribution of charge but produces an electronically excited state *M and neutral ground state R. On the other hand, the electron-transfer process creates a reactive intermediate with significant charge separation, a radical ion pair I(R$^{\bullet+}$, M$^{\bullet-}$), or I(R$^{\bullet-}$, M$^{\bullet+}$), whose energetic characteristics will be very sensitive to polar effects, such as the solvation around the charged ions. The latter structural change may be classified as "reorganization" of the solvent around the ions and requires a certain amount of structural "reorganization" energy that will be very important in determining the rate of electron transfer. We deal with the issue of solvent reorganization energy and its influence on the rate of electron transfer in detail in Section 7.14. The latter section describes the Marcus theory of electron transfer.

In solution, the rates and efficiencies of energy and electron-transfer processes will depend not only on electronic and energetic factors but also on key mechanical processes related to molecular diffusion: (1) the diffusional process that bring the reactants (*R and M) together to within a critical distance (r_c) at which energy or electron transfer can occur in competition with deactivation of *R; (2) the relative competition between reaction and nonreactive separation of the reactants after the critical distance r_c has been achieved; and (3) the successful or unsuccessful irreversible separation of the primary products produced by either energy or electron transfer.

Initially, we deal with an analysis of the interactions by which energy and electron transfer can take place, without being concerned with the details of diffusional processes by which donors and acceptors approach within the critical distances required for successful energy or electron transfer. We will next deal with the diffusional and structural processes that are responsible for bringing *R and M to within a "critical distance" r_c required for efficient energy or electron transfer. We also consider systems where electron and energy transfer occur, even when the reactants (*R and M) are separated by a molecular "spacer" framework. The spacer may be rigid and fixed during the time scale of the energy- or electron-transfer events or the spacer may be flexible and dynamic during the time scale of the electron- or energy-transfer events. Remarkably, we will see that indeed even the σ bonds of spacers can serve as "conductors" of electrons in both energy- and electron-transfer processes and that the dynamics of flexible spacers can control the rates and efficiencies of energy- and electron-transfer processes (Sections 7.19 and 7.22).

The diffusion of *R and M from random initial positions in a liquid to form a collision pair in a solvent cage is characterized by a rate constant for diffusion (k_{dif}). The latter is a very important solvent-dependent quantity since k_{dif} determines the limiting rate at which *R and M can be delivered as colliding partners in a solvent cage. Benchmarking the typical values of k_{dif} for organic solvents is therefore of great value in the study of energy and electron transfer. Fortunately, k_{dif} can be readily estimated for liquids from their viscosities using a simplified expression (known as the Debye equation, Eq. 7.1). From this expression, the rate constants (k_{dif}) for diffusion-controlled processes can be readily estimated simply from knowledge of temperature (T) and viscosity (η).

$$k_{dif} = \frac{8RT}{3000\eta} \tag{7.1}$$

In Eq. 7.1, η is the viscosity of the solvent (in units of poise, P) and R is the gas constant (8.31×10^7 erg mol^{-1}, or 1.987 cal mol^{-1} K^{-1}). Here, we note that for typical nonviscous organic solvents (benzene, acetonitrile, hexane) at room temperature, η is $\sim$ 1–10 cP, so that from Eq. 7.1, the value of k_{dif} for nonviscous solvents is in the range of 10^9–10^{10} M^{-1} s^{-1}. This range is a benchmark for diffusional processes in fluid solution, since *it represents the fastest rate at which electron and energy transfer that require substantial orbital overlap (collisions) can occur.* Diffusional processes will be discussed in more detail in Section 7.33.

Frontier orbital interactions are responsible for *both* electron and energy transfer that occurs as the result of electron exchange. Frontier interactions are weak since they represent the very beginning of the orbital overlap process. Thus, from the standpoint of quantum mechanics, we expect that for *weak electronic interactions,* perturbation theory (Section 3.5) for transitions between an initial electronic state Ψ_1 and a final electronic state Ψ_2 will serve as an appropriate theoretical basis and a good approximation for describing the rates of energy and electron transfer between *R and M. As for all electronic transitions, a resonance between *R and M and energy conservation must occur for energy or electron transfer to occur.

7.2   The Electron Exchange Interaction for Energy and Electron Transfer

The term "interaction" implies a mutual and reciprocal force or influence of two particles or waves on one another (Newton's third law). For electron transfer, electron exchange is the only significant interaction that need be considered. However, for energy transfer, indicated in Scheme 7.1, two very different interactions, electron exchange (overlap of orbitals) and dipole–dipole interactions (overlap of dipole fields), can trigger the transfer of energy from *R to M.

Energy- and electron-transfer reactions (induced by electron exchange interactions) will be covered jointly and comparatively to emphasize the conceptual similarities and equivalence of the two processes in terms of orbital interactions (orbital overlap and electron exchange). Figure 7.2a and b shows schematically the basic orbital interactions that relate these two *electron exchange* processes from the point of view of frontier orbital (HO and LU) representations. Although in an energy-transfer step, *R is always an energy donor, as shown in Scheme 7.1, in a photoinduced electron-transfer step, *R may be either an electron acceptor *R(eA) or an electron-donor *R(eD). In the physics literature, a half-filled HO is considered a "positive hole" in the electronic framework of a molecule; similarly, *R is viewed as simultaneously possessing both a positive hole (one electron in the half-filled HO) and one electron in the half-filled LU. Physicists call *R an *exciton*, that is, a positive hole (a half-filled HO) plus a nearby electron (a half-filled LU). In Fig. 7.2a, instead of using the symbols R and M, it is now convenient to use the symbols D and A to represent an electron donor (D) and an electron acceptor (A), respectively, in an electron-transfer step. In Fig. 7.2a (top left), the positive hole appears initially on A (half-filled HO, $A^{\bullet+}$). In Fig. 7.2a (top right), the positive hole $A^{\bullet+}$ has moved from its initial site to a site producing $D^{\bullet+}$. The overall process of moving a half-filled HO from one molecule to another is termed a "positive-hole transfer" $D + A^{\bullet+} \rightarrow D^{\bullet+} + A$. The same process could also be viewed as an electron transfer from D to $A^{\bullet+}$. Similarly, the orbital description of an electron transfer from $D^{\bullet-}$ (electron in a half-filled LU) to the LU of an acceptor A is shown schematically in Fig. 7.2a (bottom) for the overall $D^{\bullet-} + A \rightarrow D + A^{\bullet-}$ electron-transfer process. Note that the term "charge transfer" is often used to refer to the general cases for transfer of an electron or a positive hole.

The orbital situations shown schematically in Fig. 7.2a are summarized by Eqs. 7.2a and b. In Eq. 7.2a, $A^{\bullet+}$ (Fig. 7.2a, top left) can be considered as analogous to the orbital situation for which the electronic characteristics of only the half-filled HO of *R is emphasized. This emphasizes $D^{\bullet+}$ as a positive charge or electronic "positive hole" that is seeking a single electron to restore neutrality. The latter may be achieved by taking an electron from the HO of some electron donor (D); thus, the orbital characteristics of $D^{\bullet+}$ may be viewed as analogous to those of the half-filled HO of *R. In Fig. 7.2a (bottom), the orbital characteristics of $D^{\bullet-}$ may be viewed as analogous to those of the half-filled LU of *R, which is seeking a positive charge or hole to which an electron can be donated. Thus, the positive hole transfer between $A^{\bullet+}$ and D involves only HOs, whereas the electron transfer between $D^{\bullet-}$ and A involves only LUs. Generally, these electron-transfer processes will be plausible (Section 7.13) if the overall steps in Eqs. 7.2a and b are exothermic in terms of their total free energy change (ΔG).

$$A^{\bullet+}(HO)^1 + D(HO)^2 \rightarrow A(HO)^2 + D^{\bullet+}(HO)^1$$

Positive hole transfer, Fig. 7.2a, top

(7.2a)

$$D^{\bullet-}(LU)^1 + A(LU)^0 \rightarrow D(LU)^0 + A^{\bullet-}LU)^1$$

Electron transfer, Fig. 7.2a, bottom

(7.2b)

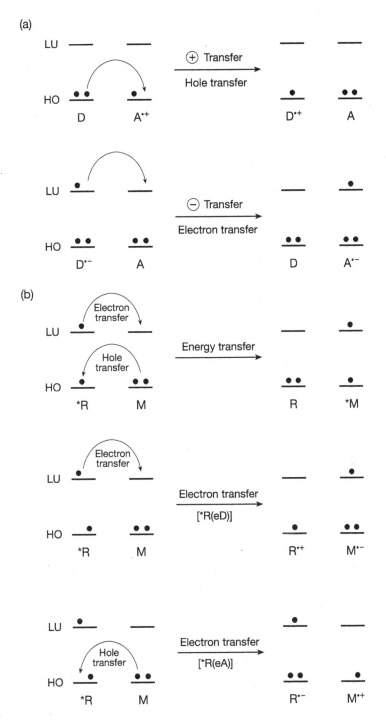

Figure 7.2　(a) Frontier orbital representation of positive hole (top) transfer from D to $A^{\bullet+}$ and an electron (bottom) transfer from $D^{\bullet-}$ to A as elementary steps of CT. (b) Schematic relationship among hole, electron, and energy transfer.[2] Energy transfer is considered to be a coordinated double electron transfer of an electron from the LU of the donor to the LU of the acceptor that occurs concurrently with a transfer of an electron from the HO of the acceptor to the HO of the donor.[2] The term electron transfer is usually employed for single electron transfers involving LUs of the donor and acceptor; the term hole transfer is usually employed for electron transfer involving the HOs of the donor and acceptor (Fig. 7.2a).

For the energy- and electron-transfer processes involving an electronically excited state, it is convenient to translate *R and M notation of Scheme 7.1 into the D and A notation of Eq. 7.3a–c.

$$^*D + A \rightarrow D + {}^*A \qquad \text{Energy transfer from } {}^*D \text{ to } A \qquad (7.3a)$$

$$^*D + A \rightarrow D^{\bullet+} + A^{\bullet-} \qquad \text{Electron transfer from } {}^*D \text{ to } A \qquad (7.3b)$$

$$D + {}^*A \rightarrow D^{\bullet+} + A^{\bullet-} \qquad \text{Hole transfer from } A^* \text{ to } D \qquad (7.3c)$$

Figure 7.2b illustrates how *both* energy- and electron-transfer processes involving *R can be viewed in terms of FMO interactions. These weak initial orbital interactions are sometimes termed CT interactions. This means that, as a result of orbital interactions, a certain amount of electric charge is transferred from *R to M (LU of *R → LU of M) or from M to *R (HO of M → HO of *R). In this description (Fig 7.2b, top), the overall *energy-transfer* process can be viewed conceptually as a sum of a more or less synchronous (negative) electron- and (positive) hole-transfer processes.[2] Thus, the theoretical formulation of energy transfer by frontier orbital overlap is described in terms of electron exchange, which at the initial stages of interaction is equivalent to partial CT.

In Fig. 7.2a, the species ($A^{\bullet+}$) that possesses a half-occupied HO is an electron acceptor and the molecule ($D^{\bullet-}$) that possesses a half-occupied LU is an electron donor. We emphasize that an electronically excited state (*R) simultaneously possesses a half-filled HO (analogous to the positive hole in $A^{\bullet+}$) and a half-filled negative LU (analogous to a high-energy electron in $D^{\bullet-}$), and *R *may serve as either an electron donor* (Fig. 7.2b, middle) *or as an electron acceptor* (Fig. 7.2b, bottom). Whether *R will serve as an electron donor or acceptor will depend on the factors that determine the exothermicity of the overall electron-transfer process, that is, on the electron-accepting or -donating ability of M. These factors will be discussed in detail in Section 7.13 and subsequent sections.

Now, we consider possible mechanisms for *delivery* of *R and M, from initial positions in a solution, to the separation r_c at which energy or electron transfer can occur with a certain efficiency. We will start with a common case for a fluid solution for which *R and M start out initially at random separations in a solvent and then undergo random diffusion until they come to within the critical distance r_c.

First, we cover the general cases for delivery mechanisms by which electron or energy transfer take place by random diffusion in solution, and then in Sections 7.17 and 7.22, we consider the special cases for which *D and A are separated by rigid and flexible spacers.

Photochemical and photophysical processes of *R are always potentially available to compete with the primary processes of energy or electron transfer. With energy transfer as an exemplar, Eqs. 7.4–7.7 illustrate steps that can be plausibly competitive with energy transfer between an excited donor (*D) and acceptor (A).

$$^*D \xrightarrow{k_D} D\ (+\ h\nu \text{ or } \Delta) \qquad (7.4)$$

$$^*D + A \xrightarrow{k_{ET}} D + {}^*A \qquad (7.5)$$

$$*D + A \xrightarrow{k_w} D + A \tag{7.6}$$

$$*D + A \xrightarrow{k_{rxn}} \text{Chemical change} \tag{7.7}$$

Equation 7.4 represents all of the *unimolecular photophysical* processes of radiationless or radiative deactivation of *D to D, which are grouped together by the rate constant k_D. Equation 7.5 specifically represents the energy-transfer (ET) step of interest with rate constant k_{ET}. Equation 7.6 represents all "energy-wasting (w) steps" in which bimolecular interaction of *D and A causes "nonreactive quenching" of *D to D with a bimolecular rate constant (k_w). Equation 7.7 represents all primary photochemical reactions of *D and A with bimolecular rate constant k_{rxn}.

Note that of the processes available to *D, only Eq. 7.5, with rate constant k_{ET}, leads to energy transfer (for convenience, we use the uppercase letters 'ET' as a shorthand to denote an *energy transfer* and the lowercase letters 'et' to denote an *electron transfer*). Thus, if we determine an experimental rate constant (k_q) for "total *bimolecular* quenching" of *D by A, the value of k_q *will incorporate all modes of bimolecular deactivation (Eqs. 7.5–7.7) of *D by A* as shown in Eq. 7.8a.

$$k_q = k_{ET} + k_w + k_{rxn} \tag{7.8a}$$

If we determine an experimental rate constant for decay of *D in the presence of A, k_{exp} is a measure of the overall unimolecular (k_D) and bimolecular (k_q) processes available to *D given by Eq. 7.8b:

$$k_{exp} = k_D + k_q[A] \tag{7.8b}$$

where k_D is the unimolecular rate constant for decay of *D in the absence of A (Eq. 7.4) and k_q is the bimolecular rate constant for the *total* quenching of *D by A (Eq. 7.8a).

The quantum efficiency ϕ_{ET} (Chapters 4 and 5) corresponds to the fraction of *D molecules that decay via energy transfer (Eq. 7.5). The quantum efficiency of energy transfer (ϕ_{ET}) is given by Eq. 7.9.

$$\phi_{ET} = \frac{k_{ET}[A]}{k_D + k_q[A]} = \frac{k_{ET}[A]}{k_D + (k_{ET} + k_w + k_{rxn})[A]} \tag{7.9}$$

The quantum *efficiency* ϕ_{ET} in Eq. 7.9 differs from the quantum *yield* of energy transfer (Φ_{ET}) in that the latter takes into account the efficiency with which *D is formed (Φ_{*D}), as shown in Eq. 7.10.

$$\Phi_{ET} = \Phi_{*D} \, \phi_{ET} \tag{7.10}$$

In Eq. 7.10, Φ_{*D} is the quantum yield for formation of *D; for example, Φ_{*D} may be the quantum yield for intersystem crossing (Φ_{ISC}) in the case of triplet states.

7.3 "Trivial" Mechanisms for Energy and Electron Transfer

Curiously, there exist energy- and electron-transfer processes that occur even when there is *no electronic interaction (no electron exchange or dipole–dipole interactions)* *between *D and A!* If there is no molecular electronic interaction between *R and M, how can electron or energy transfer occur at all? The answer is that there exist "trivial" mechanisms for energy or electron transfer. In the trivial mechanism for energy transfer, *R emits a photon (fluorescence or phosphorescence) that is subsequently absorbed by M to produce *M; in the trivial mechanism for electron transfer, *R can eject an electron (photoionization) that is subsequently attached to M to produce $M^{\bullet-}$. Such processes have been termed "trivial" probably because they are easy to visualize and understand.[3] Trivial energy transfer will be described in some detail below because the factors involved are quite analogous to those for energy transfer by the dipole–dipole mechanism and to those analogous to the absorption of light by M to form *M. In the latter case, the "energy donor" is a photon in the electromagnetic field.

A *trivial* or *radiative* emission–absorption energy-transfer mechanism consists[3] of the emission (Eq. 7.11) for a quantum of light by the excited donor (*D) followed by the absorption (Eq. 7.12) of the emitted photon by a ground-state acceptor (A). This mechanism is readily understandable in terms of the principles that determine the efficiencies of light absorption and emission, such as oscillator strengths of the donor and acceptor (Chapter 4). For energy transfer that occurs by electron exchange or dipole–dipole interactions, the lifetime of *D is *always* decreased as the result of the bimolecular interaction (Eq. 7.5). For the trivial mechanism of energy transfer, the acceptor does not at all influence the emission lifetime or the emission probability of the excited donor molecule (*D). However, the acceptor A, if it happens to be in the path of the photon, merely intercepts the photon *after* it has been emitted by *D (in essence, A is unaware whether the source of the photon it absorbs is a lamp, a laser, or an emitting molecule, such as *D). Indeed, in the case of trivial energy transfer, *D and A can even be in different physical containers! In a manner of speaking, *D serves as a "molecular lamp" that emits photons that can be absorbed by A.

In summary, trivial radiative energy transfer occurs by the two sequential steps of Eqs. 7.11 and 7.12.

$$^*D \longrightarrow D + h\nu \tag{7.11}$$

$$h\nu + A \longrightarrow {}^*A \tag{7.12}$$

Since "trivial" mechanisms for electronic energy transfer require that *D emits photons that A is capable of absorbing, the emission spectrum of *D must overlap the absorption spectrum of A.

The rate or probability per unit time of energy transfer from *D to produce *A by the trivial mechanism will depend on a number of factors, the most important of which are

1. The quantum yield (Φ_e^D) for emission of *D.
2. The number of A molecules (concentration) in the path of photons emitted by *D.
3. The light absorbing ability of A.
4. The overlap for the emission spectrum of *D and the absorption spectrum of A, with consideration given to the extinction coefficient of A at the wavelength of overlap.

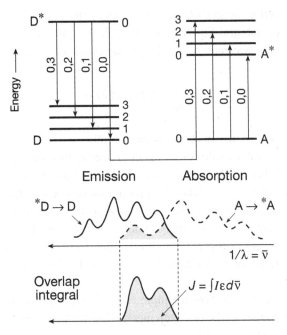

Figure 7.3   Schematic representation of the overlap integral J (see Eq. 7.14), and its relation to the experimental absorption and normalized emission spectrum of *D, and extinction coefficient of A. (This overlap integral also can be used for energy transfer.)

Trivial energy transfer will be favored when each of the parameters 1–4 is maximized, while the process will fail completely *if any one of the four criteria above is not met*. Thus, the ideal conditions for which $\Phi_e^D \sim 1$ require a high extinction coefficient of A (ε_A) and good spectral overlap of the emission spectrum of *D and of the absorption spectrum of A. The last two factors may be quantified in terms of the *normalized spectral overlap integral, J*, which is the integrated overlap of the experimental absorptions and emission curves (Fig. 7.3). Note that while we normally represent spectra in terms of wavelength (λ), wavelength is not directly proportional to either energy or frequency; however, $1/\lambda$ possesses units of wavenumber that are directly proportional to energy. Since the spectral overlap integral (J) compares energies for the *D $\rightarrow$ D and A $\rightarrow$ *A transitions that match exactly (i.e., are capable of resonance), it is necessary to calculate the overlap in units of energy. Consequently, it is convenient for the spectral overlap integral to be plotted versus $1/\lambda$ (cf. Figs. 7.3 and 7.1). The preferred unit for $1/\lambda$ is 1/cm (generally written cm^{-1}). The energy unit reciprocal centimeters (cm^{-1}) is termed a *wavenumber* and given the special symbol $\bar{\nu}$ (Eq. 7.13). The symbol $\bar{\nu}$ should not be confused with the symbol for frequency (ν) whose unit is reciprocal seconds (s^{-1}). Both $\bar{\nu}$ and ν are directly proportional to energy ($E = h\nu = hc\bar{\nu}$).

$$\bar{\nu} = \frac{1}{\lambda} \quad \text{Wavenumber} \tag{7.13}$$

The spectral overlap integral (J) shown schematically in Fig. 7.3 (bottom) is given by the mathematical integral Eq. 7.14, where I_D is the emitted light intensity for the donor as a function of wavenumber, and ε_A is the extinction coefficient of A.

$$J = \int_0^\infty I_D(\bar{\nu})\varepsilon_A(\bar{\nu})d\bar{\nu} \tag{7.14}$$

Note that both ε_A and I_D are functions of wavenumber $\bar{\nu}$.

In Eq. 7.14, note that the value of the spectral overlap integral J will be small when ε_A is small, even if there is good overlap of the *D $\rightarrow$ D and A $\rightarrow$ *A transitions. The smaller the values of ε_A, the weaker the probability of photon absorption by A, no matter what the value of the spectral overlap integral (J). Since values of ε_A for direct singlet–triplet absorption are extremely small, it is implausible for *A, the excited state of the acceptor, to be a triplet state; thus *singlet–triplet and triplet–triplet energy transfer do not take place by the trivial mechanism*, leaving singlet–singlet energy transfer as the only plausible candidate.

Equations 7.15 and 7.16 show a simple two-step mechanism for trivial electron transfer that is analogous to the two-step mechanism for trivial energy transfer given by Eq. 7.11 followed by Eq. 7.12.

$$D \xrightarrow{h\nu} {}^*D \longrightarrow D^{\bullet+} + e^{\bullet-}_{solv} \tag{7.15}$$

$$e^{\bullet-}_{solv} + A \longrightarrow A^{\bullet-} \tag{7.16}$$

Equation 7.15 is analogous to the "emission," or ejection, of a photon by *D in Eq. 7.11. However, in this case the "emitted" species is an electron, which in a liquid becomes immediately surrounded by solvent to yield a "solvated electron" ($e^{\bullet-}_{solv}$). Equation 7.16 is analogous to the "absorption" of a photon by A in Eq. 7.12. However, in the case of Eq. 7.16, an electron is "absorbed" by a suitable acceptor (A) to form a radical anion ($A^{\bullet-}$). Equation 7.15 is a primary photochemical process termed *photoionization*, which may involve a transient electronically excited state (*D). Nearly all molecules undergo photoionization when they absorb a photon whose energy ($E = h\nu$) exceeds the ionization potential (IP) of a molecule in solution; that is, $E = h\nu = E_{*R} > IP_R$. Lasers[4,5] can produce a very high concentration of photons that are absorbed by D. The concentration may be so high that absorption of a photon by D to produce *D may be followed by absorption of a second photon by *D to cause photoionization of Eq. 7.16.

"Solvated" electrons (electrons surrounded by solvent molecules) can be readily detected by ultraviolet (UV) spectroscopy in solution and are among the most important species produced in the radiolysis of water.[6] For example, in aqueous solution solvated electrons absorb with $\lambda_{max} \sim 720$ nm and an extinction coefficient[7] of $\varepsilon_{max} \sim 20,000$ M^{-1} cm^{-1}.[7] A characteristic reaction of solvated electrons involves scavenging by N_2O (Eq. 7.17), which is often used as a diagnostic test for the occurrence of a solvated electron.[6]

$$e^{\bullet-}_{aq} + N_2O + H_2O \longrightarrow N_2 + OH^\bullet + OH^- \tag{7.17}$$

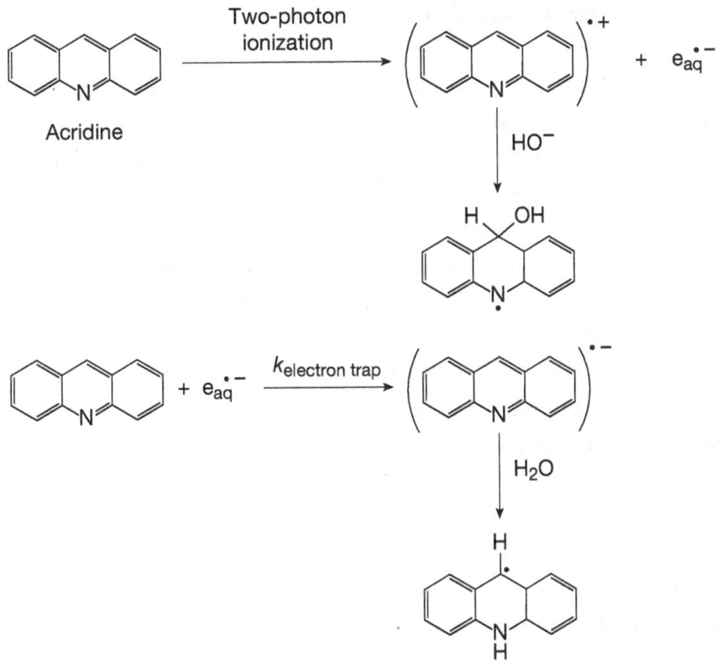

Scheme 7.2   Two photon laser absorption by acridine in aqueous solution leads to its photoionization. The hydrated electron is then trapped by ground-state acridine molecules. Both the acridine radical cation and anion react to give neutral radical products.[8]

Acridine (Scheme 7.2) provides an exemplar of an organic molecule that undergoes photoionization.[8] Upon laser excitation of acridine, photoionization occurs and an electron is ejected (symbolized as $e^{\bullet-}_{aq}$ to note that the electron is solvated by water); this electron can react with many substrates, but in this case a good scavenger present in sufficient concentration is ground-state acridine itself. Thus, a ground-state molecule of acridine that has not participated in the photoionization process reacts with the electron to give the acridine anion radical that reacts with water to produce a neutral radical, as shown in Scheme 7.2. Acridine also provides an interesting example of trivial electron transfer where the electron donor and acceptor are identical. The value of $k_{electron\ trap}$ has been measured as $4 \times 10^{10}\ M^{-1}\,s^{-1}$.

Acridine is by no means unique in its photoionization behavior; in fact, many organic molecules undergo photoionization.[4,9,10] For example, pyrene (Py) has been a frequently used molecule for photoionization studies to produce a solvated electron and the pyrene radical cation ($Py^{\bullet+}$) by two-photon absorption (Eqs. 7.18 and 7.19). The reason that Py so easily absorbs two photons is probably due to a combination of its particularly long singlet and triplet lifetime and its strong extinction coefficients for absorption of a second photon from these two states. The ejection of an electron by photoionization is not limited to highly polar media; thus, two-photon ionization of Py not only occurs in water, and alcohols, but also in alkanes.

$$Py + h\nu_1 \rightarrow {}^*Py \tag{7.18}$$

$$^*Py + h\nu_2 \rightarrow Py^{\bullet+} + e^{\bullet-}_{solv} \tag{7.19}$$

7.4   Energy Transfer Mechanisms

As described in Scheme 7.1, the most commonly encountered electronic energy-transfer processes in organic photochemistry take place by two distinct types of electronic interactions between *D and A: case 1, the electron exchange interaction (also termed in the literature as "orbital overlap mechanism," or "electron exchange mechanism" for electronic energy transfer) and case 2, the dipole–dipole interaction (also termed in the literature as the "Coulombic," or "resonance mechanism" for electronic energy transfer). Dexter[11] and Förster[12] developed the theory of energy-transfer induced electron exchange interactions and dipole–dipole interactions, respectively. In honor of the developers of the theories for energy transfer, electron exchange energy transfer is sometimes referred to in the literature as "Dexter" energy transfer (Fig. 7.4, bottom) and dipole–dipole energy transfer is sometimes referred to in the literature as "Förster" energy transfer (Fig. 7.4, top). In Section 6.18, we learned that it is the overlap of frontier orbitals that is responsible for the lowest-energy paths of photochemical reactions; the same orbital overlap is also responsible for the lowest-energy paths for exchange energy transfer. Of course, both electron exchange and dipole–dipole energy transfer require that the energy conservation condition of Fig. 7.1 be satisfied.

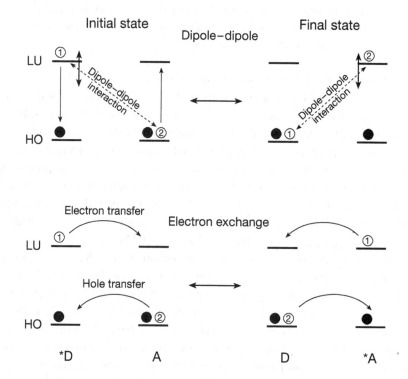

Figure 7.4   Comparison of the dipole–dipole and exchange mechanisms of electronic energy transfer. The spin of the electrons exchanged must obey the spin conservation rules. The hole transfer from *D to A could also be viewed as an electron transfer from HO(A) to HO(*D) in addition to an electron transfer from LU(*D) to LU(A).

In Fig. 7.4, for bookkeeping purposes, we labeled electrons 1 and 2 as the interacting electrons. A key difference between the two mechanisms in Fig. 7.4 is that, for the dipole–dipole mechanism, the interaction between *D and A is made through space by *the overlap of dipolar electric fields of * D with A,* while for the exchange mechanism, the interaction between *D and A is made through *the overlap of the orbitals of *D and A.* The dipole–dipole interaction operates through an oscillating electric field produced by *D (described in the following paragraph) and does not require a van der Waals contact of *D and A or an overlap of the orbitals for *D and A. From Fig. 7.4 (bottom), it is seen that electrons 1 and 2 exchange positions between A and D for electron transfer; whereas from Fig. 7.4 (top), it is seen that electron 1 stays on D and electron 2 stays on A.

How exactly do we visualize the oscillating electric field of *D, which is responsible for the dipole–dipole interaction? A simple pictorial model is available from the classical theory of electromagnetic radiation that views all of the electrons of a molecule as being harmonic oscillators that can undergo oscillation (similar to electronic vibrations) along some axis on a molecular framework (Section 4.2). In the classical model, *an electron in the ground state of the harmonic oscillator does not oscillate at all in the ground state.* However, *D possesses an excited electron, which according to classical theory corresponds to an excited state of the harmonic oscillator. For any excited state of a harmonic oscillator, *the electron undergoes periodic harmonic oscillations along the molecular framework with a certain natural frequency* (ν_0). Such oscillations along the molecular framework create an oscillating electric dipole, much the same as the oscillating electric dipole for the electric field of a passing light wave (Section 4.12). Thus, according to classical theory, *D (but not A) is imagined to possess an oscillating electric dipole that, in turn, produces an oscillating electric field in the space around *D.

The effect of this oscillating electric field of *D on a nearby A can be visualized through a familiar analogue, namely, an electrical transmitting antenna (*D) and electrical receiving antenna (A). In classical theory, A is visualized as being an initially non-oscillating electrical receiver that is potentially capable of being driven into resonance by the oscillating electric field of the transmitting antenna (*D). Suppose the frequency (ν_0) at which the electron of *D oscillates matches a natural frequency for A oscillation. This condition is the first one required for resonance between *D and A. If the oscillating electric field of *D is of sufficient strength and is close enough to A to interact with A to induce oscillations of A electrons, the conditions for classical resonance and dipole–dipole energy transfer from *D to A are met. In terms of classical antennae (or tuning forks), energy flows back and forth between *D and *A. From this classical model, the energy transfer requires the existence of a common frequency of oscillation (ν_0) for *D and *A. When this condition for a common frequency is met, the efficiency of energy transfer will be determined mainly by the distance of separation (R_{DA}) of *D and A, the strength of the oscillating field produced by *D at A at the separation distance, the ease at which A can be set into oscillation for the common frequency, and the relative orientation of *D and A. A quantum modification of a classical picture for the dipole–dipole interaction will be discussed in Section 7.5.

According to the classical model of dipole–dipole energy transfer, electrons do not "exchange molecules or orbitals" (Fig. 7.4, bottom), but rather the two transitions (*D $\rightarrow$ D and A $\rightarrow$ *A) occur simultaneously as resonance by which the oscillating dipole field of *D triggers the creation of a coupled oscillating dipole field about A and leads to *A. The oscillating dipoles of *D and *A are represented by the vertical doubled-headed arrows in Fig. 7.4, top. The excitation of A to *A by the dipole–dipole mechanism is analogous to the classical mechanism for absorption of light described in Chapter 4. In the case of light absorption, the oscillating electromagnetic field of light provides the oscillating electric dipolar field that interacts with the electrons of A. The oscillating electron of the excited molecule *D thus serves as a "virtual photon" for the production of *A. In other words, the ground state of A cannot tell whether the oscillating field that causes it to be excited to *A is due to a "real" photon from the oscillating electromagnetic field, or to the "virtual" photon from the oscillating dipolar electric field of *D!

Quantum mechanics handles both energy (electron exchange) and electron transfer in terms of interacting (overlapping) wave functions for *R and M. For any given *R and M pair, there will always be a certain degree of electron exchange interaction and dipole–dipole interaction, as they approach each other, although in general, at any given separation, one will dominate over the other. Since both the electron exchange and the dipole–dipole interaction are relatively weak, from the standpoint of quantum mechanics, electronic interactions can be analyzed in terms of the matrix elements for the interactions, as described by the golden rule for electronic transitions (Section 3.5). We can qualitatively estimate the values of the rate constants for energy transfer (k_{ET}) in terms of the corresponding matrix elements for the electron exchange and dipole–dipole interactions given by Eq. 7.20, where α and β refer to the degree of contribution for the two interactions. According to quantum mechanics, the *strength (energy)* of the interaction that triggers energy transfer is directly proportional to the magnitude of the matrix element corresponding to the interaction. However, the *rate* of energy transfer, given by k_{ET}, is proportional to the *square* of the strength (energy) of the interaction; therefore, k_{ET} is proportional to the square of the matrix element corresponding to the interaction (Eqs. 3.9 and 3.10).

In the case of electron exchange, the form of the operator H_{ex} in Eq. 7.20 is $\exp(-R_{DA})$, where R_{DA} is the separation between the energy donor *D and the energy acceptor A. This mathematical form is reasonable because, in general, the magnitude of electronic wave functions tend to fall off exponentially as a function of distance from a point of reference, such as a nucleus. Thus, the rate constant for energy transfer by the electron exchange interaction is expected to fall off exponentially as the separation R_{DA} between *D and A increases. In the case of dipole–dipole energy transfer, the operator H_{dd} has the form $\mu_{^*D}\mu_{^*A}/R_{DA}^3$, where $\mu_{^*D}$ is the strength of the oscillating dipole (the same as the transition dipole discussed in Section 4.16) due to *D in Fig. 7.4; $\mu_{^*A}$ is the strength of the oscillating dipole of *A; and R_{DA} is the separation of the donor and acceptor. Since the matrix element is squared in Eq. 7.20, the rate of energy transfer by the dipole–dipole interaction will fall off as the square of $1/R_{DA}^3$ (i.e., $1/R_{DA}^6$).

$$k_{ET}(\text{total}) \propto [\alpha \langle \Psi(^*D)\Psi(A)|H_{ex}|\Psi(D)\Psi(^*A)\rangle^2$$
$$\text{Electron exchange}$$

$$+ \beta \langle \Psi(^*D)\Psi(A)|H_{dd}|\Psi(D)\Psi(^*A)\rangle^2]$$
$$\text{Electron dipole–dipole interactions}$$

(7.20)

7.5   Visualization of Energy Transfer by Dipole–Dipole Interactions: A Transmitter–Antenna Receiver–Antenna Mechanism

As a concrete quantitative model for the dipole–dipole interaction, we assume that the oscillating electric field near *D behaves similarly to the field generated by a classical harmonic electric oscillator antenna whose frequency of oscillation is ν_0 and whose instantaneous oscillating, or *transition dipole* (Chapter 4), at any instant is μ (Fig. 7.4, top). If $|\mu_0|$ is the *maximum* value of the transition dipole μ that can be achieved, we apply the classical expression for the oscillating field of a *harmonic oscillator* (Eq. 7.21) to determine the value of μ at any instant t:

$$\mu = \mu_0 \cos(2\pi \nu t) \tag{7.21}$$

where t is time and ν is the frequency of oscillation. In molecular terms, we can identify this oscillating dipole moment as resulting from amplitude of the back-and-forth electric vibrational motion along the molecular framework of the excited electron on *D; the electric charge oscillates along the molecular framework just like the charge oscillates back and forth along an antenna. Recall that in classical theory, for the ground state of A, the electrons are assumed not to oscillate at all (the oscillation is considered to have zero amplitude). The resulting dipolar electron-charge oscillation of *D will induce the oscillation and eventually the excitation of electronic systems of nearby molecules, if the usual resonance conditions (correct frequency, finite interactions, equal energy gaps for transitions and conservation laws) are met. This dipole–dipole coupling mechanism for energy transfer is plausible only in multiplicity–conserving (spin allowed) transitions that have large transition dipoles (μ). Only singlet–singlet transitions have large oscillator strengths and are associated with large transition dipoles (Chapter 4); therefore, *only singlet–singlet energy transfer is generally plausible by the dipole–dipole mechanism*. However, we see in Section 7.9 that electron exchange provides an effective mechanism for triplet–triplet energy transfer. Thus, whenever we find an example of energy transfer from a triplet donor to produce a triplet acceptor, we can readily assume that electron exchange is involved. Dipole–dipole interactions are ruled out as an implausible mechanism for energy transfer *since neither the donor* *D, nor the acceptor A, possess significant transition dipoles, so that the dipole–dipole interaction will be very weak.*

For the radiative transition $A + h\nu \rightarrow {^*A}$, the resonance condition (energy of transition is equal to the energy for a photon of frequency, ν) is given by Eq. 7.22a.

$$\Delta E(A \rightarrow {^*A}) = h\nu \tag{7.22a}$$

From Fig. 7.1 (replacing *R with *D and M with A), we are reminded that energy conservation is an absolute requirement for energy transfer by any mechanism. For molecules, the matching of the energy for the *D → D and A → *A transitions will generally involve matching of vibrational energy levels. From Fig. 7.1, since *D will be in its lowest vibrational level ($v = 0$), we see that an excited vibrational level of *A will be produced by the energy-transfer process of Eq. 7.22b.

$$\Delta E(^*D \rightarrow D) = \Delta E(A \rightarrow {^*A}) \tag{7.22b}$$

Since the resonance condition must be met from Eq. 7.22b, we can deduce the common frequency of oscillation, since $\Delta E = hv$ so $v = \Delta E / h$.

7.6 Quantitative Aspects of the Förster Theory of Dipole–Dipole Energy Transfer

Förster[12] pointed out that in classical theory *the electrostatic interaction energy (E)* between two *electric dipoles* is directly related to the magnitude of the two interacting dipoles (μ_D and μ_A, Eq. 7.21) and inversely related to the cube of the distance between the donor and acceptor (R_{DA}), as shown in Eq. 7.23 (see Section 2.39 for an analogous discussion of magnetic dipoles):

Electrostatic interaction energy

$$E(\text{dipole–dipole}) \propto \frac{\mu_D \mu_A}{R_{DA}^3} \tag{7.23}$$

From Eq. 7.23, we see that the key parameters determining the energy (strength) of the dipole–dipole interaction are the size of the interacting dipoles μ_D and μ_A and the cube of their separation R_{DA}. Förster[12] related the electric dipoles (μ_D and μ_A) to the oscillator strengths (f_D and f_A) for radiative *D ↔ D and A ↔ *A transitions, respectively (Chapter 4). Recall that the theoretical quantities f_D and f_A are related to the experimental extinction coefficients ε_D and ε_A (Eq. 4.18). The oscillator strength (f) for the interaction of the electromagnetic field and the electrons of a molecule is based on the model of an idealized harmonic oscillator for both the oscillating electric field of a light wave and the oscillating electron for an electronically excited state. The energy (E) of classical dipole–dipole interaction can be formulated in terms of f_D and f_A (the oscillator strengths are measurable through extinction coefficients, ε) for the radiative transitions of *D → D and A → *A (see Chapter 4). Now, we can see how *factors that control the strengths of electronic radiative transitions also control the strength of dipole interactions in dipole–dipole energy transfer at any fixed distance of separation (R_{DA}).* A final theory for energy transfer by the dipole–dipole interaction for real systems must include electronic, vibrational, and spin factors in addition to solvent dielectric factors (dipole–dipole interactions are dependent on the dielectric constant of the surrounding medium).

Quantum mechanical principles can be applied to Eq. 7.20 to modify the classical model as follows. Since the dipole–dipole interactions are weak electronic interactions, the quantum mechanical golden rule for electronic transitions (Eq. 3.8), which may be applied to all weakly interactive systems, can be used to compute the rate of energy transfer (k_{ET}) by the dipole–dipole mechanism as shown in Eq. 7.24a. The matrix element (for the form of Eq. 7.20) for energy-transfer process $^*D + A \rightarrow D + {^*A}$ involves the product wave function $\Psi(^*D)\Psi(A)$ for the initial state, the product wave function $\Psi(D)\Psi(^*A)$ for the final state, and an operator $P_{D \rightarrow {^*A}}$ that corresponds to dipole–dipole interaction of Eq. 7.24a, which mixes the wave functions $\Psi(^*D)$ and $\Psi(A)$ of the initial state and causes transition to the wave function of the final states $\Psi(D)$ and $\Psi(A)$. According to the golden rule, the value of the rate constant for energy transfer (k_{ET}) depends on the *square of this matrix element* (Eq. 7.24a). In addition, there is a term ρ that is a measure of the "density of states" that have the same energy in $^*D$ and A and that are also coupled by the dipole–dipole interaction. This set of "overlapping states" of $^*D$ and A can be expressed quantitatively as an "overlap integral" whose value is given by ρ. In addition, the golden rule can be conveniently formulated as a product of the square of an electronic matrix element and the square of the Franck–Condon (FC) factors for the energy conserving transitions (Eq. 7.24b). The term $<\chi_i|\chi_f>$ is related to the term ρ and is the FC factor (Section 3.10). This latter term is important because it shows that simply conserving energy and having good positive overlap of the electronic wave functions is necessary, but not sufficient, for an energy transition to be probable. In addition, the states involved in the energy transfer must have good positive overlap for the vibrational wave function of the initial state (χ_i) and the vibrational wave function of the final state (χ_f). (The spin wave function, of course, is also important, but is ignored at this point for simplicity).

$$k_{ET} \sim <\Psi(^*D)\Psi(A)|P_{D \rightarrow {^*A}}|\Psi(D)\Psi(^*A)>^2 \rho \quad \text{Golden rule} \qquad (7.24a)$$

$$k_{ET} \sim <\Psi(^*D)\Psi(A)|P_{D \rightarrow {^*A}}|\Psi(D)\Psi(^*A)>^2 <\chi_i|\chi_f>^2 \qquad (7.24b)$$

From this analysis, we conclude that in general the magnitude of k_{ET} will be proportional to the square of the interaction energy, as shown in Eq. 7.25. A poor vibrational, FC factor, or a change in multiplicity will lead to small interaction energy because poor FC factors or changes in spin correspond to small transition dipole moments. Recall that oscillator strength is related to the inherent radiative lifetime and the extinction coefficient of a given transition (Eqs 4.20 and 4.30). By applying the golden rule (Eq. 7.24a), k_{ET} (dipole–dipole) is related to E^2 quantitatively through Eq. 7.25. Note from Eq. 7.25 that the rate of dipole–dipole energy transfer (assuming point dipoles) *falls off as the inverse sixth power of the separation of the dipoles* (R_{DA}). Thus, the rate of electronic energy transfer is expected to fall off as the separation (R_{DA}) between $^*D$ and A increases, by a factor of $1/R_{DA}^6$. *This $1/R_{DA}^6$ distance dependence, when it can be measured accurately, is a basis for distinguishing energy transfer that occurs by dipole–dipole interactions from electron exchange interactions, since the latter generally falls off exponentially with the separation R_{DA}* (Section 7.10). An exemplar that compares the dependence for rate of energy transfer by the dipole–dipole and by the electron exchange mechanism is shown in Fig. 7.5, where the falloff of $\ln k_{ET}$ compared to deactivation of the donor (k_D) is given as a function of separation of $^*D$ and A (R_{DA}). In this exemplar, it is seen that for relatively

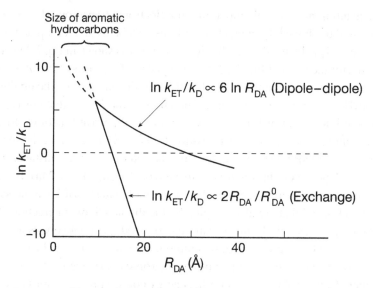

Figure 7.5   Hypothetical graphs of the rate ratio of energy transfer (k_{ET}) to decay of *D (k_D) plotted as $\ln k_{ET}/k_D$ versus $6 \ln R_{DA}$ (dipole–dipole energy transfer) and versus $2R_{DA}/R_{DA}^0$ (exchange energy transfer). R_{DA}^0 is the separation of *D and A when they are in van der Walls contact.

small separations (<10Å) for both interactions the rate of energy transfer is $\gg k_D$, and is therefore very efficient. However, for values of $R_{DA} > 10$ Å, the exponential falloff for the rate of energy transfer by the exchange mechanism (a linear falloff since ln is being plotted) is generally steeper than the $1/R^6$ falloff of the dipole–dipole interaction. Consequently, in the hypothetical exemplar example shown, even at values of $R_{DA} \sim 30$–40 Å dipole–dipole energy transfer and decay of *D are still competitive, but the energy transfer by electron exchange is not significant. This example shows that in favorable cases the range of separations for *D and A energy transfer by the dipole–dipole mechanism can be much larger than (> 30 Å) the size of typical organic molecules (5–10 Å).

$$k_{ET} \text{ (Dipole–dipole)} \propto E^2 \approx \left(\frac{\mu_D \mu_A}{R_{DA}^3}\right)^2 = \frac{\mu_D^2 \mu_A^2}{R_{DA}^6} \qquad (7.25)$$

In summary, from Eq. 7.25, Förster theory predicts that k_{ET} for an energy transfer via dipole–dipole interactions will be proportional to the following quantities:

1. The square of the transition dipole moment μ_D, corresponding to the *D $\rightarrow$ D transition.
2. The square of the transition dipole moment μ_A corresponding to the A $\rightarrow$ *A transition.
3. The inverse sixth power of the separation between *D and A (i.e., $1/R_{DA}^6$).

Förster related the theoretical quantities of oscillator strength (f) and transition dipoles (μ) to experimental quantities, such as extinction coefficients (ε) or radiative rate constants (k^0). Section 4.16 described the relationships between the transition moments and experimental quantities ε and k^0. From these relationships, we deduce that a direct proportionality exists (Eqs. 7.26 and 7.27) between μ^2 and ε (or k^0):

$$\mu_D^2(D^* \leftrightarrow D) \rightarrow \int \varepsilon_D \quad \text{or} \quad \int k_D^0 \tag{7.26}$$

$$\mu_A^2(A^* \leftrightarrow A) \rightarrow \int \varepsilon_A \quad \text{or} \quad \int k_A^0 \tag{7.27}$$

where $\int \varepsilon$ represents an integration of the experimental extinction coefficient for an absorption band over energy and $\int k^0$ represents an integration of the pure radiative rate constant over an emission band.

Since we are specifically considering an energy-transfer process for the simultaneous coupled transitions $^*D \rightarrow D$ and $A \rightarrow \,^*A$, we select k_D^0 and $\int \varepsilon_A$ as the relevant experimental terms to replace the square of the theoretical transition dipole moments of Eqs. 7.26 and 7.27 and obtain Eq. 7.28, which relates k_{ET} to experimentally measurable quantities, k_D^0 and $\int \varepsilon_A$, as a function of the separation of R_{DA}.

$$k_{ET} \text{ (dipole–dipole)} = \alpha \frac{k_D^0 \int \varepsilon_A}{R_{DA}^6} \tag{7.28}$$

Finally, the value of k_{ET} also depends on the spectral overlap requirement (Fig. 7.3). By considering the overlap of $^*D$ emission with A absorption, we obtain Eq. 7.29, from which values of k_{ET} can be computed from experimental data.

$$k_{ET} \text{ (dipole–dipole)} = \alpha \frac{\kappa^2 k_D^0}{R_{DA}^6} J(\varepsilon_A) \tag{7.29}$$

The term α in Eqs. 7.28 and 7.29 is a proportionality constant determined by experimental conditions, such as concentration and solvent index of refraction. The term κ^2 takes into account the fact that the dipoles μ_D and μ_A are vector quantities, and that the interaction between two oscillating dipole vectors will depend on the mutual orientation of the dipoles in space. For a *random* distribution of orientations for dipoles in space from geometric considerations, κ^2 turns out to be a constant equal to two-thirds. The term $J(\varepsilon_A)$, the spectral density integral, is a quantity that is similar to the overlap integral of Eq. 7.14, except that *the value of the extinction coefficient of the acceptor (ε_A) is included in the integration*. A large value of ε_A means that A has a large oscillatory strength and can readily be set into resonance to form $^*A$. The spectra density integral (Fig. 7.3) is related to the density of states (ρ) in the golden rule expression (Eq. 7.24a). We include ε_A because the number of states for the same energy, which is given by the normalized overlap integral, is important, as is the dipole–dipole interaction of each isoenergetic $^*D$ and A transition (FC factors in Eq. 7.24b). The strength of the interactions for these isoenergetic transitions requires inclusion of the oscillator strength for the acceptor or ε_A. The inclusion of k_D^0, the inherent radiative rate constant of *D, takes into account the strength of the transition dipole for the energy donor since k_D^0 is proportional to the oscillator strength of the radiative *D $\rightarrow$ D + $h\nu$ transition (Chapter 4).

7.7   The Relationship of k_{ET} to Energy-Transfer Efficiency and Separation of Donor and Acceptor R_{DA}

From Eq. 7.29, we can anticipate that the conditions for which the rate constant of energy transfer from *D to A, induced by the dipole–dipole mechanism, will be *maximal* when:

1. The *D $\rightarrow$ D and A $\rightarrow$ *A transitions correspond to a large (spectral) overlap integral, $J(\varepsilon_A)$. A large value of $J(\varepsilon_A)$ means that there are many resonant *D $\rightarrow$ D and A $\rightarrow$ *A transitions, that is, there is a high "density of isoenergetic states," ρ, in the golden rule equations Eqs. 3.8 and 7.24a.

2. The radiative rate constant (k_D^0) is as large as possible [$k_D^0 = (\tau_D^0)^{-1}$]. A large value of k_D^0 means that the *D $\rightarrow$ D transition possess a large oscillator strength (f), which in turn means that the size of the oscillating transition dipole (μ_D) due to the excited electron of *D is very large and is a strong oscillator. Therefore, the dipolar field generated by *D in the space around its vicinity is large. Now, *D is a strong and effective transmitting antenna for inducing the oscillation of accepting dipoles in the space around it when the resonance conditions are met.

3. The magnitude of ε_A is as large as possible in the overlap region (Fig. 7.3). A large value of ε means that the A $\rightarrow$ *A transition possess a large oscillator strength, which in turn means that the size of the oscillating dipole (μ_A) is large and that A is a good antenna for receiving dipole oscillations, if the resonance condition is met.

4. The spatial separation (R_{DA}) between *D and A is smaller than the critical separation required for efficient energy transfer. In actual applications, it is convenient to define a *specific critical average separation* (R_{DA}^0) for which the rate of electronic energy transfer from *D to A is equal to the rate of deactivation for *D; that is, $k_D = k_{ET}[A]$. When $R_{DA} < R_{DA}^0$, most of the *D molecules will be deactivated by energy transfer, and when $R_{DA} > R_{DA}^0$ energy transfer becomes inefficient. Pictorially, the closer *D is to A, the stronger is its oscillating force field felt by A, and therefore the more powerful its interaction with A.

5. For a given separation for which *D is interacting with A, there will be preferred relative orientations of *D and A for which energy transfer is favorable and fast, and other orientations for which energy transfer is unfavorable and slow. Recall that the interaction energy between two dipoles, electric or magnetic, is related to the angle θ between the dipoles (Fig. 2.17). Thus, for certain orientations the interaction will be small, even if *D and A are very close to one another, that is, closer than the average separation (R_{DA}^0). Indeed, near

the "magic angle" of $\theta \sim 54°$, the dipole–dipole interaction is close to zero and both have large transition dipoles, no matter how close the dipoles are in space (within the point dipole approximation). We can see that for a detailed computation, we must consider not only the "center-to-center" separation of R_{DA}^0 (based on a simple spherical model) but also the orientation of the dipoles of *D and A.

Experimentally, it is often more convenient to measure the *efficiency* of energy transfer (ϕ_{ET}) rather than the *rate constant* of energy transfer (k_{ET}) since ϕ_{ET} depends only on the spatial separation R_{DA} between *D and A for a given *D and A pair that are randomly separated in space. Conceptually, the *efficiency of energy transfer* ϕ_{ET} provides information on the fraction of initial *D molecules that succeed in transferring energy to the acceptor A. As mentioned in condition (4), it is convenient to define a separation R_{DA} for which the rate of energy transfer equals the sum of the rates for deactivation of *D, as shown in Eqs. 7.30a and b, where k_D is the reciprocal of the *experimental* lifetime of *D under the conditions of the experiment and for which [A] = 0. Note that k_D is not the inherent radiative lifetime k_D^0 that corresponds to the situation for which there is no radiationless deactivation of *D. When Eq. 7.30a is valid, since the rate of energy transfer from *D to A and the rate of deactivation of *D are equal, 50% of *D is quenched by electronic energy transfer to A and 50% of *D is quenched by the processes that deactivate *D in the absence of A. The distance at which Eq. 7.30a is valid is termed the "critical separation distance" (R_{DA}^0).

$$k_{ET}[*D][A] = k_D[*D] \quad \text{at} \quad R_{DA} = R_{DA}^0 \qquad (7.30a)$$

or

$$k_{ET}[A] = k_D = \tau_D^{-1} \qquad (7.30b)$$

Note that in Eq. 7.30b, k_D and τ_D refer to the experimental rate constant for total decay and lifetime of *D, respectively. The actual lifetime of *D under most experimental conditions will normally be *shorter* than the radiative lifetime τ_D^0 [recall from Eq. 4.30 that $\tau_D^0 = (k_D^0)^{-1}$], since generally radiationless processes will compete with radiative processes for the deactivation of *D. The *efficiency* of emission from *D is related to the measured and radiative lifetimes by Eq. 7.31.

$$\phi_{emission} = \frac{\tau_D}{\tau_D^0} \qquad (7.31)$$

When [A] is such that the equality of Eq. 7.30a is found experimentally, we may calculate R_{DA}^0, the average separation for *D and A for which energy transfer and deactivation of *D occur at equal rates. Taking into account geometric factors and assuming spherical shapes for D and A, the relationship between R_{DA}^0 and the concentration $[A]_0$ that meets the criteria of Eq. 7.30a is given by Eq. 7.32.

$$R_{DA}^0 (\text{in Å}) = 6.5[A]^{1/3} \quad \text{with [A] in M units} \qquad (7.32)$$

The *rate constant* and the *efficiency* for energy transfer by the dipole–dipole mechanism may then be related to the actual separation R_{DA} of *D and A by Eqs. 7.33a and b, respectively.[12]

Rate constant for any separation

$$k_{ET} \propto k_D \left(\frac{R_{DA}^0}{R_{DA}} \right)^6 = \frac{1}{\tau_D} \left(\frac{R_{DA}^0}{R_{DA}} \right)^6 \tag{7.33a}$$

Efficiency for any separation

$$\phi_{ET} \propto \left(\frac{R_{DA}^0}{R_{DA}} \right)^6 \tag{7.33b}$$

In Eq. 7.33a, τ_D is the experimental lifetime of *D, R_{DA} is the actual separation between centers of *D and A, R_{DA}^0 is the critical separation as defined by Eq. 7.30a, and ϕ_{ET} is the efficiency for energy transfer (see Eq. 7.10). Thus, when $R_{DA} = R_{DA}^0$, the rate of energy transfer equals the rate of deactivation (Eqs. 7.30a and b). When $R_{DA} < R_{DA}^0$, energy transfer predominates, whereas when $R_{DA} > R_{DA}^0$ deactivation of *D dominates.

Förster resonance energy transfer (commonly termed FRET) is widely used in photobiology as a ruler to determine distances between chromophores, sometimes strategically placed to examine special features or conformations in biomolecules. This interest has led to the determination of numerous R_{DA}^0 values for selected donor–acceptor pairs.[13] Typical values of R_{DA}^0 are in the 10–50-Å range.[3] For example, representative R_{DA}^0 values are tryptophan–pyrene (28 Å), pyrene–coumarin (39 Å), and naphthalene–dansyl (22 Å).

7.8   Experimental Tests for Dipole–Dipole Energy Transfer

Experimental evidence for dipole–dipole energy transfer is available from the determination of the distance dependence from the efficiency (or rate) of the energy-transfer process; a $1/R_{DA}^6$ dependence of the efficiency (or rate) rather than an exponential dependence of R_{DA} for the efficiency (or rate) of energy transfer confirms the dipole–dipole interaction (Fig. 7.5 on page 402). For systems involving random separations of *D and A, there is no single value of R_{DA}; as a result, analysis of the data can only yield an average dependence of ϕ_{ET} and k_{ET} on distance. A common procedure to determine the distance dependence of ϕ_{ET} (Eq. 7.34 on page 410) or k_{ET} (Eqs. 7.28 and 7.30b) involves designing molecular spacers that provide a way of controlling the separation of *D and A in a precise manner. An exemplar is provided by the bis-steroid N(Sp)A, where (Sp) is a spacer moiety, naphthyl–steroid–anthracyl) shown in Scheme 7.3.[14] From measurement of the quantum yield of emission for the naphthyl donor and the spectral overlap integral, an average value for R_{DA}^0 (taking into account the rotational conformers of the system) was calculated from the experimental values of Eq. 7.28. Then, from the assumed $1/R^6$ dependence of energy transfer, a value of

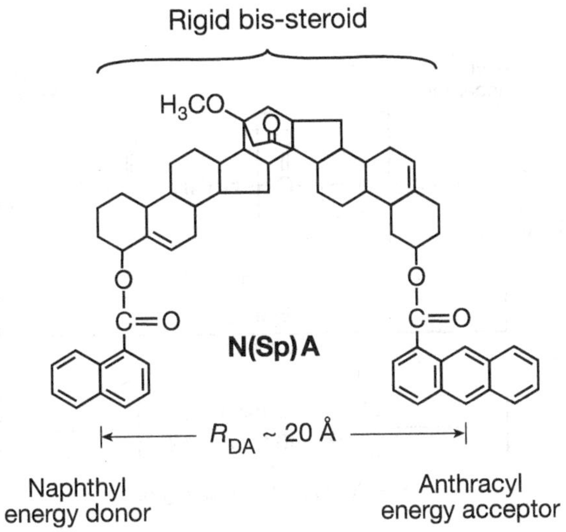

Rigid bis-steroid

N(Sp)A

$R_{DA} \sim 20$ Å

Naphthyl
energy donor

Anthracyl
energy acceptor

Scheme 7.3   Bis-steroid N(Sp)A used to test the distance dependence of dipole–dipole singlet energy transfer between naphthalene (donor) and anthracene (acceptor) chromophores.

R_{DA} was evaluated. Experimentally, the calculated value of R_{DA} was 19 ± 2 Å and the value of R_{DA} from molecular models was 21 ± 2 Å, thus providing an excellent verification for the predicted distance dependence (Eq. 7.34) for the dipole–dipole energy-transfer mechanism.

A particularly revealing exemplar of the dependence of singlet–singlet energy transfer on distance is given by the results from a series of poly-L-proline oligomers (Scheme 7.4a). In the series, the donor group (α-naphthyl energy donor) and the amino-substituted acceptor group (dansyl energy acceptor) were separated by distances from 12 (for $n = 1$) to 46 Å (for $n = 12$).[15] This system allows an important test for the Förster equation; namely, it allows evaluation of the efficiency of energy transfer as a function of distance where the key parameters for *D and A are not expected to change significantly and where a straightforward comparison of the distance dependence with theory (Scheme 7.4b) is allowed. Singlet-singlet energy transfer was 100% efficient for $n \leq 4(R_{DA} \sim 10$–20Å). The efficiency then systematically drops off as n (and R_{DA}) increases and eventually reaches a value of $\sim 15\%$ when $n = 12$. Approximately 50% transfer occurs at distances ~ 35 Å, so that for this system R_{DA}^0 is ~ 35 Å. The expected $1/R_{DA}^6$ distance dependence for dipole–dipole energy transfer holds nicely for this system. In contrast, for the exchange mechanism (see below), the efficiency is predicted to fall off exponentially and the rate of exchange energy transfer is expected to be too slow to compete with the decay of *D when the separation of *D and A is greater than ~ 15 Å.

A ln–ln plot of Eq. 7.33b for the poly-L-proline oligomers shown in Scheme 7.4b yields a slope of ~ 6; the latter corresponds to the exponent of the distance for separation predicted by Eq. 7.33b.

(a)

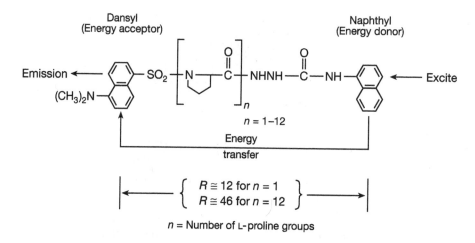

(b)

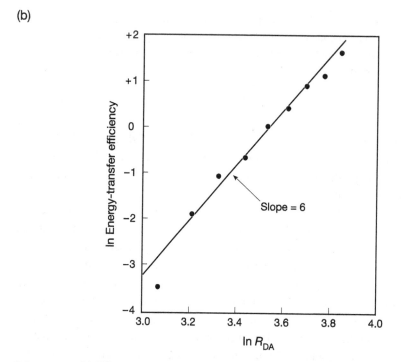

Scheme 7.4   (a) Oligomers of poly-L-proline used to test the distance dependence of singlet–singlet energy transfer by the dipole–dipole mechanism. (b) Plot of the ln for the efficiency of energy transfer versus the ln of the distance between the energy donor and acceptor of (a) according to Eq. 7.33b. The slope of the line is 5.9, which is in excellent agreement with 6.0, the value expected for a R^6 distance dependence predicted by Förster theory in terms of Eq. 7.33b.

An interesting example of the structural control of energy transfer is provided by a trichromophoric molecule **PBN**, which contains a phenanthrene, a biphenyl, and a naphthalene moiety.[16] In the example of Scheme 7.5, the chromophores are linked by adamantyl (Ad) moieties (shown as shaded spheres; see Scheme 7.6 for a more detailed structure) and are held in a semirigid conformation.

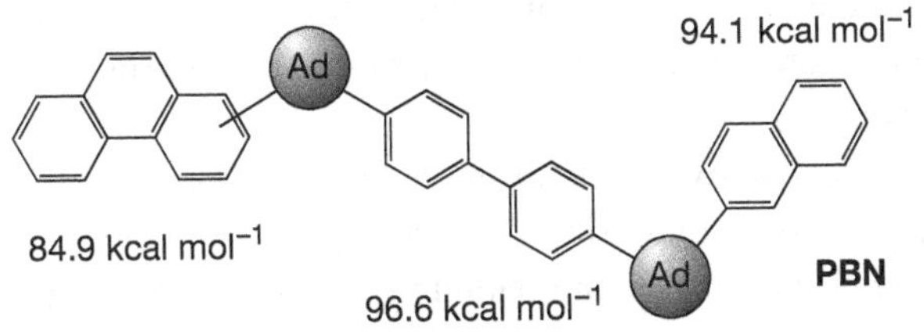

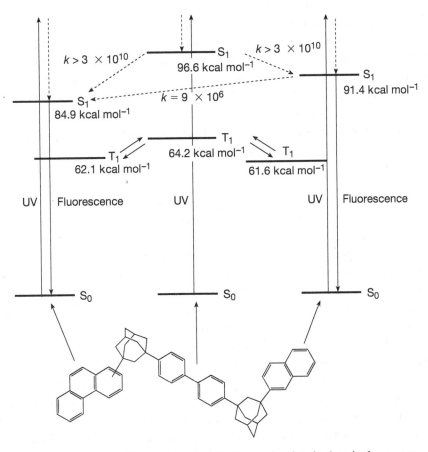

Scheme 7.5   A trichromophoric molecule, PBN. Spheres represent adamantyl spacers, as shown in Scheme 7.6.

Scheme 7.6   Energy diagram showing the intramolecular singlet–singlet energy transfer (SSET) and triplet–triplet energy transfer (TTET) processes and estimated rate constants.

The **PBN** trichromophoric system was selected for study because of its very favorable spectral characteristics and singlet energies of the three chromophores; namely, it is possible to achieve wavelength selectivity for excitation of each of the three chromophores for this system: the singlet energies (E_S) of the three chromophores are biphenyl ($E_S \sim 97$ kcal mol^{-1}), naphthalene ($E_S \sim 92$ kcal mol^{-1}), and phenanthrene ($E_S \sim 85$ kcal mol^{-1}). These characteristics provide the means of selectively exciting any one of the chromophores. The excited chromophore then becomes a potential energy donor and the two other chromophores become potential energy acceptors. Depending on the chromophore that is selectively excited, the energy transfer is either exothermic or endothermic.

For example, selective excitation of the biphenyl chromophore, which possesses the highest singlet energy of the triad, leads to rapid and efficient singlet–singlet energy transfer to *both* phenanthyl and naphthyl acceptor groups with high efficiency and a very large rate constant $> 6 \times 10^{10}$ s^{-1}. Since biphenyl as an energy donor possesses a higher energy than either energy acceptor, and since the groups are separated by < 10Å, the energy transfer from excited biphenyl is expected to occur rapidly to both acceptors. However, upon selective excitation of the naphthalene chromophore, the transfer of singlet energy to the more distant and lower-energy phenanthrene chromophore is faster and more efficient than the transfer of singlet energy to the closer, but higher-energy, biphenyl chromophore. In this case, the observed fluorescence spectra consists of a composite of emission from the naphthalene and phenantherene chromophores. By combining spectral and fluorescence lifetime data, the rate constant for singlet energy transfer from naphthalene to phenantherene was measured to be 9×10^6 s^{-1}, and to occur with an efficiency ϕ_{ET} of 37%. With this information, assuming that the long-distance transfer is due to dipole–dipole interaction, it is possible to calculate the ratio R_{DA}/R_{DA}^0, corresponding to 50% efficiency of transfer, according to Eq. 7.34.

$$\phi_{ET} = \frac{1}{1 + [R_{DA}^6/(R_{DA}^0)^6]} \qquad (7.34)$$

The calculated value of R_{DA}/R_{DA}^0 is 1.1. The value of R_{DA}^0 independently calculated from Förster's theory for dipole–dipole energy transfer is 14 Å, leading to $R_{DA} = 16$ Å.

Another important feature of the PBN system is that, in addition to singlet–singlet energy transfer, it is also possible to experimentally measure triplet–triplet electronic energy transfer (to be discussed in Section 7.10), which must occur by an exchange mechanism. The triplet energies of the three chromophores are biphenyl ($E_T \sim 64.2$ kcal mol^{-1}), naphthalene ($E_T \sim 62.1$ kcal mol^{-1}), and phenanthrene ($E_T \sim 61.6$ kcal mol^{-1}). Since the rate of triplet–triplet transfer must occur by the electron exchange interaction, its rate falls off more rapidly than the rate of dipole–dipole energy transfer at separations > 10 Å (Fig. 7.5), and the naphthalene and phenanthrene chromophores are too far apart for the exchange mechanism (the only one allowed for triplet energy transfer) to operate efficiently. However, triplet energy transfer is observed from the phenanthrene chromophore (populated directly or via singlet energy transfer) to the naphthalene chromophore, which has the lowest triplet state. The process may be an activated transfer mediated by the slightly higher biphenyl chromophore. The activated process makes up for the large distance of separation for the phenanthrene and naphthalene by achieving energy transfer in two steps; that is, step 1 is ^{3}phenanthrene + biphenyl $\rightarrow$ phenanthrene + ^{3}biphenyl, and step 2 is ^{3}biphenyl + naphthalene $\rightarrow$ biphenyl $+^3$ naphthalene. The various energy-transfer steps for this system are summarized in Scheme 7.6.

7.9   Electron Exchange Processes: Energy Transfer Resulting from Collisions and Overlap of Electron Clouds

Bimolecular chemical interactions are usually viewed as occurring via *collisions* between molecular reaction partners. By collisions, we mean that the participants in the reaction are sufficiently close that their electron clouds overlap significantly (the separation of the colliding partners is slightly smaller than their van der Waals radii). In any region of orbital overlap, electron exchange always occurs. The following processes of interest to photochemists can result from electron exchange interactions produced by molecular collisions:

1. Triplet–triplet energy transfer.
2. Singlet–singlet energy transfer.
3. Triplet–triplet annihilation.
4. Electron transfer.

We deal with energy-transfer processes that occur by electron exchange in Section 7.10, and then deal with electron-transfer processes in Section 7.13.

7.10   Electron Exchange: An Orbital Overlap or Collision Mechanism of Energy Transfer

For the simple case of two spherical orbitals of *D and A, the overlap between the orbitals falls off *exponentially* as the separation R_{DA} between *D and A increases. This exponential falloff is a characteristic distance dependence of orbital overlap. Since the degree of exchange for energy transfer is directly related to the orbital overlap of *D and A, the rate constant for exchange energy transfer (k_{ET}) is also expected to fall off as an exponential function of the distance separating *D and A. In addition to the dependence of the rate of exchange energy transfer on the separation of *D and A, the rate of energy transfer will also be directly related to J, the spectral overlap integral (Fig. 7.3), which, as expected from the golden rule for energy transfer between states (Eq. 7.24), is a measure of the number of states that are capable of satisfying the resonance condition, once *D and A are coupled by the electron exchange interaction as the result of finite orbital overlap. The rate constant of energy transfer by electron exchange[11] is given by an equation of the form of Eq. 7.35a:

$$k_{ET} \text{ (exchange)} = KJ \exp(-2R_{DA}/R_{DA}^0) \qquad (7.35a)$$

where K is a parameter related to the specific orbital interactions, such as the dependence of orbital overlap on the instantaneous orientations of *D and A. The parameter J is the normalized spectral overlap integral (Eq. 7.14), where normalized means that both the emission intensity (I_D) and extinction coefficient (ε_A) have been adjusted to unit area on the wavenumber ($\bar{\nu}$) scale. It is important to note that J, because it is normalized, *does not depend on the actual magnitude of ε_A*. This difference is an important distinction from the situation for dipole–dipole energy transfer for which the magnitude of ε_A is a direct factor in determining the overall value of J (Eq. 7.29). The overlap integral J can be identified with ρ, the density of degenerate states that couple *D and A (Eqs. 3.8 and 7.24).

Equation 7.35b is a convenient expression of the distance dependence for the rate constant for energy transfer by electron exchange, k_{ET}. The parameter R_{DA} is the separation of *D and A; R_{DA}^0 is the separation of *D and A when they are in van der Waals contact (note the definition of R_{DA}^0 is different for exchange and dipole–dipole energy transfer); and k_D represents the rate of energy transfer when *D and A are in van der Waals contact (when $R_{DA} = R_{DA}^0$, then for Eq. 7.35b, $\exp[-\beta(R_{DA} - R_{DA}^0)] = 1$). The β term in Eq. 7.35 is a parameter that represents the sensitivity of k_{ET} to a distance of separation (R_{DA}) for a given *D and A pair. Typical values of β are on the order of $1\,\text{Å}^{-1}$ and generally do not depend significantly on the electronic characteristics of *D and A when energy transfer is exothermic. This finding means that the rate constant for energy transfer falls off by $\sim 1/e$ as the value of R_{DA} increases by 1 Å. Values of $\beta < 1$ mean that β is not very sensitive to separation; for small values of β, energy transfer can occur over very large distances. The maximum value of k_D is expected to be on the order of $10^{13}\,\text{s}^{-1}$. From Eq. 7.35b, for $k_D = 10^{13}\,\text{s}^{-1}$, the rate of energy transfer is $10^{13}\,\text{s}^{-1}$ when *D and A collide and are in contact; that is, $R_{DA} = R_{DA}^0$.

$$k_{ET} \sim k_D \exp[-\beta(R_{DA} - R_{DA}^0)] \qquad (7.35b)$$

In comparing dipole–dipole interactions and electron exchange energy-transfer processes, we note the following differences in their characteristics.

1. The rate of dipole-induced energy transfer decreases as $1/R_{DA}^6$, whereas the rate of exchange-induced transfer decreases as $\exp(-R_{DA}/R_{DA}^0)$. Quantitatively, this means that k_{ET} (exchange) drops to negligibly small values (relative to the lifetime of *D) as the intermolecular distance between *D and A increases to values that are more than one or two molecular diameters (5–10 Å), as shown in Fig. 7.5.

2. The rate of dipole-induced transfer depends on the extinction coefficients (ε) of the *D → D and A → *A radiative transitions, but the rate of the exchange-induced transfer is independent of the extinction coefficients of the *D → D and A → *A transitions.

3. The efficiency of energy transfer (fraction of transfers per donor lifetime $\sim$ k_{ET}/k_D) by the dipole mechanism depends mainly on the oscillator strength of the A → *A transition (since a smaller extinction coefficient for the *D → D transition is compensated for by a slower radiative rate constant) and is directly related to the quantum yield for emission of *D, whereas the efficiency of

energy transfer by the exchange interaction cannot be as directly related to experimental quantities.

4. Both Förster[12] and Dexter[11] theories predict a direct dependence of k_{ET} on J, the spectral overlap integral, but the Förster theory includes the extinction coefficient of the A $\rightarrow$ *A transition in the computation of J.

7.11   Electron-Transfer Processes Leading to Excited States

The formation of excited states of *A via electron exchange can be imagined to occur by way of a simultaneous two-electron transfer (Scheme 7.7), requiring simultaneous orbital overlap between both HO of *D and HO of A, as well as orbital overlap between LU of *D and LU of A.

Excited states of *A can also be obtained via electron- or hole-transfer processes starting from radical ion pairs (D•+, A•−) as shown in Scheme 7.7b, hole transfer, D•+ + A•− → D + *A, and in Scheme 7.7c, electron transfer, D•+ + A•− → *D + A. Note that charge-separated structures, as illustrated on the right of Scheme 7.7, may make some contribution to energy transfer if the two-electron exchange is not "fully concerted"; that is, exciplex-like structures (Section 4.36), for which CT is a significant contributor, may be involved as intermediates. From fundamental considerations, we expect that, depending on the systems under investigation, there will be a continuum of structures with differing degrees of CT character, all the way from weak CT exciplexes to complete electron transfer.

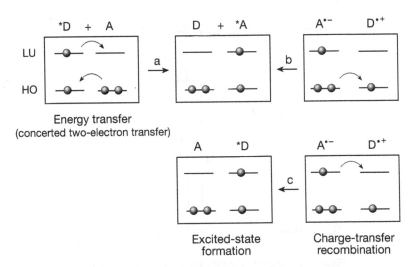

Scheme 7.7   Mechanisms for forming D and *A via electron-exchange processes. The conversion from the left to the center box corresponds to energy transfer, while those originating from the boxes on the right illustrate electron transfer, where D acts as a hole donor (top) or as an electron donor (bottom).

7.12   Triplet–Triplet Annihilation (TTA): A Special Case of Energy Transfer via Electron Exchange Interactions

The energy gap ($\Delta E_{T_1-S_0}$) between the lowest triplet state (T_1) of a molecule and its ground state (S_0) is generally (Eq. 7.36) larger than the energy gap ($\Delta E_{S_1-T_1}$) between the lowest singlet excited state (S_1) and the lowest triplet state (T_1).

$$\Delta E_{T_1-S_0} > \Delta E_{S_1-T_1} \tag{7.36}$$

If Eq. 7.36 is valid for a given system, when two triplets encounter (Fig. 7.6), generally there will be enough electronic excitation energy ($2 \times \Delta E_{T_1-S_0}$) available to promote one of the two molecules into an excited singlet state (S_1), provided the second molecule relaxes to the ground state (S_0). Equation 7.37 represents the case where the two triplet molecules are derived from the same ground-state molecule. The reaction shown in Eq. 7.37 in which two triplets interact to produce an excited singlet and a ground-state singlet is termed a *triplet–triplet annihilation*.

$$^*D(T_1) + {}^*D(T_1) \xrightarrow{k_{TTA}} {}^*D(S_1) + D(S_0) \quad \text{with} \quad \Delta H < 0 \tag{7.37}$$

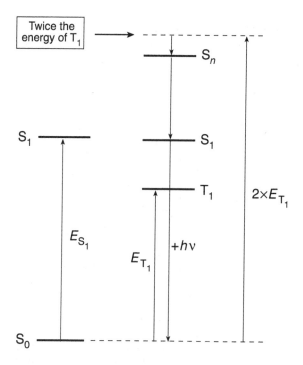

Figure 7.6   Energy diagram for triplet–triplet annihilation. The relative energy levels of S_1 and T_1 are usually such that the sum of the energy for two triplets is sufficient to produce one molecule in the S_1 state and one molecule in the S_0 state.

Table 7.1   Spin Statistics for the Interactions of Two Triplets

	T_n	$T_{n.}$	Total Spin $S = T_n + T_n$	Arrow Notation	Multiplicity $= (2S + 1)$	Final State, Symbol
Case 1	$T_+(\uparrow\uparrow)$	$T_-(\downarrow\downarrow)$	0	$\uparrow\downarrow\uparrow\downarrow$	1	Singlet, S
Case 2	$T_+(\uparrow\uparrow)$	$T_0(\downarrow\uparrow)$	+1	$\uparrow\uparrow\downarrow\uparrow$	3	Triplet, T
Case 3	$T_+(\uparrow\uparrow)$	$T_+(\uparrow\uparrow)$	+2	$\uparrow\uparrow\uparrow\uparrow$	5	Quintet, Q

The annihilation of two triplets to form two singlets may appear at first to be a violation of the spin conservation rule, which requires the orientations of electron spins to be identical before and after an elementary step. However, we find some interesting results from spin statistics for the possible allowed combination of two triplets. When two random triplet states encounter, there are three possible ways the electron spins can combine (Table 7.1). The spin combinations $T_+(\uparrow\uparrow) + T_-(\downarrow\downarrow)$ can obey the spin conservation rule by forming two singlet states. For case 1, either $S_0(\uparrow\downarrow) + S_0(\uparrow\downarrow)$ or $S_1(\uparrow\downarrow) + S_0(\uparrow\downarrow)$ can be formed, assuming that certain energetic conditions of Fig. 7.6 are met. Note that for both the initial state, $T_+(\uparrow\uparrow) + T_-(\downarrow\downarrow)$, and the final states, $S_0(\uparrow\downarrow) + S_0(\uparrow\downarrow)$ or $S_1(\uparrow\downarrow) + S_0(\uparrow\downarrow)$, the systems possess two $\uparrow\uparrow$ spins and two $\downarrow\downarrow$ spins. Therefore, for this orientation of spins, the overall orientation of spins is the same for the reactants and products and the overall process is spin allowed. From Table 7.1, we see that there are other possible combinations of two triplet spins that cannot lead to two singlet states as products and simultaneously obey the spin selection rules. For case 2, $T_+(\uparrow\uparrow) + T_0(\uparrow\downarrow)$ results in a net total spin of 1 ($\uparrow\uparrow\uparrow\downarrow$), since two of the spins cancel each other; the allowed final state would be an overall *triplet*. For case 3, $T_+(\uparrow\uparrow) + T_+(\uparrow\uparrow)$ results in a net total spin of 2 ($\uparrow\uparrow\uparrow\uparrow$) so that the allowed final state would be an overall *quintet*. For case 1, the rate constant for triplet–triplet annihilation approaches the rate constant for diffusion control (Table 7.2).

In summary, there are three possible ways the spins of two triplets can combine and obey the spin conservation rules (Table 7.1 and Fig. 7.7): one combination produces a singlet state (net total spin = 0), a second combination produces a triplet state (net total spin = 1), and a third combination produces a quintet (net total spin = 2). From the total spin, the number of possible spin states (the multiplicity) is singlet = 1, triplet = 3, and quintet = 5 for a total of 9 states.

Table 7.2   Representative Rate Constants for TTA in Solution.[20,21,22]

Substrate	Reference	Solvent	T, (K)	$k_{TTA}(10^9 \text{ M}^{-1} \text{ s}^{-1})$
Anthracene	20	Toluene	258	2.74
Anthracene	20	Toluene	298	4.10
1,2-Benzanthracene	21	n-Hexane	296	20.3
Pyrene	22	Cyclohexane	Room T	7 ± 2
Pyrene	22	Dodecane	Room T	5 ± 1
Pyrene	22	Hexadecane	Room T	1.9 ± 0.2

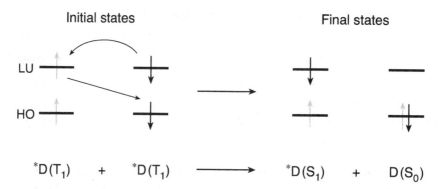

Figure 7.7 Pictorial representation of electron exchange interactions in triplet–triplet annihilation leading to $^*D(S_1) + D(S_0)$. Only one-ninth of triplet–triplet encounters can have the correct spin configuration for these interactions to be possible (Table 7.1).

Statistically, three-ninths of the encounters will result in triplet products of the type $^*D(T_1) + ^*D(T_1) \rightarrow ^*D(T_1) + D(S_0)$. This type of encounter represents a form of quenching for one of the two initial triplets. Pictorially, this process corresponds to the following spin diagram: $T_+(\uparrow\uparrow) + T_0(\downarrow\uparrow) \rightarrow T_+(\uparrow\uparrow) + S_0(\downarrow\uparrow)$. Effectively, one T_1 state induces the formation of the S_0 state from the other T_1 state; this is an example of *spin-catalyzed intersystem crossing.*[17]

If $^*D(S_1)$, produced by TTA, fluoresces with measurable efficiency, the result is a *long-lived fluorescence*. Although $^*D(S_1)$ itself has a very short lifetime, according to Eq. 7.37, this state is populated via $^*D(T_1)$, which has a relatively long lifetime; thus, the concentration of $^*D(S_1)$ will continue to be replenished as long as $^*D(T_1)$ is present. The *apparent* lifetime of this long-lived fluorescence will be of the order of magnitude of the triplet lifetime because the triplet is the immediate precursor of the fluorescence in the slow step of TTA. The fluorescence is a fast kinetic step after the rate-liming TTA has occurred. The actual value of the triplet lifetime will depend on the modes of decay that deactivate the triplet under experimental conditions.[18,19] Thus, TTA leads to a form of "delayed" fluorescence, referred to as "P-Type" *delayed fluorescence*, to differentiate it from the "E-Type" involving thermal population of the excited singlet state from the triplet state.

The rate constants for TTA (k_{TTA}) are generally very large and close to the rate constants for diffusion (see Table 7.2 for representative values).

7.13   Electron Transfer: Mechanisms and Energetics

Now, we switch gears and start a detailed discussion of electron-transfer processes.[23] Electron transfer causes charge separation, and charge separation can have several manifestations in photochemical reactions. In Chapter 4, we saw how CT can assist in the formation of excimers and exciplexes, both of which have properties that reflect the separation of charge or charge shift between reaction partners. This section is concerned with cases where *full electron transfer* takes place either in the form of *charge transfer* between an excited and a ground state, or an *electron or hole* transfer between ground-state species of different charge, as shown in Eqs. 7.38–7.41. Recall that an excited state may serve as an electron donor (Eq. 7.38) or as an electron acceptor (Eq. 7.39).

Charge transfer (electron transfer) from or to an excited state

$$^*D + A \longrightarrow D^{\bullet+} + A^{\bullet-} \tag{7.38}$$

or

$$D + {^*A} \longrightarrow D^{\bullet+} + A^{\bullet-} \tag{7.39}$$

Electron transfer

$$D^{\bullet-} + A \longrightarrow D + A^{\bullet-} \tag{7.40}$$

Hole transfer

$$D + A^{\bullet+} \longrightarrow D^{\bullet+} + A \tag{7.41}$$

As a rule of thumb, from simple thermodynamic considerations, we expect that the rates of electron-transfer reactions will be related to the overall free energy change of reaction so that electron-transfer reactions whose overall free energy is negative (exothermic) will be favored, whereas electron transfer reactions whose overall free energy is positive (endothermic) will be disfavored. At this point, intuitively we might suspect that the rate of electron transfer will depend on the magnitude of the exothermicity for the electron-transfer elementary step. We will see that the theory of electron transfer (Section 7.14) will require a rather drastic modification of this intuition!

We noted in Section 7.2 that *R is *always* both a better oxidizing and reducing agent than R. This finding is true for both S_1 and T_1 states. A simple and powerful MO basis for this generalization is presented in Fig. 7.8, where the ionization potential (IP) of the ground-state R is compared to the ionization potential (*IP) of the excited state *R and the electron affinity (EA) of the ground state R is compared to the electron affinity (*EA) of the excited state *R. From Fig. 7.8, we conclude that

1. The *EA for the excited state (*R) is higher than the EA for the ground state (R). Since addition of an electron to a molecule is generally exothermic, more energy is clearly released in going from the ionization limit to the half-filled lower energy bonding HO than in going from the ionization limit to the antibonding higher energy LU of any molecule. As a result, the addition of an electron to *R is generally more exothermic and energy releasing than the addition of an electron to R.
2. The *IP is lower for the excited state (*R) because it takes less energy to remove the antibonding electron from the LU of *R than to remove a nonbonding or bonding electron from the HO of R.

This analysis of Fig. 7.8 leads to the remarkable and important generalization that *any *R is both a better reducing agent (lower IP) and a better oxidizing agent (higher*

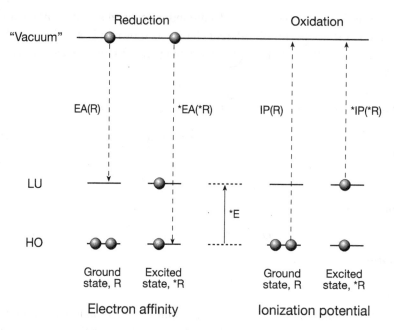

Figure 7.8   Orbital representation of oxidation and reduction processes for the ground state (R) and excited state (*R).

EA) than R. These energetic features of *R will be of great importance for the electron-transfer reactions of *R.

The energies associated with Fig. 7.8 refer to the situation in the gas phase (the term "vacuum" refers to the energy of a free electron not bound to a molecule). For the *ground-state* electron-transfer reaction (Eq. 7.42) in the gas phase, the free energy change for ground-state electron transfer (ΔG_{et}) is given by Eq. 7.43.

$$D + A \longrightarrow D^{\bullet+} + A^{\bullet-} \qquad (7.42)$$

$$\Delta G_{et} = (IP)_D - (EA)_A \qquad (7.43)$$

For an excited-state donor (*D) (Eq. 7.38), the value of the free energy change for excited-state electron transfer ($^*\Delta G_{et}$) differs from that of Eq. 7.43 by the magnitude of the electronic excitation energy (E_{*D}). The latter electronic excitation energy is available as free energy to do work on the electrons of *D and assist in moving an electron from the LU of *D to the LU of A in an electron-transfer processes. Thus, for reactions 7.38 and 7.39, we have Eqs. 7.44 and 7.45, respectively, which take the excitation energy into account (the subscript refers to the reaction being considered) in computing the overall value of $^*\Delta G_{et}$.

$$^*\Delta G_{7.38} = (IP)_D - (EA)_A - E_{*D} \qquad (7.44)$$

$$^*\Delta G_{7.39} = (IP)_D - (EA)_A - E_{*A} \qquad (7.45)$$

The subscripts in ΔG refer to the equations to which they apply.

As a convention in thermodynamics, the more negative energy a free energy change has in a reaction, the more exothermic (more energy released) is a reaction. By comparison with Eq. 7.43, we see that from Eqs. 7.44 and 7.45, CT in the excited state always will be more favorable than in the ground state, whether *R acts as an electron donor (Eq. 7.38) or an acceptor (Eq. 7.39). In another unfortunate convention that has to be taken into account, the values of electron affinity are assigned positive values when energy is released! Consequently, since addition of an electron to a molecule is generally electron releasing, the larger the positive value (by convention) of EA, the more negative is the value of the free energy of a reaction. To take this sign convention into account, in Eqs. 7.44 and 7.45, the term $-(EA)_A$ generally provides a negative contribution to the free energy. Furthermore, the $-E_{*D}$ and $-E_{*A}$ terms will always correspond to negative energies (since electronic excitation is defined as a positive energy). In summary, the larger the values of $(EA)_A$ and E_{*A} and E_{*D}, the greater the negative overall free energy of an electron-transfer reaction in the gas phase.

Because of the relatively low IP and relatively high EA of *R compared to R, it is *qualitatively* expected that from the point of view of organic photochemistry, electron-transfer processes in solution will be exothermic overall and will be commonly encountered. However, to deal with electron transfer in solution *quantitatively*, we need a means of translating the gas-phase values of ΔG_{et} for Eqs. 7.44 and 7.45 into values of ΔG_{et} for solutions. Clearly, we expect the solvation of the charged species produced by electron transfer (e.g., Eqs. 7.38 and 7.39) to produce a significant modification of the gas-phase values of ΔG. The free energy of electron-transfer processes in solution can be estimated by two distinct approaches:

1. The value of $*\Delta G$ for the gas-phase photoinduced electron-transfer reaction is calculated (e.g., Eq. 7.44 for reaction 7.38) and then corrected to take into account the solvation energies for all the participants (i.e., *D, *A, $D^{\bullet+}$, and $A^{\bullet-}$) in the electron-transfer reaction.
2. The experimental electrochemical potentials for the oxidations $E^o_{(D^+/D)}$ and reductions $E^o_{(A/A^-)}$ in solution are measured and then employed to calculate ΔG directly for the solution electron transfer.

While both approaches (1) and (2) are valid in principle, some of the parameters required for approach (1) are frequently difficult to obtain with high accuracy. For example, the solvation energies for *D or $D^{\bullet+}$ (going from the gas phase to solution) are not readily available from experiment. On the other hand, for approach (2) the key electrochemical parameters are more commonly available or can be determined using standard electrochemical techniques (e.g., cyclic voltammetry); as a result, approach (2) is most commonly used and is the approach taken in this text.[23]

The values of the electrochemical potentials are translated into the correct units by multiplying with the Faraday constant ($\mathcal{F}$). Thus, the free energy for reaction 7.38 in solution is given by Eq. 7.46:

In solution:

$$\Delta G_{7.38} \sim \mathcal{F}E^o_{D^+/D} - \mathcal{F}E^o_{A/A^-} - E_{*D} \tag{7.46}$$

where $\mathcal{F}$ is the Faraday constant (9.65×10^4 C mol^{-1}), $E^o_{(D^+/D)}$ and $E^o_{(A/A^-)}$ are the appropriate electrochemical potentials for A and D, and E_{*D} is the electronic excitation energy of *D, respectively. It is essential to note that by convention in electrochemistry, *both $E^o_{(D^+/D)}$ and $E^o_{(A/A^-)}$ are expressed as reductions*; that is, both reactions are expressed as A + e → A$^{\bullet-}$ and D$^{\bullet+}$ + e → D. Because of this convention, one must pay careful attention to the signs of the values of $E^o_{(D^+/D)}$ and $E^o_{(A/A^-)}$ when computing the overall value of ΔG. The sign "∼" in Eq. 7.46 emphasizes that this is only an approximate expression. There are two significant approximations that are commonly required for quantitative analysis of photochemical reactions involving CT:

1. The term E_{*D} normally represents PE, and therefore is always an enthalpy (ΔH), not a free energy (ΔG). The free energy includes an entropy term (ΔS); that is, $\Delta G = \Delta H - T\Delta S$. Thus, $\Delta G = E_{*D} - T\Delta S_{*D}$, where ΔS_{*D} is the excitation entropy for *D. While the entropy term is often neglected or assumed to be negligible, the entropic contribution can be significant if extensive changes in conformational freedom occur upon excitation (see entropy discussion in Section 7.29) or if solvent reorganization about the initial and final states is significantly different.[24,25]

2. There is a Coulombic energy gain (a more negative free energy change) associated with bringing two particles of opposite charge closer together. For reaction 7.38, where two neutral molecules undergo electron transfer to yield an anion (A$^{\bullet-}$) and a cation (D$^{\bullet+}$), the Coulombic correction term is proportional to $-e^2/\varepsilon r$, where e is the charge of the electron, ε is the dielectric constant of the solvent, and r is the approach distance between D$^{\bullet+}$ and A$^{\bullet-}$. As a first approximation, the distance r can be taken as the sum of the radii for the two ions. From this expression, we see that the final contribution to ΔG will decrease as the separation between the charged species increases and as the dielectric constant of the solvent increases. In nonpolar solvents (ε is small), this Coulombic term can be considerable when D$^{\bullet+}$ and A$^{\bullet-}$ are close together (r is small). However, as the solvent polarity increases (ε is large) the value decreases and may become negligible even for small values of r (see Fig. 7.9 for some data on the values of the Columbic term as a function of solvent and separation of D$^{\bullet+}$ and A$^{\bullet-}$).

Scheme 7.8 shows an exemplar for the computation of ΔG for an electron-transfer reaction:[25] the electron transfer from the excited singlet state of naphthalene (S$_1$) to the ground state of 1,4-dicyanobenzene (S$_0$). Electrochemical potentials for the pertinent $E^o_{(D^+/D)}$ and $E^o_{(A/A^-)}$ have been measured in acetonitrile against a "standard calomel electrode." (In addition to paying careful attention to the sign convention corresponding to the electron transfer, one must also be careful to specify the standard electrode to which the values of $E^o_{(D^+/D)}$ and $E^o_{(A/A^-)}$ refer. Both the hydrogen and calomel electrode are commonly used as standards, so care must be taken to know which is being used and not to mix data from the two standards unless appropriate corrections are made.) The computed value of ΔG for the reaction in Scheme 7.8 in acetonitrile solvent is -17.6 kcal mol^{-1}. Thus, the electron transfer is significantly

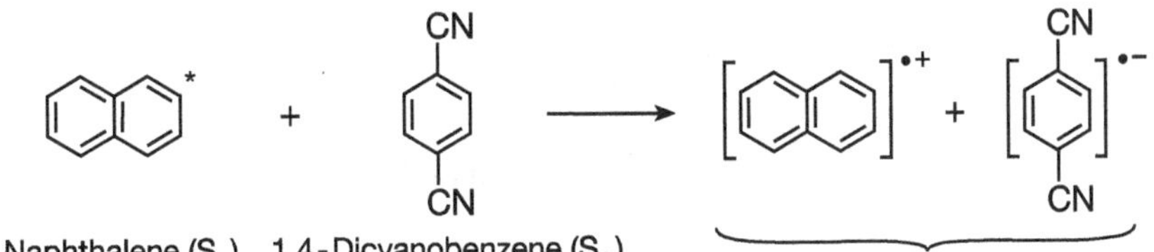

Naphthalene (S$_1$)   1,4-Dicyanobenzene (S$_0$)

$E^0_{D^+/D}$ = +1.60 V E^0_{A/A^-} = −1.64 V

Radical ions

$E(S_1)$ = 3.94 eV = 90.9 kcal mol^{-1}

$\Delta G° = \mathscr{F}E^0_{D^+/D} - \mathscr{F}E^0_{A/A^-} - E^*_D - 0.2$

$\Delta G° = 36.9 - (-37.8) - 90.9 - 0.2 = -16.4$ kcal mol^{-1}

k (electron transfer) = 1.8×10^{10} M^{-1} s^{-1}

Scheme 7.8   Electron transfer between the excited singlet state of naphthalene (donor) and 1,4-dicyanobenzene as acceptor. Data corresponds to acetonitrile.[26]

exothermic and intuitively is expected to occur with a very large rate constant. Indeed, the rate constant is $\sim 1.8 \times 10^{10}$ M^{-1} s^{-1}, the rate of diffusion in acetonitrile.

The reverse of the reaction in Scheme 7.8 is termed a "back electron transfer" reaction: naphthalene radical cation ion (D$^{•+}$) + 1,4-dicyanobenzene radical anion (A$^{•-}$) → naphthalene (D) and 1,4-dicyanobenzene (A). The back electron transfer produces *both naphthalene and 1,4-dicyanobenzene in their ground states* and has a ΔG value of about −75 kcal mol^{-1}. Thus, the back electron-transfer reaction is strongly exothermic. Scheme 7.8 is an exemplar of a common situation for photoinduced electron transfers; when the forward electron transfer *D + A → D$^{•+}$ + A$^{•-}$ is exothermic, the back electron transfer D$^{•+}$ + A$^{•-}$ → D + A is even more strongly exothermic. We will elaborate further on this aspect later in Section 7.14 when we find that beyond a certain point, higher exothermicity actually can cause a decrease in the value of the rate constant for electron transfer.

Now, let us consider the effect of the Coulombic stabilization experiences by two oppositely charged ions. By introducing the "Coulombic term" for stabilization of opposite charges ($-e^2/\varepsilon r$), Eq. 7.46 transforms to Eq. 7.47:

$$\Delta G^0_{7.38} \approx \mathscr{F}E^{°}_{D^+/D} - \mathscr{F}E^{°}_{A/A^-} - E^*_D - N_A \frac{e^2}{4\pi\varepsilon_o\varepsilon r} \qquad (7.47)$$

where N_A is Avogadro's number, e is the charge of the electron (1.60×10^{-19} C), ε_o is the permittivity of vacuum (8.85×10^{-12} C^2 N^{-1} m^{-1}), ε is the dielectric constant of the solvent, and r is the distance between the two charges. Equation 7.48a provides a means of computing the energy associated with the Coulombic term for various separations of the two charges in solvents of varying dielectric constant.

$$\text{Coulombic term} = N_A \frac{e^2}{4\pi\varepsilon_o\varepsilon r} \tag{7.48a}$$

$$= \frac{331.5}{\varepsilon r} \text{ (in kcal mol}^{-1} \text{ with } r \text{ in Å)}$$

Figure 7.9 shows the magnitude of the *Coulombic term* for representative solvents with dielectric constants ranging from 2.27 (benzene) to 80.2 (water).

From Fig. 7.9, we note that the Coulombic term will be very small (e.g., compared with the value of E_{*D}) in very polar solvents, such as water and acetonitrile, since the value of ε is very high for these solvents, even for close separations of $D^{\bullet+}$ and $A^{\bullet-}$. For example, in water the Coulombic term is < 1 kcal mol^{-1} at separations of ~ 2 Å. In contrast, in nonpolar solvents the Coulombic term may become sufficiently large to favor ionic association, rather than ion dissociation. For example, in benzene the Coulombic term is ~ 10 kcal mol^{-1} at a charge separation of ~ 2 Å. These values should be compared to the average thermal kinetic energy (KE) of a molecule at room temperature, which is ~ 1 kcal mol^{-1}. Note that from Eq. 7.47, as the ions move farther apart ($r \to \infty$), the Coulombic term drops to zero. However, when we calculate $\Delta G°$ for electron-transfer processes, we are generally interested in the free energy change associated with the formation of the *nascent ions*, which are within several angstroms or less of one another (typically resulting from encounter or collision pairs), even if after the electron transfer some or all of these may eventually separate to infinity.

Quenching of fluorescence often occurs as the result of an electron-transfer reaction; therefore, a kinetic analysis of the bimolecular electron-transfer quenching of fluorescence between appropriate *R and A pairs, termed a "Stern–Volmer analysis,"[25] provides an experimental method for the measurement of the rate constant (k_{et}). By using the Stern–Volmer analysis, values of k_{et} have been determined for a large number of such systems as a function of the free energy for the electron-transfer process.[25]

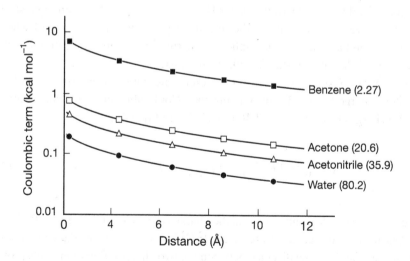

Figure 7.9   Semi-logarithmic plot of the Coulombic term calculated for solvents at various dielectric constants (ε in parentheses).

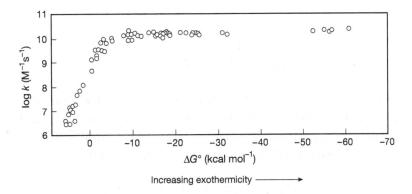

Figure 7.10   Plot of $\log(k_{et})$ for fluorescence quenching of excited states in acetonitrile.[26] The graph is generally known as the Rehm–Weller plot.

For example, the dependence of the rate constant (k_{et}) for electron-transfer fluorescence quenching in acetonitrile by a variety of quenchers on the exothermicity for the electron-transfer reaction are shown in Fig. 7.10 (such plots of k_{et} vs. ΔG° are known as "Rehm–Weller plots").[26] The measured values of k_{et} span the range from $\sim 10^6$ to $\sim 2 \times 10^{10}$ M^{-1} s^{-1}, with the latter rate constant being close to the rate constant for diffusion in acetonitrile. The free enthalpy change for the electron-transfer processes investigated varied from ~ 5 to -60 kcal mol^{-1}. Since acetonitrile possesses a high value of ε, the contribution of the Coulombic term is expected to be small (Fig. 7.9). The plot of k_{et} versus ΔG°_{et} shows a steep decrease in the rate constant for more positive ΔG°_{et} values for endothermic reactions ($\Delta G^\circ_{et} > 0$). The most striking feature of the plot in Fig. 7.10 is that the value of k_{et} reaches a plateau value of $\sim 2 \times 10^{10}$ M^{-1} s^{-1} after an exothermicity of about -10 kcal mol^{-1} and that the value of k_{et} remains the diffusion controlled value to the highest negative values of ΔG°_{et} achievable. Thus, for some reason, increasing the reaction exothermicity beyond about -10 kcal mol^{-1} does not increase or decrease for a measured value of the experimental rate constant for electron transfer. It can be concluded that when the value of ΔG° approaches about -10 kcal mol^{-1}, the process being measured is controlled by something other than electron transfer. Indeed, the value of the rate constant k_{et} in the plateau region of Fig. 7.10 approximately corresponds to the expected rate constant for diffusional processes k_{dif} in acetonitrile ($k_{dif} \sim 2 \times 10^{10}$ M^{-1} s^{-1}). Thus, it can be concluded that the rate of the electron transfer ceases to be rate limiting in the plateau region, and diffusion becomes the rate-limiting process. *This finding means that the true rate constant for electron transfer cannot be measured in the plateau region, because diffusion is the rate-limiting step, not electron transfer.*

The data in Fig. 7.10 fits Eq. 7.48b, where $\Delta G^\ddagger_{et}$ is the activation free energy for electron transfer and k_{et} is the experimental rate constant for quenching by electron transfer:

$$k_{et} = k_0 \exp\left(-\frac{\Delta G^\ddagger_{et}}{RT}\right) \qquad (7.48b)$$

where

$$\Delta G_{et}^{\ddagger} = \frac{\Delta G_{et}}{2} + \left[\left(\frac{\Delta G_{et}}{2}\right)^2 + (\Delta G_{et}^{\ddagger}(0))^2\right]^{1/2} \qquad (7.48c)$$

The term ΔG_{et} is the free energy for electron transfer of each reactive system, and $\Delta G_{et}^{\ddagger}(0)$ is the activation free energy for an isoenergetic reaction ($\Delta G_{et} = 0$).

The value of ΔG_{et} can be calculated from electrochemical reduction and oxidation (redox) potentials and excited-state energy data according to Eq. 7.48d:

$$\Delta G_{et} = E_{1/2}^{ox}(D) - E_{1/2}^{red}(A) - E_{exc}(A) + \Delta E_{Coulombic} \qquad (7.48d)$$

where $E_{1/2}^{ox}(D)$ and $E_{1/2}^{red}(A)$ are the electrochemical potentials for the donor and acceptor, $E_{exc}(A)$ is the excited-state energy for the singlet or triplet state involved, and $\Delta E_{Coulombic}$ is the Coulombic energy for the separated charges in the solvent used. Note that in Eq. 7.48d we have used the redox potentials ignoring the Faraday constant (cf. with Figs. 7.46 and 7.47), which converts the units of redox potentials to energy units. This is common practice in the literature, where it is understood that $E_{1/2}^{ox}(D)$ and $E_{1/2}^{red}(A)$ will be expressed in free energy units.

Now, we deal with the theories that allow us to understand the details of the mechanisms of electron transfer and that relate the value of k_{et} to experimental quantities, such as reaction exothermicity.

7.14   Marcus Theory of Electron Transfer

At first glance, the electron-transfer processes $D + A \rightarrow D^{\bullet+} + A^{\bullet-}$ and $^*D + A \rightarrow D^{\bullet+} + A^{\bullet-}$ appear to be two of the simplest possible chemical reactions, since in an electron-transfer reaction no bonds appear to be formed or broken. The main chemical event seems to be that an electron "changes its owner," that is, moves from an orbital on D (or $^*D$) to another orbital on A. However, this conceptually simple step involves the creation of a pair of ions ($D^{\bullet+} + A^{\bullet-}$), which will interact strongly with the solvent. Indeed, the solvent that is not considered explicitly for energy-transfer processes may have to undergo considerable structural "reorganization" to accommodate and stabilize the new charged molecular system, $D^{\bullet+} + A^{\bullet-}$. Thus, it is clear that solvent reorganization and the influence of opposite charges on themselves and the solvent (Coulombic term, Fig. 7.9) must be considered in the quantitative aspects of any theory of electron transfer. Thus, in spite of their apparent simplicity and their ubiquity and importance in many biological and technological systems, formulation of a quantitative theory of electron-transfer processes has challenged theoretical chemists for many decades. By far, the most important "application" of photoinduced electron transfer occurs in nature in the process of photosynthesis, by which solar energy is captured by plants and through a series of remarkable electron-transfer processes converts water and carbon dioxide to oxygen and carbohydrates, two essential materials for sustaining life.

The beginnings of the modern theory of electron transfer may be traced to a seminal paper, published in 1952, in which Libby[27] correctly pointed out the potential importance of "solvent reorganization" in controlling the rates of ground-state electron-transfer reactions, that is, $D + A \rightarrow D^{\bullet+} + A^{\bullet-}$ reactions. For example, consider an electron-transfer reaction in a very polar solvent (e.g., water or acetonitrile). When a neutral electron-donor molecule D (solvated) converts to a ion $D^{\bullet+}$ (solvated), the electronic structures of D and $D^{\bullet+}$ are considerably different with respect to their charge distribution. It is therefore expected that the solvation spheres of D (solvated) and $D^{\bullet+}$ (solvated) must undergo considerable reorganization of solvent dipoles around the nascent electrical charges that are being generated about $D^{\bullet+}$ and $A^{\bullet-}$, as the electron is being transferred. The change in free energy resulting from the reorganization of solvent molecules, as reactants proceed to products during an electron-transfer reaction, is termed "solvent reorganization energy." Even in a simple charge translocation between identical molecules (Eq. 7.49a), the solvent spheres need to reorganize as the $R^{\bullet+}$ (solvated) $\rightarrow$ R(solvated), and the R(solvated) $\rightarrow$ $R^{\bullet+}$(solvated) processes occur concurrently, since the solvent organization about R will be different from that about R^+. An important early and well-studied experimental example of such an identical electron exchange reaction comes from inorganic chemistry, the electron transfer between Fe^{2+} and Fe^{3+} (7.49b). Isotopic labeling of R^+ and R permits an easy distinction between otherwise identical reagents and products of Eq. 7.49b. For example, in Eq. 7.49a, the $\ddagger$ indicates an isotope of iron(II); that is, $\left[{}^{\ddagger}Fe(H_2O)_6\right]^{2+}$ transfers an electron to iron(III), that is $\left[Fe(H_2O)_6\right]^{3+}$. The self-exchange reaction eliminates one factor that can influence the rate of chemical reactions significantly: the relative free energy difference between reactants and products. In the case of an identical electron exchange reaction, the energy of the reactants and products are identical.

$$R^{\bullet+}(\text{solvated}) + R(\text{solvated}) \rightarrow R(\text{solvated}) + R^{\bullet+}(\text{solvated}) \qquad (7.49a)$$

$$\left[{}^{\ddagger}Fe(H_2O)_6\right]^{2+} + \left[Fe(H_2O)_6\right]^{3+} \rightarrow \left[{}^{\ddagger}Fe(H_2O)_6\right]^{3+} + \left[Fe(H_2O)_6\right]^{2+} \quad (7.49b)$$

In his theory, Libby[27] argued (Fig. 7.11a) that electron transfer would be a much faster process than intramolecular bond reorganization or solvent reorganization, so that the solvent structural changes would *follow* the transfer or "jump" of the electron from R to $R^{\bullet+}$. Figure 7.11a schematically shows his concept of the solvent change that must occur about R and $R^{\bullet+}$ during the electron transfer process of Eq. 7.49a, assuming that the electron jump and transfer occur first, followed by reorganization of the internal structure of the reactants and external structure of the solvent. The electron jump from R to $R^{\bullet+}$ is analogous to the electron jump from a HO to a LU that leads to formation of an electronically excited state. From this analogy, the electron jump is expected to occur "vertically" and to follow the FC principle, which states that the geometry of the products formed by an electron jump (an electron transfer) is the same as the geometry of the reactants. In Fig. 7.11a, this feature is illustrated by showing $R^{\bullet+}$ as having a smaller spherical shape (the positive charge will pull the electrons in closer to the nuclear framework) than R, which is more diffuse and is schematically given an oval shape. After the electron transfer occurs, the immediate shapes of R

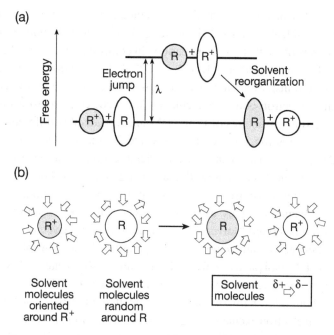

Figure 7.11   (a) An identity electron-transfer reaction (Eq. 7.49a) showing electron transfer as a stepwise process in which electron transfer *hypothetically* occurs *without* any solvent rearrangement (vertical transition along the reaction coordinate). Solvent reorganization follows as a distinct step (horizontal motion along the reaction coordinate). The different shapes illustrate the different solvation (and perhaps molecular structure) for R and R^+. The shading in the shapes is used as a marker to help us follow each species as electrons are exchanged; while we cannot "shade" our molecules and ions as we study these processes experimentally, isotopic composition can be used as a marker. (b) Visualization of the *inner* (note size change) and *outer* (note solvent reorganization) changes accompanying reaction 7.49a.

and $R^{\bullet+}$ are "structurally out of whack," since immediately after a vertical electronic transition, the newly formed $R^{\bullet+}$ still has the smaller spherical shape that it had just before electron transfer, and the newly formed R still has the larger oval shape that it had just before electron transfer. Furthermore, immediately after the electron-transfer event, the solvent molecules are still oriented about the newly formed $R^{\bullet+}$ as if it were R and about the newly formed R as if it were $R^{\bullet+}$. This situation is clearly kinetically unstable with respect to the potential and free energy of the system. Two types of reorganization will have to occur: (1) an electronic and vibrational reorganization associated with the electron transfer, termed *internal molecular reorganization*; and (2) a solvent reorganization associated with the solvent reorientation to accommodate the new electronic structures produced by electron transfer, termed *external solvent reorganization* or simply solvent reorganization.

From the representation of electron transfer in Fig. 7.11a, the "vertical" electron jump is analogous to the absorption of a photon and would require a significant amount of positive energy input if it occurs before the internal and external structural reorganizations have time to occur. The amount of *positive* energy corresponding to the jump was given the symbol λ. *This positive energy (λ) corresponds to the total internal and external reorganization energy that is required for an electron-transfer to occur in an isoenergetic electron transfer reaction*, such as that shown in Eq. 7.49a, if nuclear reorganizations did not precede the electron-transfer step. We will see that although a significant modification of the idea of FC excitation in modern electron-transfer theories has occurred, the total reorganization energy (λ) has remained an important parameter in modern theories, based on the modification of Libby's theory produced by Marcus.[28,29] Note that λ is defined as an activation energy and *is therefore always a positive energy quantity*. This will be important when discussing both λ and $\Delta G°$ since the free energy of a reaction may be a positive or a negative quantity, depending on whether the reaction is endothermic ($\Delta G° > 0$, a positive quantity) or exothermic ($\Delta G° < 0$, a negative quantity).

Marcus[28,29] pointed out that for thermal reactions a vertical jump from the ground- to the excited-state surface corresponding to electron transfer requires the very unlikely sudden input from a considerable amount of a pulse of *thermal* energy. Such a process would be plausible only if the system absorbed a photon whose energy = $\lambda = h\nu$. Thus, Marcus concluded that a vertical electron jump could not plausibly be the rate-limiting step in a *thermal* electron-transfer reaction but would be plausible for a photoinduced electron-transfer reaction.

Instead, Marcus proposed that the rate-limiting feature of an elementary *thermal* electron-transfer process requires only that the molecules and solvent involved in the electron transfer overcome an energy barrier at the crossing point of two PE surfaces (e.g., Fig. 7.12). Moving along the reaction coordinate to the crossing point corresponds to the reorganization of the reactants and solvent to achieve the lowest energy path for electron transfer. Once the reorganization of the molecular and solvent structures have taken place and the representative point is at the crossing point for the intersection of the PE surface for the reactants and products (the transition state, given the symbol ‡ on the reaction coordinate, for electron transfer), a weak electronic interaction can trigger the electron-transfer event from R to R•+.

Marcus' insight focused on an important question concerning electron-transfer processes: What is the magnitude and the nature of the total reorganization energy changes (λ) required to prepare a system for electron transfer? In other words, *what is the energy required to reorganize the reactants and the solvent so that they are "prepared" for the electron-transfer event at the crossing point along the reaction coordinate?* The early theories attempted to provide a basis for understanding experimental examples dealing with electron transfers involving metal ions; for these metal ion systems, the term "inner sphere" was used to refer to the ligands that are directly bonded to the metal ion complex and the term "outer sphere" was used to refer to the shell of solvent molecules that solvate the metal ion complex. Extending this early terminology to modern theories of electron transfer, the energies of reactants and solvent can be conceptually divided into two types, termed *inner- (molecular)* and *outer-sphere* (*supramolecular*) reorganization energies. In Marcus' theory for organic molecules, the term *inner sphere* refers to the internal molecular coordinates of the reactants and products (i.e., bond lengths and angles), while *outer sphere* refers to coordinates defining the arrangements of solvent molecules around the molecular reactants and products (supramolecular effects).

In the case of the identity reaction, Eq. 7.49a, if R is a nonpolar molecule, the orientation of the solvent molecules will be largely random. However, considerable solvent organization about R is expected if R has a significant dipole moment. In the case of $R^{\bullet+}$, the dipoles of a polar solvent will tend to be oriented with their negative end toward $R^{\bullet+}$. Thus, we can visualize reaction 7.49 for a nonpolar organic molecule, as shown in Fig. 7.11b. The solvent will be randomly oriented around the R molecule but more highly and tightly organized around $R^{\bullet+}$. For simplicity, in Fig. 7.11b, we assume spherically symmetric reactants and products.

A basic assumption of Marcus theory[29] is that only a *weak electronic interaction* of the reactants (at the crossing point of two PE curves) is needed for a simple electron-transfer process to occur. As we have seen a number of times, quantum mechanics teaches that the rate constant of processes that involve weak electronic interactions can be formulated in terms of the "golden rule" (Eq. 3.8, modified for the rate constant (k_{et}) for electron transfer).

$$k_{obs} \sim \rho \left[< \Psi_1 | P'_{1 \to 2} | \Psi_2 > \right]^2 \tag{3.8}$$

$$k_{et} \sim \rho < \Psi_1 | P_{et} | \Psi_2 >^2 \tag{3.8a}$$

We can connect the abstract quantum mechanical model given by the golden rule (Eq. 3.8) with the more physically intuitive model of the Arrhenius expressions, Eqs. 7.50a and b.

$$k_{et} = A \exp(-\Delta G^{\ddagger}/RT) \tag{7.50a}$$

$$k_{et} = \nu_N \kappa \exp(-\Delta G^{\ddagger}/RT) \tag{7.50b}$$

(A notation reminder: In this chapter, we use the subscript "et" (lowercase) for electron transfer, and "ET" (uppercase) for energy transfer.)

The "A factor" or "pre-exponential factor" in Eq. 7.50a has the units of reciprocal seconds (s^{-1}) for a unimolecular reaction (or for a bimolecular reaction for which both components are at 1 M concentration) and represents the probability that the representative point will jump from the reactants curve to the products curve when it is in the vicinity of the surface crossing point, the transition state for electron transfer (Fig. 7.12). A completely allowed crossing with probability of 1 will have a pre-exponential factor of $\sim 10^{13}$ s^{-1}. Much lower values will be observed if major electronic, vibrational, supramolecular, or spin reorganization is required for the curve crossing to occur. In Eq. 7.50b, the term ν_N is an *electronic* factor, effectively determining the maximum possible value for k_{et}, and κ, is the transmission coefficient. The student may be familiar with κ from transition state theory; κ is the probability that the reactants, once they reach the transition state for a reaction, that is, some critical arrangement of inner and outer coordinates, will successfully proceed to products.

In both Eqs. 7.50a and b, the $\exp(-\Delta G^{\ddagger}/RT)$ term is the familiar exponential dependence of rate on the free energy of activation as hypothesized by transition state reactivity theory.

Comparison of Eq. 3.8 with Eq. 7.50a shows that the A factor correlates with the $< \Psi | P_{1\to2} | \Psi >^2$ term, which will determine the ease or probability of the representative point switching states at the crossing point of two energy surfaces. The larger the value of $< \Psi | P_{1\to2} | \Psi >^2$, the more the two states will look alike and the higher the probability that the reactant state will switch to the product state at the crossing point. The $\exp(-\Delta G^{\ddagger}/RT)$ term is identified with the FC factors and density of states (ρ) in the golden rule expression. Intuitively, we can interpret the connection between $\exp(-\Delta G^{\ddagger}/RT)$ and ρ to be due to the internal and external reorganization energies that are required to make it to the crossing point. There will be both enthalpic (bond stretches and bends that occur during the internal reorganization and dipole–dipole interactions that occur during the external reorganization) and entropic (tightening or loosening of bonds and rotations that occur during the internal reorganization and tightening or loosening of solvent motion that occurs as the result of external reorganization).

Let us clarify the significance of the various terms in Eqs. 7.50 with an exemplar reaction for the situation for which $\Delta G^{\circ} = 0$; the latter corresponds to an isoenergetic electron transfer, with identical reactants and products (e.g., the reactions of Eq. 7.49). In Fig. 7.12, the PE curve of the reactants (R) is represented by the parabola on the left and the energy curve of the products (P) is represented by the parabola on the right. The x-axis represent the change in the geometry of the reactants and solvent on the way to products. We recognize these curves as having the same form as the parabolas representing the PE as a function of separation for a harmonic oscillator (Fig. 2.16).

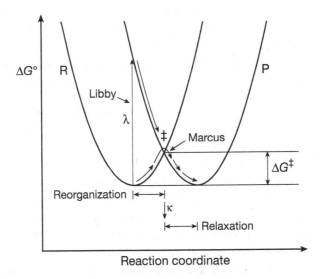

Figure 7.12   Potential energy description of an electron-transfer reaction with $\Delta G^{\circ} = 0$. The point $\ddagger$ represents the transition state. The energy λ (page 427) is defined as the energy for the vertical jump from the ground state minimum of R to the potential energy curve of P.

The harmonic oscillator approximation of the energy curves as parabolas allows for a powerful tool from geometry to make some remarkable predictions concerning the rate constants of energy-transfer reactions as a function of the inner- and outer-sphere reorganization energies involved in the electron-transfer process.

From Fig. 7.12, we can define several important parameters that will be the basis for the quantitative computation of k_{et} from Marcus theory:

1. The *reorganization energy*, λ corresponds to the vertical transition from the minimum of the ground-state parabola for the reactants (R) to the intersection with the parabola for the products (P). This thermal vertical transition is equivalent to the absorption of a photon ($h\nu = \lambda$), which causes a FC electron transfer (transfers the representative point *vertically* from the parabola for R to a position of high energy on the parabola for P). Recall from Libby theory (Fig. 7.10) that λ can be viewed as the organization energy required for an electron transfer before any inner- or outer-sphere reorganization.

2. The *free energy of activation* ($\Delta G^{\ddagger}$) represents the free energy required to reach the transition state configuration (TS‡) starting from the minimum of the reactant PE curve. The reactants (R) need to *reorganize* their nuclei and electrons both internally and externally to meet the requirements for electron transfer to be possible. The term "reaction coordinate" is used in a broader sense than just the usual nuclear coordinates of R; here it also considers not only the nuclear coordinates of R (inner sphere) but also the coordinates of the environment, specifically the solvent rearrangement (outer sphere). Once the TS‡ (crossing point of the two parabolas) has been reached, some trajectories of the representative point will proceed to the product (P) surface, while others will be "reflected" and return to the reactants.

3. The *transmission coefficient* κ gives us the probability that the reorganization required to reach TS‡ will be followed by movement of the representative point along the product surface, followed by rapid relaxation to P.

4. The *thermodynamic free energy of reaction* (ΔG°) represents the difference in free energy between the reactants and products (minima of the parabolas representing R and P) and is zero in the exemplar of Fig. 7.12.

Now, we shall show how Marcus' insight was to use information from the geometry of intersecting PE curves, described mathematically as parabolas, to quantitatively relate k_{et} to the parameters λ, ΔG°, and $\Delta G^{\ddagger}$.

In Fig. 7.12, a parabolic energy curve of the reactant R is shown as intersecting a parabolic energy curve of the product (P) at some value of the reaction coordinate, corresponding to the transition state (TS‡). The energy curves represent the *equilibrium free energy* of the reactant and product for a hypothetical isoenergetic ($\Delta G^{\circ} = 0$) *thermal* reaction, as a function of a reaction coordinate for *both* the inner- and outer-sphere reorganization of the reactants, products, and solvent. The solvent is assumed to reorganize its coordinates in a continuous fashion along the reaction coordinate.

The essential difference between the early Libby and current Marcus theories of electron transfer is highlighted in Fig. 7.13. In the Libby theory, the representative point is assumed to pay an initial steep energetic price by jumping from the reactant

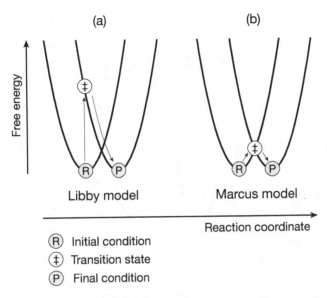

Figure 7.13   A comparison of the free energy requirements of
the electron-transfer pathways in the Libby (a) and Marcus (b)
models. See Fig. 7.12 for a comparison of (a) and (b) on the
same two potential surfaces for R and P.

to the product curve, vertically at first and then passing on to product (Fig. 7.13a).
There is no obvious source of thermal energy (equal to λ) required for the electron
jump, and therefore it is implausible. In Marcus' theory, the representative point pays
a much lower energy price to achieve the transition state $\ddagger$ for electron transfer. Thus,
thermal energy is used for reorganizing the molecular structure (inner sphere) and the
solvent structure (outer sphere) along the reaction coordinate and for overcoming the
energy barrier at the crossing point ($TS^{\ddagger}$) of the two energy curves (Fig. 7.13b). At
the surface crossing, the free energy of the system is the same whether the electron is
transferred to the reactant or to the product.

Chemists are accustomed to the intuitively attractive principle that, for a series
of structurally related reactions, as the reaction becomes more exothermic (ΔG°
becomes more negative), the activation energy ($\Delta G^{\ddagger}$) becomes smaller. The basis
for this principle can be understood by considering the behavior of the crossing point
(which corresponds to $\Delta G^{\ddagger}$) of the two parabolas (Fig. 7.14) as the minimum for the
parabola corresponding to products (conventionally the right-hand curve) is decreased
(along the y-axis) relative to the minimum for the curve corresponding to the reactants,
without displacing the two curves along the reaction coordinate (along the x-axis).

As a reaction becomes increasingly exothermic (ΔG° becomes more negative), the
values of $\Delta G^{\ddagger}$ (corresponding to the point of intersection of the parabolas representing
R and P and the transition state, $TS^{\ddagger}$) change and reflect the vertical displacement of
the reagent (R) and product (P) energy curves. Figure 7.14 shows the evolution of the
PE curves as the relative *vertical* displacement of the minimum (ΔG° becomes more
negative) for the product curve relative to the reactant curve.

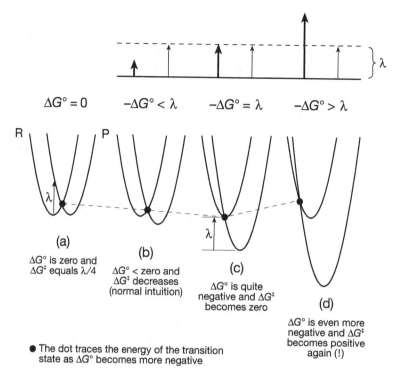

$\Delta G° = 0$ $-\Delta G° < \lambda$ $-\Delta G° = \lambda$ $-\Delta G° > \lambda$

(a)

ΔG° is zero and
ΔG‡ equals λ/4

(b)

ΔG° < zero and
ΔG‡ decreases
(normal intuition)

(c)

ΔG° is quite
negative and ΔG‡
becomes zero

(d)

ΔG° is even more
negative and ΔG‡
becomes positive
again (!)

● The dot traces the energy of the transition
state as ΔG° becomes more negative

Figure 7.14   According to Marcus theory, as the product curve (P) moves down
vertically relative to the reactant (R) curve ($\Delta G°$ becomes more negative),
the activation energy first decreases (a → b), then becomes zero (c) when
the intersection coincides with the minimum in the reactant (R) curve, and
eventually $\Delta G^{\ddagger}$ increases again as $\Delta G°$ becomes more negative (c → d). All
situations for more negative values of $\Delta G°$ after case c are grouped together
as the Marcus inverted region of electron transfer. In the scheme at the top of
the figure, the light arrows refer to λ and the heavy arrows refer to $-\Delta G$.

If we consider the geometry of the curves and relate them to PE along the y-axis, we
imagine that we increase the exothermicity of the electron-transfer reaction by starting
with the isoenergetic situation ($\Delta G° = 0$, Fig. 7.14a). Now imagine "pulling the value
of $\Delta G°$ down vertically" (like a window shade!) so that the parabola corresponding
to the products leaves the energy for the minimum of the parabola corresponding to
the reactants fixed. This amounts to keeping the minima of the two curves at the same
separation along the reaction coordinate (x-axis) but changing the energy difference
between the minima for R and P (y-axis). In the starting energy curve situation for
$\Delta G° = 0$, note that λ corresponds to the vertical separation from the reactant minimum
to the energy curve for the product. Now let us start pulling down the product curve on
the right relative to the reactant curve on the left (Fig. 7.14a–d), which corresponds to
making the reaction more and more exothermic (making $\Delta G°$ more and more negative
as we go from a to b to c to d).

For the curves in Fig. 7.14b, we see by inspection that our "chemist's intuition"
with respect to the effect of $\Delta G°$ on $\Delta G^{\ddagger}$ as $\Delta G°$ becomes more negative is correct (so
far): The energy barrier $\Delta G^{\ddagger}$ clearly decreases as $\Delta G°$ becomes more negative. The

decrease in $\Delta G^{\ddagger}$ is a simple consequence of geometry, since keeping the shape of the curves fixed on the x-axis (keeping the reactions structurally related with respect to a reaction coordinate) simply moves the left-hand portion of the product curve down in energy relative to the minimum of the reactant curve. As a result, the crossing point (TS‡) corresponding to $\Delta G^{\ddagger}$ moves closer and closer to the minimum of the reactants' potential curve and the magnitude of $\Delta G^{\ddagger}$ decreases.

Indeed, when a certain value of ΔG° is reached (Fig. 7.14c), the crossing point passes through the minimum of the reactants' curve! For this particular value of ΔG°, *there is no barrier to electron transfer and the rate of electron transfer will be at a maximum.* Indeed from geometric considerations alone, Marcus showed that the value of ΔG°, for which $\Delta G^{\ddagger} = 0$, is equal to the negative value of λ. Thus, when $\Delta G^{\circ} = -\lambda$, the value of $\Delta G^{\ddagger} = 0$ (Fig. 7.14c), and there is no activation required for electron transfer!

But what happens when the reaction exothermicity ΔG° becomes greater than $-\lambda$ (Fig. 7.14d)? Recall that ΔG° becomes an increasingly negative number as the minimum of the product curve on the right drops to increasingly larger values than the minimum of the reactant curve on the left. Continuing to pull down the product curve minimum causes a remarkable thing to happen to the crossing point, which determines the activation energy of the transition state (TS‡) for the electron-transfer reaction ($\Delta G^{\ddagger}$). The energy curve for the products now intersects *above the minimum of the reactant curve on the left as the value of ΔG° becomes more negative and the value of ΔG° becomes greater than the value $-\lambda$.* As the reaction becomes more exothermic (ΔG° becomes more negative than $-\lambda$), the point of intersection of the energy curves moves to higher and higher energies. Since the point of intersection of the energy curves corresponds to the activation energy, $\Delta G^{\ddagger}$, which determines the reaction rate, we are forced to the very nonintuitive conclusion that *the rate of electron transfer will slow down as the reaction becomes more exothermic beyond the point for which $\Delta G^{\circ} = -\lambda$!* The slowing of the electron-transfer rate with increasing value of ΔG° beyond the point that $\Delta G^{\circ} = -\lambda$ is the basis for the so-called "inverted" region of electron transfer; this is the region of free energies for which the rate of electron transfer decreases with increasing negative values of ΔG°. The region for which the rate of electron transfer increases with increasing negative values of ΔG° is termed the "normal" region of electron transfer. In this region, $\Delta G^{\circ} < -\lambda$. The region at or near $\Delta G^{\circ} = -\lambda$ is termed the "barrierless" region, since in this region $\Delta G^{\ddagger} \sim 0$.

By applying the principles of geometry to the parabolas of Fig. 7.14, Marcus showed that some important geometric relationships provide a quantitative relationship between the experimental quantities k_{et}, λ, $\Delta G^{\ddagger}$, and ΔG°.

1. Under the assumption that the shapes of the curves do not change and that the separation between the minima of the reactants and products along the x-axis is constant (similar reaction coordinates for similar reactions) we can derive Eqs. 7.51a–d, which provide a quantitative relationship among λ, $\Delta G^{\ddagger}$, and ΔG°. As a starting point, the quadratic relationship between $\Delta G^{\ddagger}$ and ΔG° and λ was derived (Eq. 7.51a).

$$\Delta G^{\ddagger} = (\Delta G^{\circ} + \lambda)^2 / 4\lambda \qquad (7.51a)$$

By algebraic rearrangement, Eq. 7.51a can be transformed to Eq. 7.51b.

$$\Delta G^{\ddagger} = (\Delta G^{\circ}/\lambda + 1)^2/4\lambda \tag{7.51b}$$

2. From the special isoenergetic case for $\Delta G^{\circ} = 0$, Eq. 7.51a, we have Eq. 7.51c.

$$\Delta G^{\ddagger} = \lambda/4 \tag{7.51c}$$

3. For the special case, where $-\Delta G^{\circ} = \lambda$, from Eq. 7.51b, we have Eq. 7.51d.

$$\Delta G^{\ddagger} = 0 \tag{7.51d}$$

4. From Eq. 7.50b and the general relationship for $\Delta G^{\ddagger}$ in Eq. 7.51a, we establish the remarkable relationship between k_{et}, λ, $\Delta G^{\ddagger}$, and ΔG° given by Eq. 7.52.

$$k_{et} = \nu_N \kappa \exp(-\Delta G^{\ddagger}/RT) = \nu_N \kappa \exp([(-\Delta G^{\circ} + \lambda)^2/4\lambda]/RT) \tag{7.52}$$

Equations 7.51a–d and Eq. 7.52 provide the theoretical link between the experimental rate constant (k_{et}), the activation energy ($\Delta G^{\ddagger}$), the reaction exothermicity (ΔG°, a negative number), and the reorganization energy (λ, a positive number) of an electron-transfer reaction. It is instructive to consider the results of the arbitrary but systematic range of values of ΔG° from $\Delta G^{\circ} = 0$ ($\Delta G^{\ddagger} = \lambda/4$) to $-\Delta G^{\circ} = \lambda$ (at which point $\Delta G^{\ddagger} = 0$) to $-\Delta G^{\circ} = 2\lambda$ (at which point $\Delta G^{\ddagger} = \lambda/4$). From a pedagogical standpoint, this range clearly shows the activation energy ($\Delta G^{\ddagger}$), starting with a value of $\Delta G^{\ddagger} = \lambda/4$ for $\Delta G^{\circ} = 0$, first decreases as long as $-\Delta G^{\circ} < \lambda$, up to $-\Delta G^{\circ} = \lambda$; and then $\Delta G^{\ddagger}$ increases as $-\Delta G^{\circ} > \lambda$ and returns to a value of $\Delta G^{\ddagger} = \lambda/4$ when $\Delta G^{\circ} = -2\lambda$.

Now, let us review the important conclusions concerning electron transfer that can be deduced from inspection of Fig. 7.14, where we see increasing reaction exothermicity (ΔG°):

1. The "normal" region from $\Delta G^{\circ} = 0$ to any negative value of ΔG° for which $-\Delta G^{\circ} < \lambda$ (Fig. 7.14 a → b). From Eq. 7.51a, in this region the rate of electron transfer will *continuously increase* as the exothermicity increases, as long as $-\Delta G^{\circ} < \lambda$, because the value of $\Delta G^{\ddagger}$ *continuously decreases* in this "normal" region of reaction exothermicity (see arrow comparing $-\Delta G^{\circ}$ to λ). From Eq. 7.51b, we can deduce the mathematical reason to explain why the free energy of activation for electron transfer ($\Delta G^{\ddagger}$) decreases and the value of k_{et} increases as the reaction becomes more exothermic in the normal region. The parameter ΔG° is always a negative number in the regions of interest and λ is defined as a positive number (energy increase). Since ΔG° becomes *more negative* as the reaction exothermicity increases, the quantity $(\Delta G^{\circ} + \lambda)^2$ of Eq. 7.51b will decrease and approach a value of 0 *until* $-\Delta G^{\circ} = \lambda$. From Eq. 7.52, $k_{et} = \nu_N \kappa \exp(-\Delta G^{\ddagger}/RT)$. Note that the value of $\exp(-\Delta G^{\ddagger}/RT)$ will become *larger* as $\Delta G^{\ddagger}$ becomes smaller; thus, at a fixed temperature, k_{et} will increase as $\Delta G^{\ddagger}$ decreases.

2. The barrierless region ($\Delta G^{\ddagger} = 0$) at which $-\Delta G^{\circ} = \lambda$ (Fig. 7.14c). We can see the mathematical basis to explain why the rate of the electron transfer is maximal and the value of $\Delta G^{\ddagger}$ is 0. When $-\Delta G^{\circ} = \lambda$, from Eq. 7.51a (and Eq. 7.51b), $\Delta G^{\ddagger} = 0$, and from Eq. 7.52, $k_{et} = \nu_N \kappa \exp(-\Delta G^{\ddagger}/RT) = \nu_N \kappa \exp(0) = \nu_N \kappa$. This result corresponds to the maximum rate of electron transfer, the inherent "zero-point" rate for crossing from reactants to products. As a result, in solution for the case $\Delta G^{\ddagger} = 0$, the rate constant for electron transfer will frequently be diffusion controlled.

3. The "inverted" region from $-\Delta G^{\circ} = \lambda$ to any value of $-\Delta G^{\circ} > \lambda$. Although this region appears to be "nonintuitive," we can see the mathematical basis to explain why the rate of electron transfer begins to decrease when $-\Delta G^{\circ} > \lambda$ from Eq. 7.51a. As ΔG° increases in this region, since its absolute value is larger than that of λ, as the value of ΔG° continues to increase in the inverted region, the difference in the term $(\Delta G^{\circ} + \lambda)$ will become more and more negative. However, since we see in Eq. 7.51a that $\Delta G^{\ddagger}$ is proportional to $(\Delta G^{\circ} + \lambda)^2$, increasing the negative value of ΔG° will make the term $(\Delta G^{\circ} + \lambda)^2$ increasingly positive. An increase in the positive value of the exponential for the rate expression, $k_{et} = \nu_N \kappa \exp(-\Delta G^{\ddagger}/RT)$, causes the value of $\exp(-\Delta G^{\ddagger}/RT)$ to decrease, and therefore the value of k_{et} decreases.

Figure 7.15 shows the effect of free and reorganization energy on the value of the free energy of activation, according to Marcus theory (Eq. 7.51a) for different values of λ. Note the use of a reverse scale in the vertical axis that plots the activation energy ($\Delta G^{\ddagger}$) in order to retain the usual appearance or Marcus-type curves. These curves generally plot the log of the rate constant (k_{et}) on the vertical axis. *Also note that the horizontal axis corresponds to increasing exothermicity from left to right.*

The $\Delta G^{\ddagger}$ values that intersect the vertical line at $\Delta G^{\circ} = 0$ correspond to $\lambda/4$, as given in Eq. 7.51c. The maxima correspond to the situations for which $\Delta G^{\circ} = \lambda$, for which $\Delta G^{\ddagger} = 0$ according to Eq. 7.51c. Note that the value for $\Delta G^{\ddagger} = 0$ moves toward the more exothermic region as the value of λ increases.

In summary, simple geometric considerations of intersecting parabolas and application of the ideas for Figs. 7.12 and 7.14 allow the connection *between the theoretical quantity λ, the reorganization energy for electron transfer, the activation free energy $\Delta G^{\ddagger}$ for electron transfer, and the free energy of the overall reaction ΔG°. Since the activation free energy of a reaction is related to the rate of a reaction through Eq. 7.52, we can relate λ to k_{et}.* These remarkable relationships are a direct geometric consequence of the use of harmonic oscillator (parabolic) functions to describe the dependence of ΔG° on the reaction coordinates. Most important from Eq. 7.50c, for $\Delta G^{\circ} > -\lambda$ the activation free energy will increase as ΔG° becomes more negative and theory predicts an "inverted" region for electron transfer. In this region, as the electron-transfer reaction becomes more negative (more exothermic), the value of k_{et} will decrease.

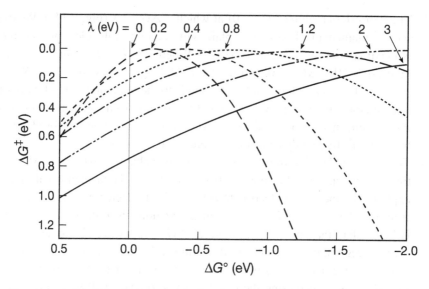

Figure 7.15   Plots for the free energy of activation ($\Delta G^{\ddagger}$), according to Eq. 7.51a for various values of the reorganization energy (λ). Note how the condition for $\Delta G^{\ddagger} = 0$ (the maximum of the curve) moves to the more exothermic region when the value of λ increases.

7.15   A Closer Look at the Reaction Coordinate for Electron Transfer

Let us take a closer look at the horizontal axis in Figs. 7.12 and 7.14 and ask, What does the reaction coordinate for electron transfer actually represent? While the reaction coordinate is usually discussed in a rather qualitative fashion, it is generally defined for a given reaction as *the lowest equilibrium energy nuclear configuration of the reactants and the molecules in the solvent surrounding the reactant as the conversion to products (R $\rightarrow$ P) proceeds.* In essence, the reaction coordinate provides a simplified way of representing in two-dimensions (2D) the evolution of many internal and external coordinates of the system as they progress from reactants to products. During this evolution, the change in internal coordinates of the reactant (inner-sphere coordinates) must be accompanied by an adequate change in the structure of the solvent (outer-sphere coordinates).

What is the significance of the crossing points of the potential curves representing R and P in Figs. 7.12 and 7.14? According to Marcus[29] "*in this atomic configuration, a hypothetical system possessing the electronic wave function (and therefore the ionic charges) of the reactants must have the same energy as that of a hypothetical system possessing the electronic wave function of the products in the same configuration.*"

Given that the curves in Figs. 7.12 and 7.14 reflect changes to both inner and outer coordinates, we can anticipate that the value of the reorganization energy λ will be determined by two contributions. One contribution comes from the reorganization of the inner coordinates of the reactant and product molecules (λ_{inner}), and one contribution comes from the reorganization of the solvent around the reactants and products (λ_{outer}) as the electron transfer proceeds (Eq. 7.53a).

$$\lambda = \lambda_{inner} + \lambda_{outer} \tag{7.53a}$$

Both contributions can be independently evaluated if sufficient information is available on the system of interest.

The value of λ_{inner} will reflect any significant changes in bond orders and angles between the structures of the reactants and the products. Examination of reactant and product structures can therefore provide an intuitive idea of the magnitude of λ_{inner}. The parameter λ_{inner} can be evaluated accurately using Eq. 7.53b:[30]

$$\lambda_{inner} = \sum_i \left(\frac{f_i^R f_i^P}{f_i^R + f_i^P} \right) \Delta q_i \tag{7.53b}$$

where Δq_i is the change in interatomic distance, f_i is the force constant for the ith vibration, and the superscripts R and P refer to the reagents and products, respectively. Typical values of λ_{inner} will be between zero and a few kilocalories per mole. It is common to express the values in units of electron volts ($1\,eV = 23\,kcal\,mol^{-1}$) because the units of energy for the electrochemical redox processes are conventionally expressed in electron volts.

The outer-reorganization energy can be estimated[28,29] from Eqs. 7.54a and b,

$$\lambda_{outer} = e^2 \left(\frac{1}{2r_D} + \frac{1}{2r_A} - \frac{1}{r_{AD}} \right) \left(\frac{1}{\varepsilon_{op}} - \frac{1}{\varepsilon_s} \right) \tag{7.54a}$$

$$r_{AD} = r_A + r_D \tag{7.54b}$$

where r_A and r_D are the radii of A and D, respectively; ε_{op} is the dielectric constant of the medium that responds to the electronic polarization (ε is the square of the refractive index); and ε_s is the static dielectric constant or relative permittivity corresponding to the solvent dipole. Values of λ_{outer} are typically a few tens of kilocalories per mole (most frequently $\lambda_{outer} < 40\,kcal\,mol^{-1}$) but may differ greatly depending on the polarity of the solvent employed.

The expressions given above for λ assume a dielectric continuum model for the solvent and a harmonic oscillator model for the vibrational terms.

As a rule, typical values of λ may be a few tens of kilocalories per mole. Recall, however, that it is $\lambda/4$ (Eq. 7.51a) that directly affects $\Delta G^{\ddagger}$; thus, for $\lambda \sim 30.0\,kcal\,mol^{-1}$ (1.30 eV), $\Delta G^{\ddagger}$ for the isoenergetic reaction would be $7.5\,kcal\,mol^{-1}$, and lower for most exothermic reactions. For common parameters involving organic molecules, ΔG° would have to be negative by more than $-60\,kcal\,mol^{-1}$ (see Eq. 7.51a) in the inverted region before $\Delta G^{\ddagger}$ exceeds $\lambda/4$.

7.16 Experimental Verification of the Marcus Inverted Region for Photoinduced Electron Transfer

Marcus' prediction of the existence of a counterintuitive inverted region for electron transfer involving electronically excited states (*R) where an increased driving force ($\Delta G°$) leads to decreased reactivity[29-31] posed a major challenge to experimentalists. Photoinduced electron transfers are obvious candidates to test the existence of the inverted region since the electronic excitation energy contributes to the overall reaction exothermicity, thereby enhancing the possibility of a system having a very negative value of $\Delta G°$ and being in the inverted region. However, before the 1980s, the experimental examples of photoinduced electron transfer demonstrated that in solution for freely diffusing donor and acceptor molecules, k_{et} initially increased with increasing driving force ($\Delta G°$), as expected intuitively, but as the driving force became very large, the values of k_{et} did not decrease. Instead, the value of k_{et} reached a limiting value: that of a diffusion controlled reaction with rate constant k_{dif} (Fig. 7.10).

The absence of experimental evidence for photoinduced electron transfer in an inverted region, meant that either the inverted region for electron transfer could not be reached or did not exist or that it was masked by rate-limiting diffusion followed by fast and irreversible electron transfer. The latter would be the case if, once the rate of electron transfer became equal to the rate of diffusion, the rate of electron transfer no longer was rate determining and one was measuring only the rate constant for diffusion (k_{dif}), not the rate constant for electron transfer (k_{et}). In effect, diffusion could set a ceiling for the maximum observable rate constant for electron-transfer quenching of *R in solution. This situation is illustrated qualitatively in Fig. 7.16 by showing the effect of the diffusional rate constant on the experimentally observed value. In the diagram, the rate constant of electron transfer is assumed to start off (Fig. 7.16, right) slower than the rate of diffusion; the rate constant of electron transfer is assumed to be in the "normal" region and to increase at first with increasing exothermicity up to a certain value of $\Delta G°$. However, after this value of $\Delta G°$ is reached, the rate of electron transfer is assumed to be *faster* than the rate of diffusion for further increases in exothermicity. Thus, *from this point on, the rate of quenching is limited by the rate of diffusion, not the rate of electron transfer.* In other words, the diffusion of *D and A together is the slow step and not the electron transfer. The experimental value is expected to blend between the two regions (as shown in the curved section), rather than simply reflect the minimum of the two possible rate constants. The details of the kinetics diffusion controlled reactions will be discussed in Section 7.34.

For many years, the problem of the diffusion limit illustrated in Fig. 7.16 frustrated all attempts to verify the Marcus inverted region for electron transfer between two freely diffusing molecules in liquids. For common parameters involving organic molecules, $\Delta G°$ would have to be negative by more than -60 kcal mol^{-1} (see Eq. 7.51a) in the inverted region before $\Delta G^{\ddagger}$ would exceed $\lambda/4$ (Fig. 7.16, far right). Consequently, systems satisfying the conditions required for the inverted region for freely diffusing D and A were difficult to find experimentally. Indeed, finding experimental examples of the inverted region evidently required a strategy that avoided

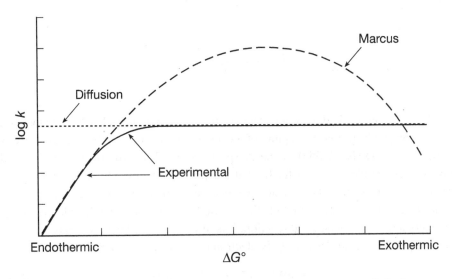

Figure 7.16   The graph illustrates how the experimental rate constant (k_{et}) electron transfer is limited by the diffusion rate constant in the solvent, and as a result effectively masks the Marcus inverted region. On the right of the plot, the reaction is endothermic and the prediction of the Marcus equation is followed. The Rehm–Weller equation (Eq. 7.48a) does not make allowance for an inverted region (value of $\Delta G°$ for which $k_{dif} \sim k_{et}$).

diffusion control limiting electron-transfer reaction. Three strategies have proved successful for the experimental observation of the inverted region:

1. The elimination of diffusion for D and A by running the electron transfer in a rigid medium where diffusion is strongly inhibited.
2. The elimination of diffusion for D and A by attaching the D and A to a rigid molecular framework that serves as a spacer, D(Sp)A, that allows electron transfer in liquids but prevents diffusion of D and A.
3. Allowing diffusion of D and A and electron transfer to form $D^{\bullet+}$ and $A^{\bullet-}$ to occur but measuring the unimolecular rate of back electron transfer from $D^{\bullet+}$ $A^{\bullet-}$ to form D and A.

The first successful approach, providing convincing evidence for the inverted region, examined electron transfer between aromatic molecules in rigid organic systems and was developed by Closs and Miller.[32] Their idea was to combine strategies (1) and (2) and investigate a D(Sp)A system in a rigid medium. Pulse radiolysis is a method to generate electrons in rigid media. By using pulse radiolysis, they were able to generate radical anions, $D^{\bullet-}$(Sp)A from a precursor D(Sp)A and then to measure the rate at which $D^{\bullet-}$ transferred an electron to an acceptor (A) as a function of reaction exothermicity, that is, the overall $D^{\bullet-}$(Sp)A to D(Sp)$A^{\bullet-}$ electron-transfer reaction.

Experimentally, D and A were separated by a rigid hydrocarbon Sp based on the steroidal 5α-androstane structure to produce a D(Sp)A system to investigate intramolecular electron transfer.

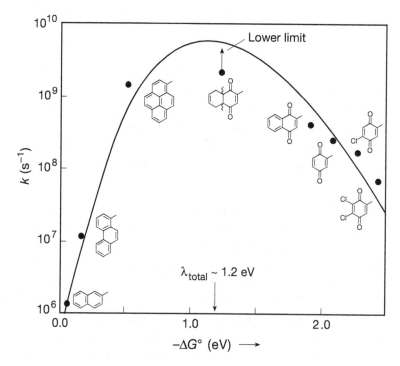

D(Sp)A

The reaction studied was an electron transfer involving charge translocation (Eqs. 7.55a and b) following capture of an electron and generated by pulse radiolysis by the moiety D (Eq. 7.55a) of the A(Sp)D system. The experimentally observed process was the electron transfer from the donor (a biphenyl moiety) to a range of electron acceptors (A). The acceptors were selected to provide a wide range of ΔG° values that would hopefully include the inverted region. Although this classic example is not a true photochemical reaction, it is discussed here because it has served as an exemplar for the strategy to study electron transfer in the inverted region.

$$D(Sp)A + e^{\bullet -} \longrightarrow D^{\bullet -}(Sp)A \qquad (7.55a)$$

$$D^{\bullet -}(Sp)A \xrightarrow{k_{et}} D(Sp)A^{\bullet -} \qquad (7.55b)$$

Figure 7.17 shows the experimental results of this seminal study for eight different acceptors. The calculated theoretical curve shown gives the best fit to date for $\lambda_{inner} = 0.45$ eV and $\lambda_{outer} = 0.75$ eV, so that the value of λ is ~ 1.2 eV (~ 27.7 kcal mol^{-1}).

Figure 7.17 Intramolecular electron-transfer rate constants (see Eq. 7.55b) as a function of ΔG° in a methyltetrahydrofuran solution at 296 K.[34] Adapted with permission of the publisher.

Recall that when $-\Delta G^{\circ} > \lambda$, the systems are predicted to be in the inverted region. Indeed, for systems for which values of $-\Delta G^{\circ} > \sim 1.2$ eV, the rate constant (k_{et}) decreases with increasing thermodynamic diving force, as predicted by Marcus theory (Fig. 7.17, right side of maximum). In electron-transfer studies, it is common to express ΔG° values in units of electon volts (as shown in Fig. 7.17), rather than the kilocalorie per mole units that are more familiar to organic chemists. This is because the experimental values used to determine ΔG° are usually derived from electrochemical measurements for which the electron volt units are the most commonly employed. The conversion to other units of energy frequently used is $1 \, \text{eV} = 23.06 \, \text{kcal mol}^{-1} = 96.48 \, \text{kJ mol}^{-1}$.

The rate constants vary by almost four orders of magnitude for the intramolecular systems shown in Fig. 7.17. Importantly, this dynamic range of rate constants is reduced to just a factor of four and is close to the diffusion rate constants when the same electron-transfer processes are studied as *inter*molecular reactions in solution, showing how diffusion masks the differences in electron-transfer rates.

After the observations of Fig. 7.17 were reported, numerous examples of both intra- and intermolecular systems, which demonstrate the existence of an inverted region in other electron-transfer processes, were reported. A few examples of these are presented in Section 7.23.

7.17 Examples of Photoinduced Electron Transfer That Demonstrate the Marcus Theory

The elegant work of Miller and Closs[32-34] discussed above demonstrated the validity of the fundamental principles of Marcus' theory of electron transfer for a nonphotochemical system. It is important to understand the way in which these principles have become part of the general paradigm of organic photochemistry, and the consequences these principles have on photochemical reactions. Interestingly, in the vast majority of examples for application of Marcus theory to photochemical reactions, the key electron-transfer steps are frequently back electron transfers from "ground-state" $D^{\bullet+}$, $A^{\bullet-}$ radical ion pairs produced by photoinduced electron transfer. In the exemplars that follow, we will see how the concepts of Marcus theory influence the efficiency of charge separation, and how the yields of final products in an organic photoreaction are controlled by electron-transfer energetics.

7.18 Long-Distance Electron Transfer

By long-distance electron transfer, we mean electron-transfer reactions that occur over distances that are significantly larger than the sum of the van der Waals radii of the electron donor and the acceptor. An understanding of long-distance electron transfer reactions is fundamental for an understanding of important electron-transfer processes in biology, such as photosynthesis. Insight into the mechanisms of these processes requires addressing issues such as how the electron-transfer rate depends on the distance of separation for the electron donor and acceptor (R_{DA}), on the relative orientation of D and A (which determines the effective orbital overlap), and on the nature of the intervening medium (solvent, rigid spacer, flexible spacer, supramolecular medium) that separates the electron donor and acceptor.

The rate constant for electron transfer depends on the electronic coupling between the donor (D) and acceptor (A) involved in an electron-transfer reaction. Electronic coupling refers to the matrix element for the electron-transfer process. Electronic coupling will generally have the form of Eq. 7.24a (in the literature the matrix element for electron transfer is often given the symbol V) and remains constant throughout the steroid series of Fig. 7.17. It is also interesting to investigate the effect of a variation in the electronic coupling on the rate of electron transfer. The magnitude of the electronic coupling depends on the overlap of the wave functions of D and A. Within the limit of weak electronic coupling, if D and A are approximated as spheres, it is expected that the electronic coupling term V should fall off exponentially as the separation between the donor and acceptor (R_{DA}) increases.

The FC factors should be kept as constant as possible throughout a series for which the rate data are being compared. The expected dependence of V on R_{DA} is given by Eq. 7.56a, where R_{DA}^0 is the separation when D and A are in van der Waals contact and R_{DA} is the actual separation, which must, of course, be equal or greater than R_{DA}^0. The β term reflects the sensitivity of the coupling to distance and V_0 is a proportionality constant or "pre-exponential" factor. The parameter β is inversely proportional to the orbital overlap of donor and acceptor, so that the magnitude of interaction (and the rate constant, k_{et}) between D and A orbitals is inversely proportional to β. From Eq. 7.56a, it can be seen that since β is positive, larger values of β will reduce the value of the (negative) exponential term.

$$V(R_{DA}) = V_0 \exp[-\beta(R_{DA} - R_0/2)] \qquad (7.56a)$$

An equivalent expression can be written for the rate of electron transfer as shown in Eq. 7.56b. Note the close similarity between the expression for electron transfer and the expression for the rate constant of energy transfer by the electron exchange mechanism (Eq. 7.35). Note also that an exponential fall off of a rate constant with separation of the reacting partners is characteristic of reactions that are rate limited by orbital overlap since the latter falls off exponentially as a function of separation of the reacting partners.

$$k_{et} = k_0 \exp[-\beta(R_{DA} - R_0)/2] \qquad (7.56b)$$

7.19 Mechanisms of Long-Distance Electron Transfer: Through-Space and Through-Bond Interactions

A remarkable series of molecules (Fig. 7.18, **1–5**) possessing a 1,4-dimethoxynaph-thalene moiety as an excited-state electron donor (*D) and a 1,1-dicyanoethylene moi-ety as an electron acceptor (A) interconnected by five different, rigid, nonconjugated bridges have been synthesized and studied.[35] The length of the bridges varies with

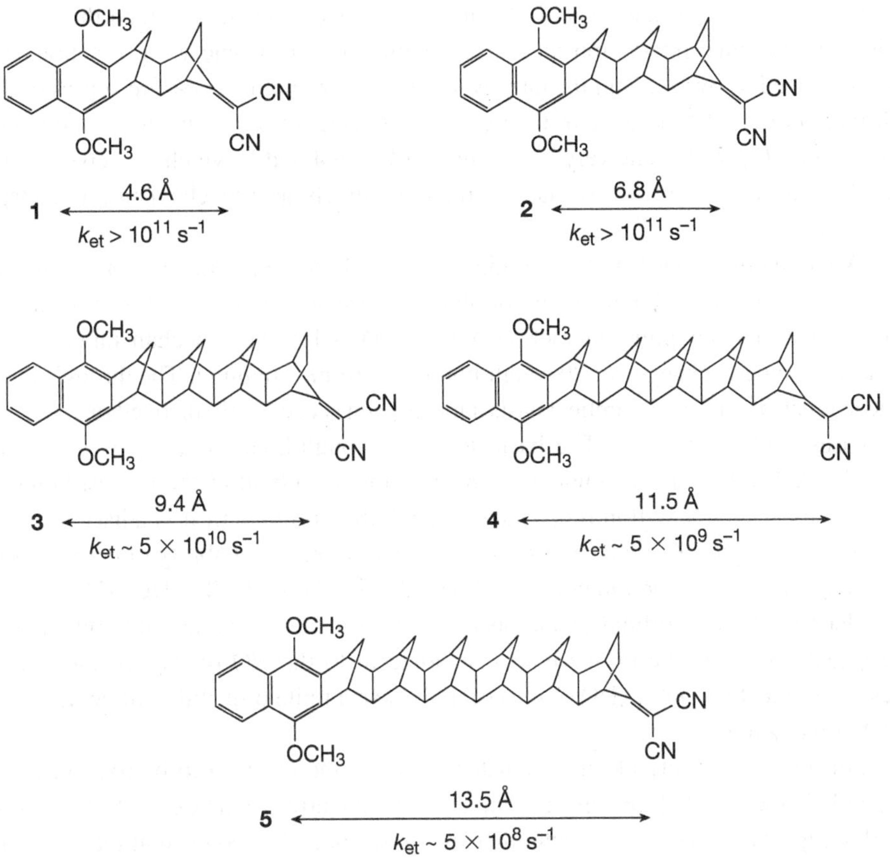

Figure 7.18   Structure of a series (1–5) of electron donors and electron-acceptor pairs separated by rigid spacers.

increments of 4 σ bonds for 1 to 12 σ bonds for 5. This corresponds to an edge-to-edge separation of ~ 5 Å for 1 to ~ 14 Å for 5. In solvents where the electron transfer was exothermic, the values of k_{et} were measured by fluorescence quenching of the donor naphthalene chromophore. The value of k_{et} for a given molecule showed only a small dependence on solvent (change by a factor of 3 or less in going from benzene to acetonitrile); however, the value of k_{et} showed a strong dependence on the separation of donor and acceptor. For 1 and 2, the values of k_{et} were $> 10^{11}$ s^{-1} (too fast to measure by the fluorescence technique); the values of k_{et} were $\sim 5 \times 10^{10}$ s^{-1} for 3, $\sim 5 \times 10^9$ s^{-1} for 4, and $\sim 5 \times 10^8$ s^{-1} for 5, respectively.

An obvious question arising from the data for electron transfer for 1–5 is How does the electron move from the donor to the acceptor when the intervening bridge is a set of σ bonds that have very little electron affinity and possess a high excitation energy for promotion of an electron to a σ^* orbital? The golden rule expression (Eq. 3.8), which requires electronic coupling between the donor and acceptor, provides a guide for understanding the results. One mechanism to achieve electronic coupling between *D and A would be through the usual electron exchange interaction that occurs from overlap of the wave functions of *D and A. This mechanism is termed the normal electron exchange interaction mechanism. However, electronic wave functions are not expected to overlap significantly when the donor and acceptor are separated by distances of ~ 10 Å because of the exponential decrease in overlap as a function of separation (Eq. 7.56). The very high values of k_{et} for **4** and **5**, which are separated by > 10 Å, makes the orbital overlap interaction unlikely as a mechanism for electron transfer.

A second mechanism for electronic coupling of the donor and acceptor is to use the wave functions of the σ and σ^* orbitals of the spacer to assist in the propagation of the electronic coupling interaction between *D and A. This mechanism is termed "electron superexchange" or "through-bond" electronic coupling. The through-bond mechanism has a firm theoretical basis and has been successfully invoked in a number of cases over the years.[36] The basic idea of through-bond coupling is shown in Fig. 7.19. The LU of the donor overlaps with the σ^* orbital of the nearest bond of the bridge. This interaction propagates through the bridge until it is adjacent to the LU of the acceptor. At this point, the electron has a finite probability of being on the acceptor, a through-bond interaction. Alternatively, the half-filled HO of the donor overlaps with the σ orbital of the nearest bond for the bridge and this interaction propagates through the bridge until it is adjacent to the HO of the acceptor. As a result, the electrons in the HO of the acceptor have a finite probability of being in the HO of the donor.

For an excited state electron transfer to occur, the electron to be transferred is initially in the LU of *D and must eventually be transferred to the LU of A. The donor and acceptor are attached to each other by a sequence of σ bonds that make up the spacer. To each σ bond, we can associate a filled HO and an empty LU. As shown in Fig. 7.19, the σ HOs of the spacer bonds have a lower energy than the HO of the donor and acceptor, and the σ^* LUs of the spacer bonds have a higher energy than that of the donor and acceptor.

Two mechanisms exist for through-bond coupling. The first mechanism involves a coupling of the LUs of *D with the LU of A through mixing wave functions of *D

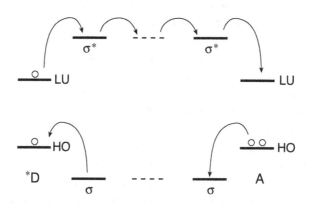

Figure 7.19   Schematic of through-bond coupling for an electron hole (or energy) transfer.

and A through the LUs of the spacer. In this mechanism, the donor LU mixes with the nearest spacer LU, which then in turn mixes with the next spacer LU closer to the acceptor; similar mixing occurs sequentially between adjacent LUs of the space until the spacer LU next to the acceptor mixes with the LU of the acceptor. This mixing provides a path for the electron to be transferred from the LU of *D to the LU of A. The second mechanism is a "hole transfer" by which the HO of the acceptor mixes with the nearest HO of the spacer, followed by sequential coupling within the HO chain of bonds in the space and finally with the HO of the donor. This mixing provides a path for the electron to be transferred from the HO of A to the HO of *D. Which pathway operates will depend on the relative energy gaps of the HOs and LUs of the donor, spacer and acceptor, and the overlap of the adjacent orbitals.

Qualitatively, the through-bond mechanism predicts that the rate of electron transfer will fall off as the number of σ bonds between the donor and acceptor increase, since the degree of overlap will fall off as the number of orbitals involved increases. Experimentally, this expectation is verified for the molecules shown in Fig. 7.18. It was found that the stereochemistry of the bonds attaching to D and A is also important, and that there is an exponential dependence of the rate constant on the number of σ bonds between D and A only for a series of the same stereochemistry.[37] For such a series, the rates were found to follow Eq. 7.57, where V_0 is the electronic coupling for separation by one σ bond, N is the number of σ bonds separating D and A, and ρ is the density of states parameter.

$$k_{et}(N) = |V_0|^2 [\exp -\beta(N - 1)]\rho \qquad (7.57)$$

7.20   A Quantitative Comparison of Triplet–Triplet Energy and Electron Transfer

Since triplet–triplet energy transfer and electron transfer both occur by electron exchange interaction, it is expected that both should exhibit similar features. The existence of the inverted region was originally demonstrated (Fig. 7.17) by the investigation of compounds for which a 4-biphenyl group served as an electron donor (D) to a series of electron acceptors (A). In a second series, biphenyl electron donor and naphthalene electron acceptors were connected at various positions of a decalin nucleus spacer (**NaBi-6** and **NaBi-7**, Table 7.3). By replacing the 4-biphenyl group with a 4-benzophenone group (**NaBz-6** and **NaBz-7**, Table 7.3), the family of structures employed for electron transfer is elegantly converted into an ideal system for investigation of *triplet energy transfer* from a triplet benzophenone moiety to a naphthalene moiety. Table 7.3 lists values for the rate constants for electron transfer (k_{et}), hole transfer (k_{ht}), and energy transfer (k_{ET}). For structures **NaBi-6** and **NaBz-6**, the donor and acceptor groups are separated by six σ bonds and for structures **NaBi-7** and **NaBz-7**, the donor and acceptor groups are separated by seven σ bonds.

A logarithmic plot of the rate constants for triplet transfer versus the rate constants for electron transfer shows a good correlation between the two processes. Interestingly, the ratio of the difference in the rates of triplet and electron transfer is a factor of ~ 2,

Table 7.3   Rates of electron transfer (k_{et}), hole transfer (k_{ht}), and triplet energy transfer (k_{ET})[a]

Compound	N	k_{et} (s^{-1})	k_{ht} (s^{-1})	k_{ET} (s^{-1})	R_{DA} (Å)
Naphthalene —〜〜— Biphenyl 1 2 3 4 5 6 **NaBi-6**	6	3×10^8	6×10^8		6
Naphthalene —〜〜— Biphenyl 1 2 3 4 5 ⁶ ⁷ **NaBi-7**	7	5×10^7	6×10^7		7
Naphthalene —〜〜— Benzophenone 1 2 3 4 5 6 **NaBz-6**	6			9×10^7	6
Naphthalene —〜〜— Benzophenone 1 2 3 4 5 ⁶ ⁷ **NaBz-7**	7			3×10^6	7

a. For electron and hole transfer, the D group is Bi and for energy transfer the D group is Bz. In all cases the electron or energy acceptor is Np. Legend: Np = 2-Naphthyl, Bi = 4-Biphenyl, Bz = 4-benzophenoyl.

which is consistent with the triplet energy transfer corresponding to a simultaneous *two*-electron transfer and the electron transfer corresponding to a single-electron transfer as implied by FMO interactions (Fig. 7.2b). Another important finding is that triplet–triplet transfer does not show strong solvent dependence, whereas electron transfer commonly shows large solvent effects. These results are expected, since for energy transfer the interacting partners remain electrically neutral overall; therefore, energy transfer is not expected to require significant solvent reorganization or show a strong dependence on solvent polarity. On the other hand, electron transfer involves formation of a fully charged positive cation and charged negative anion (radical ion pair. $D^{\bullet+}$, $A^{\bullet-}$) that are fully charged; the pair interacts strongly with solvent molecules and experiences significant solvent reorganization. Consistent with the expected solvent dependences, the rate changes only by a factor of < 3 for triplet transfer upon going from acetonitrile to hexane, while for electron transfer the rate changes by many orders of magnitude for the same solvent change.[38]

7.21   A Connection between Intramolecular Electron, Hole, and Triplet Transfer

For a D(Sp)A system, the frontier orbital diagram description of the electron exchange induced processes of Fig. 7.2b can be represented in terms of electron transfer (Eq. 7.58a), hole transfer (Eq. 7.58b), or energy transfer (Eq. 7.58c). In Fig. 7.2b, electron transfer is viewed as the transfer of an electron between the LUs of D and A; hole transfer is viewed as the transfer of an electron from the HOs of D and A; whereas energy transfer by the exchange mechanism is viewed as a double electron transfer involving both HOs and LUs.

$$D^{\bullet -}(Sp)A \rightarrow D(Sp)A^{\bullet -} \quad k_{et} \quad \text{Electron transfer} \qquad (7.58a)$$

$$D^{\bullet +}(Sp)A \rightarrow D(Sp)A^{\bullet +} \quad k_{ht} \quad \text{Hole transfer} \qquad (7.58b)$$

$$^3D(Sp)A \rightarrow D(Sp)^3A \quad k_{ET} \quad \text{Energy transfer} \qquad (7.58c)$$

For both electron and hole transfer, the electronic coupling V (Eq. 7.56a) has a similar distance dependence.[38] This result implies that since V is transmitted through the σ^* antibonding LU orbitals of the spacer for the anions (electron transfer), and through σ bonding HO orbitals for the cations (hole transfer); this result implies that there is a substantial symmetry of the σ and σ^* orbitals (Fig. 7.19). In both cases, the value of $\beta \sim 1\,\text{Å}^{-1}$ (Eq. 7.57).

7.22 Photoinduced Electron Transfer between Donor and Acceptor Moieties Connected by a Flexible Spacer

In (Section 7.16), we saw how the details of electron-transfer processes can be elucidated by the design of D(Sp)A systems for which Sp is a rigid spacer. It is also of interest to investigate systems for which Sp is a flexible spacer, such as a flexible alkane chain, with D and A attached to the ends of the chain. For the case of a flexible spacer, the kinetics of electron transfer will depend on the chain length and dynamics resulting from conformational changes.[39] As exemplars, we discuss $D(Sp)_nA$ systems for which the $(Sp)_n$ is a flexible spacer $(Sp)_n$ possessing chain consisting of n CH_2 units, where $n > 1$.

Let us consider two limiting conformations of a $D(Sp)_nA$: one conformation for which the chain is a fully extended conformation so that D and A are separated by the length of the chain and a second conformation for which the chain is looped into a ring that brings the D and A into close proximity (see Sections 3.26 and 4.41 for a discussion of flexible biradicals). The many other possible conformations may be approximated as approaching one or the other of these two limiting conformations. For short chains (say, n between 2 and 6), the possibility of planar exciplexes as precursors to electron transfer is plausible. For very large n (say, > 6), the plausibility of forming planar exciplexes decreases because of the entropic penalty for bringing the ends of the chain together.

A good exemplar of a $D(Sp)_nA$ is dimethylaniline as the electron donor and pyrene as the electron acceptor (e.g., the structure PA_n). In this case, Py is photoexcited and the dynamics of the flexible chain in controlling electron transfer can be monitored directly by time-resolved fluorescence laser flash photolysis.[40]

$$\textbf{PA}_n \qquad (6 \leq n \leq 12)$$

7.23   Experimental Observation of the Marcus Inversion Region for Freely Diffusing Species in Solution

As discussed in Section 7.14, many early attempts to demonstrate the inverted region in liquids by the reaction of *D and A in solution failed because for these systems diffusion, not electron transfer, was rate limiting. The Rehm–Weller plot (Fig. 7.10) clearly demonstrates the leveling effect of electron-transfer rates as the electron-transfer reaction rate becomes diffusion controlled. This problem, the direct observation of the inverted region in liquids, could be resolved if the rate of electron transfer in the inverted region could be slowed down to the point that k_{et} is slower than k_{dif}. The latter hypothetical possibility has not yet been achieved for the electron-transfer process, such as $D + A \rightarrow [D^{\bullet+}, A^{\bullet-}]_{gem}$ or $(*D + A \rightarrow [D^{\bullet+}, A^{\bullet-}]_{gem})$ experimentally, presumably because the exothermicity required to be in the inverted region (Fig. 7.16) is not experimentally accessible. However, the existence of the inverted region in solution has been demonstrated indirectly by observing reactions of the type $[D^{\bullet+}, A^{\bullet-}]_{gem} \rightarrow D + A$ pair rather than by directly observing the formation of the pair from *D and A. The basic strategy for the extraction of the rate constant for back electron transfer from the geminate radical ion pairs, k_{bet}, is given in Scheme 7.9. The basic strategy is to find $[D^{\bullet+}, A^{\bullet-}]_{gem}$ systems for which a measurable competition is set up between back electron transfer within the geminate ion pair (rate constant, k_{bet}) and separation of the partners of the radical ion pair (rate constant, k_{sep}).

An exemplar system[41,42] involving a strong electron-accepting molecule in its excited state (*A) and a good ground-state electron donor (D), 9,10-dicyanoanthracene (DCA) as *A, and a methyl naphthalene (MN) as A is an exemplar of such a pair. It was possible to use 4,4'-dimethoxystilbene (DMS) to scavenge the free radical ions (FRIs) that escape from the geminate ion pair formed by electron transfer from *A to D. The rate constant (k_{bet}) for the back electron transfer from the geminate ion pair $[D^{\bullet+}, A^{\bullet-}]_{gem}$, could be determined through the measurement of the quantum yield for formation of free ions and by use of an established rate of separation of the geminate ion pair into free ions (which were scavenged by DMS). When both the back electron transfer from the geminate ion pair and the separation of the ion radical pair

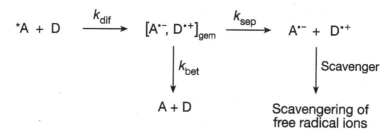

Scheme 7.9   Strategy for measuring electron-transfer rates in solution. The symbol $[A^{\bullet-}, D^{\bullet+}]_{gem}$ represents a collision complex, although in some cases electron transfer may occur in an encounter complex for which one or more solvent molecules is separating $A^{\bullet-}$ and $D^{\bullet+}$.

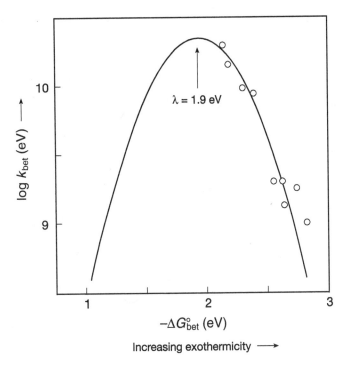

Figure 7.20   Rate constants (k_{bet}) for back electron transfer from
DCA ($A^{\bullet -}$) to aromatic hydrocarbon radical anions ($D^{\bullet +}$) in acetonitrile, plotted versus the negative of the reaction exothermicity
($-\Delta G^{\circ}_{bet}$). The curve is a fit to Marcus theory. Note that λ is the
exothermicity (~ 1.9 eV $=$ kcal mol^{-1}) at which the maximum
value of k_{bet} is observed.[41,42]

partners are competing unimolecular processes, the issue of diffusional encounter of
the donor–acceptor is eliminated.

Results[41,42] for the measurement of k for some DCA/MN systems are shown in
Fig. 7.20. The results can be fit to the Marcus theory to yield a value of $\lambda \sim 1.5$
eV. The data show the clear *decrease* in the rate constant k_{bet} from $\sim 7 \times 10^9$ s^{-1}
to $\sim 1 \times 10^8$ s^{-1} as the exothermicity is increased from ~ -2 to -3 eV (note that
$-\Delta G_{bet}$ is plotted along the energy axis).

7.24   Control of the Rate and Efficiency of Electron-Transfer Separation by Controlling Changes in the Driving Force for Electron Transfer

An elegant example of the control of photoinduced electron-transfer processes by
variation of the driving force for electron transfer is provided by investigation of the
photoinduced electron transfer for structures **6–9** shown in Fig. 7.21. The increased
understanding of the mechanism involved in photosynthetic energy conversion mechanism has led to attempts to mimic the process, in an area of study frequently called

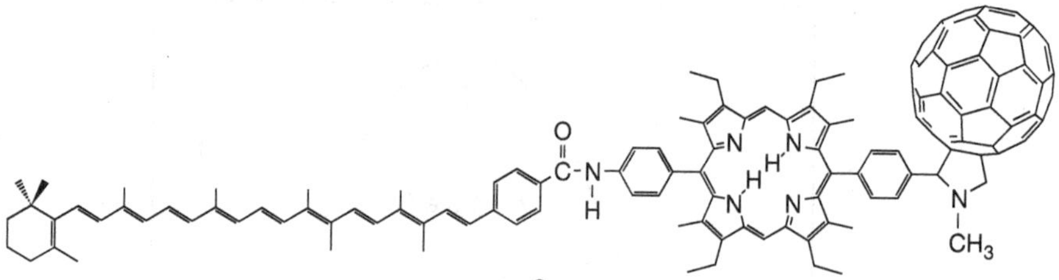

6

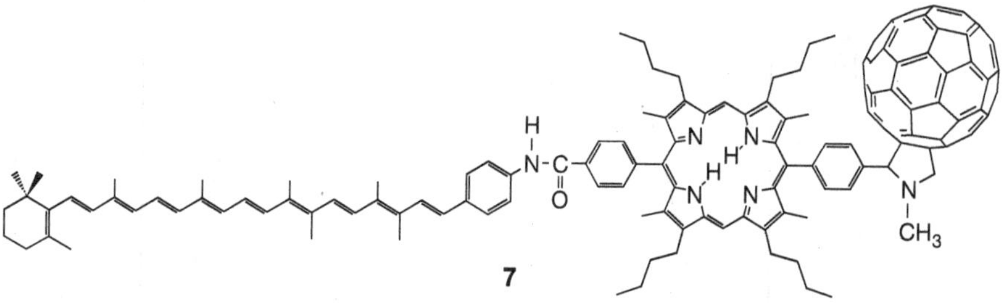

7

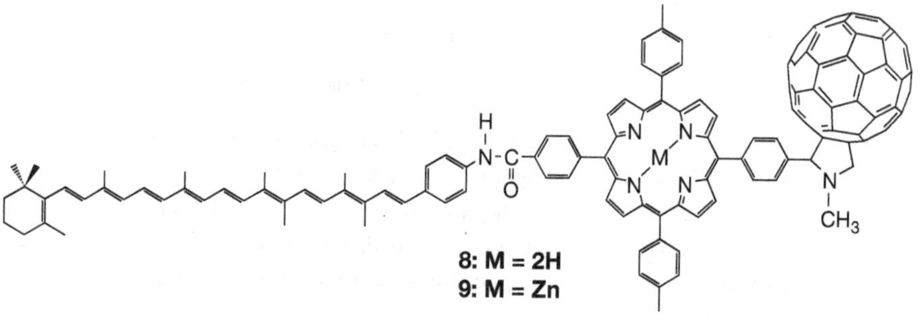

8: M = 2H
9: M = Zn

Figure 7.21   Triads showing different electron-transfer efficiencies.[46]

"artificial photosynthesis."[43–46] The photoinduced electron transfer reactions of structures **6–9** nicely illustrate the delicate balance of excited state energies and redox properties that control the efficiency and kinetics of electron transfer. Structures **6–9** in Fig. 7.21 are referred to as "triads" because they contain three unconjugated chromophores. The label $C(P)C_{60}$ refers to the following chromophores: a carotenoid (C, left), a porphyrin (P, center), and a fullerene (C_{60}, right).[46] The fullerene C_{60} is an excellent electron acceptor.

Indeed, the absorption spectra of the triads can be approximated by linear combinations of the absorption spectra of model carotenoids, porphyrins, and fullerenes, confirming the absence of significant electronic perturbations among the chromophores in their ground states. These molecules were examined by initial photoexcitation of the central porphyrin to its lowest excited singlet state, that is, by formation of $C(^1P)C_{60}$. A goal was to maximize the yield of the charge-separated form: $C^{\bullet+}(P)C_{60}^{\bullet-}$. Note that the latter is a biradical ion pair, which means that the radical cation and anion moieties cannot undergo irreversible diffusional separation. The overall reaction $C(^1P)C_{60} \rightarrow C^{\bullet+}(P)C_{60}^{\bullet-}$ involves a cascade of events, as illustrated in Scheme 7.10.

Figure 7.22 shows the energy levels and reaction paths for the reactive intermediates involved in the overall cascade processes for compound **6** of Fig. 7.21. The ener-

$$C(P)C_{60} \xrightarrow{hv} C(^*P)C_{60} \xrightarrow{(1)} C(P)^*C_{60} \xrightarrow{(2)} C(P^{\bullet+})C_{60}^{\bullet-} \xrightarrow{(3)} C^{\bullet+}(P)C_{60}^{\bullet-}$$

Scheme 7.10   Charge separation in the C-P-C_{60} triad.

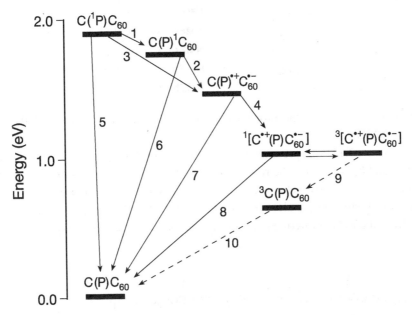

Figure 7.22   Relevant high-energy states and interconversion pathways for triad **6** (Fig. 7.21) following excitation. The energies of the charge-separated states are estimated from cyclic voltammetric data on model compounds in polar solvents. Adapted from the copyright owner.[46]

gies listed in Fig. 7.22 are based on an analysis of emission spectroscopy and knowledge of the electrochemical reduction and oxidation potentials of the components of the triad. Figure 7.22 also shows a singlet–triplet interconversion in the charge-separated biradical, $C^{\bullet+}(P)C_{60}^{\bullet-}$. The latter intermediate has been detected directly by time-resolved spectroscopic methods, lives 170 ns in 2-methyltetrahydrofuran at room temperature, and is formed with a quantum yield of 0.22. Decay of this species yields predominantly the triad triplet state, localized on the carotenoid moiety, $^3C(P)C_{60}$.

7.25   Application of Marcus Theory to the Control of Product Distributions

It is important to note that rates of electron transfer can influence not only the dynamics of a reaction but also the final outcome in terms of products and their distribution. Among various examples that have been recognized over the last few years, we will discuss the photochemistry of benzylic esters as an exemplar.[47] Photolysis of a series of 1-naphthylmethyl esters (**NMEs**) yields products resulting from both homolytic (formation of radicals with unpaired spins) and heterolytic cleavage (formation of electron spin paired anions and cations) from the singlet state (S_1).

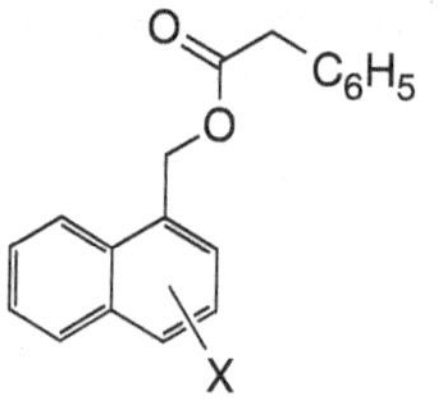

NMEa	X = H
NMEb	X = 4-CN
NMEc	X = 3-CH_3O
NMEd	X = 4-CH_3
NMEe	X = 4-CH_3O
NMEf	X = 4-C_2H_5
NMEg	X = 4,8-$(CH_3)_2$
NMEh	X = 4,7-$(CH_3O)_2$

NME

Scheme 7.11 shows the process that takes place upon irradiation of the esters **NMEa–NMEh**.

In Scheme 7.11, k_I represents the rate of the heterolytic process to yield a radical ion pair and k_R represents the rate of the homolytic process to yield radical pairs. In this case, $k_R \gg k_I$, so that caged radical ion pairs are formed in a two-step process from radical pairs via the electron process described by k_{et}. Since the rate of loss of CO_2 from the radical intermediates (k_{CO_2}) can be estimated from experimental data

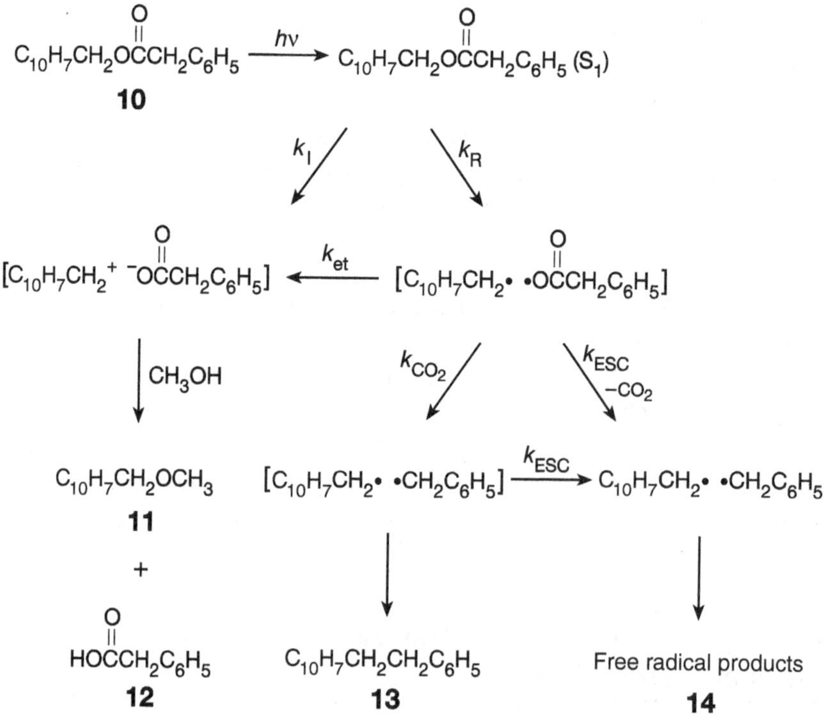

$$C_{10}H_7CH_2OCCH_2C_6H_5 \xrightarrow{h\nu} C_{10}H_7CH_2OCCH_2C_6H_5 \ (S_1)$$

10

$$[C_{10}H_7CH_2^+ \ ^-OCCH_2C_6H_5] \xleftarrow{k_{et}} [C_{10}H_7CH_2 \bullet \ \bullet OCCH_2C_6H_5]$$

$$\downarrow CH_3OH \qquad k_{CO_2} \swarrow \qquad k_{ESC} \searrow -CO_2$$

$$C_{10}H_7CH_2OCH_3 \qquad [C_{10}H_7CH_2 \bullet \ \bullet CH_2C_6H_5] \xrightarrow{k_{ESC}} C_{10}H_7CH_2 \bullet \ \bullet CH_2C_6H_5$$

11

$+$

$$HOCCH_2C_6H_5 \qquad C_{10}H_7CH_2CH_2C_6H_5 \qquad \text{Free radical products}$$

12   **13**   **14**

Scheme 7.11 Mechanism of photolysis of 1-naphthylmethyl esters in methanol.[47] Square brackets denote geminate species. See Scheme 12.45 for a discussion of the overall photochemistry. Illustrated for NMEa.

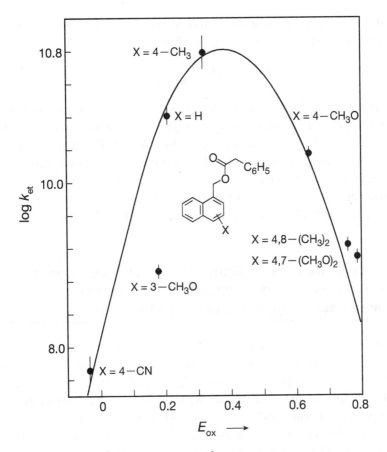

Figure 7.23   Plot of log k_{et} versus E° of electron transfer for the conversion of the radical pair to the ion pair. The curve is a fit according to the Marcus theory using $\lambda = 0.39\,\text{eV}$. Adapted from the copyright owner.[47]

$(k_{CO_2} \sim 4.8 \times 10^9\ \text{s}^{-1})$,[47] it is possible to determine k_{et} from the product ratios (that represent the competition between k_{et} and k_{CO_2}) as shown in Eq. 7.59.

$$k_{et} = \frac{\text{Yield}\ (\text{ArCH}_2\text{OCH}_3)}{\text{Yield}\ (\text{ArCH}_2\text{CH}_2\text{C}_6\text{H}_5)}\ (k_{CO_2}) \tag{7.59}$$

Figure 7.23 shows that a plot of the rate constants derived from Eq. 7.59, when plotted as a function of the free energy change for the reaction, yields a bell-shaped curve that is analogous to the systems involved in both the normal and Marcus inverted regions of electron transfer (Fig. 7.17).

In the electron-transfer systems of Fig. 7.23, the free energy change can be calculated as indicated in Eq. 7.47, which in this case corresponds to Eq. 7.60:

$$\Delta G_{et} = \mathcal{F}E^\circ_{(\text{ArCH}_2^+/\text{ArCH}_2^\bullet)} - \mathcal{F}E^\circ_{(\text{C}_6\text{H}_5\text{CH}_2\text{CO}_2^\bullet/\text{C}_6\text{H}_5\text{CH}_2\text{CO}_2^-)} - N_A \frac{e^2}{4\pi\varepsilon_o\varepsilon r} \tag{7.60}$$

where the term E_{*D} has been omitted, since a ground state reaction is involved in the electron transfer. Equation 7.60 can be simplified to Eq. 7.61:

$$\Delta G_{et} = \mathcal{F} E^{\circ}_{(ArCH_2^+/ArCH_2^{\bullet})} + B \tag{7.61}$$

which reflects that the only change along the series is in the substituent in $ArCH_2$, and B incorporates all of the other terms. Figure 7.23 has therefore been plotted as a function of the electrode potential, E° (rather than ΔG), a common practice in studies of this type. This system provides a clear example of how electron transfer can control product distribution (see Eq. 7.59) and how Marcus theory can explain an unusual dependence for variation of product formation on the free energy of reaction when the free energy change results in both the normal and Marcus inverted regions.

7.26   The Continuum of Structures from Charge Transfer to Free Ions: Exciplexes, Contact Ion Pairs, Solvent Separated Radical Ion Pairs, and Free Ion Pairs

It is conceptually useful to examine a plausible hypothetical sequence of steps that might occur along a reaction coordinate, as a donor approaches an acceptor (for which either the donor or acceptor is electronically excited), and that concludes with the electron-transfer event to form a radical ion pair. This exercise will review an expanded array of *possible* intermediates that might exist along the reaction coordinate. As a specific exemplar (Eq. 7.62a), we will assume that the donor is the electronically excited species (*D) and that *D encounters A and first forms an encounter complex, symbolized for situations where the excitation is localized on D, $(*D,A)_{ex}$. The latter then forms an exciplex with delocalized excitation, symbolized as $*(D, A)_{ex}$. The latter often will possess significant contributions from CT stabilization.

Next, we imagine that electron transfer occurs in the exciplex, which is converted to a geminate *contact radical ion pair* (CRIP, symbolized as $D^{\bullet+}$, $A^{\bullet-}$), followed by insertion of one or more solvent molecules between the partners of the ion pair to form a geminate *solvent separated radical ion pair* (SSRIP, symbolized as $D^{\bullet+}(S)A^{\bullet-}$). Finally, we imagine that the ions separate to become free radical ions (FRIs) in the bulk solvent (FRI, $D^{\bullet+} + A^{\bullet-}$). At various stages, each step along this hypothetical pathway will compete with other possible pathways, such as chemical reactions or back electron transfer. Thus, the radical ion pairs, CRIP, SSRIP, and FRI appear to be similar chemical species, since they both formally consist of a $D^{\bullet+}$ and $A^{\bullet-}$ unit; however, they may possess vastly different rates for back electron transfer to form D + A because of the distance dependence of electron transfer and the reorganization energies involved.

$$
\begin{array}{ccccccc}
{}^{*}D + A & \rightarrow & {}^{*}(D,A) & \rightarrow & D^{\bullet+},A^{\bullet-} & \rightarrow & D^{\bullet+}(S)A^{\bullet-} & \rightarrow & D^{\bullet+} + A^{\bullet-} \\
& & \text{Exciplex} & & \text{Contact radical} & & \text{Solvent separated} & & \text{Free ion pair} \\
& & \text{(EX)} & & \text{ion pair (CRIP)} & & \text{radical ion} & & \text{(FIP)} \\
& & & & & & \text{pair (SSRIP)} & &
\end{array}
$$

$$\tag{7.62a}$$

In Section 7.13, we saw that energetic terms associated with electron transfer have a strong dependence on the distance between charges (see, e.g., Eqs. 7.47 and 7.55). One could easily conceive that nascent radical ion pairs (e.g., reaction 7.38) may have different initial structures (e.g., those shown in Eq. 7.62a) depending on the nature of the reactions that generated them and the solvent properties for stabilizing the development of charges. In addition, in certain cases, D and A may be able to form a *ground-state electron-transfer stabilized complex* and electron transfer may occur directly following excitation of this complex, as shown in Eq. 7.62b.

$$D + A \; \rightleftharpoons \quad D,A \quad \xrightarrow{h\nu} \quad D^{\bullet+}, A^{\bullet-}$$

Ground-state   Contact

CT complex   radical ion pair   (7.62b)

(GSCTC)   (CRIP)

In this case, the partners of the initial radical ion pair ($D^{\bullet+}, A^{\bullet-}$) must be a geminate CRIP, since they were in contact in order to form a ground-state CT complex (GSCTC); thus, the initial geminate radical ion pair ($D^{\bullet+}, A^{\bullet-}$) is formed with no solvent molecules separating the partners of the pair.

Alternatively, a compositionally identical radical ion pair could be formed by excitation of D to form *D, followed by diffusion of *D and A together, to form an encounter complex from which electron transfer occurs. Two possible situations are illustrated in reactions 7.63 and 7.64, where (S) represents a solvent molecule located between the donor–acceptor pair. The electrostatic interactions involved in each case will be different and controlled by Eq. 7.48.

$$^*D + A \; \rightleftharpoons \; ^*D(S)A \; \rightleftharpoons \; ^*D,A \; \longrightarrow \; D^{\bullet+}, A^{\bullet-}$$
(CRIP)   (7.63)

$$^*D + A \; \rightleftharpoons \; ^*D(S)A \; \longrightarrow \; D^{\bullet+}(S)A^{\bullet-}$$
(SSRIP)   (7.64)

The product of reaction 7.64 is a SSRIP, $D^{\bullet+}(S)A^{\bullet-}$ and is distinct from the CRIP, $D^{\bullet+}, A^{\bullet-}$, formed in reaction 7.63. The solvent molecules (S) of $D^{\bullet+}(S)A^{\bullet-}$ "shield" the charges of $D^{\bullet+}$ and $A^{\bullet-}$ from one another: This *shielding effect* is highest in polar solvents (see Fig. 7.9). By comparison, the effects of solvent are not nearly as large with neutral molecules in energy-transfer processes, where only minor effects (e.g., due to dipolar orientation) can be expected.

As an exemplar of the photochemistry of ground-state CT complexes, the ion pairs formed by photolysis of ground-state complexes between tetranitromethane (**A**) and anthracenes substituted at the 9-position (**D**, Y = H, CH$_3$, NO$_2$) have been investigated in detail.[48a]

A   **D**

Equation 7.65 illustrates two types of nomenclatures currently in use.

$$D,A \xleftarrow{k_1} CRIP \underset{k_{-2}}{\overset{k_2}{\rightleftharpoons}} SSRIP \qquad (7.65a)$$

$$D,A \xleftarrow{k_1} D^{\bullet+},A^{\bullet-} \underset{k_{-2}}{\overset{k_2}{\rightleftharpoons}} D^{\bullet+}/A^{\bullet-} \qquad (7.65b)$$

In dichloromethane at room temperature for the parent anthracene (D, $Y = H$), the values of the rate constants for the steps in Eq. 7.65b are $k_1 = 9 \times 10^8$ s^{-1}; $k_2 = 9 \times 10^8$ s^{-1}; $k_{-2} = 4.3 \times 10^7$ s^{-1}. In addition, it was found that $\Delta G_2 = 1.8$ kcal mol^{-1}, where ΔG_2 corresponds to the free energy for the conversion between CRIP and SSRIP. The positive value of ΔG_2 indicates that the CRIP is more stable than the SSRIP. The values of k_2 were largely insensitive to the nature of Y, but k_1 values changed between 3×10^8 s^{-1} (D, $Y = CH_3$) and 3.6×10^9 s^{-1} (D, $Y = NO_2$). Values of k_1 were also strongly dependent on the solvent; charge recombination is faster in nonpolar solvents. Thus, for $Y = H$, k_1 changes from 9×10^8 s^{-1} in dichloromethane, to 7.1×10^9 in benzene, and $> 4 \times 10^{10}$ s^{-1} (i.e., diffusion controlled) in n-hexane.

For simplicity in Eq. 7.62b, the ground-state complex is shown as yielding the CRIP directly upon excitation. However, the process can also be viewed as occurring in two steps, where the initial excitation produces an unrelaxed (or Franck–Condon) CRIP that then undergoes a configurational and solvent reorganization to the lowest energy or relaxed configuration of the CRIP (Eq. 7.66). Relaxation times (τ_d) are between 0.5 and 2 ps for CRIPs derived from the reaction of 1,2,4,5-tetracyanobenzene with benzene, toluene, and mesitylene.[48b]

$$D,A \xrightarrow{h\nu} (CRIP)_{FC} \xrightarrow{\tau_d} (CRIP)_{relaxed} \qquad (7.66)$$

The steps of Eq. 7.66 correspond to the processes represented schematically by the PE surfaces of Fig. 7.24. Note the offset minimum of the nuclear coordinates between the ground-state CT complex and excited-state CRIP. In other words, the two states have different relaxed geometries, as well as different vibrational frequencies (i.e., the parabolas have different widths). Excitation leads to a FC state that then relaxes. Note the similarity of Fig. 7.24 with Libby's model of electron transfer in Fig. 7.13 (left). However, in the case of Fig. 7.24, there is no conflict with energy conservation, since the energy required for electron transfer is supplied by the absorption of a photon (whose energy is the reorganization energy, λ, of Marcus theory). Figure 7.24 also points out that the *CRIP is an electronically excited state and not a ground state*! That the CRIP is an excited state could be anticipated from Eq. 7.62a, which shows an EX, *(D,A), as a precursor to the CRIP if *D approaches A by diffusion. The most compelling evidence that the CRIP is an excited state is that it has the possibility to emit a photon, even though this process may be of low probability because of poor FC factors.

Both CRIPs and SSRIPs differ in their electronic characteristics.[48b] Thus the electronic coupling between partners in CRIPs is higher than in SSRIPs because in the CRIP there is greater overlap of the orbitals for the ions; this difference in overlap can approach two orders of magnitude.[48b] On the other hand, the solvent reorganization energy in the CRIP is generally lower than in SSRIP, an effect that is

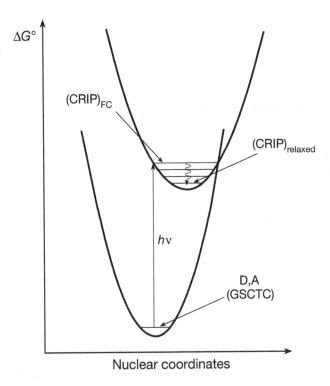

Figure 7.24   Excitation and relaxation of a ground-state CT complex.

mainly due to the solvent reorganization term (λ_{outer}). In general, one expects that the solvent shell of the SSRIP will be more loosely held and therefore more dynamic than a more tightly held solvent shell about the CRIP. Therefore, the solvent shell about a SSRIP will possess a less well defined structure. Another important difference between a CRIP and its associated SSRIP is that of the two, only the CRIP ($D^{\bullet+}, A^{\bullet-}$) is expected to possess an emission spectrum, because the FC factors are much more favorable for emission to the ground-state D,A complex. The SSRIP ($D^{\bullet+}/A^{\bullet-}$) does not possess a corresponding well-defined FC ground state to which it can emit since the solvent molecules separate $D^{\bullet+}$ and $A^{\bullet-}$, leading to poor electronic overlap and a low oscillator strength (Chapter 4). Conversely, the excitation of a ground-state CT complex (D,A) will directly produce a CRIP ($D^{\bullet+}, A^{\bullet-}$).

Both CRIPs and SSRIPs are clearly transient species that under normal conditions will either collapse to the D,A pair, or separate into the free ions $D^{\bullet+} + A^{\bullet-}$. If the D,A pair does not form a ground-state complex, they will be simple collision partners in a solvent cage and will rapidly separate into free D and A. Similarly, SSRIPs will also rapidly separate into FRI, typically in 10^{-7}–10^{-9} s. Later, we will examine some of these processes in the context of solvent cage effects. Now it is useful to have some idea of the separation kinetics for the radical ion pairs, as shown in Eq. 7.67.

$$D^{\bullet+}(S)A^{\bullet-} \quad \xrightarrow{k_{sep}} \quad D^{\bullet+} + A^{\bullet-} \qquad (7.67)$$

SSRIP   Free radical ions

Typical values[48b–51] for k_{sep} are between 10^7 and 10^9 s^{-1}. The values are usually higher in more polar solvents but are not very sensitive to details of molecular structure.

7.27 Comparison between Exciplexes and Contact Radical Ion Pairs

Earlier (Chapter 4) we described exciplexes as excited-state complexes stabilized by CT interactions. Charge-transfer stabilization may range from weak (as for CT complexes) to very strong (as involved in full electron transfer). As a result, depending on the extent of the CT, a CRIP may be alternatively described[44] as equivalent to an exciplex exhibiting partial or nearly complete CT from the donor to the acceptor. Radiative and nonradiative back electron-transfer processes are spontaneous transitions in which the energy difference between the D,A and the D$^{\bullet+}$, A$^{\bullet-}$ pairs is dissipated as the emission of light or as nuclear motions (heat) of D,A and the solvent, respectively. Thus, pure CRIP (exciplex) fluorescence emission represents back electron transfer to regenerate the starting materials in a contact ground-state pair (D,A). After transition from the excited state, the partners may be on a repulsive ground-state surface and separate (true exciplex) or remain in contact depending on the tendency of the pair to form ground-state CT complexes.

Now, let us consider the case of a CRIP undergoing a radiative transition to the ground-state contact pair D,A, that is, CRIP $\rightarrow$ D,A $+ h\nu$. This process is illustrated in Fig. 7.25, where the heavy lines at the bottom of the vibrational ladder represent the lowest vibrational levels of D,A and D$^{\bullet+}$,A$^{\bullet-}$, and the thin lines are low-frequency modes relating largely to solvent motion; that is, the lines representing vibrational motion are related to the same motions that determine λ_{outer}. While under most conditions only the lowest vibrational state of CRIP will be active, in the case of the final state (D,A), other vibrational modes may participate as a result of the transition to the ground state. For simplicity, we start with a hypothetical example where only the lowest vibrational level ($v = 0$) of D,A is active.

For this hypothetical system, each emission frequency in Fig. 7.25 corresponds to a different energy, or different ΔG_f, the free energy associated with the transition between the numerous low-frequency modes available for each state. The probabilities for each one of these combinations between low-frequency modes of CRIP and of D,A will be given by the usual Marcus type of dependence, potentially with a "normal" and an "inverted" region. Thus, *the emission spectrum at low frequencies represents the normal region and the emission spectrum at high frequencies represents the Marcus inverted region.*[52]

In a real system corresponding to Fig. 7.25, a number of vibrational modes of D,A can be expected to be active, each one associated with a set of different solvent low-frequency modes, as shown in Fig. 7.26.

Each Gaussian curve (numbered traces) in Fig. 7.26 represents a transition from the lowest vibronic level of CRIP to a vibronic level (identified by $j = 0$, 1, or 2, 3 or 4) of D,A. Each of these traces is a Marcus-type free energy curve (Fig. 7.15). Their

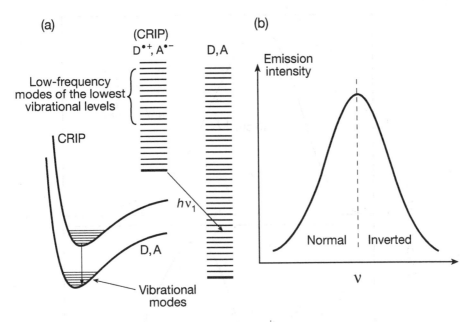

Figure 7.25   (a) Schematic representation of radiative back electron transfer from a CRIP to the ground-state pair D,A, and (b) emission intensity as a function of frequency.

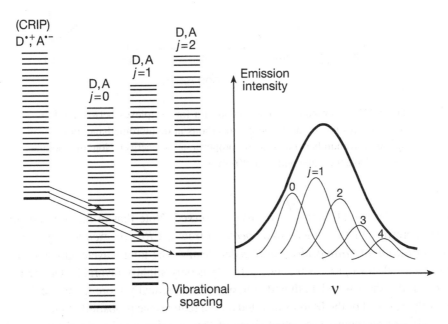

Figure 7.26   Radiative back electron transfer from CRIP to several vibrational levels (j) of D,A. Each vibrational level has a quasicontinuum of low-frequency modes associated with it. The arrows illustrate a few possible transitions, but many of these are possible. Each vibrational level leads to an emissive curve similar to that in Fig. 7.25 and adds up to give the overall emission illustrated by the heavy curve.[52]

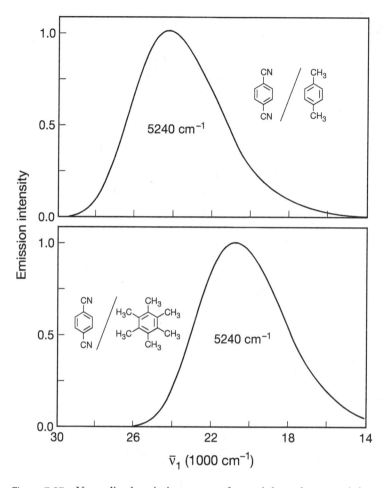

Figure 7.27   Normalized emission spectra for exciplexes between 1,4-dicyanobenzene and aromatic hydrocarbons in dichloromethane, plotted against wavenumber (i.e., directly proportional to energy). Note that both spectra have the same width (5240 cm^{-1}).

sum gives the overall emission intensity observed. Note that in this interpretation it is interesting that *each emission spectrum from a CRIP or exciplex represents a demonstration of Marcus electron-transfer theory.* The long-wavelength side of the emission reflects a process that energetically is only slightly favorable. On the other hand, at the short-wavelength end emission is energetically more favorable, but the underlying effect of the Marcus inverted region reduces the probability (and therefore intensity) of emission. The actual shape of the emission band is a reflection of the interplay of these factors. A classic example of this behavior is shown in Fig. 7.27 for two related systems leading to somewhat different emission energies.

The concepts in Figs. 7.25 and 7.26 are fundamental to understanding the formation of excited states, frequently leading to chemiluminescence or bioluminescence, phenomena that are discussed in Section 7.31. Figures 7.25 and 7.26 have not included radiationless electron transfer from CRIP to D,A. Such processes usually will occur in competition with emission and can be represented by horizontal lines from CRIP to upper solvent modes of D,A.

7.28   Electron-Transfer Equilibria

Now, we consider the issue of reversibility in electron-transfer processes. So far, we have written energy and electron-transfer processes $^*D + A \rightarrow D^{\bullet+} + A^{\bullet-}$ as *irreversible,* since the reverse process $D^{\bullet+} + A^{\bullet-} \rightarrow {}^*D + A$ is generally endothermic after vibrational relaxation has occurred. The cases of *back electron transfer* presented above represent a very exothermic, and therefore irreversible $D^{\bullet+} + A^{\bullet-} \rightarrow D + A$ process, where the ground state of D and A are produced. However, although electron transfer is generally irreversible, under certain favorable conditions, electron-transfer processes can be reversible.

7.29   Energy-Transfer Equilibria

For an energy-transfer equilibrium to be established, it is necessary to have concentrations of an energy acceptor and rate constants for energy transfer sufficiently large that the dominant mode of decay for the excited states will be via energy transfer (Eq. 7.70) and not first- or second-order decays of $^*D$ or $^*A$ (reactions 7.68 and 7.69).

$$^*D \xrightarrow{k_D} D \ (+h\nu \text{ or } \Delta) \tag{7.68}$$

$$^*A \xrightarrow{k_A} A \ (+h\nu' \text{ or } \Delta) \tag{7.69}$$

$$^*D + A \underset{k_{-ET}}{\overset{k_{ET}}{\rightleftharpoons}} D + {}^*A \tag{7.70}$$

For reversible energy transfer to be important, it is necessary for Eqs. 7.71 and 7.72 to be valid.

$$k_{ET}[A] > k_D \tag{7.71}$$

$$k_{-ET}[D] > k_A \tag{7.72}$$

For singlet excited states, k_D and k_A may frequently have values $\geq 10^8 \text{s}^{-1}$, and the largest values of k_{ET} in fluid solvents will be on the order of 10^9–$10^{10} \text{ M}^{-1} \text{ s}^{-1}$ or less in the exothermic direction and much smaller in the endothermic direction. These boundary conditions suggest that equilibration in energy-transfer reactions between singlet states will be rare at the low concentrations of D and A typically employed in photochemical experiments. Thus, while singlet-state energy-transfer equilibration in solution is possible, it is not likely for intermolecular energy transfer because of the short lifetime of singlet states. However, the situation may be quite different if chromophores are linked by some type of *molecular spacer*. In this case, singlet energy transfer may be fast enough for equilibration to occur. We have already covered an example illustrating fast singlet energy transfer in tethered systems (Section 7.8).

In contrast to the case for singlet states, intermolecular equilibration of triplet states undergoing energy transfer can be readily achieved in a number of cases, since k_D and k_A values of Eqs. 7.71 and 7.72 for relatively long-lived triplets are frequently three to four orders of magnitude lower than the values for singlet states.[53a]

The energy-transfer equilibrium constant for equilibration between excited states undergoing energy transfer is given by Eq. 7.73.

$$K_{ET} = \frac{[D][A^*]}{[D^*][A]} = \frac{k_{ET}}{k_{-ET}} \tag{7.73}$$

The free energy for energy transfer (ΔG_{ET}) is related to the equilibrium constant (K_{ET}) by Eq. 7.74.

$$\Delta G_{ET} = -RT \ln K_{ET} \tag{7.74}$$

The values of ΔG_{ET} are in turn related to ΔH_{ET} and ΔS_{ET} by Eq. 7.75.

$$\Delta G_{ET} = \Delta H_{ET} - T \Delta S_{ET} \tag{7.75}$$

The enthalpic change for the overall energy-transfer step (ΔH_{ET}) refers to the energy minimum of *D and *A, that is, the relaxed triplet energies for donor and acceptor. This contrasts with energies derived from radiative processes (emission or absorption) that refer to vertical or FC transitions (see Chapters 4 and 5).

Since the reorganization energy changes for energy-transfer processes are usually small, it is common practice to assume that $\Delta S_{ET} \sim 0$, so that Eq. 7.76 is generally a good approximation.

$$\Delta G_{ET} \approx \Delta H_{ET} \tag{7.76}$$

When Eq. 7.76 is valid, it is possible to make a direct comparison of the energies for *D and *A from spectroscopic measurement (e.g., phosphorescence) and energy-transfer equilibria. This assumption is reasonable only for molecules where there is little change in shape or in the degree of conformational freedom. The triplet states of rigid polyaromatic molecules, such as naphthalene or chrysene, meet this criterion. For molecules where considerable changes in conformational freedom take place upon excitation (e.g., biphenyl, which loses its internal mobility in the triplet state), the assumption of $\Delta S \sim 0$ breaks down.

Naphthalene   Chrysene

For the naphthalene–chrysene system, ΔS_{ET} (chrysene as donor) was found to be 0.04 G mol^{-1}, as expected for two nearly rigid molecules. In contrast for ben-zophenone (donor) and biphenyl (acceptor),[53b] $\Delta S_{ET} = -1.8$ G mol^{-1}, and for 4-

methylbenzophenone and 4-methylbiphenyl,[53b] $\Delta S_{ET} = -2.0 \, \text{G mol}^{-1}$, indicating that biphenyl loses more entropy than is regained by benzophenone on triplet energy transfer.

Benzophenone   Biphenyl

7.30 Electron-Transfer Equilibria in the Ground State

There are numerous examples of electron-transfer equilibration between ground-state species. For example, there exists a ground-state equilibrium for the radical anion of D reacting and ground-state electron acceptor A (Eq. 7.77):

$$D^{\bullet-} + A \underset{k_{-et}}{\overset{k_{et}}{\rightleftharpoons}} D + A^{\bullet-} \tag{7.77}$$

The equilibrium constant for reaction 7.77 can be readily estimated if the corresponding electrochemical potentials, as given by Eq. 7.78, are available.

$$E^{\circ}_{(D/D^-)} - E^{\circ}_{(A/A^-)} = -\frac{RT}{\mathcal{F}} \ln K_{ET} \tag{7.78}$$

The equilibrium constant for reaction 7.77 can also be determined if the appropriate kinetic data, as given by Eq. 7.79, are available.

$$K_{et} = \frac{k_{et}}{k_{-et}} \tag{7.79}$$

Equation 7.80 provides an example of a ground-state electron-transfer equilibrium from the abundant literature available on the chemistry of superoxide $(O_2^{\bullet-})$.[54]

$$k_{et} = 5 \times 10^6 \, \text{M}^{-1} \, \text{s}^{-1}; \; k_{-et} = 2 \times 10^8 \, \text{M}^{-1} \, \text{s}^{-1}; \; K_{et} = 0.023, \text{ at } 22 \, ^\circ\text{C} \tag{7.80}$$

7.31 Excited-State Electron-Transfer Equilibria

Expressions similar to Eq. 7.78 can also be written for equilibrium electron-transfer reactions involving electronically excited states. For example, we can write the same equation used in previous examples, but now as a reversible process, as given by Eq. 7.81.

$$^*D + A \underset{k_{-et}}{\overset{k_{et}}{\rightleftharpoons}} D^{\bullet+} + A^{\bullet-} \tag{7.81}$$

Equation 7.46 provides an estimate for ΔG_{et}, although it ignores entropic terms. A more accurate expression (Eq. 7.82) includes the entropy terms.

$$\Delta G_{et} = \mathscr{F} E^{\circ}_{(D^+/D)} - \mathscr{F} E^{\circ}_{(A/A^-)} - \Delta H_{et} + T \Delta S^{\circ}_{et} \tag{7.82}$$

Energy-transfer equilibration, as shown in Eq. 7.81 (i.e., an excited and ground state with free ions), is not generally observed. The reason is that the back reaction to give the excited donor (rate constant k_{-et}), as shown in Eq. 7.81, is generally much less favorable energetically than back electron transfer (rate constant k_{bet}) to yield the ground state, as illustrated in Eq, 7.83.

$$D^{\bullet+} + A^{\bullet-} \xrightarrow{k_{bet}} D + A \tag{7.83}$$

Thus, if $k_{bet} >> k_{-et}$, Eq. 7.83 will dominate Eq. 7.82, and the equilibration of reaction 7.81 will not occur.

7.32   Excited-State Formation Resulting from Electron-Transfer Reactions: Chemiluminescent Reactions

There are situations where ion recombination in solution can lead to an excited state, as shown in Eq. 7.84:

$$D^{\bullet+} + A^{\bullet-} \xrightarrow{k_{bet}} {}^*D + A \tag{7.84}$$

where $^*D$ is an electronically excited state, which may or may not be the same one in which the reactants were initially formed. These are examples of *chemiexcitation reactions* for which chemical energy of recombination is converted into electronic excitation energy. When $^*D$, which is produced by Eq. 7.84, emits a photon, such reactions are termed *chemiluminescent reactions*.

Consider the back electron transfer reaction of Scheme 7.8. The back reaction between the radical ions from naphthalene and 1,4-dicyanobenzene is $\sim 75 \text{ kcal mol}^{-1}$ exothermic. This result is substantially more than the triplet energy of naphthalene ($\sim 60 \text{ kcal mol}^{-1}$). Thus, triplet formation via back reaction in Scheme 7.8 is energetically plausible.

Two situations that can contribute to the formation of an excited state from an electron transfer between $D^{\bullet+}$ and $A^{\bullet-}$ are

1. When back electron transfer to the ground state ($D^{\bullet+} + A^{\bullet-} \rightarrow D + A$) is in the Marcus inverted region (very large ΔG°_{et}) and is therefore inhibited; as a result, the formation of the excited products (e.g., $D^{\bullet+} + A^{\bullet-} \rightarrow {}^*D + A$) may

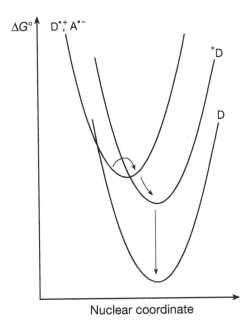

Figure 7.28   Potential energy curves for ion recombination (see reaction 7.84) leading to chemiluminescence. Note that ground state formation occurs in the inverted region.

be kinetically preferred because of the smaller ΔG°_{et} change involved in the formation of the excited state.

2. When a triplet radical ion pair undergoes back electron transfer, spin selection rules forbid the formation of $D + A$ (singlet products) but allow the formation of an excited triplet state of D or A, that is, when $^3(D^{\bullet+}, A^{\bullet-}) \rightarrow {}^1(D,A)$ is spin forbidden, while $^3(D^{\bullet+}, A^{\bullet-}) \rightarrow {}^3 D, A$ is spin allowed, and therefore is plausible if the process is exothermic.[55]

Figure 7.28 illustrates the formation of an excited state from back electron transfer by case (1) for which the system is in the Marcus inversion region. The energy curves show that the activation required for the ground-state formation (for the more exothermic reaction) is higher than excited-state formation.

Among the many systems where these effects have been characterized, the pyrene-N,N-dimethylaniline system has figured prominently. Here the excited triplet state of pyrene (^{3}Py) is formed by reaction of its radical anion (Py$^{\bullet-}$) with the radical cation from N,N-dimethylaniline (DMA$^{\bullet-}$), as shown in Eq. 7.85.[56,57]

$$Py^{\bullet-} + DMA^{\bullet+} \longrightarrow {}^{3*}Py + DMA \tag{7.85}$$

Only the triplet state of Py is energetically accessible from the radical ion pair precursor; formation of the Py singlet state is considerably endothermic. The occurrence of reaction 7.85 is detected through the delayed fluorescence of Py that results from triplet–triplet annihilation (Section 7.12), as shown in reactions 7.86 and 7.87.

$$^{3*}Py + {}^{3*}Py \xrightarrow{k_{TTA}} {}^{1*}Py + Py \qquad\qquad (7.86)$$

$$^{1*}Py \xrightarrow{k_F^0} Py + h\nu \qquad\qquad (7.87)$$

Recall that reactions, such as those given by Eq. 7.86 (i.e., triplet–triplet annihilation), generally occur close to the diffusion limit (Table 7.3).

Reaction 7.85 occurs not only between free radical ions in solution but also in compounds for which the pyrene and aniline are covalently connected by a flexible linker. In this case, the process effectively involves a *biradical* or *biradical ion pair*. Systems linked by long flexible spacer hydrocarbon chains allow the pyrene and aniline to approach each other closely and effectively overlap their electron clouds. They are also generally most efficient[58,59] (e.g., as in structure PN_n in Section 7.22, where n is between 6 and 12).

7.33 Role of Molecular Diffusion in Energy and Electron-Transfer Processes in Solution

The preceding sections provide us with a background on the possible interaction types and characteristic mechanisms for energy and electron transfer. Whether the mechanism of the interaction is an electron exchange or a dipole–dipole process, a key factor determining which one of these will be most effective for *R (where the * may refer to either D or A in a photoinduced process) is the mechanism for *delivery* of energy or electrons from donor to acceptor. The delivery may take advantage of a molecular property, such as a flexible intramolecular link of covalent bonds between D and A (as in the chemiluminescent systems described in Section 7.32) or, as in the trivial mechanism described in Section 7.3, the ability of D to release and of A to capture the energy of a photon or electron.

Delivery processes can be conveniently divided into three types:

1. *Structurally preorganized proximity of D and A.* Both D and A may be "preorganized" so that they are in the proximity of one another when one or the other absorbs a photon. For example, D and A may form a CT complex (D,A) for which a D molecule is always in the proximity of an A molecule as the result of intermolecular (supramolecular) bonding. If D and A are covalently linked at the ends of a flexible spacer (e.g., a chain of CH_2 groups), there will be a certain probability that D and A will be in the proximity of one another. This probability will be determined by the conformational equilibrium at the ends of the chain. In nature, the photosynthetic unit takes advantage of this delivery process, by preorganizing the components (light absorbing electron-donating chlorophyll molecules and electron-accepting quinones) necessary for photosynthesis.

2. *Diffusional processes that bring D and A into proximity.* When D and A start at random positions in a liquid, a fluid medium allows enough mobility for reaction partners to approach each other and to come into one another's proximity. We can think of this case as one of *material or molecular transport.* Most commonly, the motion of D and A is modeled by a "random walk" by which the displacements of D and A are viewed as a series of "jumps" of more or less fixed length from one initial site to another.

3. *Conducting medium that can transport energy or charge.* In these cases, the medium (solvents, bonds, space) can transport either electronic excitation or electrons. The trivial mechanisms for energy or electron transfer discussed earlier exemplify an example for which empty space provided the transport medium for a photon. In some cases, spacers may prove to be part of the delivery or communication between donor and acceptor. In addition to energy migration through a spacer, overlap of the orbitals of *D with the orbitals of the space can weakly couple *D with A through σ bonds (Section 7.19)

7.34   An Exemplar Involving Energy Transfer Controlled by Diffusion

Our first exemplar of diffusional control deals with energy transfer where *R = *D and A is the ground-state acceptor. The overall energy-transfer process is shown in Eq. 7.88.

$$^*D + A \longrightarrow D + {}^*A \tag{7.88}$$

For our analysis, the following assumptions are made: reaction 7.88 is exothermic (Fig. 7.1), and the lifetime of *D and concentration of A are such that *D sees a "random" distribution of A molecules in its immediate surroundings. These assumptions are valid for typical situations in fluid solutions when the lifetime of $^*D > 10^{-8}$ s and the concentration of A is < 0.1 M.

Now, we discuss more precisely what we mean by an "encounter" and an "encounter complex" involving *D and A. When *D and A approach each other through diffusion and come within a separation of 2–5 Å, there is a statistical probability (according to random walk theory) that because of their proximity they will remain in the same region of space for a certain period of time, even though one or more solvent molecules separate them. When *D and A have reached this distance, they are "statistically bonded" in the same region for a certain period of time and we say that *D and A have become an "encounter complex." In Section 7.1, the notation *D/A was introduced to denote an encounter complex, where *D and A are separated by one (or several) solvent molecule and the notation (*D,A) denotes a pair of molecules in a solvent cage and uses the term "collision complex," or "caged pair," to describe this situation. The *D/A pair may be considered as "statistically bonded" for a period of time simply by having diffused into each other's proximity. If *D and A diffuse even closer to one another and come into contact by collision, we have a special situation in which the pair is now in a solvent cage. For simplicity, the following discussion will refer to situations for which *D and A have become collision partners (*D,A) in a solvent cage. An encounter complex (*D/A) is a precursor of a collision complex (*D,A).

A successful encounter of *D and A will result in energy transfer (Eq. 7.88) through the following sequence of events:

1. Both *D and A diffuse through the solution until they encounter and form the encounter complex (*D/A) and eventually become partners in a collision complex (*D,A).

2. During collisions between *D and A in the collision complex (*D,A) the orbitals of *D and A overlap and electron exchange interactions occur between *D and A; these interactions eventually lead to energy transfer and generation of a new collision complex (D,*A) for which the electronic excitation resides on A, not on D. The energy transfer is in principle reversible; that is, *D,A and D,*A will pass the excitation back and forth in the collision complex. However, if *A possesses a lower excitation energy than *D, the energy transfer will be exothermic (Fig. 7.1). Consequently, the collision complex D,*A will eventually be the major species because of vibrational relaxation to the lowest vibrational level of the D,*A collision complex and there will be a significant activation energy required to re-form *D,A. In this case, the energy transfer is irreversible.

3. The collision complex (D,*A) dissociates into the free molecules D + *A, which diffuse apart via a diffusional random walk.

The steps for irreversible energy transfer are summarized in Scheme 7.12.

Note that point (2) introduces the possibility that when *D and A meet they do not undergo just one collision and exchange interaction and separate, but rather that the pair may undergo *multiple collisions and exchange interactions* during the lifetime of a collision complex. The phenomena of multiple collisions between partners (*D,A) in a solvent cage is called the *cage effect*. Cage effects are quite important for radical pairs undergoing a competition between spin evolution and escape.

Diffusional encounters always occur in competition with the spontaneous or induced decay of *D through photophysical or photochemical processes. For simplicity, we assume that photochemical primary processes are not important and lump all competing photophysical deactivations into an overall unimolecular rate constant k_D

$$*D + A \underset{\text{Diffusion}}{\rightleftharpoons} *D,A \underset{\text{ET}}{\rightleftharpoons} D,*A \underset{\substack{\text{Diffusional} \\ \text{separation}}}{\rightleftharpoons} D + *A$$

Scheme 7.12   Representation processes (1)–(2) above leading to energy transfer and separation of donor and acceptor.

(Eq. 7.89 is equivalent to Eq. 7.4 discussed in Section 7.2).

$$^*D \xrightarrow{k_D} , D \tag{7.89}$$

$$^*D + A \xrightarrow{k_{dif}} {}^*D,A \xrightarrow{k_{ET}} D,{}^*A \xrightarrow{k_{-dif}} D + {}^*A \tag{7.90}$$

Note that all steps in Eq. 7.90 were written as irreversible; this will therefore be a "true" example of diffusion-controlled energy transfer, where *every encounter between *D and A that results in the formation of a caged pair leads to successful energy transfer before the separation of the pair can occur*. Equation 7.90 can be viewed as a detailed expansion of steps involved in Eq. 7.5. Each individual elementary step may also compete with other energy wasting processes, although these are somewhat less common when energy transfer approaches diffusion control. The only requirement for the overall energy-transfer process to be highly efficient is that the concentration of A be such that essentially all of the *D molecules decay via energy transfer rather than by the decay process of Eq. 7.89; that is, the condition of Eq. 7.91 is valid.

$$k_{dif}[A] >> k_D \tag{7.91}$$

Again, we emphasize that not every *collision* between *D and A in a solvent cage (*D,A) necessarily leads to energy transfer. The only requirement is that energy transfer and vibrational relaxation of *A occur before the encounter complex dissociates irreversibly; that is, energy transfer must occur within the lifetime of the "cage." This concept is illustrated in the sequence of events of Eq. 7.92. Thus, a diffusion control of energy transfer does not prove that energy transfer is collision controlled. This distinction is important because as we have seen in Section 7.13, diffusion control masked the occurrence of the Marcus inversion region of electron transfer for many years. Only when diffusion was eliminated as a factor in determining the formation of collision complexes (rigid media, rigid molecular spacers that inhibit diffusion) was the existence of the Marcus inverted region convincingly demonstrated experimentally.

$$\begin{array}{ccccc} \text{Free diffusing} & & \text{One} & & \text{Many} & & \text{One favorable collision leads to} \\ \text{molecules} & \rightarrow & \text{encounter} & \rightarrow & \text{collisions} & \rightarrow & \text{successful and irreversible} \\ & & & & & & \text{electron transfer} \end{array} \tag{7.92}$$

7.35 Estimation of Rate Constants for Diffusion Controlled Processes

The value of k_{dif} for a diffusion-controlled energy-transfer process between *D and A can be estimated from information on the viscosity of the solvent and on the diffusional properties of molecules involved in the energy-transfer process. Equations

7.93 and 7.94 represent a simplified equation that is frequently employed to estimate the magnitude of k_{dif}.

$$k_{dif} = 4\pi N_A \sigma D \times 10^{-3} \tag{7.93}$$

$$\sigma = r_A + r_D \tag{7.94}$$

In Eq. 7.93, N_A is Avogadro's number, σ is the interaction distance, which is frequently taken as the sum of radii for the two reactants (r_A and r_D in Eq. 7.94); D is the diffusion coefficient and is equal to the sum of the individual diffusion coefficients D_D and D_A, for D and A, respectively (the factor of 10^{-3} takes care of unit conversion, m^3 to L).

For relatively large solute molecules diffusing among relatively small solvent molecules, it is possible to estimate the diffusion coefficients D_A or D_D from an equation of the form of Eq. 7.95.

$$D_A = \frac{kT}{6\pi \eta_A} \tag{7.95}$$

In Eq. 7.95, k is the Boltzmann constant and η is the solvent viscosity. Assuming $r_A \sim r_D$ (similarly sized molecules) and combining Eq. 7.93 with Eq. 7.95 leads to Eq. 7.96, which was introduced earlier in Section 7.1.

$$k_{dif} = \frac{8RT}{3000\eta} \tag{7.96}$$

Equation 7.96 is known as the Debye equation[60] and allows for a convenient estimate of diffusion controlled rate constants at a given temperature based simply on knowledge of the viscosity for the solvent (it is assumed that the size of small molecules is not a significant factor).

Viscosities are normally expressed in units of poise; since this is a cgs unit, it requires that the gas constant (R) be expressed in the same units ($R = 8.31 \times 10^7$ erg mol^{-1}). Examples of the viscosities and values of k_{dif} for common organic solvents are given in Table 7.4. Typical values for k_{dif} for "nonviscous solvents" are in the $3 \times 10^9 - 4 \times 10^{10}$ $M^{-1} s^{-1}$ range (Table 7.4).

When actual diffusion coefficients (D) and critical distances (σ) are known, a better estimate of k_{dif} can be obtained using Eq. 7.93. Figure 7.29 gives plots of expected values of k_{dif} for a common range of solvent viscosities. In general, when we talk about typical *nonviscous solvents* or *fluid solvents*, we mean that the viscosity of the solvents is in the range 0.2 cP $\leq \eta \leq$ 2 cP.

Experimental criteria for diffusion-controlled energy-transfer processes usually fall into one or more of the following categories:

1. The observed value of the experimental bimolecular rate constant (k_{obs}) is close to that calculated with Eqs. 7.93 or 7.96.
2. The experimental value of k_{obs} is a function of T/η, as expected from Eq. 7.96.
3. The value of k_{obs} is essentially invariant for *D quenchers of widely varying structure; that is, the magnitude of the k_{obs} value is a property of the solvent and not of the detailed molecular structure of D and A.

Table 7.4 Representative Rate Constants for Diffusion in Various Solvents at 25 °C, Calculated According to Eq. 7.96

Solvent	Viscosity (cP)	k_{dif} (M^{-1} s^{-1})
Hydrocarbons		
Pentane	0.24	2.7×10^{10}
Heptane	0.42	1.5×10^{10}
Octane	0.55	1.2×10^{10}
Cyclohexane	0.98	6.6×10^{9}
Dodecane	1.51	4.3×10^{9}
Hexadecane	3.3	1.9×10^{9}
Alcohols		
Methanol	0.55	1.2×10^{10}
Ethanol	1.20	5.4×10^{9}
Ethylene glycol	20	3.3×10^{8}
Aromatic hydrocarbons		
Toluene	0.59	1.1×10^{10}
Benzene	0.65	1.0×10^{10}
Miscellaneous solvents		
Acetonitrile	0.34	1.9×10^{10}
Methylene chloride	0.41	1.6×10^{10}
Tetrahydrofuran	0.46	1.4×10^{10}
Chloroform	0.54	1.2×10^{10}
Water	0.89	7.4×10^{9}
Carbon tetrachloride	0.90	7.3×10^{9}

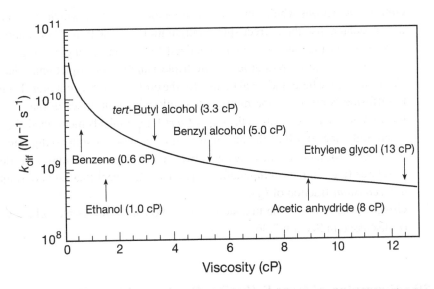

Figure 7.29 Calculated values of k_{dif} at 25 °C based on Debye's equation (Eq. 7.96) for a range of common viscosities. Note that viscosities are usually expressed in centipoise, cP (1 cP = 0.01 P). The poise is the cgs unit of viscosity.

4. The values of k_{obs} for different quenchers reach a limiting value that corresponds to the fastest bimolecular rate constant measured for that solvent. This "leveling" of the k_{obs} value for energy transfer is analogous to the issue of the masking of the Marcus inversion region in electron transfer.

Examples of truly diffusion-controlled energy-transfer processes are not common. The most compelling experimental criterion for a truly diffusion-controlled process is probably a strict correlation of the rate constant for the process investigated with the solvent viscosity, as expected from Eq. 7.96. Such an example is the observation[61] that the rate constants of singlet–singlet energy transfer from the quenching of naphthalene fluorescence by biacetyl in solvents of viscosity ranging from 0.34 (hexane) to 17.2 cP (liquid paraffin) were proportional to η^{-1}. In this case, quenching is probably due to singlet–singlet energy transfer via a simple overlap electron-exchange mechanism. In this case, we conclude that every encounter of *D and A leads to energy transfer. The data are consistent with a quenching range of ~ 11 Å for effective transfer. It is important to note that, although each encounter leads to quenching of *D, it may take more than one collision of *D and A in the collision complex to result in energy transfer. Indeed, the ~ 11 Å quenching range indicates that some energy transfer can occur even when *D and A are separated by a molecule (or two).

Diffusion control reactions are not limited to reactions involving electronically excited states. For example, the self-reaction of *tert*-butyl radicals in low-viscosity solvents shows diffusion controlled behavior and also follows (within 0.5 kcal mol^{-1}) the activation energies expected for simple diffusion through the solvent. Many reactions "approach," but do not reach, diffusion control; these reactions can be divided into two groups.

Group 1: Reactions where spin statistical factors determine which fraction of the encounters has the correct spin configuration for reaction. This situation is common in the case of triplet sensitized formation of singlet oxygen from molecular oxygen (a ground state triplet) and triplet–triplet annihilation (Section 7.12), where statistical factors for the probability of encounters leading to diffusion control are one-ninth.[20] A similar situation arises in the self-reaction of free radicals (as in the case of *tert*-butyl radicals mentioned above), where only one-fourth of the encounters have the correct spin configuration for product formation. These systems behave as typical diffusion-controlled processes (e.g., in their temperature dependence), except that rate constants are a *constant* fraction of k_{dif}.

Group 2: Cases where the first step in Eq. 7.90 can be reversible, which will be discussed in Section 7.36.

7.36   Examples of Near-Diffusion-Controlled Reactions: Reversible Formation of Collision Complexes

The scheme of reactions 7.89 and 7.90 can be readily modified to include the possibility of reversibility in the energy-transfer process, as shown in Eqs. 7.97 and 7.98.

$$^*D \longrightarrow D \tag{7.97}$$

$$^*D + A \underset{k_{-dif}}{\overset{k_{dif}}{\rightleftharpoons}} {}^*D,A \xrightarrow{k_{ET}} D,{}^*A \underset{k_{-dif}}{\overset{k_{dif}}{\rightleftharpoons}} D + {}^*A \tag{7.98}$$

When the reversible steps of Eq. 7.98 occur, the observable rate constant for reaction between *D and A is given by Eq. 7.99.

$$k_{obs} = \frac{k_{dif} k_{ET}}{k_{-dif} + k_{ET}} \tag{7.99}$$

Note that if $k_{-dif} << k_{ET}$, then Eq. 7.99 reduces to the case discussed in Section 7.35 and $k_{dif} = k_{obs}$. All other situations will lead to $k_{obs} < k_{dif}$ and the reactions are not diffusion controlled, although they may approach this limit. Note also that Eq. 7.99 is the basis for absence of the observation of Marcus inversion regions in the Rehm–Weller plot (Fig. 7.10), since the rate measured in solution is limited by diffusion, when $k_{-dif} << k_{ET}$.

If we assume that k_{ET} is approximately independent of the solvent viscosity, then it is easy to see from Eq. 7.99 that a reaction that is not diffusion controlled in a low-viscosity solvent may become diffusion controlled in high-viscosity solvents, that is, when $k_{-dif} << k_{ET}$. Note that as the solvent becomes more viscous at the macroscopic level, the "walls" of the solvent cage increasingly hinder separation of the caged radical pair. This finding is an important *supramolecular* effect of the solvent. Effectively, as the lifetime of the pair (*D,A) increases, more collisions between *D and A in the solvent cage occur before separation of the pair, and the energy transfer step can approach or reach an efficiency of 1.

An example of the relationship of viscosity to the rate of energy transfer is illustrated in Fig. 7.30 for the case of exothermic triplet energy transfer from the

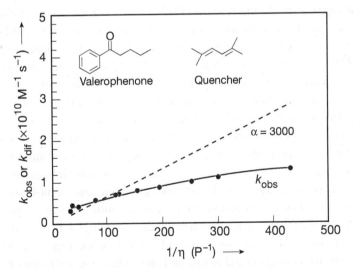

Figure 7.30   Comparison of theoretical rate constants (dashed lines) calculated using Eq. 7.96 and experimental rate constants at 25.5 °C as a function of $1/\eta$.[62] for triplet energy transfer from valerophenone to 2,5-dimethyl-2,4-hexadiene.

valerophenone triplet (ET $\sim$ 73 kcal mol^{-1}) to 2,5-dimethyl-2,4-hexadiene ($E_T \sim$ 58 kcal mol^{-1}). The measured k_{obs} for triplet energy transfer is not a linear function of reciprocal viscosity for a series of inert solvents.[62] Note that the experimental rate constant approaches diffusion controlled only for $\eta \geq 2$ cP, in a range of viscosities that we could regard as *viscous fluids*.

7.37   The Cage Effect

It is instructive to compare the collision behavior of two molecules in the gas phase to the collisions of the same two molecules in solution. Molecules in the gas phase are visualized as moving mostly in uninterrupted straight line trajectories through the vast empty space of a gas. Only on rare occasions is the straight line trajectory interrupted by an "encounter" with another molecule. For neutral molecules, an encounter is generally followed by an "elastic collision," which means that the two partners in the "collision complex" immediately recoil from each other, separate, take on new trajectories, and essentially never re-encounter or collide again. This classic "billiard ball" model is for collisions between molecules in the gas phase. In this model, a gas-phase encounter of two molecules consists of a single collision that typically lasts $\sim 10^{-12}$ s (i.e., on the time scale of the vibration of a very weak bond between two heavy particles). So, for the gas phase the rule is *one encounter, one collision, followed by irreversible separation*.

The situation for collisions of two molecules is very different in solution. In solution, D and A first form an encounter complex (D/A) and eventually a collision complex (D,A) in a solvent "cage". Compared to the gas phase, the two colliding partners find themselves in a very "crowded environment" and are surrounded completely by solvent molecules, that is, (D,A) is "caged" by solvent molecules. The "walls" of the solvent cage confine D and A to be collision partners for a certain period of time. As a result, the pair in the collision complex may undergo many collisions before separating from the solvent cage and begin a random walk that will lead to irreversible diffusion. Effectively, the solvent "walls" around this pair of molecules cause the pair to become collisionally correlated and to "bounce" against the walls and each other many times before they find a "hole" in the solvent wall through which one of the colliding partners can diffuse and become a solvent separated pair, D(S)A. This solvent cage model suggests that collisions between encountering molecules in solution will occur "in sets" before the colliding partners separate; this contrasts with the situation for the gas phase, for which collisions are followed by immediate separation of the colliding partners after a single collision. The same solvent molecules that tend to hinder the formation of a collision complex of two molecules in solution (viscous solvents) also hinder their separation once a collision complex is formed, thus leading to a "long-lived" cage encounter. In a collision complex, there is a much higher probability that the partners D and A will find and collide with each other than collide with other molecules of D and A that are randomly distributed in the solution; this tendency for D and A to undergo multiple collisions in a collision complex in solution compared to the gas phase is termed the solvent "cage effect."

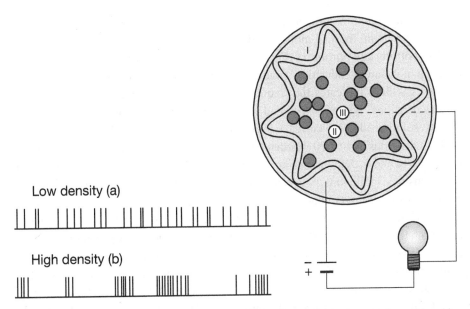

Low density (a)

High density (b)

Figure 7.31   Schematic representation of the collision frequency in the gas phase (a) and in solution (b). Note that the average number of collisions is the same in both cases (based on Rabinowitch & Wood[63] and the type of representation given in Laidler[64]). Here I is a zigzag boundary that keeps the balls confined to the internal area of the device; II is a metallic conducting ball among a set of nonconducting wooden balls; an electrical circuit is closed when II collides with III and a lamp flashes to indicate the collision.

The basic idea of the cage effect and its control of the collisions between a pair of molecules in a solvent cage was cleverly demonstrated[63] by a simple 2D mechanical model of the molecules in a liquid, and it showed convincingly that collisions between two "solute" molecules occur in sets that increase in size as the density of the "solvent" particles is increased (Fig. 7.31).

The device consisted of a flat electrically conducting brass plate with a zigzag border (I) and a metal knob (white circle, III) that is isolated from the brass plate. The metal knob is connected to one of the poles of a battery and the brass plate itself is connected to the other pole. A number of nonconducting wooden balls (filled circles in Fig. 7.31) are distributed on the plate and a single conducting metal ball (white circle, II) is added to the set of wooden balls. The nonconducting wooden balls represent the molecules of the solvent and the single conducting ball together with the central metal knob represent two dissolved molecules that eventually form a collision complex. When the collision complex is formed (the ball II strikes the central metal knob), the number of collisions between II and the central metal knob can be tracked in real time, since every time the conducting ball and the central knob come into contact, the electrical circuit is closed through the battery and the collision is registered electrically (by a lamp flashing or through a recording device). The "experiment" demonstrating the cage effect consisted of placing the brass plate on a shaking machine capable of producing strong agitation of the plate. The zigzag border of the device served to transform the regular motion of the shaking machine into chaotic agitation of the balls, simulating random thermal motion of the solute and solvent.

The bottom left portion of Fig. 7.31 shows the results of the experiment as a function of different "densities" (number) of added insulated balls. Figure 7.31a (low density, gas-like) indicates the frequencies of electrical impulses due to collisions at "low densities" when 25 insulated wooden balls are on the plate; Fig. 7.31b (high density, solution-like) indicates the frequencies of electrical impulses due to collisions at "high densities" when 50 wooden balls are on the plate. Comparison of the results in Fig. 7.31a and b, demonstrate an important feature of the cage effect: at sufficiently high densities, collisions between "solute" molecules occur in "sets." Depending on the density, the sets contained an average of about one collision per set at low density (Fig. 7.31a), representative of the gas phase, while at high solvent density, representative of a liquid, the sets contained an average of five or more collisions per set. When the percent coverage of the tray surface was 72%, approximately 20 collisions took place during each encounter. The important conclusion is that if two molecules do become nearest neighbors and collide in a solvent cage, they can remain as colliding neighbors for a significant period of time; that is, a solvent cage effect exists.

The model of "inert" balls mechanically encountering each other is an appropriate model for *D and A undergoing energy transfer in a liquid. However, in studies of electron transfer it is important to distinguish whether the reaction partners are in *contact* within a solvent cage or if the partners are separated by one or a few solvent molecules.

7.38   Distance–Time Relationships for Diffusion

It is important when considering processes competing with deactivation of *D in solution to have benchmarks for the time it takes for a molecule to diffuse a given distance from an initial point as the result of a random walk. The diffusion coefficient (D) is a parameter (Eq. 7.95) that provides this information. Eq. 7.95 shows the dependence of D on temperature (T), molecular size (r_A), and solvent viscosity η.

As described in Section 7.2, the term "random walk" refers to a motion that consists of random steps or jumps of various lengths and angles starting from some point of origin.[64] The motion of a diffusing molecule in solution follows a complex pattern, largely determined by its random collisions with other molecules, especially those of the solvent, which cause the motion of the molecule to make random jumps of various lengths at various angles. Thus, the motion of a molecule in solution is expected to follow the diffusional features of a random walk. Random walk theory predicts that the average displacement x of a diffusing particle from an origin during a time t is related to the diffusion coefficient D of the particle according to Eq. 7.100.

$$x = \sqrt{2Dt} \qquad (7.100)$$

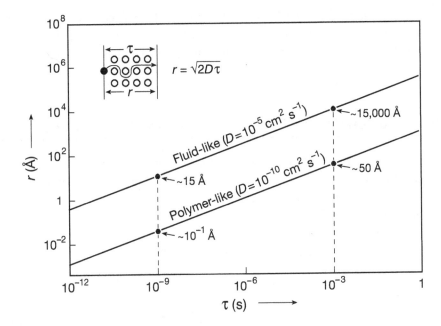

Figure 7.32   Graphs of the relationship between diffusional distance and time for a small organic molecule in a fluid (nonviscous) and polymer-like (viscous) solvent.

This remarkably simple expression takes into account the complex zigzag pattern involved in the random walk displacement of a molecule in solution. Figure 7.32 shows the displacement achieved by a small molecule in a nonviscous ($D \sim 10^{-5}\,cm^2\,s^{-1}$) and in a viscous ($D \sim 10^{-10}\,cm^2\,s^{-1}$) solvent; the latter represents a relatively viscous system, such as a polymer solution.

As an exemplar, consider the mean distances traveled by oxygen in water at room temperature with a diffusion coefficient of $\sim 10^{-5}\,cm^2\,s^{-1}$. This diffusion coefficient corresponds to a rate constant on the order of $5 \times 10^9\,M^{-1}\,s^{-1}$. After 1 ps, the oxygen molecule has diffused only 1–2 Å, which is less than the size of a solvent molecule from its original point of origin. However, after a few microseconds, the oxygen molecule has diffused $> 10,000$ Å, a distance equal to hundreds of solvent molecules!

It is also informative to estimate how long a solvent cage persists. Let us compute the time t from Eq. 7.100: a typical diffusion coefficient (D) in a nonviscous solvent, such as benzene, at 25 °C is $\sim 2 \times 10^{-5}\,cm^2\,s$. If we assume the encounter to be over when one of the molecules has traveled a distance equivalent to the size of several solvent molecules (e.g., $x \sim 10$ Å, equivalent to about two benzene molecules), then, by applying Eq. 7.100, we obtain an estimate of the time required to be $\sim 2.5 \times 10^{-10}\,s$ or ~ 250 ps. During this time, the partner molecules undergo many collisions. It has been suggested[64] that, since even displacements of molecular dimensions are achieved on a complicated zigzag path (random walk) consisting of a great number of single segments, then Eq. 7.100 can be applied even to such tiny displacements. Also, we have a benchmark for the lifetime of a solvent cage from this computation. If we take the origin as the collision complex and say that the pair has become random when the pair has separated by ~ 10 Å, then the lifetime of the solvent cage or encounter complex for benzene is computed to be ~ 100 ps. This benchmark is important for the maximum time that a geminate pair remains geminate in a nonviscous liquid, such as benzene.

7.39   Diffusion Control in Systems Involving Charged Species

It is important to realize that in the cases of electron transfer the effect of charge may be quite different on the reactants and on the products. For example, let us assume that an initially *uncharged* donor and acceptor pair interact by transferring a single electron from *D to A (Eq. 7.101).

$$\text{*D} + \text{A} \underset{k_{-\text{dif}}}{\overset{k_{\text{dif}}}{\rightleftharpoons}} \text{*D,A} \xrightarrow{k_{\text{et}}} \text{D}^{\bullet+},\text{A}^{\bullet-} \xrightarrow{k^c_{-\text{dif}}} \text{D}^{\bullet+} + \text{A}^{\bullet-}$$

$$\downarrow k_{\text{bet}}$$

$$\text{D,A} \xrightarrow{k_{-\text{dif}}} \text{D} + \text{A} \tag{7.101}$$

In the case of Eq. 7.101, Coulombic effects do not play a significant role on the initial encounter of the uncharged reactants *D and A but would contribute an attractive interaction between the radical ion products $\text{D}^{\bullet+}$ and $\text{A}^{\bullet-}$, which will then separate more slowly than they would if they had no charge; that is, as shown in Eq. 7.102, where the superscript "c" indicates a charged species.

$$k^c_{\text{dif}} < k_{\text{dif}} \tag{7.102}$$

As discussed in Section 7.23, a common reaction following photoinduced electron transfer is <u>back electron transfer</u>, for which we use the subscript "bet" in Eq. 7.103. Note that back electron transfer (k_{bet}) from $\text{A}^{\bullet-}$ to $\text{D}^{\bullet+}$ regenerates D and A in their ground states, unless there is a low-energy excited state of D or A (e.g., a triplet state) whose energy is between the radical ion pair and the ground states of D + A (see Section 7.32). As the time scale of the encounter increases (due to Coulombic attraction), then the rate of diffusion out of the encounter complex decreases and the *probability* of back electron transfer will increase (Eq. 7.103).

$$\text{Probability of bet} = \frac{k_{\text{bet}}}{k_{\text{bet}} + k^c_{\text{dif}}} \tag{7.103}$$

A fast back electron transfer may result in an inefficient overall reaction of *D and A, even if the initial electron transfer occurs with very high efficiency.

The diffusion equations presented above need to be modified to accommodate the electrostatic forces resulting from the interactions of electrically charge particles. The electrostatic potential (U_{es}) is given by Eq. 7.104[60,64]:

$$U_{\text{es}} = \frac{Z_A Z_B e^2}{4\pi \varepsilon_o \varepsilon r} \tag{7.104}$$

where Z_A and Z_B are the charges of the ions involved, e the charge of the electron, ε_o the permittivity of vacuum, ε the dielectric constant of the solvent, and r the separation between particle centers of charge Z_A and Z_B. The value of U_{es} is zero if either Z_A or Z_B are zero.

For example, for a critical separation distance of 3 Å, the correction terms in water ($\varepsilon = 80$), acetonitrile ($\varepsilon = 36$), and dichloromethane ($\varepsilon = 8.9$) are 0.25, 0.03, and 1.6×10^{-8} for $Z_A Z_B = 1$ and 2.6, 5.2, and 21, respectively, for $Z_A Z_B = -1$. In other words, bringing together two ions of different charge is easier (by the amount of the correction) than for two neutral species; in contrast, like charges hinder the approach since repulsive interactions are involved ($Z \geq 1$).

7.40 Summary

Electron and electronic energy transfer are of great importance in photochemistry. There are two important mechanisms for electronic energy transfer, the dipole–dipole interaction and the electron exchange interaction. Electron transfer occurs by orbital overlap interactions and may occur through one or more solvent molecules or, in systems with a rigid spacer between the electron donor and acceptor, through the σ bonds of the spacer. Energy and electron transfer require delivery mechanisms of the donor and acceptor molecules. The encounter of donor and acceptor and their subsequent separation are crucial in determining the efficiency of electronic energy transfer and electron transfer. From the theory of electron transfer, it is predicted, and found, that when an electron transfer becomes highly exothermic the process may be in the "inverted" region for energy transfer. In this region, the reaction rate decreases as the electron transfer becomes more exothermic.

Nature has taken advantage of the inverted region of photoinduced electron transfer in designing the mechanism of photosynthesis. The key photochemical step in photosynthesis is a photoinduced electron transfer. In order to proceed to produce energy storing products, the initially produced radical ion pair is inhibited from undergoing a nonproductive back electron transfer because the process is highly exothermic and in the inverted region.

References

1. N. J. Turro, *Modern Molecular Photochemistry,* Chapter 9, University Science Books, Mill Valley, CA, 1991.

2. G. L. Closs, M. D. Johnson, J. R. Miller, P. Piotrowiak, *J. Am. Chem. Soc.* **111** (10), 3751 (1989).

3. J. B. Birks, *Photophysics of Aromatic Molecules.* Wiley-Interscience, New York, 1970.

4. P. L. Piciulo and J. K. Thomas, *J. Chem. Phys.* **68**, 3260 (1978).

5. H. Kawazumi, Y. Isoda, and T. Ogawa, *J. Phys. Chem.* **98**, 170 (1994).

6. R. V. Bensasson, E. J. Land, and T. G. Truscott, *Flash Photolysis and Pulse Radiolysis.* Pergamon Press, New York, 1983.

7. G. L. Hug, *Optical Spectra of Nonmetallic Inorganic Transient Species, in Aqueous Solution.*

National Bureau of Standards, Washington, 1981, Vol. NSRDS-NBS 69, p. 160.

8. A. Kellmann and F. Tfibel, *J. Photochem.* **18**, 81 (1982).

9. A. Kellman and F. Tfibel, *Chem. Phys. Let.* **69**, 61 (1980).

10. J. T. Richards, G. West, and J. K. Thomas, *J. Phys. Chem.* **74**, 4137 (1970).

11. D. L. Dexter, *J. Chem. Phys.* **21**, 836 (1953).

12. T. Förster, *Fluorenzenz Organische Verbindungen.* Vandenhoech and Ruprecht: Göttingen, 1951.

13. P. G. Wu, L. Branch, *Anal. Biochem.* **218**, 1 (1994).

14. S. A. Latt, H. T. Cheung, and E. R. Blout, *J. Am. Chem. Soc.* **87**, 995 (1965).

15. L. Stryer and R. P. Haugland, *Proc. Natl. Acad. Sci. U.S.A.* **58**, 720 (1969).

16. Z. Tan, R. Kote, W. N. Samaniego, S. J. Weininger, and W. G. McGimpsey, *J. Am. Chem. Soc.* **103**, 7612 (1999).

17. A. L. Buchachenko and V. L. Berdinsky, *Chem. Rev.* **102**, 603 (2002).

18. J. B. Birks, *Chem. Phys. Lett.* **2**, 417 (1968).

19. C. A. Parker and C. G. Hatchard, *Trans. Faraday Soc.* **59**, 284 (1963).

20. J. Saltiel and B. W. Atwater, *Adv. Photochem.* **14**, 1 (1988).

21. B. Stevens and M. I. Ban, *Trans. Faraday Soc.* **60**, 1515 (1964).

22. C. Bohne, E. B. Abuin, and J. C. Scaiano, *J. Am. Chem. Soc.* **112**, 4226 (1990).

23. G. J. Kavarnos, *Fundamentals of Photoinduced Electron Transfer*, VCH Publishers, New London, CT, 1993, p. 359.

24. D. Zhang, G. L. Closs, D. D. Chung, and J. R. Norris, *J. Am. Chem. Soc.* **115**, 3670 (1993).

25. G. Kavarnos, *Topics Curr. Chem.* **156**, 21 (1990).

26. D. Rehm and A. Weller, *Isr. J. Chem.* **8**, 259 (1970).

27. W. F. Libby, *J. Phys. Chem.* **56**, 863 (1952).

28. R. A. Marcus, *J. Chem. Phys.* **24**, 966 (1956).

29. R. A. Marcus, *Can. J. Chem.* **37**, 155 (1959).

30. G. J. Kavarnos and N. J. Turro, *Chem. Rev.* **86**, 401 (1986).

31. R. A. Marcus, *Disc. Faraday Soc.* **29**, 21 (1960).

32. J. R. Miller, L. T. Calcaterra, and G. L. Closs, *J. Am. Chem. Soc.* **106**, 3047 (1984).

33. G. L. Closs and J. R. Miller, *Science* **240**, 440 (1988).

34. G. L. Closs, L. T. Calcaterra, N. J. Green, K. W. Penfield, and J. R. Miller, *J. Phys. Chem.* **90**, 3673 (1986).

35. H. Oevering, M. N. Paddonrow, M. Heppener, A. M. Oliver, E. Cotsaris, J. W. Verhoeven, and H. S. Husch, *J. Am. Chem. Soc.* **109**, 3258 (1987).

36. R. Hoffmann, *Acc. Chem. Res.* **4**, 1 (1971).

37. G. L. Closs, P. Piotrowiak, J. M. MacInnis, and G. R. Fleming, *J. Am. Chem. Soc.* **110**, 2652 (1988).

38. M. D. Johnson, J. R. Miller, N. S. Green, and G. L. Closs, *J. Phys. Chem.* **93**, 1173 (1989).

39. M. Winnik, *Chem. Rev.* **81**, 491 (1981).

40. Y. Hirata, Y. Kanda, and N. Mataga, *J. Phys. Chem.* **87**, 1659 (1983).

41. I. R. Gould and F. Farid, *Acc. Chem. Res.* **29**, 522 (1996).

42. I. R. Gould, J. E. Moser, B. Armitage, and S. Farid, *J. Am. Chem. Soc.* **111**, 1917 (1989).

43. D. Kuciauskas, P. A. Liddell, A. L. Moore, T. A. Moore, and D. Gust, *J. Am. Chem. Soc.* **120**, 10880 (1998).

44. P. A. Liddell, D. Kuciauskas, J. P. Sumida, B. Nash, D. Nguyen, A. L. Moore, T. A. Moore, and D. Gust, *J. Am. Chem. Soc.* **119**, 1400–1405 (1997).

45. D. Kuciauskas, P. A. Liddell,; S. Lin, S. G. Stone,; A. L. Moore, T. A. Moore, and D. Gust, *J. Phys. Chem. B* **104**, 4307 (2000).

46. J. L. Bahr, D. Kuciauskas, P. A. Liddell, A. L. Moore, T. A. Moore, D. Gust, *Photochem. Photobiol.* **72**, 598 (2000).

47. D. P. DeCosta and J. A. Pincock, *J. Am. Chem. Soc.* **111**, 8948 (1989).

48a. T. Yabe and J. K. Kochi, *J. Am. Chem. Soc.* **114**, 4491–4500 (1992).

48b. I. R. Gould, D. Ege, J. E. Moser, and S. Farid, *J. Am. Chem. Soc.* **112**, 4290 (1990).

49. B. R. Arnold, D. Noukakis, S. Farid, J. L. Goodman, and I. R. Gould, *J. Am. Chem. Soc.* **117** (15), 4399 (1995).

50. H. Masuhara and N. Mataga, *Acc. Chem. Res.* **14**, 312 (1981).

51. I. R. Gould, R. H. Young, R. E. Moody, and S. Farid, *J. Phys. Chem.* **95**, 2068 (1991).

52. I. R. Gould and S. Farid, *J. Photochem. Photobiol. A: Chem.* **65**, 133 (1992).

53a. A. Kira and J. K. Thomas, *J. Phys. Chem.* **78**, 196 (1974).

53b. F. Gessner and J. C. Scaiano, *J. Am. Chem. Soc.* **107**, 7206–7207 (1985).

54. K. B. Patel and R. L. Willson, *J. Chem. Soc., Faraday Trans. I* **69**, 1597 (1973).

55. A. Weller and K. Zachariasse, *J. Chem. Phys.* **46**, 4984 (1967).

56. H. J. Werner, H. Staerk, and A. Weller, *J. Chem. Phys.* **68**, 2419-2426 (1978).

57. A. Weller, F. Nolting, and H. Staerk, *Chem. Phys. Lett.* **96**, 24-27 (1983).

58. A. Weller, H. Staerk, and R. Treichel, *Faraday Discuss. Chem. Soc.* **78**, 271 (1984).

59. Y. Tanimoto, N. Okada, M. Itoh, K. Iwai, K. Sugioka, F. Takemura, R. Nakagaki, and S. Nagakura, *Chem. Phys. Lett.* **136**, 42-46 (1987).

60. P. Debye, *Trans. Electrochem. Soc.* **82**, 265 (1942).

61. J. B. Birks, M. S. S. C. P. Leite, *J. Phys. (B), Atom. Molec. Phys.* **3**, 417 (1970).

62. P. J. Wagner and I. Kochevar, *J. Am. Chem. Soc.* **90**, 2232 (1968).

63. E. Rabinowitch and W. C. Wood, *Trans. Faraday Soc.* **32**, 1381 (1936).

64. K. J. Laidler, *Chemical Kinetics*. 3rd ed., Harper and Row, New York, 1987, p. 531.

Index